D0075812

Probability Models and Applications

Probability

Models and

Applications

Ingram Olkin *Stanford University*

Leon J. Gleser *Purdue University*

Cyrus Derman *Columbia University*

Macmillan Publishing Co., Inc.
New York

Collier Macmillan Publishers
London

Copyright © 1980, Macmillan Publishing Co., Inc.
Printed in the United States of America

A portion of this book is reprinted from
A Guide to Probability Theory and Application,
copyright © 1978 by Cyrus Derman, Leon J. Gleser, and Ingram Olkin.

Macmillan Publishing Co., Inc.
866 Third Avenue, New York, New York 10022

Collier Macmillan Canada, Ltd.

Library of Congress Cataloging in Publication Data

Olkin, Ingram.
   Probability models and applications.

   Includes index.
   1. Probabilities.   I. Gleser, Leon J., joint
author.   II. Derman, Cyrus, joint author.
III. Title.
QA273.04   1980   519.2   79-15729
ISBN 0-02-389230-7

Printing: 4 5 6 7 8   Year: 4 5 6

ISBN   0-02-389230-7

To my mother, Clara, on her 90th birthday
Ingram Olkin

To my parents
Leon J. Gleser

In memory of Manuel Derman
Cyrus Derman

Preface

There are at present many textbooks on probability theory written for under-graduate and graduate students in the physical, biological, and social sciences, in engineering, and in business. Why, then, another probability text?

Our primary goal in writing a new textbook in probability theory is to attempt to bridge the gap between the theory necessary to construct and describe probability models, both discrete and continuous, and the application (and applicability) of these models to real phenomena. Feller's classic *An Introduction to Probability and Its Applications* (Vol. I) shares some of our goals, but limits itself largely to discrete models; the second volume covers continuous models, but with a different orientation.

In the present textbook we give the usual theory concerning the axioms, probability calculus, distributional representations, moments, generating functions, and so on, found in other probability texts. However, the unifying theme of our discussion is the role probability models can play in scientific and practical applications, and how to judge the fit of such models to data. We provide numerous examples, with actual data, of situations where probability models have been used in practice, and in Chapters 6 and 8 give detailed discussions of the properties and uses of certain probability models (binomial, hypergeometric, gamma, Weibull), other than the normal distribution, that have been successful in modeling real phenomena.

The development of fast and convenient computer facilities has made it possible to consider complex models in which several variables are measured jointly (multivariate models). Although most current probability textbooks mention multivariate distributions and briefly define selected indices of such distributions, few go into the depth provided in Chapter 9, particularly as concerns fitting such models, computing probabilities, and prediction.

A chapter on Markov chains has been included to introduce students to an important simple discrete process. This provides an alternative model for dependencies over time.

Finally, more than in most purely theoretical textbooks on probability theory, we have attempted to build a firm foundation for the subsequent study of statistics. Distribution theory is the heart of statistical methodology, providing the theoretical tools by which the variability of sample indices is seen to reflect properties of the underlying population. Since correct application of probability models involves techniques of statistical inference (point and interval estimation of parameters, tests of fit), the material on data transformations found in Section 6 of Chapter 4 and in Chapters 10 and 11 can be valuable even to those students who do not go on to take courses in statistics.

In short, we feel that the present textbook, while covering the usual topics met in a first textbook on probability theory, provides a unique emphasis on where and how to apply probability models to real phenomena and an unusually strong preparation in the tools needed for such applications.

Prerequisites

Most of the material in this book will be comprehensible to readers who have had a one-year undergraduate course in differential and integral calculus. Some exposure to partial derivatives and integrals in at least two variables is recommended for the discussion in Chapter 9. The proofs in Chapter 11 will be best understood by readers who have been exposed to a rigorous treatment of limits, sequences, and series, but this background is not needed to comprehend the results (and their applications) obtained in that chapter. An exposure to elementary concepts of matrix algebra will make some parts of Chapter 12 more easily read.

How to Use This Book

This book is intended for either a one-semester or a two-quarter course in probability theory. Two classes of user are envisioned: those who are taking a probability course for its own sake, or to lead to more advanced courses in probability theory, and those who intend to use probability theory as a background to courses in statistics. It is unlikely that any course using this book can cover all of the topics included, particularly if too much time is spent on the early chapters (particularly Section 5 of Chapter 3). Chapters 1–5 should be covered by all users. It is recommended that at least Sections 1–4 of Chapter 6, Chapter 7, and Sections 1, 2, 4, and 5 of Chapter 8 be covered, with students being invited to skim omitted sections of Chapters 6 and 8 on their own. The material in Chapters 6–8 can serve as a useful handbook of models for later use in applications, as well as providing examples of distributions for use in later courses. Students planning to take further probability courses will then find Chapters 9, 11, and 12 of greatest use, while Chapters 9–11 are recommended for students intending to take courses in statistics. If time is pressing, Section 6 of Chapter 9 and Section 3 of Chapter 11 can be omitted. If additional time is available, as in a one-year sequence, the material in Chapter 12 could be discussed in greater detail.

The exercises in this book were chosen to illustrate the textual material and to provide additional learning material.

Necessary tables appear at the end of the text. Blank entries are either equal to 0 or correspond to undefined quantities.

Because of roundoff errors, some totals that should be 1.00 may be 0.98–1.02, and similarly for other tables. Where these discrepancies are minor and recognizable, we have made no attempt to correct them.

Acknowledgments

This book has been through a number of drafts. We are greatly indebted to Carolyn Knutsen and Nancy Steege for maintaining their equilibrium during the many changes while still efficiently preparing the manuscript for publication.

Ingram Olkin
Leon J. Gleser
Cyrus Derman

Contents

1

Introduction

1. HISTORY

The growth of probability theory in its modern form as a unified theory of chance phenomena began in the middle of the seventeenth century when a French nobleman, the Chevalier de Méré, consulted the mathematician Blaise Pascal (1623–1662) about a problem that had arisen in a game of chance. Pascal and his fellow mathematicians, most notably Pierre de Fermat (1601–1665), developed a lively interest in mathematically analyzing odds and strategies for various games of chance then popular in French society. Starting from about the year 1700, these fragmentary efforts began to be unified into a more general theory of chance by several European mathematicians, outstanding among whom were the Swiss mathematician Jakob (James) Bernoulli (1654–1705) and the French mathematician Abraham de Moivre (1667–1754).

Although much of the early work on probability theory concentrated on games of chance, eighteenth- and nineteenth-century students of natural phenomena noted close analogies between the laws of uncertainty proposed by the mathematicians for games of chance and the laws of variation observed in various natural phenomena. For example, in the eighteenth century it was observed that the sequential record of sexes (male or female) from successive births in city hospitals exhibited patterns similar to the patterns of heads and tails resulting from successive tosses of a coin. As a consequence, scientists developed models to study the variability of natural phenomena that treated such variability as if it resulted from a game of chance played by nature. Increasingly, these *probabilistic* (or *stochastic*) models[1] assumed important roles in the study and practical utilization of natural phenomena.

The success of probabilistic models in so many diverse contexts clearly

---

[1] The word *stochastic* derives from Greek στόχοσ (*stochos* a target); in ancient Greece a *stochastiches* was a person who forecast a future event (in the sense of aiming at the truth). In this sense the word occurs in sixteenth-century writing. Bernoulli in the *Ars conjectandi* (1713) refers to the "ars conjectandi sive stochastice."

demonstrated a need for a unified approach to a study of chance phenomena. Several such approaches were proposed. Among these, the axiomatic approach, based on "the stability properties of relative frequencies" (Section 3), proved to be the most popular. It is this approach that is emphasized in the following chapters.

Some examples of today's widespread use of stochastic models may be of interest. Life insurance policies take into account the chance that a person with a certain medical history will die within the year (actuarial tables); public opinion experts gauge the reactions of the total populace by questioning a few people selected through a lottery (random sampling); theorists in physics conceive of elementary particles moving, colliding, and splitting according to chance; psychologists analyze learning behavior as if it were a chance phenomenon; some sociologists view mobility of populations as being governed by a probabilistic mechanism; hereditary characteristics of biological organisms are hypothesized to be assigned by chance; and inventory systems are devised to meet demands fluctuating in a random fashion. From this short list, it should be apparent that the use of stochastic models (and thus probability theory) pervades most sophisticated attempts to explore, explain, and control our physical and social environment.

2. MODELS

A *scientific model* is an abstract and simplified description of a given phenomenon. Certain basic aspects of this phenomenon are isolated as being of primary interest, and an analogy is drawn between these aspects and some logical structure concerning which we already have detailed information. Scientific models are most often based on mathematical structures; but organic models have been used in sociology [e.g., see Spencer (1877)], physical models in psychology and economics, and economic models in engineering. Figure 2.1 illustrates how a scientific model may deepen our understanding of natural phenomena.

When an investigator builds a mathematical model for a particular natural phenomenon (say, the motion of an asteroid), important elements of this phenomenon (the position, mass, shape, and speed of the asteroid) are

Figure 2.1: *Use and development of a scientific model.*

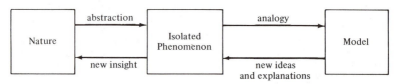

identified with the basic elements of some mathematical structure (numbers). Certain fundamental facts connecting the important elements of the phenomenon are restated as axioms relating the analogous mathematical entities. Finally, the more complex relationships between the basic elements of the natural phenomenon are made to correspond to laws or theorems in the mathematical structure. If this correspondence is reasonably valid, the investigator does not have to experiment haphazardly with the phenomenon to find new facts; instead, logical arguments based on the mathematical axioms can lead to a theorem that presumably has an analogy to a law of nature. Experimentation can now be directed toward verifying this law.

The fact that an investigator need only concentrate on the few axioms that define the mathematical structure of his model leads to a simplification and unification of his knowledge concerning the natural phenomenon. Every fact known to him can be reproduced by starting from the axioms and using mathematical logic. Thus, his discipline becomes a cohesive whole in which all facts are logically interrelated, rather than merely a list of isolated facts.

For more than a century after Newton's time, the prevalent scientific and philosophical viewpoint was that all variability in any observed phenomenon could be attributed to a failure to identify and control certain causal variables. Once all relevant causal variables were identified and measured, any phenomenon could be *exactly* predicted. From this point of view, even the face on a tossed coin could be predicted once the magnitude and direction of all forces acting on the coin were known. In short, nature was seen as being a fully predictable machine. Such *mechanistic models* proved very fruitful in physics, chemistry, and engineering, and their success led many scientists to deny both the existence of chance (unpredictable) phenomena and the need for stochastic models as descriptions of nature.

However, even in the eighteenth century it had become apparent that models incorporating uncertainty (stochastic models) were needed. Some of the reasons for this need are

(1) Even if nature is mechanistic, to list and measure all the variables involved in complex phenomena is an impossible task.

(2) Some scientists and philosophers [e.g., the German physicist Heisenberg (1901–1976)] argue that there is a limit to man's ability to measure natural phenomena because our measurements themselves produce unpredictable variations in the phenomena.

(3) There are occasions when a deliberate *introduction* of uncertainty provides a cheaper (and often more accurate) way of investigating natural phenomena. For example, instead of taking a complete census of a population, information can be obtained more cheaply, in less time, and more accurately, by interviewing a random sample (a sample chosen by lottery) of the population and then extrapolating to the population by means of probability theory.

In consequence, starting with the work of Gauss (1777–1855) and Laplace (1749–1827) on errors in measurement, and continuing with the nineteenth-century actuaries and demographers, even believers in an inherently mechanistic nature have used stochastic models of natural phenomena. A model need not be completely true to be useful; the decisive criteria for choosing a model are practical, not metaphysical: Does the model provide a simple, yet comprehensive, explanation of known phenomena and at the same time have a strong potentiality for providing further insight into the natural world? Stochastic models meet these criteria. As a consequence, the assumption that chance phenomena exist and can be described, whether true or not, has proved valuable in almost every discipline.

3. THE EMPIRICAL BASIS OF PROBABILITY THEORY

The applicability of probability theory is a consequence of the observed fact that chance phenomena exhibit statistical regularities. A clear example of this fact is the ancient game of dice throwing. Although we cannot say what the result of any given throw of a die will be, we can predict in a large number of throws of the die what proportion of the throws will result in (say) a six. We use this information to place our bets for each throw. A more practical application of the statistical regularity of chance phenomena occurs in the field of insurance. Although it may not be possible to predict the life span of a given individual, precise statements concerning longevity can be made about large populations of individuals. Such a statement may be, for example, that 50% of the present population will die before age 65. For the purposes of life insurance or annuities, these statements are entirely adequate to enable the insurance company to "place its bet"—that is, to decide what premium to charge for a given amount of insurance. In offering a policy to a large group of individuals, it is not important to the company to know what will happen to any one individual; only what happens to the group as a whole is of concern.

Let us discuss this subject in a slightly more formal way. A basic concept for all discussions of chance phenomena is the notion of a random experiment. A random experiment is a procedure that involves certain actions under specified conditions and that has as its outcome one (and only one) of a collection of possible simple results. The particular simple result that occurs cannot be predicted with certainty before the experiment is performed.

Example 3.1 (Sample Survey). A phone number chosen by means of a lottery from a phone book is called and the person who answers is asked whether he or she is listening to a certain radio station. The possible simple results are "yes" or "no" responses (or perhaps "refuses to respond"). Before the person is chosen there is no way of accurately predicting what his or her answer will be.

Example 3.2 (Physical Measurement). A physicist attempts to find the weight of a given object. Here the possible simple results are all possible weights. The experiment is random because minute changes in moisture content, air pressure, and temperature in the room affect the scale and the actual weight of the object in unknown and unpredictable ways and because there may be slight visual or adjustment errors on the part of the physicist.

We can think (at least ideally) of repeating a random experiment under identical conditions, a so-called *repeated trial*. Since the experiment is random, the result of each trial is unpredictable. However, let us assume that we make many—say, N—repeated trials of the experiment and record the (simple) result for each trial. In the physical measurement example (Example 3.2), we might have

Trial	1	2	3	4	$\cdots$	$N-1$	N
Result (in pounds)	4.5	4.4	4.6	4.4		4.2	4.4

Consider some statement that relates to the result of an individual trial. In Example 3.2, the statement might be "The weight of the object is less than 4.4 pounds." Each individual trial has as its outcome a single simple result, but out of all possible simple results, more than one may satisfy the statement in which we are interested. Thus, in Example 3.2 each of the weights 4.1 lb, 4.2 lb, and 4.3 lb will satisfy the statement "The weight of the object is less than 4.4 pounds."

Let us call the collection of all possible simple results that satisfy the statement of interest to us the *composite result* defined by this statement, and denote this composite result by the letter E. If the trial yields a simple result that is a member of the composite result E, then we say that E has occurred. Among the N trials that we observe, we count the number, $\#E$, of times that the composite result E occurs and divide this number by N. The result of this operation is a fraction, $\#E/N$, that gives the proportion of times in N trials that the composite result E occurs. We call this fraction the *relative frequency* (in N trials) of E, and denote this fraction by r.f.(E).

As we continue to make more and more trials, let us keep a record to see how r.f.(E) varies with the number of trials. Thus, after the first trial, we compute r.f.(E) (it is 0 if E did not occur; otherwise it is 1); after the second trial, we again compute r.f.(E) over the two trials (this number is 0 if E has not occurred on either trial, $\frac{1}{2}$ if E has occurred on exactly one trial, or 1 if E occurred on both trials); after the third trial, we compute r.f.(E) for the three trials; and so on. We graph the relative frequencies so obtained against the

number of trials for which they were obtained. Thus, if we have observed a list
of happenings such as the following:

Trial	1	2	3	4	5	$\cdots$
Result	E	Not E	Not E	E	Not E	$\cdots$
r.f.(E)	1	$\frac{1}{2}$	$\frac{1}{3}$	$\frac{2}{4}$	$\frac{2}{5}$	$\cdots$

(where "not E" indicates that the composite result E did not occur), we could
construct a table like Table 3.1 and from this table obtain Figure 3.1. Note
that a logarithmic scale is used in order to more easily exhibit a large number
of trials. As N grows larger and larger, the number r.f.(E) should grow closer
and closer to a certain value, which we shall call p. In Figure 3.1, $p = \frac{2}{5}$.

For small N, r.f.(E) may vary widely over the range of numbers from 0
to 1; but as N gets large, r.f.(E) approaches p. This phenomenon of conver-
gence is what we call the *statistical regularity of chance phenomena*; it is also
known as the *stability of relative frequencies*. Such stability may be all that we
need in order to describe and utilize chance phenomena (a point noted earlier
in connection with life insurance policies). *When the property of stability of
relative frequencies is assumed to hold for all composite results of interest in a
given experiment, we can use probability theory as a mathematical model for this
chance phenomenon.*

Before we turn to an exposition of probability theory, we should make
two points. First, we do not always test the stability of relative frequencies in
order to justify the use of probability theory. In many random experiments, it
may be impossible or impractical to perform many repeated trials (although
we can perhaps conceive of such trials in our imagination). For example, if we

Figure 3.1: *Relative frequency of the composite result E.*

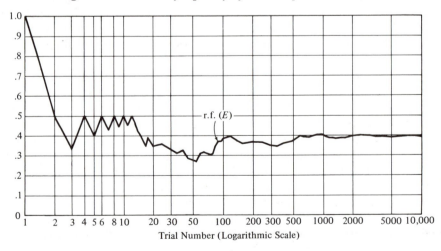

Trial Number (Logarithmic Scale)

Table 3.1: *Record of How the Relative Frequency r.f.(E) of the Composite Result E Varies with the Number N of Trials*

Trial	Relative frequency	Trial	Relative frequency	Trial	Relative frequency	Trial	Relative frequency
1	1.0000	30	0.3333	300	0.3567	1700	0.3912
2	0.5000	35	0.3143	350	0.3514	1800	0.3961
3	0.3333	40	0.3250	400	0.3625	1900	0.3974
4	0.5000	45	0.2889	450	0.3689	2000	0.3985
5	0.4000	50	0.2800	500	0.3760	2500	0.4032
6	0.5000	55	0.2727	550	0.3927	3000	0.4003
7	0.4286	60	0.3167	600	0.4000	3500	0.3986
8	0.5000	65	0.3231	650	0.3985	4000	0.3973
9	0.4444	70	0.3143	700	0.3929	4500	0.3953
10	0.5000	75	0.3067	750	0.3947	5000	0.3956
11	0.4545	80	0.3125	800	0.4025	5500	0.3958
12	0.5000	85	0.3529	850	0.4059	6000	0.3978
13	0.4615	90	0.3778	900	0.4078	6500	0.3983
14	0.4286	95	0.3789	950	0.4084	7000	0.3994
15	0.4000	100	0.3900	1000	0.4060	7500	0.4019
16	0.3750	120	0.4000	1100	0.3973	8000	0.4000
17	0.3529	140	0.3786	1200	0.3917	8500	0.4016
18	0.3889	160	0.3625	1300	0.3938	9000	0.4016
19	0.3684	180	0.3667	1400	0.3893	9500	0.3994
20	0.3500	200	0.3700	1500	0.3880	10000	0.4001
25	0.3600	250	0.3680	1600	0.3919		

allow an electric light bulb to burn until it burns out and measure the time it takes for the bulb to expire, we have performed one trial of a random experiment that cannot be repeated (once a bulb burns out, it is gone forever). In Example 3.2 we might find that taking repeated trials is an expensive or time-consuming task. Even when repeated trials are not performed, however, we may have theoretical reasons for assuming that the relative frequencies are stable. The only test of this assumption then becomes the success of the model in describing and explaining the phenomenon. This brings us to our second point: probability theory has proved itself successful whether or not relative frequencies ever really stabilize.

NOTES AND REFERENCES

The reader interested in the history of probability and in the use of scientific models may find the books by Pearson and Kendall (1970), David (1962), Ore (1933), Todhunter (1865), and Kline (1962) and the articles by Ore (1960) and

Kendall (1963) to be of interest. A series of original essays on probability by Pierre Laplace, Charles S. Peirce, John Maynard Keynes, Henri Poincaré, and Ernest Nagel is reproduced in Newman (1956). Another interesting original source is an English translation of Laplace's popular exposition of probability theory [Laplace (1951)].

EXERCISES

1. A study of purchasing intentions and actions of members of a consumers organization was made by Columbia University in 1962. Table E.1 gives the annual family income reported by each person in a sample of 49 members of the organization. Regard the results reported in this table as if they arose from 49 repeated trials of the random experiment in which a member of the consumers organization is chosen by lottery from among all members of the organization, and in which the annual family income of the person so chosen is recorded.

Table E.1: *Annual Family Income (in hundreds of dollars) for a Sample of 49 Members of a Consumers Union*

Member number	Annual family income	Member number	Annual family income	Member number	Annual family income
1	$085	18	$065	34	$075
2	135	19	147	35	083
3	200	20	084	36	092
4	050	24	120	37	060
5	070	22	092	38	105
6	076	23	090	39	088
7	105	24	100	40	055
8	065	25	074	41	085
9	121	26	070	42	110
10	105	27	110	43	096
11	130	28	100	44	100
12	065	29	075	45	066
13	120	30	069	46	072
14	100	31	067	47	060
15	127	32	060	48	095
16	150	33	120	49	090
17	105				

(a) What are the simple results of the random experiment?

(b) What is the relative frequency (in these 49 trials) of the composite result *E* that the annual family income of the individual chosen in the experiment is not greater than $10,000 and not less than $8500? What simple results are contained in this composite result?

2. The following experiment is part of the conceptual basis of a theory of statistical linguistics.

Table E.2 [taken from Herdan (1964)] gives the frequency with which the letters A through Z of the English alphabet occur as the first letter of the nouns used in a sample from the works of the English author John Bunyan (1628–1688).

Table E.2: *Frequency of Occurrence of the Letters of the English Alphabet as the Initial Letters of the Nouns in a Sample of the Writing of John Bunyan*

Initial letter of noun	Frequency of occurrence	Initial letter of noun	Frequency of occurrence
A	111	N	40
B	147	O	41
C	210	P	188
D	153	Q	7
E	69	R	133
F	112	S	256
G	72	T	112
H	110	U	16
I	72	V	43
J	22	W	100
K	18	X	0
L	84	Y	5
M	124	Z	1

(a) What is the relative frequency of the composite result "The letter chosen is a vowel"?

(b) What letter has the largest relative frequency?

(c) Construct a table (similar to Table E.2) based on a sample (two or three typewritten pages in length) of your own writing, and answer questions (a) and (b) for your own writing style.

3. Consider the following experiment. With a tape measure, a person measures the length of the longest wall in a given room to the nearest $\frac{1}{8}$ inch.

(a) What are the possible simple results of this experiment?

(b) Twenty-five people were asked to perform this experiment. The results obtained are shown in Table E.3. Regard each person's measurement as a repeated trial of the measuring experiment described above. Find the relative frequency of each simple result that you actually observe in these repeated trials.

Table E.3: *Measured Length of a Given Wall in a Given Room Obtained by 25 Different People*

Person	Length obtained	Person	Length obtained
1	20 ft $8\frac{1}{8}$ in.	14	20 ft $7\frac{3}{8}$ in.
2	20 ft $8\frac{1}{2}$ in.	15	20 ft $8\frac{1}{8}$ in.
3	20 ft $8\frac{3}{8}$ in.	16	20 ft 8 in.
4	20 ft $7\frac{7}{8}$ in.	17	20 ft $8\frac{1}{4}$ in.
5	20 ft 8 in.	18	20 ft $7\frac{7}{8}$ in.
6	20 ft $8\frac{1}{8}$ in.	19	20 ft $7\frac{3}{4}$ in.
7	20 ft $7\frac{3}{4}$ in.	20	20 ft 8 in.
8	20 ft $7\frac{7}{8}$ in.	21	20 ft $8\frac{1}{8}$ in.
9	20 ft $8\frac{1}{2}$ in.	22	20 ft $7\frac{1}{2}$ in.
10	20 ft $8\frac{1}{4}$ in.	23	20 ft $8\frac{1}{4}$ in.
11	20 ft $7\frac{5}{8}$ in.	24	20 ft 8 in.
12	20 ft 8 in.	25	20 ft $8\frac{3}{8}$ in.
13	20 ft $7\frac{3}{4}$ in.		

4.   A standard deck of cards is well shuffled several times and cut by a blindfolded individual. After the deck has been cut, the top card in the deck is turned up and the suit (clubs, spades, hearts, or diamonds) and the value (ace, deuce, ..., jack, queen, king) of the card are noted. The result of this process is one trial of a certain random experiment.

(a) List the simple results of this experiment.
(b) Consider the composite result E_1 consisting of simple results in which the card drawn is a heart. List the simple results belonging to E_1.
(c) If you conducted many trials of this experiment, at what number p do you think the relative frequency r.f.(E_1) would stabilize?
(d) Consider the composite result E_2 consisting of simple results in which the card drawn is a "face card" (king, queen, jack). List the simple results belonging to E_2.

(e) If you conducted many trials of this experiment, to what number p do you think r.f.(E_2) would tend as the number of trials becomes large?

5. In a cloud-seeding experiment, clouds are randomly chosen and categorized as

Category A: The cloud dissipates before producing rain.

Category B: The cloud produces rain somewhere else but not on the area over which it was first observed by the meteorologist.

Category C: The cloud produces rain on the area over which it was first observed.

Table E.4 gives the results of observations on 100 clouds, listed in the order in which these clouds were observed.

Table E.4: *Results of Observation of 100 Clouds*

(Read across)

A A A A B B C B A B C A C C B B A A B C
B A A A A B A A A C A A B C A B C A A B
A B B A B A A C C A C A A A B B B A A A
C C A B A A A B C C C A A A A B A A A B
C B A A B A B A B B C A B B A B A C A B

(a) Consider the composite result "The observed cloud produces rain somewhere." Construct a table showing how the relative frequency of this composite result varies with the number of trials. Graph the relative frequencies so obtained against the number of trials for which they were obtained (see Figure 3.1). Around what number p do the relative frequencies stabilize?

(b) Consider the result "The cloud does not produce rain on the area over which it was first observed." Construct a table and a graph for this result similar to the table and graph constructed in part (a). Around what number p do the relative frequencies stabilize?

6. A diagnostic computer receives a medical history and a list of symptoms from a patient and makes a diagnosis of the patient's problem. For each patient, a team of physicians also makes a diagnosis. If the diagnoses agree, an A is recorded; otherwise a D is recorded. The results for 80 patients are shown in Table E.5.

(a) The table can be regarded as giving the results of 80 trials of a certain random experiment. What are the simple results of that experiment?

(b) List all the different composite results for this experiment. How many such composite results are there?

(c) For each of the composite results that you listed in part (b), determine whether or not the relative frequencies for this result stabilize. Based on your conclusions, can the random experiment that you described in part (a) be modeled using probability theory?

Table E.5: *Results of the Diagnosis of 80 Patients*
(Read across)

A A A D A A A A A D A A A A A A A D D A
A A A A A A A A A D A A A A A A A A A D
A A A A D A A D A A A A A A D A A A A A
A A A A A A D A A A A A D A A A D A A A

2

The Elements

of Probability Theory

1. PRELIMINARIES

Our discussion has isolated three basic elements that are of major common interest in all random experiments: simple results, composite results, and relative frequencies. We now seek a mathematical structure with entities that correspond to and describe the properties of these basic elements.

Corresponding to simple results of a random experiment, we conceive of basic mathematical entities in our model called *outcomes*. We denote individual outcomes in our model by ω_1, ω_2, ω_3, and so on.

We have remarked that composite results of a random experiment are collections of simple results. In our model, we define an *event* to be a collection of outcomes. Corresponding to every composite result E of a random experiment, there is an event defined in the model for that experiment—namely, the event consisting of the outcomes identified with the simple results of E.

Example 1.1. In Exercise 2 of Chapter 1, we discussed a random experiment in which the simple results are the capital letters of the English alphabet. To form a model for this experiment, we start by assigning outcomes ω_1, ω_2, ..., ω_{25}, ω_{26} to the letters (simple results) A, B, ..., Y, Z of the experiment. One composite result in the experiment is the collection of vowels (A, E, I, O, U). The corresponding event in the model is the collection $\{\omega_1, \omega_5, \omega_9, \omega_{15}, \omega_{21}\}$, since A corresponds to ω_1, E to ω_5, I to ω_9, O to ω_{15}, and U to ω_{21}.

Both composite results and events are special cases of a *set*. A set is a well-defined collection of entities. Thus, a composite result is a set of simple

13

results, while an event is a set of outcomes. There is a mathematical theory of sets (*set theory*) that allows us to analyze the properties of complicated sets in terms of the properties of more simple collections of entities, and vice versa. In the following section, we provide a brief outline of some of the more useful concepts, rules, and methods of analysis of set theory. However, because we intend to use this theory only to deal with events, all of our discussion will be given in terms of events and outcomes.

2. A BRIEF OUTLINE OF SET THEORY

We have seen that events E are collections of outcomes ω. We can define an event E by giving a proposition or property $\mathscr{P}$ that is to be satisfied by every outcome in the event. In such a case, we write

(2.1) $E = \{\omega: \quad \omega \text{ satisfies property } \mathscr{P}\},$

or, more succinctly,

(2.2) $E = \{\omega \text{ satisfies property } \mathscr{P}\}.$

Another way to describe an event E is by enumeration of all the outcomes that belong to E. Thus, if E consists of the outcomes ω_1, ω_2, and ω_6, we write

$$E = \{\omega_1, \omega_2, \omega_6\}.$$

The event consisting of all the outcomes defined for a given model is called the *sample space* for that model and is denoted Ω. The event Ω is often called the *sure event* (or certain event) because Ω corresponds to the composite result that contains all the simple results of the random experiment and thus is sure to occur on every trial.

The event $\varnothing$ that has no outcomes is called the *impossible event* (or the *null event*), since $\varnothing$ corresponds to a composite event in the random experiment that has no simple results as members and thus cannot occur on any trial of the experiment.

Example 2.1. In Example 1.1, we considered a model in which the outcomes $\omega_1, \omega_2, \ldots, \omega_{26}$ corresponded to the capital letters A, B, ..., Z, respectively, of the English alphabet. A particular event in this model is the event E_1 consisting of outcomes that correspond to letters that are vowels. Thus, one way of characterizing the event E_1 is to write

$$E_1 = \{\omega: \quad \omega \text{ satisfies property } \mathscr{P}_1\},$$

where $\mathscr{P}_1$ is the proposition "ω corresponds to a letter (simple result) that is a vowel." Because it is very cumbersome to remind ourselves constantly of the correspondence between our model and the random experiment, *we adopt the convention that outcomes and events in probability models of random experiments will be described in terms of the language used to identify the corresponding simple results and composite results in the experiment.* Thus, we describe E_1 by

$$E_1 = \{\omega: \quad \omega \text{ is a vowel}\} = \{\omega \text{ is a vowel}\}.$$

We can also describe E_1 by listing its outcomes:

$$E_1 = \{\omega_1, \omega_5, \omega_9, \omega_{15}, \omega_{21}\}$$

or, again using our convention,

$$E_1 = \{A, E, I, O, U\}.$$

Another event in our model is the event E_2, consisting of outcomes corresponding to letters that make up the word "probability." Thus,

$$E_2 = \{\omega: \quad \omega \text{ is a letter in the word "probability"}\}$$
$$= \{\omega_1, \omega_2, \omega_9, \omega_{12}, \omega_{15}, \omega_{16}, \omega_{18}, \omega_{20}, \omega_{25}\}$$
$$= \{A, B, I, L, O, P, R, T, Y\}.$$

The sure event Ω of this example can be described as

$$\Omega = \{\omega_1, \omega_2, \ldots, \omega_{26}\}$$

or, alternatively, as

$$\Omega = \{\omega: \quad \omega \text{ is a letter in "The quick brown fox jumped over the lazy dogs."}\}.$$

The null event $\varnothing$ in this example can be described as

$$\varnothing = \{\omega: \quad \omega \text{ is } not \text{ a letter in "The quick brown fox jumped over the lazy dogs."}\}.$$

Example 2.2. Consider a model whose outcomes correspond to possible rankings on a scale from 1 through 10. Thus, the outcomes $\omega_1, \omega_2, \ldots, \omega_{10}$ in this model correspond to the numbers 1, 2, ..., 10. One event E_1 in this model consists of outcomes corresponding to ranks (numbers) larger than 5; that is,

$$E_1 = \{\omega: \quad \omega \text{ corresponds to a number larger than 5}\}$$
$$= \{\omega: \quad \omega > 5\}$$
$$= \{\omega_6, \omega_7, \omega_8, \omega_9, \omega_{10}\}.$$

Another event E_2 consists of outcomes corresponding to even ranks (even integers):

$$E_2 = \{\omega: \quad \omega \text{ is an even integer}\}$$
$$= \{\omega_2, \omega_4, \omega_6, \omega_8, \omega_{10}\}.$$

The sure event Ω is $\{\omega_1, \omega_2, \ldots, \omega_{10}\}$, or can be described as $\{\omega: \omega$ is either an even integer or an odd integer$\}$. The null event $\varnothing$ can be described as $\{\omega: \omega$ is both an even and an odd integer$\}$.

Inclusion of Events

An event E_1 is said to include another event E_2 if every outcome belonging to E_2 also belongs to E_1. This relation is illustrated in Figure 2.1 [called a *Venn diagram* in honor of the English logician John Venn (1834–1923)]. In the figure, the dots represent outcomes ω, the event E_1 is represented by the larger oval, and the event E_2 is represented by the smaller oval (contained in E_1). Notice that every dot (outcome) contained in E_2 is also contained in E_1.

If the event E_1 includes the event E_2, we write $E_2 \subset E_1$ or $E_1 \supset E_2$. If two events E_1 and E_2 each include the other, then these events must have the same outcomes as members, and we say that they are *equal* to one another (and write $E_1 = E_2$). Thus,

$$(2.3) \qquad E_1 = E_2 \quad \text{if and only if } E_1 \subset E_2 \text{ and } E_2 \subset E_1.$$

One way of showing that two complicated propositions $\mathscr{P}_1$ and $\mathscr{P}_2$ describe the same event (and thus are equivalent descriptions of the same composite result) is to show that $E_1 \subset E_2$ and $E_2 \subset E_1$, where

$$E_i = \{\omega: \omega \text{ satisfies } \mathscr{P}_i\}, \quad i = 1, 2.$$

The Complement of an Event

The *complement* E^c of an event E is the event consisting of all outcomes in the sample space Ω that are not members of E (see Figure 2.2); that is,

$$(2.4) \qquad E^c = \{\omega: \omega \text{ is not a member of } E\}.$$

Figure 2.1: *Venn diagram illustrating inclusion.*

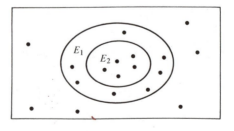

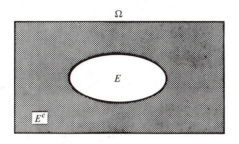

Figure 2.2: *Venn diagram illustrating the complement of an event.*

The Union of Two Events

The *union* of two events E_1 and E_2 is defined to be the event $E_1 \cup E_2$ (sometimes written $E_1 + E_2$) consisting of all outcomes that are members of E_1, of E_2, or of both E_1 and E_2 (represented by all shaded regions in Figure 2.3); that is,

(2.5) $E_1 \cup E_2 = \{\omega: \ \omega$ is a member of E_1, of E_2, or of both E_1 and $E_2\}$.

We observe that it follows directly from the definition of $E_1 \cup E_2$ that

(2.6) $$E_1 \subset (E_1 \cup E_2), \qquad E_2 \subset (E_1 \cup E_2).$$

Figure 2.3: *Venn diagram illustrating union and intersection of two events.*

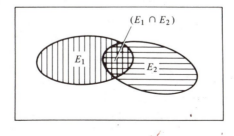

The Intersection of Two Events

The *intersection* of two events E_1 and E_2 is defined to be the event $E_1 \cap E_2$ (sometimes written $E_1 E_2$) whose outcomes are members of *both E_1 and E_2* (see Figure 2.3); that is,

(2.7) $\qquad E_1 \cap E_2 = \{\omega: \ \omega$ is a member of E_1 *and* of $E_2\}$.

Again, it follows from the definition of $E_1 \cap E_2$ that

(2.8) $\qquad\qquad (E_1 \cap E_2) \subset E_1, \qquad (E_1 \cap E_2) \subset E_2$.

We remark that there is a direct relationship between the operations of complementation, union, and intersection and the logical connectives between propositions (assertions) describing composite results. For example, if E is an event consisting of outcomes corresponding to simple results for which the assertion $\mathscr{P}$ is true, then E^c is the event whose outcomes correspond to simple results for which the assertion $\mathscr{P}$ is false (*not* $\mathscr{P}$).

Example 2.2 (continued). The operations of union and intersection on the events $E_1 = \{\omega: \ \omega$ exceeds 5$\}$ and $E_2 = \{\omega: \ \omega$ is even$\}$ described in Example 2.2 yield

$$E_1 \cup E_2 = \{\omega: \ \omega \text{ exceeds 5 } or \ \omega \text{ is even}\}$$
$$= \{\omega_2, \omega_4, \omega_6, \omega_7, \omega_8, \omega_9, \omega_{10}\},$$
$$E_1 \cap E_2 = \{\omega: \ \omega \text{ exceeds 5 } and \ \omega \text{ is even}\}$$
$$= \{\omega_6, \omega_8, \omega_{10}\}.$$

The complement of E_1 is

$$E_1^c = \{\omega: \ \omega \text{ does not exceed 5}\} = \{\omega: \ \omega \le 5\}$$
$$= \{\omega_1, \omega_2, \omega_3, \omega_4, \omega_5\},$$

while the complement of E_2 is

$$E_2^c = \{\omega: \ \omega \text{ is not even}\} = \{\omega: \ \omega \text{ is odd}\} = \{\omega_1, \omega_3, \omega_5, \omega_7, \omega_9\}.$$

Union and Intersection of Three or More Events

We can define the union of three or more events. For example, the union $E_1 \cup E_2 \cup E_3$ of the events E_1, E_2, E_3 is the event whose outcomes each belong to one (or more) of the events E_1, E_2, E_3 (see Figure 2.4); that is,

(2.9) $\quad E_1 \cup E_2 \cup E_3 = \{\omega: \ \omega$ is a member of E_1, of E_2, *or* of $E_3\}$.

In general, the union $E_1 \cup E_2 \cup \cdots \cup E_k$ of k events is defined to be the event whose outcomes belong to one or more of the events $E_1, E_2, \ldots, E_k$.

Analogously, we can define the intersection of three or more events.

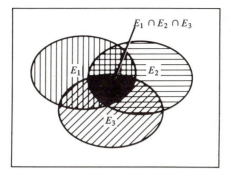

Figure 2.4: *Venn diagram illustrating union and intersection of three events.*

Given E_1, E_2, E_3, their intersection $E_1 \cap E_2 \cap E_3$ is the event whose outcomes are members of *all* the events E_1, E_2, E_3 (see Figure 2.4); that is,

$$(2.10) \qquad E_1 \cap E_2 \cap E_3 = \{\omega: \ \omega \text{ is a member of } E_1, E_2, \text{ and } E_3\}.$$

In general, the intersection $E_1 \cap E_2 \cap \cdots \cap E_k$ of k events is defined to be the event whose outcomes are members of all the events $E_1, E_2, \ldots, E_k$.

Mutually Exclusive Events

If two events have no outcomes in common, then we say that they are *mutually exclusive* or *disjoint*. If E_1 and E_2 are mutually exclusive events, then

$$E_1 \cap E_2 = \varnothing,$$

since the fact that E_1 and E_2 have no outcomes in common means that $E_1 \cap E_2$ contains no outcomes as members, and thus is the null event.

The relationship of mutual exclusiveness between events corresponds to the mutual logical incompatibility of the propositions defining these events.

Example 2.3. Suppose that the outcomes of a probability model correspond to all the chemical elements. Let E_1 be the event consisting of outcomes corresponding to chemical elements belonging to the family of halogens; that is,

$$E_1 = \{\text{bromine, chlorine, fluorine, iodine, astatine}\}.$$

Let E_2 be the event consisting of outcomes corresponding to all metals, and let E_3 be the event consisting of outcomes that correspond to solids. Then

E_1 and E_2 are mutually exclusive because no chemical element can be both a halogen and a metal, while the events E_1 and E_3 are not mutually exclusive since they have the outcome corresponding to the element iodine in common.

Some Useful Rules and Methods of Derivation of Rules

From the definitions of union, intersection, and complementation, we can derive the following frequently useful rules for analyzing events.

Rule 2.1. For any event E,

(2.11)   (a) $E \cap E^c = \varnothing$,   (b) $E \cup E^c = \Omega$.

Rule 2.2. $\Omega^c = \varnothing$, $\varnothing^c = \Omega$.

Rule 2.3. For any event E,

(a) $\varnothing \subset E \subset \Omega$,   (b) $E \cap \Omega = E$,
(c) $E \cap \varnothing = \varnothing$,   (d) $E \cup \Omega = \Omega$,
(e) $E \cup \varnothing = E$,   (f) $(E^c)^c = E$.

The reader may prove these rules either by drawing appropriate Venn diagrams or by the use of logical arguments. To use a logical argument to prove Rule 2.3(f), for example, we could show that $(E^c)^c \subset E$ and that $E \subset (E^c)^c$, and then use the definition (2.3) of what it means for two events to be equal. [If an outcome ω is in E, then it is not in E^c by definition of E^c, and thus ω is in $(E^c)^c$. It follows that $E \subset (E^c)^c$. On the other hand, if ω is in $(E^c)^c$, then it is not in E^c, so that ω must be in E. Hence, $(E^c)^c \subset E$, and we have shown that $E = (E^c)^c$.]

The following two rules relate the operations of union and intersection to one another.

Rule 2.4. For any events E, E_1, E_2,

(2.12) $E \cup (E_1 \cap E_2) = (E \cup E_1) \cap (E \cup E_2)$.

Rule 2.5. For any events E, E_1, E_2,

(2.13) $E \cap (E_1 \cup E_2) = (E \cap E_1) \cup (E \cap E_2)$.

We note that it follows from the definitions for the union and intersection of three events that

(2.14) $E \cup (E_1 \cup E_2) = (E \cup E_1) \cup E_2 = E \cup E_1 \cup E_2$,

(2.15) $E \cap (E_1 \cap E_2) = (E \cap E_1) \cap E_2 = E \cap E_1 \cap E_2$.

Equations (2.12) through (2.15) can be proved with the kind of argument used above to demonstrate Rule 2.3(f). For example, to prove (2.12), note that if an outcome ω is in $E \cup (E_1 \cap E_2)$, then ω is in E or ω is in $(E_1 \cap E_2)$. If ω is in E, then ω is in $E \cup E_1$ and also in $E \cup E_2$, and thus ω is in $(E \cup E_1) \cap (E \cup E_2)$. If ω is in $E_1 \cap E_2$, then ω is in both E_1 and E_2, and thus ω is in $E \cup E_1$ and in $E \cup E_2$, and consequently in $(E \cup E_1) \cap (E \cup E_2)$. It follows that

$$E \cup (E_1 \cap E_2) \subset (E \cup E_1) \cap (E \cup E_2).$$

In similar fashion, we can show that

$$(E \cup E_1) \cap (E \cup E_2) \subset E \cup (E_1 \cap E_2),$$

and then apply (2.3) to complete the proof.

The next rule comprises *de Morgan's Laws*, named for the English mathematician Augustus de Morgan (1806–1871), and relates the operation of complementation to the operations of union and of intersection.

Rule 2.6. For any two events E_1, E_2,

(2.16) $$(E_1 \cup E_2)^c = E_1^c \cap E_2^c,$$

(2.17) $$(E_1 \cap E_2)^c = E_1^c \cup E_2^c.$$

We could prove these results by means of the basic inclusion arguments used to prove Rules 2.3(f) and 2.4. However, the rules we have already given provide a direct way of proving (2.17), once (2.16) has been established. To see this, note that by (2.16) and Rule 2.3(f),

$$(E_1^c \cup E_2^c)^c = (E_1^c)^c \cap (E_2^c)^c = E_1 \cap E_2.$$

Thus, using Rule 2.3(f) once again, we obtain

$$(E_1 \cap E_2)^c = ((E_1^c \cup E_2^c)^c)^c = E_1^c \cup E_2^c,$$

which is the result of (2.17).

Inductive Derivations

Rules 2.4 through 2.6, and Equations (2.15) and (2.16), can be generalized to cover any number of events $E, E_1, E_2, \ldots, E_k$. For example, it follows from the definitions of union and intersection for k events that

(2.18) $$E_1 \cup E_2 \cup \cdots \cup E_k = E_1 \cup (E_2 \cup \cdots \cup E_k)$$

and

(2.19) $$E_1 \cap E_2 \cap \cdots \cap E_k = E_1 \cap (E_2 \cap \cdots \cap E_k).$$

Rule 2.4 has the following generalization: For any events $E, E_1, E_2, \ldots, E_k$,

(2.20)
$$E \cup (E_1 \cap E_2 \cap \cdots \cap E_k) = (E \cup E_1) \cap (E \cup E_2) \cap \cdots \cap (E \cup E_k).$$

We can verify Equation (2.20) by making repeated use of Rule 2.4 and Equation (2.18),

$$E \cup (E_1 \cap E_2 \cap \cdots \cap E_k) = E \cup (E_1 \cap (E_2 \cap \cdots \cap E_k))$$
$$= (E \cup E_1) \cap (E \cup (E_2 \cap \cdots \cap E_k))$$
$$= (E \cup E_1) \cap (E \cup E_2) \cap (E \cup (E_3 \cap \cdots \cap E_k)),$$

and so on, proceeding as shown above until (2.20) is obtained. This same kind of *inductive argument* can be used to obtain the generalization

(2.21)
$$E \cap (E_1 \cup \cdots \cup E_k) = (E \cap E_1) \cup (E \cap E_2) \cup \cdots \cup (E \cap E_k)$$

of Rule 2.5, and the generalizations

(2.22) $$(E_1 \cup E_2 \cup \cdots \cup E_k)^c = E_1^c \cap E_2^c \cap \cdots \cap E_k^c,$$

(2.23) $$(E_1 \cap E_2 \cap \cdots \cap E_k)^c = E_1^c \cup E_2^c \cup \cdots \cup E_k^c$$

of (2.16) and (2.17) of Rule 2.6.

The Correspondence Between the Probability Model and the Random Experiment

In Table 2.1, we summarize the correspondence established so far between our mathematical model and the random experiment that we wish to model. We have seen how the correspondences we have created between composite results and events can allow us to use the theory of sets, as applied

Table 2.1: *Relationships Between Model and Experiment*

Random experiment	Probability model
Simple Result: The basic occurrences of the experiment.	*Outcome:* The basic elements of the model.
Totality of Simple Results: All simple results that can occur in the experiment.	*Sample Space:* All possible outcomes.
Composite Result: A collection of simple results having some property.	*Event:* A collection of outcomes obeying some proposition.
Relative Frequency: The proportion of times a given composite result is observed in repeated trials of the random experiment.	?

to the events of our model, to analyze complicated logical assertions about the results of the random experiment. We now need a mathematical entity to fill the blank opposite "relative frequency" in our table. This entity is the probability of an event.

3. FORMAL STRUCTURE: PROBABILITIES

We have already assumed that in a large number of repeated trials of our random experiment the relative frequency of every composite result stabilizes to a certain number, the long-run relative frequency of that composite result. Correspondingly, we assume that for every event E, there exists a number $P(E)$, called the *probability* of the event E, that is the idealization of the long-run relative frequency of the composite result corresponding to E. In order to make the analogy between the concepts of probability and relative frequency useful, we would like the probability of an event to have all the important properties of the relative frequency of a composite result.

We first recall that the relative frequency r.f.(E) of a composite (or simple) result E in N trials is $\#E/N$, where $\#E$ is the number of times that the composite result E occurred in the N trials. Since $\#E$ and N are nonnegative numbers, and $\#E$ cannot exceed N, we must have $0 \leq \text{r.f.}(E) \leq 1$. We have already remarked that the sure event Ω corresponds to a composite result that is certain to occur (has relative frequency 1), and that the impossible event $\varnothing$ corresponds to a composite result that has relative frequency 0. Finally, if two composite results E_1 and E_2 cannot happen together on the same trial (they have no simple results in common, they are mutually exclusive), then

$$\#(E_1 \text{ or } E_2) = \#E_1 + \#E_2$$

and it follows that

(3.1) $$\text{r.f.}(E_1 \cup E_2) = \text{r.f.}(E_1) + \text{r.f.}(E_2).$$

Remark. Equation (3.1) is not, in general, true if the two composite results E_1 and E_2 have simple results in common.

The three properties that we have just described are some basic properties obeyed by relative frequencies. Corresponding to these basic properties, we demand that the probabilities of events obey the following axioms.

Axiom 1. For every event E, $0 \leq P(E) \leq 1$.

Axiom 2. $P(\Omega) = 1, \qquad P(\varnothing) = 0.$

Axiom 3. If E_1 and E_2 are mutually exclusive events, then

$$P(E_1 \cup E_2) = P(E_1) + P(E_2).$$

The null event $\varnothing$ need not be the only event that has probability 0. Certain events may be such that they correspond, loosely speaking, to composite results that are so rare that they "happen only once in a lifetime." The relative frequencies of such composite results in a large number of trials will be almost indistinguishable from 0. The conceptual importance of this fact will become apparent when we discuss random variables in Chapter 4.

4. THE PROBABILITY CALCULUS

If a sample space consists of 15 outcomes, then there are $2^{15} = 32,768$ distinct events in the probability model. The task of determining the probabilities for each of these events seems quite formidable. Fortunately, using Axioms 1, 2, and 3, we can construct a probability calculus that allows us to evaluate the probability of every event once we know the probabilities of certain basic events.

Although there are many rules of the probability calculus, the following are among the most useful for practical calculations.

Law of Inclusion

Recall that an event E_2 is said to be included in an event E_1 (denoted by $E_2 \subset E_1$) if every outcome that belongs to E_2 also belongs to E_1.

Figure 4.1: *Comparison of probabilities for included events.*

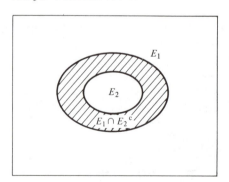

Rule 4.1. If $E_2 \subset E_1$, then $P(E_2) \le P(E_1)$.

We could have guessed that Rule 4.1 must hold for probabilities by considering the corresponding fact for relative frequencies. Suppose we have two composite results E_1 and E_2 such that every simple result in E_2 also belongs to E_1. In N repeated trials, E_1 will occur every time that E_2 occurs (and also, possibly, in cases when E_2 does not occur). Hence, $\#E_2 \le \#E_1$. Dividing both sides of this inequality by N gives r.f.$(E_2) \le$ r.f.(E_1).

We may formally derive Rule 4.1 from the axioms of probability through use of the Venn diagram shown in Figure 4.1. Note from the diagram that

$$E_1 = E_2 \cup (E_1 \cap E_2^c)$$

and that E_2 and $E_1 \cap E_2^c$ have no outcomes in common (they are mutually exclusive). From Axiom 3,

$$P(E_1) = P(E_2 \cup (E_1 \cap E_2^c)) = P(E_2) + P(E_1 \cap E_2^c) \ge P(E_2),$$

since $P(E_1 \cap E_2^c) \ge 0$ by Axiom 1.

Law of Complementation

Rule 4.2. For any event E, $P(E) = 1 - P(E^c)$.

Formal verification of this rule follows by noting from Rule 2.1 that $E \cup E^c = \Omega$ and $E \cap E^c = \varnothing$, and from Axiom 2 that $P(\Omega) = 1$, and then applying Axiom 3 to obtain $P(E) + P(E^c) = 1$.

Law of Addition

When E_1 and E_2 are mutually exclusive, then Axiom 3 tells us that $P(E_1 \cup E_2) = P(E_1) + P(E_2)$. However, in general, two events E_1 and E_2 need not be mutually exclusive (see Figure 4.2 for an illustration), and in this case the values of $P(E_1)$ and $P(E_2)$ do *not* provide enough information to allow us to compute $P(E_1 \cup E_2)$.

Indeed, we see from the Venn diagram in Figure 4.2 that

(4.1) $$E_1 \cup E_2 = E_1 \cup (E_1^c \cap E_2),$$

where E_1 and $(E_1^c \cap E_2)$ are mutually exclusive events. Thus, from Axiom 3, we have

(4.2) $$P(E_1 \cup E_2) = P(E_1) + P(E_1^c \cap E_2).$$

Again, from Figure 4.2, we see that

(4.3) $$E_2 = (E_1 \cap E_2) \cup (E_1^c \cap E_2),$$

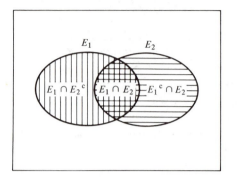

Figure 4.2: *The probability of the union of two events.*

where $(E_1 \cap E_2)$ and $(E_1^c \cap E_2)$ are mutually exclusive events. Thus, Axiom 3 yields

(4.4) $$P(E_2) = P(E_1 \cap E_2) + P(E_1^c \cap E_2),$$

from which we obtain

(4.5) $$P(E_1^c \cap E_2) = P(E_2) - P(E_1 \cap E_2).$$

Plugging the results of Equation (4.5) into Equation (4.2) yields the following result.

Rule 4.3. For any two events E_1 and E_2,

$$P(E_1 \cup E_2) = P(E_1) + P(E_2) - P(E_1 \cap E_2).$$

From Figure 4.2, we see that adding $P(E_1)$ and $P(E_2)$ counts the overlap $P(E_1 \cap E_2)$ between E_1 and E_2 twice, whereas in $P(E_1 \cup E_2)$ this overlap is counted only once. Rule 4.3 corrects for this counting.

Rule 4.3 can be extended to more than two events. For example, the following is the corresponding rule for calculating the probability of the union of three events E_1, E_2, E_3.

Rule 4.4. For any three events E_1, E_2, E_3,

$$P(E_1 \cup E_2 \cup E_3) = P(E_1) + P(E_2) + P(E_3) - P(E_1 \cap E_2)$$
$$- P(E_1 \cap E_3) - P(E_2 \cap E_3) + P(E_1 \cap E_2 \cap E_3).$$

One way of verifying Rule 4.4 is to use a Venn diagram (see Figure 2.4). By breaking up the total area into seven mutually exclusive areas

$(E_1 \cap E_2 \cap E_3, E_1 \cap E_2 \cap E_3^c,$ etc.) and using an argument similar to the one used to derive Rule 4.3, we may verify Rule 4.4.

An alternative method is to proceed recursively, applying Rule 4.3 first to the events E_1 and $E_2 \cup E_3$ [see Equation (2.14)] and then to the events E_2 and E_3. We obtain

(4.6)
$$
\begin{aligned}
P(E_1 \cup E_2 \cup E_3) &= P(E_1 \cup (E_2 \cup E_3)) \\
&= P(E_1) + P(E_2 \cup E_3) - P(E_1 \cap (E_2 \cup E_3)) \\
&= P(E_1) + P(E_2) + P(E_3) \\
&\quad - P(E_2 \cap E_3) - P(E_1 \cap (E_2 \cup E_3)).
\end{aligned}
$$

However, from Rule 2.5,

$$
E_1 \cap (E_2 \cup E_3) = (E_1 \cap E_2) \cup (E_1 \cap E_3),
$$

so that a third application of Rule 4.3 yields

(4.7) $P(E_1 \cap (E_2 \cup E_3)) = P(E_1 \cap E_2) + P(E_1 \cap E_3) - P(E_1 \cap E_2 \cap E_3).$

Substituting (4.7) into (4.6) yields the result of Rule 4.4.

An important special case of Rule 4.4 occurs when the events E_1, E_2, E_3 have no outcomes in common (i.e., $E_1 \cup E_2 = \varnothing$, $E_1 \cup E_3 = \varnothing$, $E_2 \cap E_3 = \varnothing$). From Equation (2.8) and Rule 2.3(a), we have

$$
\varnothing \subset E_1 \cap E_2 \cap E_3 = E_1 \cap (E_2 \cap E_3) \subset E_2 \cap E_3 = \varnothing,
$$

so that $E_1 \cap E_2 \cap E_3 = \varnothing$. Since $P(\varnothing) = 0$, it follows from Rule 4.4 that when E_1, E_2, and E_3 have no outcomes in common,

$$
P(E_1 \cup E_2 \cup E_3) = P(E_1) + P(E_2) + P(E_3).
$$

Using a recursive argument, we can generalize this result as follows.

Rule 4.5. If E_1, E_2, ..., E_k are k events, no two of which have outcomes in common, then

$$
P(E_1 \cup E_2 \cup \cdots \cup E_k) = P(E_1) + P(E_2) + \cdots + P(E_k).
$$

Remark. Suppose that we consider a probability model in which the outcomes ω_i correspond to the positive integers 1, 2, ... If we want to determine the probability of the event $\{\omega$ is an even integer$\}$ and we know the probabilities of the mutually exclusive events $E_i = \{\omega_i\}$, $i = 1, 2, \ldots$, consisting of a single outcome, then it seems reasonable that

$$
P\{\omega \text{ is an even integer}\} = P(\omega = 2 \text{ or } 4 \text{ or } 6 \text{ or } 8 \text{ or } \cdots)
$$

$$
= P(E_2 \cup E_4 \cup E_6 \cup E_8 \cup \cdots)
$$

should be equal to the sum $\sum_{i=1}^{\infty} P(E_{2i})$ of the probabilities of the events E_2,

E_4, E_6, E_8, ... Unfortunately, this result cannot be proved with Axioms 1, 2, and 3. In particular, Rule 4.5 only applies to a union of a *finite* number k of mutually exclusive events, whereas the event $\{\omega$ is an even integer$\}$ is a union of an infinite number of events E_i.

A mathematically correct and completely general theory of probability requires that we replace Axiom 3 by the following:

Axiom 3′. For any countable collection E_1, E_2, E_3, ... of events, no two of which have outcomes in common,

$$P(E_1 \cup E_2 \cup E_3 \cup \cdots) = \sum_{i=1}^{\infty} P(E_i).$$

This axiom corresponds to dropping the requirement in Rule 4.5 that k be finite; in the remainder of this book our only use of Axiom 3′ will be to generalize Rule 4.5. However, the reader should be aware that a rigorous treatment of probability theory (in particular, the material in Chapters 4, 5, and 7 through 10) requires that we assume that Axiom 3′ holds for the probability model.

We now give some examples of the uses of Rules 4.1 through 4.5.

Example 4.1. Police statistics in a certain city indicate that of all cars stolen, 9% are 10 or more years old, 11% are 5–9 years old, 15% are 3–4 years old, 20% are 2 years old, 20% are 1 year old, and 25% are less than 1 year old.

In this example, our sample space can consist either of outcomes corresponding to all possible ages for the stolen cars or of the outcomes $\omega_1 = 10$ years or older, $\omega_2 = 5$–9 years, $\omega_3 = 3$–4 years, $\omega_4 = 2$ years, $\omega_5 = 1$ year, and $\omega_6 = $ less than 1 year. We choose the latter sample space because the police statistics give us probabilities for the events $E_i = \{\omega_i\}$, $i = 1, 2, \ldots, 6$, as follows:

Age of cars (in years)	Over 10	5–9	3–4	2	1	Less than 1
Event	E_1	E_2	E_3	E_4	E_5	E_6
Probability	0.09	0.11	0.15	0.20	0.20	0.25

The probability that a stolen car is at least 1 year old is the probability of the event $A = E_1 \cup E_2 \cup E_3 \cup E_4 \cup E_5$, where no two of the events E_1, E_2, E_3, E_4, E_5 have outcomes in common. Thus, from Rule 4.5,

$$P(A) = P(E_1 \cup E_2 \cup E_3 \cup E_4 \cup E_5) = \sum_{i=1}^{5} P(E_i)$$

$$= 0.09 + 0.11 + 0.15 + 0.20 + 0.20 = 0.75.$$

However, it is far easier to note that $A = E_6^c$ and apply Rule 4.2:

$$P(A) = P(E_6^c) = 1 - P(E_6) = 1 - 0.25 = 0.75.$$

Remark. Notice that in Example 4.1, we have used the letter A to denote an event. In the following section, we will often use capital English letters other than E to denote events.

Example 4.2. The results of a nationwide survey show that 75% of individuals surveyed obtain current news information from television, 35% obtain current news from radio, and 25% obtain current news from both radio and television. What percent of all individuals surveyed obtained current news either from television *or* from radio?

Although the information given us is in terms of relative frequencies (percentage divided by 100), we know that relative frequencies and probabilities obey similar rules. Thus, if A is the composite result "the individual obtains current news from television" and B is the composite result "the individual obtains current news from radio," we know that r.f.$(A) = 0.75$, r.f.$(B) = 0.35$, r.f.$(A$ *and* $B) = 0.25$, and from Rule 4.3,

$$\text{r.f.}(A \text{ } or \text{ } B) = \text{r.f.}(A) + \text{r.f.}(B) - \text{r.f.}(A \text{ } and \text{ } B)$$

$$= 0.75 + 0.35 - 0.25 = 0.85.$$

Thus, $100(0.85) = 85\%$ of all individuals surveyed obtained current news either from television or from radio.

Example 4.3. An engineer is concerned that natural-gas-pipeline breaks might occur and shut off the flow of gas to certain sections of the country. Figure 4.4 illustrates a network of pipelines servicing various areas of the country. We assume that a break in the network, when it occurs, can occur in one and only one segment of the network. Let ω_i correspond to a break in segment i, $i = 1$, 2, ..., 10. The event $E_i = \{\omega_i\}$ corresponds to the composite result "pipeline segment i breaks." From previous experience the engineer has calculated the probabilities that appear in the following table.

Event	E_1	E_2	E_3	E_4	E_5	E_6	E_7	E_8	E_9	E_{10}
Probability	0.096	0.066	0.109	0.049	0.027	0.149	0.147	0.163	0.092	0.102

Note that no two of the events $E_1, \ldots, E_{10}$ have outcomes in common.

Let A be the event corresponding to the composite result "a pipeline segment leading to region I breaks" and let B be the event corresponding to

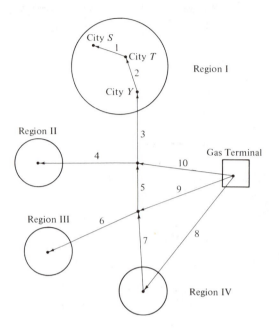

Figure 4.3: *Network of pipelines serving the various regions of the country. The branch serving region I consists of segments 1, 2, and 3.*

the composite result "a segment leading to region II breaks." From Figure 4.3, applying Rule 4.5, we find that

$$P(A) = P(E_1 \cup E_2 \cup E_3 \cup E_5 \cup E_7 \cup E_8 \cup E_9 \cup E_{10})$$

$$= P(E_1) + P(E_2) + P(E_3) + P(E_5) + P(E_7)$$
$$+ P(E_8) + P(E_9) + P(E_{10})$$

$$= 0.096 + 0.066 + 0.109 + 0.027 + 0.147$$
$$+ 0.163 + 0.092 + 0.102$$

$$= 0.802,$$

and, similarly,

$$P(B) = P(E_4 \cup E_5 \cup E_7 \cup E_8 \cup E_9 \cup E_{10}) = 0.580,$$

$$P(A \cap B) = P(E_5 \cup E_7 \cup E_8 \cup E_9 \cup E_{10}) = 0.531.$$

Thus, the probability that a segment leading to either region I or region II breaks, by Rule 4.3, is

$P(A \cup B) = P(A) + P(B) - P(A \cap B) = 0.802 + 0.580 - 0.531 = 0.851.$

5. FITTING A PROBABILITY MODEL

Having defined the probability of an event, we have completed the general construction of probability models. To the simple results, composite results, totality of simple results, and the long-run relative frequencies of a random experiment, we have assigned the outcomes, events, sample space, and probabilities, respectively, of our probability model. We can now discuss the random experiment completely in terms of the probability model. To do so, we need to determine the probability associated with each event in the model. The rule or list that assigns a probability to every event is called the *probability measure* for the model.

In Section 4, we noted that in order to determine the probability measure for a model, only probabilities for certain basic events need to be found. In practice, such values are determined in one of two different ways. We may actually take a large number of trials of the random experiment, calculate the resulting relative frequencies, and set the probabilities equal to these relative frequencies. Alternatively, we may employ theoretical arguments to specify that certain mathematical relationships are to hold between the probabilities of the basic events. To conclude this section, we give an example of each of these two methods for determining the probability measure of a random experiment.

Example 5.1 (Experimental Method). Insurance companies have traditionally been concerned with records of human life spans. These are usually summarized in *life tables*, the earliest example of which appears to have been given by Captain John Graunt (1620–1674) in 1662. (The foundation of a theory of life annuities is due to the English astronomer Edmund Halley in 1693.) An example of a life table is given in Table 5.1. The simple results are the ages at death (last birthday); the probabilities are actually relative frequencies calculated over thousands of observed life spans.

Example 5.2 (Theoretical Approach). Suppose that we are concerned with the number of fatal automobile accidents that occur over a typical weekend. This is a random experiment whose simple results are nonnegative integers 0, 1, 2, ... Using certain theoretical assumptions about the nature of automobile accidents, the probability of d deaths on the weekend can be shown to have the form

$$P\{d \text{ deaths}\} = \frac{e^{-\lambda}\lambda^d}{d!},$$

Table 5.1: *Life Table for Individuals Living in the City of Breslau (Wroclaw),*
Poland, 1693

Age	Probability	Age	Probability	Age	Probability	Age	Probability
0	0.145	21	0.007	42	0.010	63	0.010
1	0.057	22	0.006	43	0.010	64	0.010
2	0.038	23	0.006	44	0.010	65	0.010
3	0.028	24	0.007	45	0.010	66	0.010
4	0.022	25	0.007	46	0.010	67	0.010
5	0.018	26	0.007	47	0.010	68	0.010
6	0.012	27	0.007	48	0.011	69	0.011
7	0.010	28	0.008	49	0.011	70	0.011
8	0.009	29	0.008	50	0.011	71	0.011
9	0.008	30	0.008	51	0.011	72	0.011
10	0.007	31	0.008	52	0.011	73	0.010
11	0.006	32	0.008	53	0.010	74	0.010
12	0.006	33	0.009	54	0.010	75	0.010
13	0.006	34	0.009	55	0.010	76	0.010
14	0.006	35	0.009	56	0.010	77	0.009
15	0.006	36	0.009	57	0.010	78	0.008
16	0.006	37	0.009	58	0.010	79	0.007
17	0.006	38	0.009	59	0.010	80	0.006
18	0.006	39	0.009	60	0.010	81	0.005
19	0.006	40	0.009	61	0.010	82	0.023
20	0.006	41	0.010	62	0.010	and above	

where λ is a positive number (called the *parameter*) depending on the rate at which accidents occur.

Of course, one might challenge the theoretical arguments that led to this probability measure. In this case, our only recourse is to take experimental data and show that the probabilities computed from our model agree to a close enough approximation with the observed relative frequencies.

NOTES AND REFERENCES

One of the advantages of the mathematical conceptualization of a real phenomenon is that a given conceptualization may serve, with suitable translations in language, as a model for more than one reality. Probabilities were conceptualized to correspond to relative frequencies in repeated trials of a random experiment. Such probabilities are sometimes called *objective probabilities*. However, a number of present-day statisticians believe that the probability calculus can also serve as a model for the way in which a "rational" man assesses his own belief in the truth of a given statement. In most cases the statement is, objectively, either true or false. However, the rational man may not know what is the case (because he has not been informed, because the event to which the statement applies will happen in the future, and so on). Nevertheless, he may be able to make an assessment as to whether the statement is true or false. His assessment is a personal number (a probability), based on the evidence available to him, signifying the degree to which he *believes* the given statement to be true. When probabilities are given this kind of interpretation, they are referred to as *subjective probabilities*. The probability calculus to be presented in the remainder of this chapter, and in the succeeding chapters, is valid *no matter which interpretation of the probabilities* (objective or subjective) *is used*.

The reader interested in further discussion of the meaning, construction, and uses of probability models is encouraged to read the *Scientific American* survey articles by Ayer (1965), Kac (1964), and Weaver (1950, 1952). The elementary textbooks of Derman, Gleser, and Olkin (1973), Dwass (1967), Goldberg (1960), Guenther (1968), Hodges and Lehmann (1970), and Mosteller, Rourke, and Thomas (1961), and the more advanced textbooks of Feller (1968), Kolmogorov (1933), and Von Mises (1957) provide alternative approaches to this subject.

Some idea of the history, philosophy, and theoretical foundations of the subjectivist approach to probability theory can be obtained by reading the books by Savage (1954), Kyburg and Smokler (1964), and Winkler (1972).

EXERCISES

1. In Example 2.1, describe each of the following events by listing the outcomes that belong to that event.
   (a) $E_1 \cup E_2$,   (b) $E_1 \cap E_2$,   (c) E_1^c,   (d) E_2^c,
   (e) $E_1^c \cup E_2^c$,   (f) $E_1^c \cap E_2^c$,   (g) $E_1 \cap E_2^c$,   (h) $E_1^c \cap E_2$.
   Describe in words the composite results corresponding to each event.

2. In Example 2.2, describe each of the following events by listing the outcomes that belong to that event.
   (a) $E_1^c \cap E_2$,   (b) $E_1^c \cap E_2^c$,
   (c) $E_1^c \cap E_2 \cap E_2^c$,   (d) $E_1^c \cup E_2$,
   (e) $E_1^c \cup E_2^c$,   (f) $E_1^c \cup E_2 \cup E_2^c$,
   (g) $(E_1^c \cap E_2) \cup E_2^c$,   (h) $(E_1^c \cup E_2^c) \cap (E_2 \cup E_2^c)$,
   (i) $(E_1^c \cup E_2)^c$.
   Describe in words the composite results corresponding to each event.

3. A sample space Ω consists of the 52 outcomes $\omega_1, \omega_2, \ldots, \omega_{52}$. The event E_1 consists of outcomes $\omega_1, \omega_2, \ldots, \omega_{13}$. The event E_2 consists of outcomes $\omega_{11}, \omega_{12}, \omega_{13}, \omega_{24}, \omega_{25}, \omega_{26}, \omega_{37}, \omega_{38}, \omega_{39}, \omega_{50}, \omega_{51}, \omega_{52}$. What outcomes belong to
    (a) $E_1 \cup E_2$,   (b) $E_1 \cap E_2$,   (c) $E_1^c \cap E_2$,
    (d) $E_1 \cap E_2^c$,   (e) $E_1^c \cap E_2^c$,   (f) $(E_1^c \cup E_2)^c$?
    Consider a random experiment in which a blindfolded individual picks one card from a deck of 52 cards. The simple results of this experiment are "ace of clubs," "two of clubs," ..., "king of clubs," "ace of diamonds," ..., "king of diamonds," "ace of hearts," ..., "king of hearts," "ace of spades," ..., "king of spades." Show that the sample space Ω provides a model for this experiment by indicating how you would assign outcomes to simple results. Then, describe in words the composite results corresponding to E_1, E_2, and the events (a) through (f).

4. (a) Show that $E_1 \subset E_2$ if and only if $E_1^c \supset E_2^c$.
    (b) Show that $E_1 = E_2$ if and only if $E_1^c = E_2^c$.
    (c) Show that $E_1 \subset E_2$ and $E_2 \subset E_3$ implies that $E_1 \subset E_3$.
    (d) Show that if $E_1 \subset E_2$, then $E_1 \cap E_2 = E_1$ and $E_1 \cup E_2 = E_2$.

5. (a) Complete the proof of Rule 2.4 by showing that
    $$(E \cup E_1) \cap (E \cup E_2) \subset E \cup (E_1 \cap E_2).$$
    (b) Prove Rule 2.5 by the method used in the text to prove Rule 2.4.
    (c) Prove Equation (2.16) of Rule 2.6 by the method used in the text to prove Rule 2.4.

6. Use Rule 2.3(f), Rule 2.4, and Rule 2.6 to prove Rule 2.5. [Hint: Start by showing that $(E^c \cup (E_1^c \cap E_2^c))^c$ equals $(E^c \cup E_1^c) \cap (E^c \cup E_2^c)$.]

7. Use inductive arguments to verify Equations (2.21), (2.22), and (2.23).

8. A family with exactly three children is selected by lottery, and the sexes (M = male, F = female) of the children are recorded in sequence of birth. Altogether there are eight simple results of this experiment: MMM, MFF, MMF, MFM, FMM, FFM, FMF, and FFF. Let the outcomes $\omega_1, \omega_2, \omega_3, \omega_4, \omega_5, \omega_6, \omega_7, \omega_8$ correspond respectively to these simple results.
    (a) What event E_1 corresponds to the composite result that exactly two of the three children are male?
    (b) What event E_2 corresponds to the composite result that the first-born child is a female?
    (c) Suppose that when you observe the family chosen in your random experiment you only count the *number* of male children. What are the simple results of this experiment? What information do you lose by only counting the number of male children?

9. Let E_1, E_2, E_3 be three events associated with a random experiment. Express the following verbal statements in set notation.
   (a) At least one of the events occurs.
   (b) Exactly one of the events occur.
   (c) Exactly two of the events occur.
   (d) Not more than two of the events occur simultaneously.
   (e) All of the events occur.
   (f) None of the events occurs.

10. Which of the following pairs of events A and B are mutually exclusive?
    (a) A: being the son of a lawyer,
        B: being born in Chicago.
    (b) A: being under 18 years of age,
        B: voting in a Presidential election.
    (c) A: owning a Chevrolet,
        B: owning a Ford.

11. Suppose that in a survey concerning the smoking habits of adult males, it is found that (i) 50% of the men smoke cigarettes, (ii) 35% of the men smoke pipes, and (iii) 30% of the men smoke both cigarettes and pipes. Which of the following assertions is false? Explain why.
    (a) 45% of the adult males smoke neither cigarettes nor pipes.
    (b) 25% of the adult males smoke exactly one of the choices: cigarettes or pipes.
    (c) 85% of the adult males either smoke cigarettes or smoke pipes.
    (d) 70% of the adult males do not smoke *both* cigarettes *and* pipes.

12. Let the sample space of a probability model have as outcomes all schools in the United States. According to the 1965 *Statistical Abstract of the United States* (p. 106), there were 81,910 public elementary schools, 14,762 nonpublic elementary schools, 25,350 public secondary schools, 4129 nonpublic secondary schools, 721 public schools of higher education, and 1316 nonpublic schools of higher education.
    (a) How many schools were public schools?
    (b) How many schools were elementary schools?
    (c) How many schools were either private schools or schools of higher education?
    (d) How many schools were either private schools or elementary schools, but not both?
    (e) How many schools were *not* secondary schools?
    (f) How many schools had no more than one of the following characteristics: private, school of higher education, elementary school?
    (g) How many schools had *none* of the following characteristics: private, elementary, secondary?

13.   A certain city of population 100,000 has three newspapers: A, B, and C.
      The proportions of people who read these papers are:

          A: 0.10   A *and* B: 0.08   A *and* B *and* C: 0.01
          B: 0.30   A *and* C: 0.02
          C: 0.50   B *and* C: 0.04

      (a) Find the proportion and the number of people who read only one
          newspaper.
      (b) How many people read at least two newspapers?
      (c) If A and C are morning papers, and B is an evening paper, how
          many people read *at least* one morning paper and also an evening
          paper?
      (d) How many people read only *one* morning paper and *one* evening
          paper?

14.   Prove the following inequality.

$$P(E_1 \cup E_2) \le P(E_1) + P(E_2).$$

      Now use recursive arguments to show that

$$P(E_1 \cup E_2 \cup \cdots \cup E_k) \le P(E_1) + P(E_2) + \cdots + P(E_k).$$

      In the statistical literature, this last inequality is called the *Bonferroni
      Inequality* [after Carlo Emilio Bonferroni (1892–1960)].

15.   Note that

$$P(E_1 \cup E_2) = P(E_1) + P(E_2) \quad \text{if and only if } P(E_1 \cap E_2) = 0.$$

      (a) Use this fact to prove that $P(E_1 \cup E_2 \cup E_3) = P(E_1) + P(E_2) +
          P(E_3)$ if and only if $P(E_1 \cap E_2) = P(E_1 \cap E_3) = P(E_2 \cap E_3) = 0$.
      (b) Use recursive arguments (mathematical induction) to show that
          $P(E_1 \cup E_2 \cup \cdots \cup E_k) = P(E_1) + P(E_2) + \cdots + P(E_k)$ if and only if
          $P(E_i \cap E_j) = 0$ for all $i \ne j$.

16.   Refer to Example 4.1.
      (a) Find the probability that a stolen car is 3 years old or older.
      (b) Find the probability that a stolen car is 2 years old or younger.
      (c) Find the probability that a stolen car is at least 1 year old, but no
          older than 2 years of age.

17.   Refer to Example 4.2.
      (a) Find the percentage of individuals surveyed who obtain current
          news from television, but *not* from radio.
      (b) Find the percentage of individuals surveyed who obtain current
          news from radio, but *not* from television.
      (c) Find the percentage of individuals surveyed who obtain current
          news *neither* from radio *nor* from television.
      (d) Of the individuals surveyed, 25% obtained current news from

periodicals, 10% obtained current news from both television and periodicals, 6% obtained current news from both radio and periodicals, and 4% obtained current news from radio, television, and periodicals. What percentage of individuals obtained current news from radio, television, *or* periodicals?

18. Refer to Example 4.3.
(a) Find $P(A \cap B^c)$. (b) Find $P(A^c \cap B)$. (c) Find $P(A^c \cup B^c)$.
(d) Find $P(A^c \cap B^c)$. (e) Find $P(A^c \cup B)$. (f) Find $P(A \cup B^c)$.

19. Suppose that we are given a probability model in which two mutually exclusive events A and B have the respective probabilities $P(A) = 0.30$ and $P(B) = 0.40$. Find
(a) the probability that either A or B occurs;
(b) the probability that both A and B occur;
(c) the probability that exactly one of the events A and B occurs;
(d) the probability that neither A nor B occurs;
(e) the probability that B occurs but A does not occur;
(f) the probability that either B does not occur or A does not occur.

20. In the experiment of Exercise 8, suppose that the probability of the event E_1 "the first-born child is a male" is 0.50, that the probability of the event E_2 "the second-born child is a male" is 0.51, and that the probability of the event E_3 "the third-born child is a male" is 0.52. Also suppose that $P(E_1 \cap E_2) = 0.25$, $P(E_1 \cap E_3) = 0.25$, $P(E_2 \cap E_3) = 0.26$, and $P(E_1 \cap E_2 \cap E_3) = 0.13$.
(a) What is the probability of the event "at least 1 of the children is male"?
(b) What is the probability of the event "all 3 children are females"?
(c) What is the probability of the event "exactly 2 of the children are male"?

21. A study is being made of the tendency of husbands and wives to come from similar social groups. A hierarchy of four social groups is defined in which social group I is the most prestigious and social group IV is the least prestigious.
(a) The probability that both husband and wife come from the same social group is 0.45. What is the probability that a husband and wife will come from different social groups?
(b) The probabilities that the husband comes from social groups I, II, III, and IV are 0.07, 0.25, 0.45, and 0.23, respectively. What is the probability that the husband will come from either social group I or social group IV?

(c) It is asserted that the probability that husband and wife both come from social group I is 0.08. Prove that this statement must be false.

22. Suppose that in 1693 an individual was randomly chosen from the city of Breslau, Poland. Use the life table (Table 5.1) to answer the following questions.
   (a) What is the probability that this individual lived past his or her 60th birthday?
   (b) What is the probability that this individual did not live to reach his or her 18th birthday?
   (c) What is the probability that this individual lived past his or her 15th birthday, but died before his or her 70th birthday?

23. In Exercise 20 let the events S_1, S_2, ..., S_8 correspond to the results MMM, MFF, MMF, MFM, FMM, FFM, FMF, and FFF, respectively. Show that the information given in Exercise 20 about the probabilities of E_1, E_2, E_3, and their various intersections allows you to compute $P(S_1)$, $P(S_2)$, ..., $P(S_8)$. Do so by deriving formulas and by using these formulas to compute $P(S_i)$, $i = 1, 2, ..., 8$. [Hint: A Venn diagram may help you in answering this question.]

3

Further Concepts of

Probability Theory

In this chapter, we show how the probability calculus enables us to use probabilities of less complicated events to obtain probabilities of quite complex events. In Section 1, we discuss probability models whose sample spaces contain only a finite number of outcomes. When every outcome is equally probable (the *uniform probability model*), then the probabilities of events can be determined by counting outcomes. Section 2 shows how uniform probability models arise in statistical sampling problems. An appendix, Section 5, provides rules of combinatorial analysis useful for counting outcomes for probability calculations.

In Section 3, we introduce the new and important concept of *conditional probability* and illustrate how knowledge of conditional probabilities aids us in computing probabilities of complex events and in prediction. Finally, in Section 4, we introduce the concept of *probabilistic* (stochastic, statistical) *independence* among events.

1. FINITE PROBABILITY MODELS

There are a variety of random experiments whose sample spaces are finite. Examples of such experiments include population sampling experiments in the social and biological sciences, experiments in genetics, and physical experiments dealing with microscopic structures. Most popular games of chance also have sample spaces that are finite.

A sample space Ω of a probability model is finite (and the model is said to be a *finite probability model*) if Ω contains a finite number M of outcomes $\omega_1, \omega_2, \ldots, \omega_M$. We define the *simple events* $S_i = \{\omega_i\}$, $i = 1, 2, \ldots, M$, of such a probability model to be the events consisting of exactly one outcome. Figure 1.1 illustrates the sample space Ω of a finite probability model in which there are $M = 7$ outcomes $\omega_1, \omega_2, \ldots, \omega_7$ (represented by points in Figure 1.1). Corresponding to these seven outcomes are seven simple events $S_1, S_2, \ldots, S_7$, represented by the small circles in Figure 1.1. The entire box in Figure 1.1 represents Ω, while one particular event E, represented by the large circle, contains the outcomes ω_1, ω_2, and ω_4. Note that

$$E = \{\omega_1, \omega_2, \omega_4\} = S_1 \cup S_2 \cup S_4.$$

Figure 1.1 illustrates the following facts about finite probability models.

(1) Any two simple events S_i and S_j, $i \neq j$, are mutually exclusive.
(2) Every event E is the union of simple events corresponding to the outcomes contained in E.

These two facts, together with Rule 4.5 of Chapter 2, lead to a fundamental rule for computing probabilities in finite probability models.

Rule 1.1. The probability $P(E)$ of any event E of a finite probability model is equal to the sum of the probabilities of the simple events whose union is E.

From Rule 1.1, it follows that any finite probability model can be completely described by a table such as

Simple event	S_1	S_2	$\cdots$	S_{M-1}	S_M
Probability	p_1	p_2	$\cdots$	p_{M-1}	p_M

Figure 1.1: *Sample space of a finite model.*

where $p_j = P(S_j)$, $j = 1, 2, \ldots, M$. Note that since $\Omega = S_1 \cup S_2 \cup \cdots \cup S_M$ and $P(\Omega) = 1$, Rule 1.1 implies that

(1.1) $$\sum_{j=1}^{M} p_j = P(\Omega) = 1.$$

That is, the sum of the probabilities corresponding to all simple events must always be equal to 1.

Example 1.1. For the sample space illustrated by Figure 1.1, suppose that we have defined the probability model through the following table of probabilities.

Simple event	S_1	S_2	S_3	S_4	S_5	S_6	S_7
Probability	0.08	0.25	0.33	0.04	0.16	0.05	0.09

Since the event E pictured in Figure 1.1 is the union $S_1 \cup S_2 \cup S_4$ of the simple events S_1, S_2, and S_4, Rule 1.1 implies that

$$P(E) = P(S_1) + P(S_2) + P(S_4) = 0.08 + 0.25 + 0.04 = 0.37.$$

Example 1.2. A random experiment consists of spinning a roulette wheel. The outcomes of this experiment are the numbers 1, 2, $\ldots$, 35, 36, 0, 00 printed on the wheel; there are $M = 38$ outcomes in all. The simple events are $S_j = \{j$ is observed$\}$, $j = 1, \ldots, 36$, and $S_{37} = \{0$ is observed$\}$, $S_{38} = \{00$ is observed$\}$. Assuming that every number on the wheel has an equal probability of being observed when the wheel is spun, we have the following table of probabilities.

Simple event	S_1	S_2	$\cdots$	S_{36}	S_{37}	S_{38}
Probability	$\frac{1}{38}$	$\frac{1}{38}$	$\cdots$	$\frac{1}{38}$	$\frac{1}{38}$	$\frac{1}{38}$

If E is the event consisting of outcomes that are even numbers (excluding the zeros), then $E = S_2 \cup S_4 \cup \cdots \cup S_{36}$, and

$$P(E) = P(S_2) + P(S_4) + \cdots + P(S_{36}) = \frac{18}{38} = 0.474.$$

Uniform Probability Models

In many random experiments, such as Example 1.2, there is a symmetry (or interchangeability) among the outcomes. Thus, there is a symmetry between the two faces of a tossed coin, among the numbers on a balanced roulette wheel, and among the cards in a well-shuffled deck. Such symmetry among outcomes is often assumed for the choice of sex at birth, for guessing in multiple-choice situations, for the choice of magnetic polarity in a given electron, and in many other experiments of importance in science and technology. It is a *logical* consequence of this assumption of symmetry that the simple events of the probability model have *equal* probabilities. Any finite model for which the simple events S_i have equal probabilities is called a *uniform probability model*.

In a uniform probability model each of the M simple events S_i has the same probability $p_i = p$, $i = 1, 2, \ldots, M$. It then follows from Equation (1.1) that $p = 1/M$.

Rule 1.2. In a uniform probability model, the probability of any simple event S_i equals $1/M$, where M is the number of outcomes in Ω.

If an event E consists of K outcomes, then E is the union of K simple events, each having probability $1/M$. From Rule 1.1, it follows that $P(E) = K/M$.

Rule 1.3. In a uniform probability model, the probability of any event E is $P(E) = K/M$, where K is the number of outcomes in E and M is the number of outcomes in Ω.

Example 1.3. In a learning experiment, a rat is placed in a cell. There are 10 identical doors leading from the cell, 2 of which lead to rewards of food. If no learning has taken place, all doors have equal probability of being chosen by the rat, and the probability of the event that the rat is rewarded is

$$P\{\text{rat is rewarded}\} = \frac{\text{number of doors leading to rewards}}{10}$$

$$= \tfrac{2}{10} = 0.2.$$

It is clear from Rule 1.3 that probability calculations in uniform probability models reduce to problems of counting outcomes. In an appendix (Section 5), rules of counting are given that can be helpful in situations more complicated than those described in Examples 1.2 and 1.3.

2. SIMPLE RANDOM SAMPLING

Perhaps the single most important use of uniform probability models is in survey sampling. Here the randomness in the experiment is deliberately introduced by the experimenter. This permits the experimenter to sample a population in a controlled random way, examine these chosen members, and then make inferences about the entire population through the use of probability theory.

The basic probability model for survey sampling is that of *random sampling*. Suppose that we have a population of N units $u_1, u_2, \ldots, u_N$. This population may consist of people, domiciles, radio tubes, chemical distillates, items on an examination, numbers, and so on. In choosing a random sample of n of these N units, we perform a random experiment in which the outcomes are *samples* of n units and in which each particular sample (outcome) is chosen with a specified probability.

Formally, a *sample* is an array of units listed in a definite order. For example, one possible sample of $n = 5$ units from the population consisting of the units $u_1, u_2, \ldots, u_N$ is the sample $(u_1, u_4, u_5, u_7, u_3)$. This sample would be distinguished from the sample $(u_5, u_1, u_4, u_7, u_3)$ because even though the same units appear in both samples, the *order* in which these units appear is different. The definition of a sample should be interpreted to allow a given unit (say, u_2) to appear $0, 1, 2, \ldots, n$ times in an (ordered) sample of size n. For example, the ordered array of units $(u_3, u_2, u_2, u_7, u_1)$ and the ordered array of units $(u_1, u_1, u_1, u_2, u_2)$ are both permissible samples of five units.

Example 2.1. In the population consisting of the three units u_1, u_2, u_3, the possible ordered samples of two units are

$$\omega_1 = (u_1, u_1), \qquad \omega_4 = (u_2, u_1), \qquad \omega_7 = (u_3, u_1),$$

$$\omega_2 = (u_1, u_2), \qquad \omega_5 = (u_2, u_2), \qquad \omega_8 = (u_3, u_2),$$

$$\omega_3 = (u_1, u_3), \qquad \omega_6 = (u_2, u_3), \qquad \omega_9 = (u_3, u_3).$$

One type of random sampling is *simple random sampling with replacement*. This is a random experiment in which the sample space Ω consists of all possible ordered samples and in which each sample has equal probability of being selected. Another type of random sampling is *simple random sampling without replacement*. This random experiment is described by a probability model that gives zero probability to all ordered samples in which any unit appears more than once and equal probability to all other ordered samples.

Example 2.2. For the population of three units discussed in Example 2.1, the probability model for *simple random sample with replacement* is

Simple event	$\{\omega_1\}$	$\{\omega_2\}$	$\{\omega_3\}$	$\{\omega_4\}$	$\{\omega_5\}$	$\{\omega_6\}$	$\{\omega_7\}$	$\{\omega_8\}$	$\{\omega_9\}$
Probability	$\frac{1}{9}$	$\frac{1}{9}$	$\frac{1}{9}$	$\frac{1}{9}$	$\frac{1}{9}$	$\frac{1}{9}$	$\frac{1}{9}$	$\frac{1}{9}$	$\frac{1}{9}$

and the probability model for *simple random sampling without replacement* is

Simple event	$\{\omega_1\}$	$\{\omega_2\}$	$\{\omega_3\}$	$\{\omega_4\}$	$\{\omega_5\}$	$\{\omega_6\}$	$\{\omega_7\}$	$\{\omega_8\}$	$\{\omega_9\}$
Probability	0	$\frac{1}{6}$	$\frac{1}{6}$	$\frac{1}{6}$	0	$\frac{1}{6}$	$\frac{1}{6}$	$\frac{1}{6}$	0

Simple random sampling *with replacement* gives rise to a uniform probability model for the sample space Ω whose outcomes are all ordered samples.

Simple random sampling without replacement defines a uniform probability model over only those ordered samples in which no unit appears more than once.

Under either sampling model, to find the probability that any particular ordered sample will be observed, we need to compute $1/M$, where M is the total number of possible samples for that sampling model. The following facts are established in Section 5.

Rule 2.1. The total number of ordered samples of size n from a population of N units is N^n.

Rule 2.2. The total number of ordered samples of size n in which no unit appears more than once is $N(N - 1) \cdots (N - n + 1)$, where N is the number of units in the population.

Example 2.3. Suppose that the dates of birth of n individuals assembled in a room are a simple random sample *with replacement* of size n from among the $N = 365$ days of the year (excluding February 29). What is the probability of the event E that no two people in the room have the same birthday? To obtain the probability of this event, we count the number K of ordered samples of size n from among the $N = 365$ days of the year in which no day (birthdate) appears more than once, and divide this number K by the total number M of ordered samples of size n with replacement from among the $N = 365$ days of the year. From Rule 2.1, $M = 365^n$, while from Rule 2.2, $K = (365)(364) \cdots (365 - n + 1)$. For example, if $n = 2$,

$$P(E) = \frac{K}{M} = \frac{(365)(364)}{(365)^2} = \frac{364}{365} = 0.997,$$

while if $n = 5$,

$$P(E) = \frac{K}{M} = \frac{(365)(364)(363)(362)(361)}{(365)^5} = 0.973.$$

Interestingly enough, when $n = 23$,

$$P(E) = \frac{K}{M} = \frac{(365)(364) \cdots (344)(343)}{(365)^{23}} = 0.493.$$

Thus, if the assumptions of this example hold, and you are at a party in which $n = 23$ people are present, if you offer to bet that at least two people at the party have the same birthdate, your chances of winning are

$$P(E^c) = 1 - P(E) = 1 - 0.493 = 0.507.$$

Considering that there are 365 days to choose from, the result that there is better than a 50–50 chance of a match in birthdays among 23 people appears counterintuitive.

Example 2.4. A political caucus decides to choose a subcommittee to work out rules of procedure. The committee consists of $n = 3$ individuals chosen by a simple random sample without replacement from the $N = 60$ individuals attending the caucus. Suppose that 35 members (say, individuals $u_1, u_2, \ldots,$ u_{35}) of the caucus favor candidate A, and the remaining 25 members $(u_{36}, \ldots,$ $u_{60})$ favor candidate B. What is the probability that all three members of the subcommittee will favor candidate A?

Since the subcommittee is chosen without replacement, the total number M of possible outcomes is, by Rule 2.1, equal to $M = (60)(59)(58)$. If all three individuals drawn are to favor candidate A, then the sample will be drawn without replacement from among the subpopulation of 35 members who favor candidate A. Again, from Rule 2.2, the total number of ways that this can happen is $K = (35)(34)(33)$. Thus, by Rule 1.2,

$P\{$all 3 members of the subcommittee favor candidate A$\}$

$$= \frac{K}{M} = \frac{(35)(34)(33)}{(60)(59)(58)} = 0.191.$$

Similarly,

$P\{$all 3 members of the subcommittee favor candidate B$\}$

$$= \frac{(25)(24)(23)}{(60)(59)(58)} = 0.067.$$

We can also calculate the probability of the event E that two members of the subcommittee favor candidate A and one member favors candidate B. Recalling that our outcomes are *ordered* samples, we must consider three possibilities, AAB, ABA, BAA, where the ordering of the letters indicates the location of the supporter of candidate B in the ordered sample. To count the number of ways that the possibility AAB can occur, we see that the two A's must be drawn as an ordered sample without replacement from among the 35 supporters of candidate A. Using Rule 2.2 again, we see that we can do this in $(35)(34)$ ways. For every such way of drawing the two A's, there are 25 ways of choosing the supporter of candidate B; thus, there are $(35)(34)(25)$ ways of obtaining the possibility AAB. Similarly, there are $(35)(34)(25)$ ways of obtaining the possibility ABA, and $(35)(34)(25)$ ways of obtaining the possibility BAA. Since these three possibilities are mutually exclusive, the total number of ways we can obtain an outcome (sample) satisfying the definition of event E is

$K = $ (number of ways of obtaining AAB)

$\quad +$ (number of ways of obtaining ABA)

$\quad +$ (number of ways of obtaining BAA)

$= (35)(34)(25) + (35)(34)(25) + (35)(34)(25)$

$= (3)(35)(34)(25)$

and

$$P(E) = \frac{K}{M} = \frac{(3)(35)(34)(25)}{(60)(59)(58)} = 0.435.$$

Example 2.5. Suppose that in Example 2.4, the members of the subcommittee are drawn *with* replacement. (This supposition is unrealistic, but is made to illustrate relevant probability calculations.) Then arguments similar to those used in Example 2.4, but using Rule 2.1 instead of Rule 2.2, yield

$$P\{\text{all 3 subcommittee members favor candidate A}\} = \frac{(35)^3}{(60)^3} = 0.198,$$

$$P\{\text{all 3 subcommittee members favor candidate B}\} = \frac{(25)^3}{(60)^3} = 0.072,$$

$P\{2 \text{ members favor candidate A, 1 member favors candidate B}\}$
$$= \frac{(3)(35)^2(25)}{(60)^3} = 0.425.$$

3. CONDITIONAL PROBABILITY

There are many situations where we obtain partial knowledge concerning the outcome of a random experiment before the complete result becomes known. The availability of this partial knowledge suggests that we might improve our predictions concerning the outcome of the experiment if we could re-evaluate the probabilities of events of interest taking into account such partial knowledge.

For example, consider the following condensed (and fictitious) life table.

	Probability of dying before age 65	Probability of living to age 65
Men	0.26	0.74
Women	0.20	0.80
Both men and women	0.23	0.77

If actuaries in an insurance company do not know the sex of a particular applicant, they will undoubtedly base their insurance premium on a probability of death for both sexes before age 65 of 0.23. However, if the actuaries know that the applicant is a woman, they should lower their premium since the chance that a person will die before age 65, given that the person is a female, is only 0.20.

This process of re-evaluating probabilities in the light of partial information about the outcomes of an experiment is made precise by the concept of *conditional probability*. Formally, let A be an event such that $P(A) > 0$, and let B be another event. The conditional probability of B occuring *given* that A has occurred is defined by the equation

$$(3.1) \qquad P(B \mid A) = \frac{P(A \cap B)}{P(A)}.$$

If $P(A) = 0$, the conditional probability of B given A is undefined.

To distinguish the conditional probability $P(B \mid A)$ of the event B *given* the event A from the probability $P(B)$ of B evaluated before the experiment begins, we sometimes speak of $P(B)$ as being the *unconditional, marginal,* or *absolute* probability of B. Of course, *all* probabilities are conditional in the sense that we assign (or calculate) these probabilities in the light of our present knowledge of the properties of whatever random experiment we are observing.

Example 3.1. Two reports, "Doctoral Scientists and Engineers in the United States, 1975 Profile" and "Employment Status of Ph.D. Scientists and Engineers: 1973 and 1975," published by the National Research Council suggest that for the events

$A =$ "a doctoral scientist or engineer is a woman,"

$B =$ "a doctoral scientist or engineer is totally unemployed in 1975,"

the following probabilities held:

$$P(A) = 0.099, \qquad P(B) = 0.009, \qquad P(A \cap B) = 0.003.$$

The conditional probability that a doctoral scientist or engineer is unemployed in 1975 given that she is a woman is

$$P(B \mid A) = \frac{P(A \cap B)}{P(A)} = \frac{0.003}{0.099} = 0.030.$$

The conditional probability that a doctoral scientist or engineer is a woman given that the individual is unemployed in 1975 is

$$P(A \mid B) = \frac{P(A \cap B)}{P(B)} = \frac{0.003}{0.009} = 0.333.$$

Comparing $P(A)$ to $P(A \mid B)$ indicates that female doctoral scientists or engineers in 1975 had a disproportionately high rate of unemployment.

Frequency Interpretation of Conditional Probability

The frequency interpretation for the probability $P(E)$ of an event E states that the proportion r.f.(E) of trials in which E occurs in a large number of repetitions of a random experiment tends to be close to $P(E)$. In a similar manner, a frequency interpretation can be given for $P(B|A)$. Let $\#A$ and $\#(A \cap B)$ be the number of times that the events A and $A \cap B$, respectively, occur in N repetitions of a random experiment. Then, when N is large,

$$\frac{\#(A \cap B)}{\#A} = \frac{\#(A \cap B)/N}{\#A/N} = \frac{\text{r.f.}(A \cap B)}{\text{r.f.}(A)},$$

which by the frequency interpretation of the measure $P(E)$ tends to be approximately $P(A \cap B)/P(A)$. *Thus, the conditional probability of B given A is the abstraction of the long-run proportion of times that B occurs among those trials for which A also occurs.*

Example 3.2. Consider a table of relative frequencies (Table 3.1) [taken from Hollingshead (1949)], which resulted from a study of the impact of social class structures on 390 high school students in a midwestern community.

Table 3.1 can be regarded as resulting from 390 replications of the random experiment in which a high-school-age youth (from Elmtown) is observed and his social class and choice of high school curriculum are recorded. The outcomes (simple results) of this experiment are the pairs in which, say, the first position gives the social class of the individual and the second position gives the choice of curriculum for the individual.

Let A be the event that an observed individual is in social class V and let B be the event that an observed individual chooses a general high school curriculum. Note that the relative frequency of event A is given by the sum of the entries in the last column of Table 3.1; that is, $\text{r.f.}(A) = \frac{26}{390}$. The proportion of times that an individual chooses a general high school curriculum among the trials in which the individual's social is observed to be class V is then

$$\frac{\text{r.f.}(A \cap B)}{\text{r.f.}(A)} = \frac{\frac{14}{390}}{\frac{26}{390}} = \frac{7}{13}.$$

Table 3.1: Relative Frequency of Five Social Classes of Elmtown Youth in Three Alternative High School Curricula

Curriculum	I and II	III	IV	V
		Social class		
College preparatory	$\frac{23}{390}$	$\frac{40}{390}$	$\frac{16}{390}$	$\frac{2}{390}$
General	$\frac{11}{390}$	$\frac{75}{390}$	$\frac{107}{390}$	$\frac{14}{390}$
Commercial	$\frac{1}{390}$	$\frac{31}{390}$	$\frac{60}{390}$	$\frac{10}{390}$

If we interpret Table 3.1 as being a table of probabilities for the random experiment described above, then we would say that the conditional probability that an individual Elmtown youth (of high school age) chooses a general high school curriculum, *given* that he belongs to social class V, is $\frac{7}{13}$.

The Probability Calculus for Conditional Probabilities

Another useful way of thinking of conditional probabilities is the following. Knowing that event A has occurred changes the sample space of our experiment. Rather than expecting the outcome ω of the random experiment to belong to the sample space Ω, we now know that ω must belong to the event A, which now becomes the sample space of our new probability model.

In Figure 3.1, the old sample space Ω is replaced by a new sample space A, and each event B is replaced by the event $B \cap A$. Because the relative frequency of the event B given that event A has occurred is proportional to r.f.$(B \cap A) = $ r.f.$(A \cap B)$, the new model must have a probability measure P^* that assigns a probability to the event B proportional to the probability of the event $A \cap B$ under the old (unconditional) model; that is,

$$(3.2) \qquad P^*(B) = cP(A \cap B),$$

where c is a constant. Since A is the new sample space, $P^*(A)$ must equal 1, so that

$$1 = P^*(A) = cP(A \cap A) = cP(A),$$

and hence for any event B, $P^*(B) = P(A \cap B)/P(A) = P(B|A)$.

Before we can interpret $P(B|A)$ as being a probability measure defined

Figure 3.1: *Given that A occurs, the event A replaces the sample space* Ω, *and* $B \cap A$ *replaces B.*

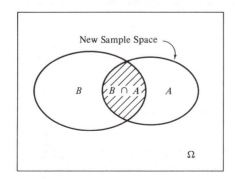

on the new sample space consisting only of outcomes belonging to the event
A, we must show that $P(B|A)$, as defined by Equation (3.1), satisfies Axioms
1, 2, and 3 of Chapter 2. Since

$$0 \leq P(\emptyset) \leq P(A \cap B) \leq P(A),$$

dividing by $P(A)$ in the foregoing inequality shows that $P(B|A)$ satisfies
Axiom 1. We have already remarked that $P(A|A) = 1$. Also, since
$\emptyset \cap A = \emptyset$, $P(\emptyset|A) = 0$. Hence, Axiom 2 is satisfied. Finally, if the events
E_1 and E_2 are mutually exclusive, then $E_1 \cap A$ and $E_2 \cap A$ are also mutually
exclusive, so that

$$P((E_1 \cup E_2) \cap A) = P((E_1 \cap A) \cup (E_2 \cap A))$$
$$= P(E_1 \cap A) + P(E_2 \cap A).$$

Dividing this equality by $P(A)$ gives

$$P(E_1 \cup E_2|A) = P(E_1|A) + P(E_2|A),$$

which establishes that $P(B|A)$ satisfies Axiom 3.

Since $P(B|A)$ is a legitimate probability measure, it must therefore obey
the following rules.

Rule 3.1 (Law of Inclusion for Conditional Probabilities). If $E_1 \subset E_2$, then
$P(E_1|A) \leq P(E_2|A)$.

Rule 3.2 (Law of Complementation for Conditional Probabilities). For any
event E, $P(E^c|A) = 1 - P(E|A)$.

Rule 3.3 (Law of Addition for Conditional Probabilities). For any two events
E_1 and E_2,

$$P(E_1 \cup E_2|A) = P(E_1|A) + P(E_2|A) - P(E_1 \cap E_2|A).$$

Other results of the probability calculus for unconditional probabilities
(Section 4 of Chapter 2) can be converted to corresponding results for condi-
tional probabilities by simply replacing $P(E)$ by $P(E|A)$ in the relevant for-
mulas. For example,

$$P(E_1 \cup E_2 \cup E_3|A) = P(E_1|A) + P(E_2|A) + P(E_3|A)$$
(3.3)
$$- P(E_1 \cap E_2|A) - P(E_1 \cap E_3|A)$$
$$- P(E_2 \cap E_3|A) + P(E_1 \cap E_2 \cap E_3|A).$$

Example 3.3. An economist proposes a probability model relating gross price
movements for various types of investments. Let A be the event that common

stock prices fall, let E_1 be the event that bond prices rise, and let E_2 be the event that commodity prices rise. The model states that

$$P(E_1 \mid A) = P(\text{bond prices rise} \mid \text{stock prices fall}) = 0.63,$$

$$P(E_2 \mid A) = P(\text{commodity prices rise} \mid \text{stock prices fall}) = 0.57,$$

$$P(E_1 \cap E_2 \mid A) = P(\text{bond } and \text{ commodity prices rise} \mid \text{stock prices fall})$$
$$= 0.50.$$

From this information, the conditional probability of a rise in bond prices *or* a rise in commodity prices *given* a drop in stock prices is

$$P(E_1 \cup E_2 \mid A) = P(E_1 \mid A) + P(E_2 \mid A) - P(E_1 \cap E_2 \mid A)$$
$$= 0.63 + 0.57 - 0.50 = 0.70,$$

while the conditional probability of a drop in both bond prices *and* commodity prices *given* a drop in stock prices is

$$P(E_1^c \cap E_2^c \mid A) = P((E_1 \cup E_2)^c \mid A) = 1 - P(E_1 \cup E_2 \mid A)$$
$$= 1 - 0.70 = 0.30.$$

Law of Multiplication

If we know $P(A)$ and $P(B \mid A)$, then Equation (3.1) tells us that we can obtain $P(A \cap B)$.

Rule 3.4 (Law of Multiplication). Given the values of $P(B \mid A)$ and $P(A)$, we can compute $P(A \cap B)$ by means of the formula

$$P(A \cap B) = P(A)P(B \mid A).$$

Example 3.4. Long-term records of the performance of a particular type of electric generator show that the probability $P(A)$ of failure in the first 10 years of life of such a generator is 0.25. Given that a generator has failed, the conditional probability $P(B \mid A)$ that the failure cannot be repaired is 0.40. Thus, the probability that such a generator will experience an irreparable failure within its first 10 years of life is

$$P(A \cap B) = P(A)P(B \mid A) = (0.25)(0.40) = 0.10.$$

We may generalize Rule 3.4 to any number of events. For example, if we have three events E_1, E_2, E_3, then repeated application of Rule 3.4 yields

$$(3.4) \qquad P(E_1 \cap E_2 \cap E_3) = P(E_1 \cap E_2)P(E_3 \mid E_1 \cap E_2)$$
$$= P(E_1)P(E_2 \mid E_1)P(E_3 \mid E_1 \cap E_2).$$

By induction, a similar result for k events $E_1, E_2, \ldots, E_k$ is

$$(3.5) \quad P(E_1 \cap E_2 \cap \cdots \cap E_k)$$
$$= P(E_1)P(E_2|E_1)P(E_3|E_1 \cap E_2) \cdots P(E_k|E_1 \cap E_2 \cap \cdots E_{k-1}).$$

Example 3.5. In Example 2.4, counting rules were used to obtain probabilities connected with a simple random sampling situation. However, any random sample of size n can be conceived of as arising from n consecutive draws from the population. Before each draw is observed, we have the opportunity to reassess our probabilities for that draw and all future draws in the light of the results of the draws already observed. For example, the situation considered in Example 2.4 was that of a political caucus of 60 individuals, 35 of whom favored candidate A and 25 of whom favored candidate B. A subcommittee (sample) of $n = 3$ members is drawn by simple random sampling *without* replacement from the 60 individuals in the caucus. Let E_j be the event that the jth individual drawn for the subcommittee favors candidate A, $j = 1, 2, 3$. Since every individual in the caucus has an equal chance to be drawn on the first draw, and since 35 of these 60 individuals favor candidate A, $P(E_1) = \frac{35}{60}$. Given that an individual favoring candidate A has been drawn on the first draw (E_1 occurs), each of the $60 - 1 = 59$ remaining individuals has an equal *conditional* probability to be drawn on the second draw, and $35 - 1 = 34$ of these individuals favor candidate A. Thus, $P(E_2|E_1) = \frac{34}{59}$. Finally, *given* that the first two individuals drawn favor candidate A ($E_1 \cap E_2$ occurs), the *conditional* probability $P(E_3|E_1 \cap E_2)$ that the third individual drawn favors candidate A is

$$P(E_3|E_1 \cap E_2) = \frac{35 - 2}{60 - 2} = \frac{33}{58}.$$

We conclude from (3.4) that

$P\{$all 3 members favor candidate A$\}$

$$= P(E_1 \cap E_2 \cap E_3) = P(E_1)P(E_2|E_1)P(E_3|E_1 \cap E_2)$$

$$= \left(\frac{35}{60}\right)\left(\frac{34}{59}\right)\left(\frac{33}{58}\right) = 0.191,$$

which is the result we obtained by counting in Example 2.4. Similarly,

$$P(E_1^c) = \frac{25}{60},$$

$$P(E_2^c|E_1^c) = \frac{25 - 1}{60 - 1} = \frac{24}{59},$$

$$P(E_3^c|E_1^c \cap E_2^c) = \frac{25 - 2}{60 - 2} = \frac{23}{58},$$

so that

$P\{$all 3 members favor candidate B$\}$

$$= P(E_1^c \cap E_2^c \cap E_3^c) = P(E_1^c)P(E_2^c \mid E_1^c)P(E_3^c \mid E_1^c \cap E_2^c)$$

$$= \left(\frac{25}{60}\right)\left(\frac{24}{59}\right)\left(\frac{23}{58}\right) = 0.067,$$

and finally,

$P\{2$ members favor candidate A and 1 member favors candidate B$\}$

$$= P((E_1 \cap E_2 \cap E_3^c) \cup (E_1 \cap E_2^c \cap E_3) \cup (E_1^c \cap E_2 \cap E_3))$$

$$= P(E_1 \cap E_2 \cap E_3^c) + P(E_1 \cap E_2^c \cap E_3) + P(E_1^c \cap E_2 \cap E_3)$$

$$= P(E_1)P(E_2 \mid E_1)P(E_3^c \mid E_1 \cap E_2) + P(E_1)P(E_2^c \mid E_1)P(E_3 \mid E_1 \cap E_2^c)$$

$$\quad + P(E_1^c)P(E_2 \mid E_1^c)P(E_3 \mid E_1^c \cap E_2)$$

$$= \left(\frac{35}{60}\right)\left(\frac{34}{59}\right)\left(\frac{25}{58}\right) + \left(\frac{35}{60}\right)\left(\frac{25}{59}\right)\left(\frac{34}{58}\right) + \left(\frac{25}{60}\right)\left(\frac{35}{59}\right)\left(\frac{34}{58}\right)$$

$$= \frac{3(35)(34)(25)}{(60)(59)(58)} = 0.435.$$

Example 3.6. A systems engineer is interested in assessing the reliability of a rocket composed of three stages. At takeoff, the engine of the first stage of the rocket must lift the rocket off the ground (event E_1). If that engine accomplishes its task, the engine of the second stage must now lift the rocket into orbit. Once the engines in both stages 1 and 2 have performed successfully (event E_2), the engine of the third stage is used to complete the rocket's mission (event E_3). The reliability of the rocket is thus $P(E_3)$. Suppose that the probabilities of successful performance for the engines of stages 1, 2, and 3 are 0.99, 0.97, and 0.98, respectively. From this information, $P(E_1) = 0.99$. Since the second engine can only come into action after the first engine fires successfully, $P(E_2 \mid E_1) = 0.97$. Finally, since the third engine can only come into action after the engines of stages 1 and 2 perform successfully, we know that $P(E_3 \mid E_2) = 0.97$. However, note that $E_3 \subset E_2 \subset E_1$, since the success of all three stages implies the success of the first two stages, which in turn implies the success of the first stage. Thus [Exercise 4(d) of Chapter 2],

$$E_2 = E_1 \cap E_2, \qquad E_3 = E_2 \cap E_3 = E_1 \cap E_2 \cap E_3,$$

and we can apply (3.5) to conclude that

Reliability of rocket $= P(E_3)$

$$= P(E_1 \cap E_2 \cap E_3) = P(E_1)P(E_2 \mid E_1)P(E_3 \mid E_1 \cap E_2)$$

$$= P(E_1)P(E_2 \mid E_1)P(E_3 \mid E_2)$$

$$= 0.99(0.97)(0.98) = 0.941.$$

Partitions

The events E_1, E_2, ..., E_k are said to *partition* the sample space Ω if every outcome ω of Ω belongs to one and only one of the events E_1, E_2, ..., E_k, that is, if $E_i \cap E_j = \emptyset$ for all $i \neq j$, and $E_1 \cup E_2 \cup \cdots \cup E_k = \Omega$. Figure 3.2 illustrates the partitioning of the sample space Ω (the entire box in Figure 3.2) by the events E_1, E_2, E_3, E_4. Note that the event B can be represented as

$$B = (E_1 \cap B) \cup (E_2 \cap B) \cup (E_3 \cap B) \cup (E_4 \cap B),$$

where $(E_i \cap B)$ and $(E_j \cap B)$ are mutually exclusive for $i \neq j$. In general, *a partition of the sample space Ω also partitions any event B:* The event B is the union of the mutually exclusive events $(E_1 \cap B)$, $(E_2 \cap B)$, ..., $(E_k \cap B)$. It follows that for any partition $E_1, E_2, ..., E_k$ of Ω, and any event B,

$$(3.6) \qquad P(B) = P((E_1 \cap B) \cup (E_2 \cap B) \cup \cdots \cup (E_k \cap B))$$

$$= \sum_{i=1}^{k} P(E_i \cap B).$$

However,

$$P(E_i \cap B) = P(E_i)P(B \mid E_i),$$

so that

$$(3.7) \qquad P(B) = \sum_{i=1}^{k} P(E_i)P(B \mid E_i).$$

Rule 3.5. If the events E_1, E_2, ..., E_k partition the sample space, then the probability of any event B can be calculated as follows.

$$P(B) = \sum_{i=1}^{k} P(E_i \cap B) = \sum_{i=1}^{k} P(E_i)P(B \mid E_i).$$

We remark that an event E and its complement E^c always partition Ω.

Figure 3.2: *The events E_1, E_2, E_3, E_4 partition the sample space.*

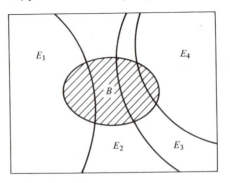

Example 3.7. A medical test is designed to give a positive response when a patient has a certain disease, disease D. Previous experience shows that the conditional probability of a positive response (event B) *given* that the patient has disease D (event E) is $P(B|E) = 0.80$, while the conditional probability of a positive response *given* that the patient does not have disease D is $P(B|E^c) = 0.15$. The probability that a randomly chosen patient has disease D is $P(E) = 0.25$. What is the probability $P(B)$ that the medical test, when applied to a randomly chosen patient, yields a positive response? Since E and E^c partition the sample space, and $P(E^c) = 1 - P(E) = 0.75$, we can apply Rule 3.5 to obtain

$$P(B) = P(E)P(B|E) + P(E^c)P(B|E^c)$$
$$= (0.25)(0.80) + (0.75)(0.15) = 0.3125.$$

Example 3.8. A social scientist is worried that randomly selected individuals in a survey will not respond to questions (event B). A previous study provides conditional probabilities $P(B|E_i)$, $i = 1, 2, 3$, of nonresponse for individuals in three different income classes and unconditional probabilities of membership in these income classes.

Income class (event)	$P(B|E_i)$	$P(E_i)$
High income (E_1)	0.08	0.25
Middle income (E_2)	0.10	0.50
Low income (E_3)	0.04	0.25

Applying Rule 3.5,

$P\{$person drawn does not respond$\} = P(B)$

$$= P(E_1)P(B|E_1) + P(E_2)P(B|E_2) + P(E_3)P(B|E_3)$$
$$= (0.25)(0.08) + (0.50)(0.10) + (0.25)(0.04)$$
$$= 0.08.$$

Bayes' Rule

Before an individual is chosen in Example 3.8, the probability of the event E_3 that the individual comes from the low-income class is $P(E_3) = 0.25$. If after an individual is chosen, the individual does not respond to questions (event B), then we have additional information about the individual and can reassess the probability of event E_3 in the light of this new information. The method for such reassessment has been known since the seventeenth century when the Reverend Thomas Bayes (1702–1761) first employed the rule that bears his name.

Rule 3.6 (Bayes' Rule). If the events $E_1, E_2, \ldots, E_k$ partition the sample space Ω and if B is any other event for which $P(B) > 0$, then the conditional probability of any partitioning event E_i given that event B has occurred is

$$P(E_i \mid B) = \frac{P(E_i \cap B)}{P(B)} = \frac{P(E_i)P(B \mid E_i)}{\sum_{j=1}^{k} P(E_j)P(B \mid E_j)}, \quad i = 1, 2, \ldots, k.$$

Example 3.9. A randomly chosen student takes an examination that is passed by 80% of all "good" students, 60% of all "average" students, and 30% of all "poor" students. In the population, 25% of all students are good students (event E_1), 50% are average (event E_2), and 25% are poor (event E_3). Given that the chosen student passes the exam (event B), what is the conditional probability that the student is a good student? Since we are given the information that $P(B \mid E_1) = 0.8$, $P(B \mid E_2) = 0.6$, $P(B \mid E_3) = 0.3$, and that $P(E_1) = P(E_3) = 0.25$, $P(E_2) = 0.50$, we can use Bayes' Rule to find

$$P\{\text{student is good} \mid \text{passes exam}\} = P(E_1 \mid B)$$

$$= \frac{P(B \mid E_1)P(E_1)}{\sum_{i=1}^{3} P(B \mid E_i)P(E_i)}$$

$$= \frac{(0.8)(0.25)}{(0.8)(0.25) + (0.6)(0.5) + (0.3)(0.25)}$$

$$= 0.35.$$

Since the student passed the exam, the odds have increased that the student is "good."

Example 3.8 (continued). Using Bayes' Rule, we obtain

$$P(E_3 \mid B) = \frac{P(E_3)P(B \mid E_3)}{P(E_1)P(B \mid E_1) + P(E_2)P(B \mid E_2) + P(E_3)P(B \mid E_3)}$$

$$= \frac{(0.25)(0.04)}{(0.25)(0.08) + (0.50)(0.10) + (0.25)(0.04)} = \frac{1}{8} = 0.125.$$

Knowing that the individual drawn did not respond to questions (event B), we have reassessed the probability that the individual came from the low-income class (event E_3). Our new conditional probability $P(E_3 \mid B) = 0.125$ is exactly half of the unconditional probability $P(E_3)$ that we assigned to event E_3 before we knew whether or not the chosen individual would respond to questions.

Example 3.7 (continued). Using Bayes' Rule and the results of the medical test, we can improve our diagnosis of disease D. Before the results of the medical test are known, the probability that an individual has disease D is

$P(E) = 0.25$. However, if the medical test shows a positive result, the new *conditional* probability that the individual has disease D is

$$P(E \mid B) = \frac{P(E)P(B \mid E)}{P(E)P(B \mid E) + P(E^c)P(B \mid E^c)} = \frac{(0.25)(0.80)}{0.3125}$$

$$= \frac{0.2000}{0.3125} = 0.64.$$

On the other hand, suppose that the medical test does not yield a positive response. Then, since $P(B^c \mid A) = 1 - P(B \mid A)$ for any event A,

$$P(E \mid B^c) = \frac{P(E)P(B^c \mid E)}{P(E)P(B^c \mid E) + P(E^c)P(B^c \mid E^c)}$$

$$= \frac{(0.25)(0.20)}{(0.25)(0.20) + (0.75)(0.85)} = 0.0727.$$

Note that $P(E \mid B^c) < P(E) < P(E \mid B)$.

4. PROBABILISTIC INDEPENDENCE

Conditional probabilities are used to reassess probabilities of events in the light of partial information concerning the outcome of a random experiment. In some instances, there may be no new probabilistic information about an event B that can be obtained from knowing that an event A has occurred. In this case, the conditional probability $P(B \mid A)$ of B *given* A is the same as the unconditional probability $P(B)$ of B; that is,

(4.1) $$P(B \mid A) = P(B).$$

If (4.1) is true, we say that *the event B is probabilistically independent of the event A*. Similarly, *the event A is probabilistically independent of the event B* if

(4.2) $$P(A \mid B) = P(A).$$

However, if (4.1) holds, then

$$P(A \cap B) = P(A)P(B \mid A) = P(A)P(B),$$

and thus

$$P(A \mid B) = \frac{P(A \cap B)}{P(B)} = \frac{P(A)P(B)}{P(B)} = P(A),$$

so that (4.2) holds. Similarly, (4.2) implies that $P(A \cap B) = P(A)P(B)$, which in turn can be used to show that (4.1) holds. We have therefore shown that B

is probabilistically independent of A if and only if A is probabilistically independent of B. Summarizing our findings, we can now define the probabilistic independence of two events A and B.

Probabilistic Independence of Two Events. Two events A and B are *probabilistically independent* of one another if either (4.1), (4.2), or

(4.3) $P(A \cap B) = P(A)P(B)$

is a true equality.

Although (4.3) is a less conceptually meaningful definition of probabilistic independence than either (4.1) or (4.2), it is computationally useful, and also has the advantage that it clearly reveals the symmetry of the events A and B in the definition of probabilistic independence. Using an obvious terminology, if A and B are not probabilistically independent events, then we say that A and B are *probabilistically dependent* events.

While we are discussing terminology, it should be mentioned that the adverbs "statistically" and "stochastically" are often used in place of "probabilistically" when discussing probabilistic independence or dependence of events. Indeed, it is customary in the literature to omit these adverbs entirely whenever it is clear what kind of "independence" or "dependence" is being discussed. We will adopt this last custom in the remainder of this book. Thus, if we say that two events A and B "are independent," we will mean that these events are probabilistically independent.

Is it possible for A and B to be independent events, and yet for A^c and B, A and B^c, or A^c and B^c to be dependent? Intuitively, the answer should be "no," and this is indeed the case. For example, if A and B are independent events, then using (4.3) yields

(4.4) $P(A^c \cap B) = P(B) - P(A \cap B) = P(B) - P(A)P(B)$

$= P(B)(1 - P(A)) = P(B)P(A^c) = P(A^c)P(B).$

Thus, A^c and B are independent events. Similar arguments show that if A and B are independent events, then A and B^c and also A^c and B^c are independent events, and

(4.5) $P(A \cap B^c) = P(A)P(B^c),$

$P(A^c \cap B^c) = P(A^c)P(B^c).$

Example 4.1. Consider the experiment of randomly selecting two units *with* replacement from a population consisting of the two units u_1 and u_2. The probability model for this experiment is

Simple event	$\{(u_1, u_1)\}$	$\{(u_1, u_2)\}$	$\{(u_2, u_1)\}$	$\{(u_2, u_2)\}$
Probability	$\frac{1}{4}$	$\frac{1}{4}$	$\frac{1}{4}$	$\frac{1}{4}$

Let A be the event that u_1 is chosen on the first draw, so that $A = \{(u_1, u_1), (u_1, u_2)\}$. Let B be the event that u_2 is chosen on the second draw, so that $B = \{(u_1, u_2), (u_2, u_2)\}$. Then $A \cap B = \{(u_1, u_2)\}$, and

$$P(A \cap B) = \tfrac{1}{4} = \tfrac{1}{2}(\tfrac{1}{2}) = P(A)P(B).$$

Hence, the events A and B are independent.

In Example 4.1, a heuristic argument can be given to explain why events A and B are independent. Note that event A refers to the first draw and event B refers to the second draw. Because the sampling is *with replacement*, knowledge of the result of the first draw yields no information as to the probabilities governing what will take place on the second draw; the second draw is from the same population as for the first draw, since the unit drawn on the first draw is replaced in the population before the second draw is made. This lack of information is what is reflected in the mathematical notion of probabilistic independence.

In contrast, if in Example 4.1, we sampled two units *without* replacement, then our probability model would be

Simple event	$\{(u_1, u_1)\}$	$\{(u_1, u_2)\}$	$\{(u_2, u_1)\}$	$\{(u_2, u_2)\}$
Probability	0	$\tfrac{1}{2}$	$\tfrac{1}{2}$	0

and

$$P(A \cap B) = P\{(u_1, u_2)\} = \tfrac{1}{2} \neq \tfrac{1}{2}(\tfrac{1}{2}) = P(A)P(B),$$

so that A and B are dependent events. Indeed, once event A has occurred on the first draw, the fact that we are sampling without replacement means that only the unit u_2 is left in the population for the second draw, and thus $P(B \mid A) = 1 \neq \tfrac{1}{2} = P(B)$. Knowing that u_1 was drawn on the first draw, we are *certain* that u_2 will appear on the second draw.

Mutually Exclusive Events and Probabilistically Independent Events

There is a tendency to equate the concepts "mutually exclusive" and "probabilistically independent." This is a fallacy. Indeed, suppose that A and B are mutually exclusive events and $P(B) \neq 0$. Then, if we know that B has occurred, A cannot occur and $P(A \mid B) = 0$. If $P(A) \neq 0$, then $P(A \mid B) = 0 \neq P(A)$, and thus A and B are dependent events. That is, two possible mutually exclusive events are always dependent (not independent). On the other hand, if A and B are independent events, and A and B both are possible ($P(A) \neq 0$, $P(B) \neq 0$), then A and B cannot be mutually exclusive for $P(A \cap B) = P(A)P(B) \neq 0$, whereas if A and B were mutually exclusive, then $A \cap B = \varnothing$ and $P(A \cap B) = P(\varnothing) = 0$.

Mutually Probabilistically Independent Events

We have given two equivalent types of definition of the (probabilistic) independence of two events, E_1 and E_2: an intuitive definition based on conditional probabilities, and the computational definition based on the equality $P(E_1 \cap E_2) = P(E_1)P(E_2)$. Both kinds of definition can be generalized to provide equivalent definitions of the *mutual probabilistic independence* of more than two events.

Intuitive Definition of Mutually Probabilistically Independent Events

The n events $E_1, E_2, \ldots, E_n$ are mutually probabilistically independent (mutually independent) if for every event E_k, the conditional probability of E_k given the intersection of *every* subcollection of the remaining events $E_1, \ldots, E_{k-1}, E_{k+1}, \ldots, E_n$ is equal to the unconditional probability of E_k, $k = 1, \ldots, n$.

For $n = 3$, the definition of the mutual independence of E_1, E_2, and E_3 requires that *all* the following inequalities be true:

$$P(E_1) = P(E_1 \,|\, E_2) = P(E_1 \,|\, E_3) = P(E_1 \,|\, E_2 \cap E_3),$$

(4.6) $$P(E_2) = P(E_2 \,|\, E_1) = P(E_2 \,|\, E_3) = P(E_2 \,|\, E_1 \cap E_3),$$

$$P(E_3) = P(E_3 \,|\, E_1) = P(E_3 \,|\, E_2) = P(E_3 \,|\, E_1 \cap E_2).$$

Alternatively, we can use the following equivalent definition of the mutual independence of n events.

Computational Definition of Mutually Probabilistically Independent Events

The n events $E_1, E_2, \ldots, E_n$ are mutually independent if

(4.7) $$P(E_{i_1} \cap E_{i_2} \cap \cdots \cap E_{i_k}) = P(E_{i_1})P(E_{i_2}) \cdots P(E_{i_k})$$

for all subcollections $E_{i_1}, E_{i_2}, \ldots, E_{i_k}$ of size $k \leq n$ of the collection $E_1, E_2, \ldots, E_n$.

For example, the events E_1, E_2, and E_3 are mutually independent if *all* the following equalities hold:

$$P(E_1 \cap E_2) = P(E_1)P(E_2),$$

(4.8a) $$P(E_1 \cap E_3) = P(E_1)P(E_3),$$

$$P(E_2 \cap E_3) = P(E_2)P(E_3),$$

(4.8b) $$P(E_1 \cap E_2 \cap E_3) = P(E_1)P(E_2)P(E_3).$$

We have claimed that the intuitive and computational definitions of mutually independent events are equivalent. We now demonstrate this fact in the case of three events E_1, E_2, E_3 by proving that (4.6) is equivalent to (4.8).

The arguments needed to prove equivalence in the case of more than three events are similar. First, recall that

$$(4.9) \qquad P(E_i) = P(E_i \mid E_j) \quad \text{if and only if} \quad P(E_i \cap E_j) = P(E_i)P(E_j)$$

for $1 \le i \ne j \le 3$. Using (3.4) and applying (4.6), we obtain

$$P(E_1 \cap E_2 \cap E_3) = P(E_1)P(E_2 \mid E_1)P(E_3 \mid E_1 \cap E_2)$$
$$= P(E_1)P(E_2)P(E_3).$$

This result and (4.9) show that (4.6) implies (4.8). On the other hand, if (4.8) holds, then

$$P(E_1 \mid E_2 \cap E_3) = \frac{P(E_1 \cap E_2 \cap E_3)}{P(E_2 \cap E_3)} = \frac{P(E_1)P(E_2)P(E_3)}{P(E_2)P(E_3)} = P(E_1),$$

and similarly, $P(E_2 \mid E_1 \cap E_3) = P(E_2)$, $P(E_3 \mid E_1 \cap E_2) = P(E_3)$. These results and (4.9) show that (4.8) implies (4.6). Hence, (4.6) and (4.8) are equivalent.

Suppose that E_1, E_2, E_3 are mutually independent events. Then

$$P(E_2)P(E_3) = P(E_2 \cap E_3)$$
$$= P((E_1 \cap E_2 \cap E_3) \cup (E_1^c \cap E_2 \cap E_3))$$
$$= P(E_1 \cap E_2 \cap E_3) + P(E_1^c \cap E_2 \cap E_3)$$
$$= P(E_1)P(E_2)P(E_3) + P(E_1^c \cap E_2 \cap E_3),$$

from which

$$(4.10) \qquad P(E_1^c \cap E_2 \cap E_3) = [1 - P(E_1)]P(E_2)P(E_3)$$
$$= P(E_1^c)P(E_2)P(E_3).$$

Arguments similar to those given above can be used to verify the following useful rule.

General Law of Multiplication

If E_1, E_2, ..., E_n are mutually independent events, then

$$(4.11) \qquad P(F_1 \cap F_2 \cap \cdots \cap F_n) = P(F_1)P(F_2) \cdots P(F_n),$$

where F_i can be either E_i or E_i^c, $i = 1, \ldots, n$.

It is clear from either of the two equivalent definitions of mutual independence, that if E_1, E_2, ..., E_n are mutually independent events, then every subcollection E_{i_1}, E_{i_2}, ..., E_{i_k}, $2 \le k \le n$, of E_1, E_2, ..., E_n are mutually independent events. In particular, every pair E_i, E_j, $i \ne j$, of events in a collection E_1, E_2, ..., E_n of mutually independent events are independent. For example, if the events E_1, E_2, E_3 are mutually independent, then so are the pairs E_1 and E_2, E_1 and E_3, and E_2 and E_3.

Although mutual independence of the events E_1, E_2, ..., E_n implies the independence of every pair E_i, E_j, $i \neq j$, the converse of this assertion is not true in general; that is, *pairwise independence of all pairs* E_i, E_j, $i \neq j$, $i, j = 1$, 2, ..., *n, does not imply mutual* independence of the events E_1, E_2, ..., E_n, except trivially when $n = 2$. For example, suppose that the sample space Ω of a random experiment has four outcomes ω_1, ω_2, ω_3, ω_4, and that $P(\{\omega_i\}) = \frac{1}{4}$, $i = 1, 2, 3, 4$. Suppose that

$$E_i = \{\omega_i, \omega_4\}, \quad i = 1, 2, 3.$$

Then

$$P(E_i \cap E_j) = P(\{\omega_4\}) = \tfrac{1}{4} = (\tfrac{1}{2})(\tfrac{1}{2}) = P(E_i)P(E_j)$$

for all $i \neq j$, $i, j = 1, 2, 3$. Thus, E_i and E_j are independent events for all $i \neq j$. However,

$$P(E_1 \cap E_2 \cap E_3) = P(\{\omega_4\}) = \tfrac{1}{4} \neq \tfrac{1}{8} = P(E_1)P(E_2)P(E_3),$$

so that (4.8) fails to hold, and E_1, E_2, E_3 are not mutually independent events.

Because of the fact that every pair E_i, E_j, $i \neq j$, of the events E_1, E_2, ..., E_n can be independent, even when E_1, E_2, ..., E_n are not mutually independent, we must be careful in establishing terminology for the case when n events are not mutually independent. Thus, if E_1, E_2, ..., E_n are not mutually independent, we say that they are *mutually dependent*.

Probabilistically Independent and Mutually Probabilistically Independent Components of an Experiment

In many cases, a random experiment can be regarded as if its outcomes are the consequence of combining the outcomes of two or more *component random experiments*. For example, if we are randomly selecting one individual, and measuring the number of years of schooling and the index of social status of this individual, then our experiment $\mathscr{E}$ can be regarded as the composition or *composite* of two *component random experiments* $\mathscr{E}^{(1)}$ and $\mathscr{E}^{(2)}$, where in $\mathscr{E}^{(1)}$ we measure an individual's years of schooling and in $\mathscr{E}^{(2)}$ we measure the index of social status of the individual. As another example, each of the n units in a random sample of n units can be thought of as being drawn as a sample of one from a population of units. The n draws that create the sample are then *component random experiments* $\mathscr{E}^{(1)}$, $\mathscr{E}^{(2)}$, ..., $\mathscr{E}^{(n)}$, whose outcomes are combined to form the outcome of the *composite random experiment*, $\mathscr{E}$, which consists of drawing the entire sample of n units. This last example is a special case of the experiment in which the same random experiment is repeated n times. When the outcomes of these n repetitions (trials) are summarized, that summary is an outcome of a composite random experiment $\mathscr{E}$ whose components are the n trials $\mathscr{E}^{(1)}$, $\mathscr{E}^{(2)}$, ..., $\mathscr{E}^{(n)}$.

We say that an event E defined for a composite random experiment $\mathscr{E}$ with component random experiments (*components*) $\mathscr{E}^{(1)}$, $\mathscr{E}^{(2)}$, ..., $\mathscr{E}^{(n)}$ *belongs to component* $\mathscr{E}^{(j)}$ [*is determined by* $\mathscr{E}^{(j)}$] if the occurrence or nonoccurrence of the event E can be investigated simply by observing only the component $\mathscr{E}^{(j)}$. For example, if we are observing the years of education and index of social status of a randomly chosen individual, and $E = \{$the individual had 16 years of schooling$\}$, then the event E belongs to the component $\mathscr{E}^{(1)}$ in which the individual's years of schooling is observed. Similarly, if we observe n trials of a certain random experiment, and E is the event that a certain happening occurs on the fifth trial, then the event E is determined by the fifth trial (component) $\mathscr{E}^{(5)}$ of the composite random experiment.

The assumption that the components of a composite random experiment are mutually independent can greatly simplify computations of probabilities of events. Two components $\mathscr{E}^{(1)}$ and $\mathscr{E}^{(2)}$ of a composite random experiment $\mathscr{E}$ are said to be *independent* if for any event A belonging to component $\mathscr{E}^{(1)}$ and any event B belonging to component $\mathscr{E}^{(2)}$, A and B are independent events. The k components $\mathscr{E}^{(1)}$, $\mathscr{E}^{(2)}$, ..., $\mathscr{E}^{(k)}$ of a composite random experiment $\mathscr{E}$ are said to be *mutually independent* if for every collection $E_1, E_2, \ldots, E_k$ of events, where E_i belongs to (is determined by) $\mathscr{E}^{(i)}$, $i = 1$, $2, \ldots, k$, the events $E_1, E_2, \ldots, E_k$ are mutually independent.

If we are given a probability model for a composite random experiment $\mathscr{E}$, then we can attempt to verify that the components $\mathscr{E}^{(1)}$, $\mathscr{E}^{(2)}$, ..., $\mathscr{E}^{(n)}$ are mutually independent. A composite random experiment consisting of k mutually independent repetitions of the same experiment is said to consist of *independent and identically distributed* trials. However, unless there is some simple functional way (see Chapter 9) of representing probabilities for the random experiment $\mathscr{E}$, the task of establishing mutual independence from a probability model for $\mathscr{E}$ can be a difficult task. Instead, we frequently construct the probability model for $\mathscr{E}$ on the *assumption* that its components are mutually independent, basing that assumption upon the nature of the process that gives rise to the joint results of these component experiments. For example, in Example 4.1 we could argue that every draw in a random sample *with replacement* of n items from a population of N items is unaffected by the results of other (previous) draws, since all previous draws are replaced in the population before each new draw. Thus, these draws are made independently (in a physical sense), and the results of these draws should be mutually probabilistically independent. Hence, the draws in a random sample of size n with replacement from a given population are independent and identically distributed trials of the random experiment of drawing a sample of size 1 from that population.

Example 4.2. Suppose that there are $N = 100$ fish in a pond, of which 30 fish are of legal size. When we fish from the pond, assume that fish are caught as if we take a random sample *with replacement* from the fish in the pond. If we catch 3 fish, what is the probability that all 3 fish caught are of legal size? Let E_i equal the event that the ith fish caught (ith draw) is of legal size, $i = 1, 2, 3$. Then $P(E_i) = \frac{30}{100} = 0.3$. Since we sample with replacement, our draws (catches) are mutually independent, and thus

$$P\{\text{all 3 fish are of legal size}\} = P(E_1 \cap E_2 \cap E_3)$$

$$= P(E_1)P(E_2)P(E_3)$$

$$= (0.3)(0.3)(0.3) = 0.027.$$

Similarly,

$$P\{\text{at least 1 fish is of legal size}\} = P(E_1 \cup E_2 \cup E_3)$$

$$= P((E_1^c \cap E_2^c \cap E_3^c)^c) = 1 - P(E_1^c \cap E_2^c \cap E_3^c)$$

$$= 1 - P(E_1^c)P(E_2^c)P(E_3^c) = 1 - (0.7)^3 = 0.657.$$

5. APPENDIX: COMBINATORIAL ANALYSIS

For some simple experiments we can enumerate all the possible outcomes by counting. For example, if we wish to choose a committee of 2 people from a group of 5 people, A, B, C, D, E, the 10 possible committees are {A, B}, {A, C}, {A, D}, {A, E}, {B, C}, {B, D}, {B, E}, {C, D}, {C, E}, and {D, E}. Now suppose that we need to choose a committee of 10 from a group of 100; the enumeration of all possible committees in this case, while possible, becomes a long and tedious task.

The preceding problem is one of a variety of counting problems for which *counting rules* would be helpful. In particular, counting rules can be useful in dealing with uniform probability models. However, such rules are merely shortcuts to computation and are not intrinsic to the understanding of the theory of probability.

Orderings and Permutations

An arrangement of n distinct objects in a given order is called a *permutation*. Suppose that we have n distinct objects; in how many distinguishable ways can we order them (that is, how many permutations can we form)?

If we have 2 distinct objects A and B, the possible permutations are AB, BA. If we have 3 distinct objects A, B, C, we may append C at any of the positions indicated by the arrows:

$$\downarrow\text{AB} \quad \text{A}\downarrow\text{B} \quad \text{AB}\downarrow \quad \downarrow\text{BA} \quad \text{B}\downarrow\text{A} \quad \text{BA}\downarrow.$$

For each of the 2 permutations of A and B, we have three choices of where to put C, thus giving a total of $2(3) = 6$ permutations. For 4 distinct objects, A, B, C, and D, we may repeat the argument. That is, for each of the $2 \times 3 = 6$ possible permutations of A, B, and C, we have 4 positions in which to place D. Thus, we have $2(3)(4) = 24$ permutations of 4 distinct objects. If we continue arguing in this way, we obtain the following rule.

Counting Rule 5.1. The number of distinguishable permutations (arrangements) of n distinct objects is $1 \times 2 \times 3 \times \cdots \times (n - 1) \times n$, denoted by $n!$

Thus,

$$2! = 1(2) = 2, \qquad 3! = 1(2)(3) = 6, \qquad 4! = 24,$$

and so on. For mathematical convenience in writing formulas, we define $0!$ to be 1.

Example 5.1. Six diplomats are asked to line up for a group picture. The number of ways in which they could order themselves is

$$6! = 1 \times 2 \times 3 \times 4 \times 5 \times 6 = 720.$$

Orderings with Similar Elements

Consider a shelf upon which we are to place 5 books, each of which has a dust jacket with a different design. If we wish to pick the most aesthetic arrangement of dust jackets, we have to view $5! = 1 \times 2 \times 3 \times 4 \times 5 = 120$ distinguishable arrangements of the books. Now, however, suppose that 3 of the books have identical dust jackets. How many arrangements of the books can we distinguish now?

Assume that books A, B, and C have identical jackets, whereas books D and E have jackets different from those of A, B, and C, and from each other. Consider any single permutation of the books, say ABDCE. Keeping books D and E fixed in place, rearrange A, B, and C in this ordering in all possible ways. There are 3! possible orderings of A, B, C. Since A, B, and C have the same dust jackets, and J denotes their common jacket design, then all 6 of the foregoing arrays look to us like JJDJE. This same argument can be carried out for every positioning of the books D and E. Thus, rather than $5! = 120$ distinguishable orderings, there are really only $5!/3! = \frac{120}{6} = 20$ different orderings.

Now suppose that D and E have dust jackets identical to one another but different from the common dust jacket of books A, B, and C. By the same argument as that given above, the number of distinguishable orderings of the book jackets is $(5!/3!)$ divided by $2!$ or $5!/(3!\,2!) = 10$.

Extending this argument to cover more general situations we obtain the following rule.

Counting Rule 5.2. The number of distinguishable arrangements of n items of which n_1 are of type 1 and n_2 are of type 2 is $n!/(n_1!n_2!)$. Here $n_1 + n_2 = n$, and types 1 and 2 are assumed to be different.

The expression

$$\binom{n}{k} = \frac{n!}{k!(n-k)!} = \binom{n}{n-k}$$

is called the kth *binomial coefficient*. The terminology "bionomial coefficient" arose historically because $\binom{n}{k}$ is the coefficient of $a^k b^{n-k}$ in the binomial expansion

$$(a + b)^n = \sum_{j=0}^{n} \binom{n}{j} a^j n^{n-j}.$$

Values of $\binom{n}{k}$ for $n = 1, 2, \ldots, 10$ and various values of k are presented in Table 5.1. The binomial coefficients can be arranged to form a triangle-like figure called Pascal's triangle, where each coefficient is obtained as the sum of the two coefficients directly above it (see Figure 5.1).

Example 5.2. One of the uses of Counting Rule 5.2 is the case in which each of the n items belongs to one and only one of two distinct and exhaustive categories, and we are interested in the number of distinguishable ways in which these items can be arranged. For example, suppose that we have 10 machine parts, $n_1 = 8$ of which are "defective," and $n_2 = 10 - n_1 = 2$ of

***Table 5.1:** Values of the kth Binomial Coefficient* $\binom{n}{k}$ *for* $k \le n$, $n = 1, 2, \ldots, 10$

	k										
n	0	1	2	3	4	5	6	7	8	9	10
1	1	1									
2	1	2	1								
3	1	3	3	1							
4	1	4	6	4	1						
5	1	5	10	10	5	1					
6	1	6	15	20	15	6	1				
7	1	7	21	35	35	21	7	1			
8	1	8	28	56	70	56	28	8	1		
9	1	9	36	84	126	126	84	36	9	1	
10	1	10	45	120	210	252	210	120	45	10	1

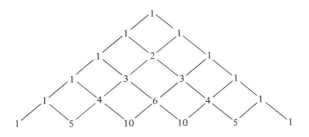

Figure 5.1: *Pascal's triangle.*

which are "not defective." The number of distinguishable ways in which these 10 parts can be arranged is

$$\binom{10}{8} = \binom{10}{2} = \frac{10!}{8!\,2!} = 45.$$

Counting Rule 5.2 can be extended to situations where the n items to be arranged belong to k distinct types and consist of n_j items of type j, $j = 1, 2, \ldots, k$.

Counting Rule 5.3. The number of distinguishable arrangements of n items, n_1 of which are of type 1, n_2 of which are of type 2, $\ldots$, n_k of which are of type k is

$$\binom{n}{n_1, n_2, \ldots, n_k} = \frac{n!}{n_1!\,n_2! \cdots n_k!}.$$

Here $n_1 + n_2 + \cdots + n_k = n$ and the types $1, 2, \ldots, k$ are distinct.

Example 5.3. In attempting to find a pleasing abstract design, we have available 20 positions in which to place patterned tiles. We have available 4 distinct tile patterns, A, B, C, and D. Seven of our tiles are of pattern A, 5 are of pattern B, 4 are of pattern C, and 4 are of pattern D. From Counting Rule 5.3, we thus have a total of $20!/(7!\,5!\,4!\,4!) = 6,983,776,800$ designs from which to choose.

Ordering of r Items from n

Consider again the problem of ordering 5 distinct books on a shelf. However, now suppose that we may choose these 5 books from a bin holding 12 different books. This problem may be viewed as one in which we attempt to fill the 5 vacant places ① ② ③ ④ ⑤ with items chosen from among 12 distinct items. For each of the 12 ways we can fill the first vacant place, there are 11 ways to fill the second. For each of the $12(11) = 132$ ways in which we can fill the first 2 places, there are 10 ways to fill the third. Continuing in this way, we see that there are $12(11)(10)(9)(8)$ or 95,040 ways of filling the 5 vacant places.

This calculation in general is given by the following rule.

Counting Rule 5.4. The number of ways of ordering r items from among n items is

$$n_{(r)} = n(n - 1)(n - 2) \cdots (n - r + 2)(n - r + 1).$$

Thus,

$$5_{(3)} = (5)(4)(3) = 60 \quad \text{and} \quad 5_{(4)} = (5)(4)(3)(2) = 120.$$

Notice that $n_{(n)} = n!$ and $n_{(1)} = n$.

The drawing of an ordered sample of size n *without replacement* from a population consisting of the N units $u_1, u_2, \ldots, u_N$ is directly analogous to the ordering of n of the units from all N of the units. Consequently, *the total number of ordered samples without replacement of size r from a population of N units is $N_{(r)} = N(N - 1)(N - 2) \cdots (N - r + 1) = N!/(N - r)!$*.

Counting of Sets of Items

In a variety of counting problems we do not distinguish between orderings containing the same items; that is, we are only interested in *sets* or *combinations* of items. For example, the number of ways in which a jury of 12 can be chosen from a panel of 300 eligible jurors does not depend on the order in which the jurors are chosen but only on *which* jurors are chosen. The number of ways a *set* (unordered sample) of 12 jurors can be chosen without replacement from among a panel of 300 is $300_{(12)}/12! = \binom{300}{12}$.

Counting Rule 5.5. The number of ways of choosing a set (combination, unordered sample) of r items from n distinct items is $\binom{n}{r}$.

Again, consider the problem of choosing a jury from a panel of 300 eligible jurors, of which 200 are men and 100 are women. The law requires that the proportion of men and women on the jury should be the same as the proportion in the panel; in this case, we need 8 men and 4 women. How many possible jurors can we choose? The answer to this question is $\binom{200}{8}\binom{100}{4}$ since for every one of the $\binom{200}{8}$ ways in which we can choose 8 men, we can choose any one of $\binom{100}{4}$ groups of 4 women.

We can extend these arguments to the following more general result.

Counting Rule 5.6. Suppose that we must choose a set of r items from a population of n items in which there are k distinct categories, with n_j items in category j, $j = 1, 2, \ldots, k$. Suppose further that our set must be chosen so that r_j items appear in category j, $j = 1, 2, \ldots, k$, $\sum_{j=1}^{k} n_j = n$, $\sum_{j=1}^{k} r_j = r$. The number of distinct ways to choose this set of r items is

$$\binom{n_1}{r_1}\binom{n_2}{r_2}\cdots\binom{n_k}{r_k}.$$

Example 5.3. In a university student council, there are 4 freshmen, 6 sophomores, 6 juniors, and 14 seniors. The number of ways of choosing a subcommittee with 1 freshmen, 1 sophomore, 2 juniors, and 3 seniors is

$$\binom{4}{1}\binom{6}{1}\binom{6}{2}\binom{14}{3} = 131{,}040.$$

The drawing of an ordered sample with replacement from a population consisting of N units $u_1, u_2, \ldots, u_N$ is directly analogous to choosing n items from n distinct categories in such a way that N items belong to each category and one item is chosen from each category. Since each category has size N, we have $n_1 = n_2 = \cdots = n_n = N$; since one item is drawn from each category, $r_1 = r_2 = \cdots = r_n = 1$. Thus, the number of possible ordered samples with replacement is

$$\binom{n_1}{r_1}\binom{n_2}{r_2}\cdots\binom{n_n}{r_n} = \left(\binom{N}{1}\right)^n = N^n.$$

The total number of ordered samples with replacement of size n from a population of N units is N^n.

NOTES AND REFERENCES

A number of books contain extensive tables of binomial coefficients. For example, in the *Handbook of Mathematical Functions*, edited by M. Abramowitz and I. A. Stegun, binomial coefficients for $n = 1(1)50$ are provided. The notation $n = 1(1)50$ means that the first value of n is 1, the last value of n is 50, and the steps are in units of 1.

A table of the factorials is given in *Tables of n! and $\Gamma(n + \frac{1}{2})$ for the First Thousand Values of n*, National Bureau of Standards, Applied Mathematics, Series 16.

EXERCISES

1. In a game of roulette, find probabilities of the following events.
   (a) $E_1 = $ "The wheel stops at a number that strictly exceeds 18."
   (b) $E_2 = $ "The wheel stops at a number (other than 0 and 00) that is divisible by 3."
   (c) $E_1 \cap E_2$.
   (d) $E_1 \cup E_2$.

2. Families with three sons are studied with respect to color blindness.
   (a) Taking account of order of birth, list the possible outcomes of the experiment in which a family is randomly drawn and it is recorded for each son whether or not he is colorblind.
   (b) Suppose that we have a uniform probability model for this experiment. Find (i) the probability that the oldest son is colorblind, (ii) the probability that the middle son is colorblind, and (iii) the probability that the youngest son is colorblind. What is the probability that all three sons are colorblind?

3. Two members of a class of 10 children are told to clean the blackboard after class. The teacher claims that the choice of the students was made by simple random sampling without replacement, yet the students actually chosen are the only two students in class who write for an underground newspaper. If the teacher is telling the truth, what is the probability of the event "both children chosen to clean the blackboard write for the underground newspaper?"

4. An urn contains 5 red balls and 5 black balls. Individual A takes a simple random sample of $n = 2$ balls *with* replacement from the urn, and then replaces the balls in the urn. Individual B then takes a simple random sample *without* replacement of $n = 2$ balls. Let E be the event "the sample contains 1 red and 1 black ball." Find $P(E)$ for (a) individual A's sample, and (b) individual B's sample.

5. A botanist studying the process of fertilization in a certain kind of plant chooses a random sample of $n = 2$ plants *without* replacement from a greenhouse containing 30 plants of the desired kind. Unknown to the botanist, 3 of the 30 plants have damaged buds and cannot flower. Find the following.
   (a) The probability that none of the chosen plants flower.
   (b) The probability that all of the chosen plants flower.
   (c) The probability that exactly 1 of the chosen plants flower.

6. Repeat Exercise 5 assuming that $n = 3$ plants are sampled.

7. A random sample of $n = 2$ units is taken *without* replacement from a population of N units, Q of which have a certain characteristic (characteristic C). Assume that the two units are drawn one at a time, and let E_j be the event that the jth unit drawn has characteristic C, $j = 1, 2$. Show that

(a) $P(E_1 \cap E_2) = \dfrac{Q(Q-1)}{N(N-1)}$, $P(E_1^c \cap E_2) = \dfrac{Q(N-Q)}{N(N-1)}$,

(b) $P(E_1) = P(E_2) = \dfrac{Q}{N}$,

(c) $P(E_1 \cup E_2) = \dfrac{Q(2N-Q-1)}{N(N-1)}$.

8. Smith and Suchman (1955) are concerned with the question "Do people know why they buy?" In a telephone survey made in March 1938 in Syracuse, New York, 764 people were called during the time Boake Carter (who was then advertising Philco radios) was on the air, and asked for the program to which they were listening at the moment of the call and the make of the radio they owned. The following relative frequencies were obtained:

		Owned a Philco radio	
		Yes	No
Listened to Boake Carter	Yes	0.136	0.170
	No	0.272	0.422

Assume that these relative frequencies are actually probabilities for the events described.
(a) Given that someone owns a Philco radio, what is the conditional probability that this person listened to Boake Carter?
(b) Given that someone listened to Boake Carter, what is the conditional probability that this person owns a Philco radio?

9. A pebble is selected at random from the shore of Lake Michigan, and the color (brown, black, green) of this pebble is observed. Based on a group of 1000 pebbles studied by Miller and Kahn (1962), an appropriate probability model for this experiment is the following:

Simple event	{brown}	{black}	{green}
Probability	0.852	0.093	0.055

(a) Find the probability that a randomly selected pebble is either black or brown.

(b) Find the conditional probability that the pebble is brown *given* that the pebble is either black or brown. Compare the conditional probability to the unconditional probability that the pebble is brown.

10. Based on studies described by Gregory (1963, p. 78), the following probability model describes the number of floods in a given wet season.
  (i) The outcomes are the numbers 0, 1, 2, 3, ... of floods. The simple events are {0 floods} = {0}, {1}, {2}, {3}, and so on.
  (ii) The probability model is determined by

Simple event	{0}	{1}	{2}	{3}	{4}	{5}	{6}
Probability	0.2466	0.3452	0.2417	0.1127	0.0395	0.0110	0.0033

Simple events corresponding to more than 6 floods have probability 0. Find the following.
  (a) The probability of 2 or more floods in a wet season.
  (b) The conditional probability of 4 or more floods in a wet season given that 2 floods have already been observed.
  (c) The conditional probability of at least 1 *additional* flood in a wet season given that a flood has already been observed.

11. Table E.1 shows data obtained by Lazarsfeld and Thielens (1958) on the relationships among age, productivity, and party vote in 1952 for a sample of 2117 social scientists studied in 1954–1955.

Table E.1: Social Scientists, Classified by Age, Productivity Score, and Party Vote in 1952

	Productivity score					
	Low		Middle		High	
Age	Demo-crats	Others	Demo-crats	Others	Demo-crats	Others
40 or younger	260	118	226	60	224	60
41–50	60	60	78	46	231	91
51 or older	43	60	59	60	206	175

Calculate the following:
  (a) The percentage of social scientists who voted Democratic in 1952.
  (b) The percentage of social scientists in the age group 41–50 who are classified in the middle of the productivity score.
  (c) The percentage of social scientists 51 years or older and with a low productivity score among all who voted Democratic in 1952.

(d) The percentage of social scientists 40 years or younger among all who did not vote Democratic in 1952.

(e) The percentage of social scientists 51 years or older and with a low productivity score among all who did not vote Democratic in 1952.

(f) The percentage of social scientists low on the productivity scale and voting Democratic in 1952 among all who are in the age group 41–50 years.

12. Studies of the number of children in families provide experimental evidence to support the following table of probabilities [*Statistical Abstract of the United States* (1965), p. 36]:

Event	{0 children}	{1 child}	{2 children}	{3 children}	{4 or more children}
Probability	0.431	0.173	0.174	0.113	0.109

(a) What is the probability that a randomly chosen family has 1 or more children?

(b) Given that a randomly chosen family has children (1 or more), what is the conditional probability that they have exactly 1 child?

(c) What is the probability that a randomly chosen family has at least 2 children? If it is learned that the randomly chosen family lives in an apartment complex where families with more than 3 children are not allowed, what is the *conditional* probability that the family has at least 2 children?

13. A marketing study of TV viewing habits concentrated on two types of program: news programs and sports programs. Suppose that it was found that 80% of college-educated adults watched the news, 50% watched sports, and 40% watched both news and sports. On the other hand, 65% of adults who had not completed college watched news, 85% watched sports, and 55% watched both. Finally, assume that 35% of all adults are college educated.

(a) What percent of all college-educated adults watch neither news nor sports on TV?

(b) What percent of all adults watch neither news nor sports on TV?

(c) What percent of all non-college-educated adults watch sports but not news on TV?

(d) What percent of all adults watch sports but not news on TV?

14. For any probability model, show that
(a) $P(E_1 \cap E_2^c \,|\, A) = P(E_1 \,|\, A) - P(E_1 \cap E_2 \,|\, A)$,
(b) $P(E_1 \cap E_2 \,|\, A) = P(E_1 \,|\, A)P(E_2 \,|\, E_1 \cap A)$,
(c) $E_2 \subset E_1$ and $P(E_1 \,|\, A) = 0$ implies that $P(E_2 \,|\, A) = 0$,
(d) $(E_1 \cap E_2) \cap A = \varnothing$ implies that $P(E_1 \cup E_2 \,|\, A) = P(E_1 \,|\, A) + P(E_2 \,|\, A)$.

15. Glass and Hall (1954) obtained occupational status data on male residents of England and Wales and their fathers. Three occupation states were distinguished: $S_1 =$ upper-level occupations (professional or executive), $S_2 =$ middle-level occupations (supervisory, nonmanual, and skilled manual), and $S_3 =$ lower-level occupations (semiskilled and unskilled manual). Given that the father had occupation state S_1, the conditional probabilities of the occupational states for the sons were

$$P(S_1 \,|\, \text{father is } S_1) = 0.45,$$
$$P(S_2 \,|\, \text{father is } S_1) = 0.48,$$
$$P(S_3 \,|\, \text{father is } S_1) = 0.07.$$

Similarly,

$$P(S_1 \,|\, \text{father is } S_2) = 0.05, \qquad P(S_1 \,|\, \text{father is } S_3) = 0.01,$$
$$P(S_2 \,|\, \text{father is } S_2) = 0.70, \qquad P(S_2 \,|\, \text{father is } S_3) = 0.50,$$
$$P(S_3 \,|\, \text{father is } S_2) = 0.25, \qquad P(S_3 \,|\, \text{father is } S_3) = 0.49.$$

If the probabilities for the various states for the fathers are $P(S_1) = 0.50$, $P(S_2) = 0.40$, and $P(S_3) = 0.10$, find the following.
(a) The unconditional probability that both father and son have occupational state (i) S_1, (ii) S_2, and (iii) S_3.
(b) The conditional probability that a son has occupational state at least as high as his father, *given* that the father has occupational state (i) S_1 and (ii) S_2.
(c) The unconditional probability that a son's occupational state is at least as high as his father's.

16. On a 3-part examination, the probability is $\frac{1}{2}$ of correctly answering the first part of the exam. The probability of correctly answering the second part of the exam given that the first part of the exam is correctly answered is $\frac{3}{4}$; while the probability of correctly answering the second part of the exam given that the first part of the exam is incorrectly answered is $\frac{1}{4}$. Finally, the probability of correctly answering part 3 of the exam is

$\frac{1}{5}$, given that both previous parts of the exam are missed;

$\frac{2}{5}$, given that part 2 is missed but part 1 is correctly answered;

$\frac{3}{5}$, given that part 1 is missed, but part 2 is correct;

$\frac{4}{5}$, given that parts 1 and 2 are correct.

(a) What is the probability that the person misses all 3 parts of the exam?
(b) What is the probability that the person misses the first 2 parts, but answers the third part correctly?

(c) What is the probability that the person misses the first and last parts, but answers the second part correctly?

(d) What is the probability that the person misses the last 2 parts, but answers the first part correctly?

(e) What is the probability that the person answers exactly 1 part correctly?

(f) What is the probability that the person answers exactly 2 parts correctly?

(g) Given that the person answers the last part correctly, what is the conditional probability that 2 or more parts are answered correctly?

17. In a probability model, events $E_1, E_2, \ldots, E_k$ are said to be decreasing or *nested* if $E_k \subset E_{k-1} \subset \cdots \subset E_2 \subset E_1$. Show that if $E_1, E_2, \ldots, E_k$ are nested events, then

$$P(E_k) = P(E_1)P(E_2|E_1)P(E_3|E_2) \cdots P(E_{k-1}|E_{k-2})P(E_k|E_{k-1}).$$

18. A rat is running a maze. After the rat enters the maze, there are 3 doors that can be chosen, only 1 of which is correct. If the rat chooses correctly, it enters a cell where there are 4 doors, 2 of which are correct. Each of these 2 correct doors leads to a cell where there are 5 doors, 2 of which lead to an exit. In each cell, assume that the rat guesses which door to choose. Use conditional probability calculations to find the probability that the rat will successfully traverse the maze.

19. A diagnostic test is used by a clinic to determine whether or not a person has a certain disease. If the test is positive (event T), the person is assumed to have the disease, while if the test is negative (event T^c), the person is assumed not to have the disease. Let D be the event that a person has the disease. Suppose that $P(D) = 0.10$, $P(T|D) = 0.90$, and $P(T^c|D^c) = 0.95$.

(a) Find $P(T)$.

(b) A mistaken diagnosis occurs when either (i) the test is negative and the person has the disease, or (ii) the test is positive and the person does not have the disease. What is the probability that the clinic will make a mistaken diagnosis?

(c) Find $P(D|T)$ and $P(D^c|T^c)$.

20. A study relating education level to income level shows that the population can be adequately represented by two income levels, "high" and "low," having relative sizes $\frac{1}{3}$ and $\frac{2}{3}$, respectively. The proportion of people in the high-income group who have a college education is 0.75, while the proportion of people in the low-income group who are college educated is 0.36. (a) Making use of Bayes' Rule, find the conditional probability that a person chosen at random will be from the high-income group, given that he is college educated. (b) If the person is college educated, what is the probability that he is from the low-income group?

21. In a study of marriages, both husbands and wives were classified as being either good tempered (G) or bad tempered (B). The following probability model (based on observations made on 300 couples) was found to hold.

Temper of wife

		G	B
Temper	G	0.22	0.24
of husband	B	0.31	0.23

(a) What is the conditional probability that a husband is good tempered given that his wife is good tempered?
(b) What is the conditional probability that a husband is good tempered given that his wife is bad tempered?
(c) Are the tempers of husband and wife probabilistically independent of one another? Based on the foregoing table of probabilities, if you want to find a good-tempered husband, should you look for a good-tempered wife?

22. The proportions of people falling into blood groups O, A, B, and AB are

Blood group	O	A	B	AB
Proportion	0.45	0.42	0.10	0.03

Assume that choice of a marriage partner is independent of blood group. A married couple is chosen by a simple random sample and the blood groups of the spouses are observed.
(a) What is the probability that the husband has blood group A and his wife has blood group B?
(b) What is the probability that one spouse has blood group A and the other spouse blood group B?
(c) What is the probability that both spouses have the same blood group?

23. In electrical systems, components may be inserted into a circuit in "parallel." If 2 or more components appear in parallel in a circuit, electric current will flow through the circuit if any component is capable of conducting the current. If 2 or more components appear in series in a circuit, all components must conduct current if electricity is to flow through the circuit. The accompanying figure shows an electric circuit in which the components B, D, and E, and the subcircuits A and C, are

inserted in series. Subcircuit A consists of 2 components, A_1 and A_2, in parallel; while subcircuit C consists of 3 components, C_1, C_2, and C_3, in parallel. The abilities to conduct electricity of the components A_1, A_2, B, C_1, C_2, C_3, D, and E are assumed to be mutually independent.

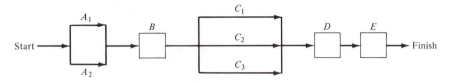

(a) Write out a list of the ways in which the circuit can conduct electric current.
(b) Assume that the following probabilities hold:

$$P(A_1 \text{ conducts}) = P(A_2 \text{ conducts}) = 0.7,$$
$$P(C_1 \text{ conducts}) = P(C_2 \text{ conducts}) = P(C_3 \text{ conducts}) = 0.6,$$
$$P(B \text{ conducts}) = P(D \text{ conducts}) = P(E \text{ conducts}) = 0.9.$$

Find the probability that the entire circuit conducts.
(c) Find the probability that current passes at least through subcircuit A and component B.
(d) Find the probability that current can pass through the part of the circuit consisting of subcircuit C and components D and E.
(e) Suppose that the entire circuit does not conduct electricity. What is the conditional probability that component B is not conducting electricity?

24. Suppose that a random experiment $\mathscr{E}$ has two component random experiments $\mathscr{E}^{(1)}$ and $\mathscr{E}^{(2)}$ and that the sample spaces $\Omega^{(1)}$ and $\Omega^{(2)}$ of $\mathscr{E}^{(1)}$ and $\mathscr{E}^{(2)}$ each have a finite number of outcomes. Show that $\mathscr{E}^{(1)}$ and $\mathscr{E}^{(2)}$ are independent if and only if

$$P(S_1 \cap S_2) = P(S_1)P(S_2)$$

for every simple event S_1 of $\mathscr{E}^{(1)}$ and every simple event S_2 of $\mathscr{E}^{(2)}$.

25. Miller and Kahn (1962, p. 6) report an experiment in which samples of rocks were collected and classified both according to type and according to modal size. Based on the data they report (2000 rock samples), the probability model shown in Table E.2 can be constructed.
(a) Note that this experiment is a composite random experiment in which there are two component random experiments: the observation of the rock type of the chosen rock sample, and the observation of the modal size of the rock sample. Find the probability models for each of the component experiments (considered separately) of this composite random experiment.

Table E.2: *Rock Samples Classified by Type and Modal Size*

Rock type	Modal size			
	Coarse	Medium	Fine	Very fine
Arkose	0.160	0.075	0.065	0.025
L. R. Gray Woeke	0.025	0.250	0.225	0.050
Quartzite	0.040	0.050	0.010	0.025

(b) Are the two component experiments probabilistically independent? That is, is rock type independent of rock size?

26. Francis Galton in his book *Finger Prints* (1892) discusses the prints of the right forefingers of 101 pairs of schoolchildren chosen at random from a large collection. In addition, 105 pairs of children having a fraternal relation are also chosen. The frequency tables for the two sets of data are given as Table E.3.

Table E.3: *Characteristics of Fingerprints of Fraternal and Nonfraternal Schoolchildren*

Nonfraternal Children

		A children			
		Arches	Loops	Whorls	Totals
B Children	Arches	5	12	8	25
	Loops	8	18	8	34
	Whorls	9	20	13	42
	Totals	22	50	29	101

Fraternal Children

		A children			
		Arches	Loops	Whorls	Totals
B Children	Arches	5	12	2	19
	Loops	4	42	15	61
	Whorls	1	14	10	25
	Totals	10	68	27	105

If the measurements on the A and B children were actually independent experiments, what distributions would we expect?

27. A fair coin is tossed 100 times, the tosses being independent. (Remember that a "fair" coin has an equal chance of landing "heads" or of landing "tails" at any given toss.) Given that the first 99 tosses result in "heads," what is the probability of obtaining a "head" on the 100th toss?

In each of the following problems, state explicitly the counting rules that you use to solve the problem.

28. How many five-place numbers can be made using each of the digits
    (a) 3, 4, 5, 8, 9;
    (b) 0, 4, 5, 8, 9;
    (c) 3, 4, 5, 5, 9?
    How many *even* five-place numbers can be made using each of the digits 2, 3, 5, 7, and 9?

29. If 13 diplomats are asked to line up for a group picture with the senior diplomat always to be in the center, in how many distinguishable ways can they be arranged?

30. If 7 types of meat are available for dinners, 1 meat for every day of the week, can we vary (permute) our choices of meats so that the meats appear in a different order each week of the year?

31. A wine taster claims to be able to discriminate among 5 different varieties of wine by taste. The wine taster is blindfolded and given the wine varieties one at a time.
    (a) In how many different possible orders could the 5 wine varieties be presented to the wine taster?
    (b) The wine taster was really boasting. When the actual experiment is run, the order in which the 5 varieties are named is chosen at random from among all possible orders. What is the probability that the wine taster guesses correctly?
    (c) What is the probability that the wine taster guesses 3 right out of 5?

32. Of 7 chest X-rays, two show a disease, while the others are normal. The radiologist only pays attention to the distinction between diseased and normal X-rays. In how many possible distinguishable orders could the X-rays be presented to the radiologist?

33. A child has 3 black blocks and 3 white blocks to arrange in one line.
    (a) How many possible distinguishable patterns can be made?
    (b) If the child arranges the blocks in 2 rows of 3 blocks each, how many distinguishable patterns can be made?

34.   Nine skaters, 3 from the United States, 3 from the USSR, and 3 from China compete. At the end of the contest, the skaters will be ranked from best to worst (no ties are permitted), but the scoring will only take account of what countries these skaters represent, not their individual identities.
   (a) For the purpose of scoring, how many possible distinguishable outcomes are there to this contest?
   (b) How many outcomes correspond to results in which the U.S. skaters are ranked 1, 2, 3?

35.   A certain stained-glass window is made of 25 panes (5 rows of 5 panes each). The artist who constructs the window has 7 blue panes, 5 red panes, 6 green panes, and 7 yellow panes. In how many different distinguishable ways can the panes be arranged to make an abstract design for the window?

36.   A family has a choice of 5 vacation spots. They decide to visit 2 of these spots on their vacation, spending part of their time at each. How many different choices of vacations do they have if
   (a) we distinguish the order in which they visit the vacation spots that they choose?
   (b) we do not distinguish the order in which they visit the chosen vacation spots, but only name which vacation spots were chosen?

37.   A dairy has 5 flavors, and a family always likes to buy 2 different flavors. How many times can the family go to the dairy and bring home a different pair of flavors?

38.   A telephone dial has a finger hole for each of the 10 digits.
   (a) How many telephone numbers, each with 7 digits but with no digit repeated, are possible?
   (b) How many telephone numbers, each with 7 digits, are possible?

39.   A signalman has 6 flags. The emblems on the various flags are a stripe, a dot, a triangle, a rectangle, a bar, and a circle. By showing 2 different flags, one after the other, the signalman can designate a signal. How many different signals are possible?

40.   A certain jury venire consists of 300 individuals of whom 135 are opposed to the death penalty and 165 favor the death penalty.
   (a) If an initial list of 12 individuals are picked from this venire by a simple random sample without replacement, what is the probability that 6 of them will favor the death penalty and 6 will oppose the death penalty?
   (b) What is the probability that a majority (7 or more) of these 12 individuals will oppose the death penalty?

41. Bicycle locks frequently are arranged with 4 adjacent disks, each disk having 5 numbers. The correct combination is an ordered list of 4 numbers.

(a) If we wish to open the lock, how many combinations do we have to try?

(b) Suppose that the digit on the third disk represents a geographical area where the lock is made. Thus, 1 may denote the Northeastern states, 2 the Southeastern states, 3 the Midwest, 4 the Northwest, and 5 the Southwest. If we know the bicycle comes from a western state (Northwest or Southwest), how many combinations do we have to try?

4

Random Variables

Most experiments of scientific interest are complex, with outcomes that are likely to be difficult to describe completely. A common device for reducing the complexity of a random experiment, and of its associated probability model, is that of confining attention to one or more quantitative aspects of the results of the experiment. For example, in an experiment in medicine, we could list all the symptoms displayed by each patient. However, if our interest is in the frequency and severity of undetected diabetes, we might simplify our analysis of the experiment by recording only the numerical measure, X, of the concentration of sugar in the blood of the patient.

The quantification of the outcomes of a random experiment in such a way that each outcome is represented by exactly one numerical value involves the notion of a *random variable* (chance variable, stochastic variable). A random variable X, then, is a prescription or rule by which every outcome ω in the sample space Ω of the random experiment is assigned a number $X(\omega)$, called the *value* of the random variable X for the outcome ω. We may view a random variable as being a coding whereby every outcome ω is given a numerical code $X(\omega)$. Alternatively, stated in mathematical terms, a random variable X is a *function* that assigns a real number $X(\omega)$ to each outcome ω in the sample space Ω. We use capital letters X, Y, Z to denote the different random variables, while possible numerical values that may be assigned by these random variables (functions of outcomes) to a particular outcome are denoted by corresponding lowercase letters x, y, z.

1. PROBABILITY MODELS FOR RANDOM VARIABLES

In some experiments, we may start with a probability model for an original description of the experiment in terms of possibly nonnumerical outcomes ω, and then simplify the model by reducing attention to the values of one or

83

more random variables. In this case, we create a new probability model in which the new outcomes refer to the values of the random variables that we are using, in which the new events refer to statements about the values of the random variables, and in which probabilities for these new events are obtained from the probabilities for simple events in the original probability model.

Example 1.1. Three infants are given a simple puzzle. Records are kept whether each of the three infants completes the solution of the puzzle during the period of observation. The eight possible outcomes of this experiment can be denoted as $\omega_1 = (S, S, S)$, $\omega_2 = (S, S, F)$, $\omega_3 = (S, F, S)$, $\omega_4 = (S, F, F)$, $\omega_5 = (F, S, S)$, $\omega_6 = (F, S, F)$, $\omega_7 = (F, F, S)$, and $\omega_8 = (F, F, F)$. Here, for example, the notation (S, F, S) means that the first and third infants succeeded (S) in completing the puzzle, while the second infant failed (F) to complete the puzzle. Suppose that we are interested only in the total number, X, of infants who succeed in completing the puzzle. By inspection, we see that $X(\omega_1) = 3$, $X(\omega_2) = 2$, $X(\omega_3) = 2$, $X(\omega_4) = 1$, $X(\omega_5) = 2$, $X(\omega_6) = 1$, $X(\omega_7) = 1$, and $X(\omega_8) = 0$. The possible values 0, 1, 2, and 3 of the random variable X now become the new outcomes of a new probability model in which only the number of infants completing the puzzle is of importance.

Suppose that a probability measure P was given for the original model in the form of a table such as the following:

Simple event	S_1	S_2	S_3	S_4	S_5	S_6	S_7	S_8
Probability	p_1	p_2	p_3	p_4	p_5	p_6	p_7	p_8

where $S_i = \{\omega_i\}$ and $p_i = P(S_i)$, $i = 1, 2, \ldots, 8$. To obtain the probabilities of the simple results $\{X = 0\}$, $\{X = 1\}$, $\{X = 2\}$, $\{X = 3\}$ for the new model, we simply note that

$$\{X = 0\} = \{\omega_8\} = S_8, \qquad \{X = 1\} = \{\omega_4, \omega_6, \omega_7\} = S_4 \cup S_6 \cup S_7,$$

$$\{X = 2\} = \{\omega_2, \omega_3, \omega_5\} = S_2 \cup S_3 \cup S_5, \qquad \{X = 3\} = \{\omega_1\} = S_1.$$

Hence, the probabilities of the simple events for the new model for the random variable X can now be given in terms of a table:

Simple event	$\{X = 0\}$	$\{X = 1\}$	$\{X = 2\}$	$\{X = 3\}$
Probability	p_8	$p_4 + p_6 + p_7$	$p_2 + p_3 + p_5$	p_1

For example, if the original model was a uniform probability model, so that $p_1 = p_2 = \cdots = p_8 = \frac{1}{8}$, then the probability model for X could be summarized as follows:

Simple event	$\{X = 0\}$	$\{X = 1\}$	$\{X = 2\}$	$\{X = 3\}$
Probability	$\frac{1}{8}$	$\frac{3}{8}$	$\frac{3}{8}$	$\frac{1}{8}$

From this table, we could find the probability of the event $\{X \geq 2\}$ that two or more infants completed the puzzle:

$$P\{X \geq 2\} = P\{X = 2\} + P\{X = 3\} = \tfrac{3}{8} + \tfrac{1}{8} = \tfrac{1}{2}.$$

Example 1.1 illustrates several important points about the use of random variables in random experiments. Notice first that there were eight possible nonnumerical outcomes ω_1, ω_2, ..., ω_8 in the original probability model, but that the new derived probability model for X had only four outcomes: 0, 1, 2, and 3. In general, when we quantize the outcomes ω of a random experiment, *the total number of possibilities that we must consider in our new model never exceeds the number of outcomes in the original model*, and may be considerably less. Thus, use of random variables may simplify the analysis of complicated random experiments.

Second, once we obtain the new probability model for the random variable X, we do not have to return to the original probability model in order to compute probabilities of events such as "the number of infants who completed the puzzle is 2 or greater." Instead, such probabilities are obtained directly from the probabilities for the simple events $\{X = 0\}$, $\{X = 1\}$, $\{X = 2\}$, $\{X = 3\}$ of the new probability model for X.

Third, suppose that we are interested in the *proportion* Y of infants who completed the puzzle rather than in the number X of infants completing the puzzle. We could use the fact that $Y(\omega_1) = 1$, $Y(\omega_2) = \tfrac{2}{3}$, and so on, to obtain a probability model for the simple events $\{Y = 0\}$, $\{Y = \tfrac{1}{3}\}$, $\{Y = \tfrac{2}{3}\}$, $\{Y = 1\}$ for Y. However, $Y = X/3$, so that $P\{Y = 0\} = P\{X = 0\}$, $P\{Y = \tfrac{1}{3}\} = P\{X = 1\}$, and so on. That is, Y and the probability model for Y can be obtained directly from X and the probability model for X. Note that Y is a function of ω (so that Y is a random variable) but that Y is also a function of X. We see that *given one random variable X we can form any number of new random variables Y by means of mathematical operations on X* (e.g., $Y = X/3$, $Y = X^2$, $Y = 1/X$, etc.). Further, *if the random variable Y is known to be a function of the random variable X, then the values and probabilities for Y can be calculated directly from the values and probabilities for X.*

For the foregoing reasons, in this chapter we discuss probability models for random variables without referring to any original nonnumerical description of the experiment that might have led to these random variables. That is, we act as if the values x of the random variable X are the original outcomes of the experiment, and that our goal was to construct and analyze probability models based on these numerical outcomes.

2. DISCRETE AND CONTINUOUS RANDOM VARIABLES

For the most part, random variables fall into two categories: discrete and continuous. *Discrete random variables* are random variables that take on only "isolated" values. That is, if we mark the *possible values* on the real line, there

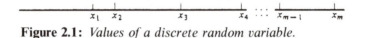

Figure 2.1: *Values of a discrete random variable.*

are gaps between the marks (see Figure 2.1). Examples of discrete random variables include (i) the number of traffic accidents over a given weekend, (ii) the number of children born to a given family, (iii) the number of dots on the upper face of a tossed die, and (iv) the income of a randomly chosen family (to the nearest dollar).

A *continuous random variable* is a random variable that (at least conceptually) can be measured to any desired degree of accuracy. The collection of all possible values of a continuous random variable usually consists of one or more intervals of real numbers.

The concept of a continuous random variable is a mathematical abstraction. Imagine a person being weighed on a scale that can give the weight to whatever number of decimal places we care to specify. Obviously, such accuracy can never be achieved, but our conception (or model) of such accuracy turns out to provide us with great mathematical simplification in many real problems. Examples of such problems are measurements of the amount of oil obtained from an oil well, the amount of alcohol distilled in a given distillation process, a child's intelligence, the distance a car can travel on 1 gallon of gas, and the blood pressure of a patient.

One important distinction between discrete and continuous variables may prove troublesome. For discrete random variables, $P\{X = x\}$ is not zero for some numbers x; however, for continuous random variables, it is necessary to assume that for all possible values x, $P\{X = x\} = 0$.

To justify this statement, recall first that a probability of zero for an event does *not* mean that the event is impossible, only that it is very very rare. Consider repeated weighings of a person on a perfectly accurate scale. The first time we obtain the value x, but the next time our measurement might differ by a very small amount from x, and it is unlikely that we would ever obtain the same weight twice. Therefore, if we calculate the relative frequency of any one value, that relative frequency would appear to be zero, or a very small number that would get smaller as we performed more and more weighings. On the other hand, if we measure the person's weight accurate only to the first decimal place, such information describes an *interval* of possible values, and we are likely to see certain readings (intervals) fairly frequently. Thus, although individual values of a continuous random variable are assumed to have zero probability of occurring, *intervals* of possible values can have positive probability. (It is worth noting that a similar, seemingly paradoxical, fact is true in geometry, where points have zero length, but lines, which are made up of points, have nonzero length.)

Because of the distinction that we have shown exists between discrete and continuous random variables, each of these types of model requires a different mathematical apparatus for its description and analysis. One descriptive tool common to both models is the notion of the cumulative distribution function.

3. CUMULATIVE DISTRIBUTION FUNCTIONS

Suppose that we have restricted ourselves to some single quantitative aspect of a given random experiment. Thus, we are interested in the probability model defined by a random variable X, the totality of whose possible values x is the sample space x. Through empirical or theoretical constructions, we wish to determine the probability measure for X.

One way to describe the probability model of *any* random variable is the *cumulative distribution function* (abbreviated c.d.f.). The cumulative distribution function $F_X(x)$ for the random variable X is defined for all numbers x by

$$(3.1) \qquad F_X(x) = P\{X \le x\}.$$

The domain of the function $F_X(x)$ is the real line $-\infty < x < \infty$, while the range of $F_X(x)$ is the interval $[0, 1]$. This last assertion follows since probabilities are numbers between 0 and 1. Note that for $x_1 < x_2$, we have

$$(3.2) \qquad F_X(x_2) = P\{X \le x_2\} = P\{X \le x_1\} + P\{x_1 < X \le x_2\}$$
$$= F_X(x_1) + P\{x_1 < X \le x_2\},$$

from which it follows that $F_X(x)$ is increasing (accumulating probability) as x increases; this is why $F_X(x)$ is called the *cumulative* distribution function of X.

When we discuss more than one random variable at a time, say the random variables X, Y, and Z, we write the c.d.f. (cumulative distribution function) of X as $F_X(x)$ so as to distinguish it from the c.d.f.'s $F_Y(y)$ and $F_Z(z)$ of Y and Z, respectively. *Every random variable has its associated cumulative distribution function.* When it is clear from the context which random variable and cumulative distribution function we are considering, the subscript X on the c.d.f. $F_X(x)$ is omitted.

We now show that when the random variable X is a discrete random variable having only a finite number of possible values, then the c.d.f. $F_X(x)$ of X determines its probability measure. Suppose that the possible values of X taken in order are $x_1 < x_2 < \cdots < x_M$. Then

$$F_X(x_i) - F_X(x_{i-1}) = P\{X \le x_i\} - P\{X \le x_{i-1}\}$$
$$(3.3) \qquad = \sum_{j=1}^{i} P\{x_j\} - \sum_{j=1}^{i-1} P\{x_j\}$$
$$= P\{x_i\}, \quad i = 1, 2, \ldots, M,$$

so that $P\{X = x_i\} = F_X(x_i) - F_X(x_{i-1})$. Since the probabilities of simple events determine the probability model of any random experiment having only a finite number of outcomes, knowledge of the values of the c.d.f. $F_X(x)$ determines the probability model of X.

Of course, as Example 3.1 illustrates, the converse of this assertion is also true.

Example 3.1. A linguist studying syllabication in the German language defines the random variable X to be the number of syllables in a randomly chosen German word. Based on some 20,453 words taken from Viëtor-Meyer's *Deutches Aussprachwörterbuch* [see Herdan (1966, p. 301)], the linguist constructs the probability model for X given in Table 3.1.

To determine the c.d.f. $F(x)$ of X, note that every event $\{X \le x\}$ can be written as the union of simple events $\{X = j\}$ corresponding to values of j less than or equal to x. Thus,

$$F(x) = P\{X \le x\} = \sum_{1 \le j \le x} P\{X = j\}.$$

In particular, if $x = 4.3$,

$$\{X \le 4.3\} = \{X = 1\} \cup \{X = 2\} \cup \{X = 3\} \cup \{X = 4\}$$

and

$$F(4.3) = \sum_{1 \le j \le 4.3} P\{X = j\} = P\{X = 1\} + P\{X = 2\} + P\{X = 3\} + P\{X = 4\}$$

$$= 0.110 + 0.313 + 0.341 + 0.178 = 0.942.$$

If $x < 1$, then $\{X \le x\}$ contains no values of X and thus has probability 0; if $x \ge 9$, then $\{X \le x\}$ contains all values of X and has probability 1. Using these facts, we obtain the following formula for $F(x)$:

Table 3.1: *Probability Model for Number, X, of Syllables in a Randomly Chosen German Word*

x	$P\{X = x\}$	x	$P\{X = x\}$
1	0.110	6	0.010
2	0.313	7	0.002
3	0.341	8	0.001
4	0.178	9	0.000[1]
5	0.045		

[1] If calculation had been carried out to five decimal places, $P\{X = 9\}$ would have been seen to equal a nonzero number. Since the values in Table 3.1 were rounded off to three decimal places, $P\{X = 9\}$ appears to be equal to zero.

$$(3.4) \qquad F(x) = \begin{cases} 0.000, & \text{if } x < 1, \\ 0.110, & \text{if } 1 \le x < 2, \\ 0.423, & \text{if } 2 \le x < 3, \\ 0.764, & \text{if } 3 \le x < 4, \\ 0.942, & \text{if } 4 \le x < 5, \\ 0.987, & \text{if } 5 \le x < 6, \\ 0.997, & \text{if } 6 \le x < 7, \\ 0.999, & \text{if } 7 \le x < 8, \\ 1.000, & \text{if } 8 \le x < 9, \\ 1.000, & \text{if } 9 \le x. \end{cases}$$

The graph of $y = F(x)$ is shown in Figure 3.1.

The graph of $F(x)$ looks like a series of steps rising from left to right; thus, $F(x)$ is called a *step function*. Notice also that the graph jumps at the numbers $x = 1, 2, \ldots, 9$, which are the possible values of X, and that the jump at $x = j$ equals $F(j) - F(j-1) = P\{X = j\}$ for $j = 1, 2, \ldots, 9$. For example, $F(5) - F(4) = 0.987 - 0.942 = 0.045 = P\{X = 5\}$. In general, *for any discrete random variable X, the c.d.f. $F(x)$ of X is a step function; the height of a jump of the graph at a point x is equal to $P\{X = x\}$.*

In contrast to the c.d.f.'s of discrete random variables, *the c.d.f.'s of continuous random variables are smooth* (see Figure 3.2), reflecting the fact that all events $\{X = x\}$ are assigned zero probability. Speaking intuitively, the probabilities of continuous random variables are not "lumped" at distinct values but are "smoothed" over all numbers.

Whether they arise from discrete or continuous random variables, all c.d.f.'s have some properties in common. We have already noted that $0 \le F(x) \le 1$ for all x. From Equation (3.2) it also follows that $F(x_1) \le F(x_2)$

Figure 3.1: *Graph of the c.d.f. $F_X(x)$ of the random variable X, which is the number of syllables in a randomly chosen German word.*

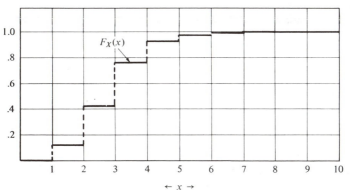

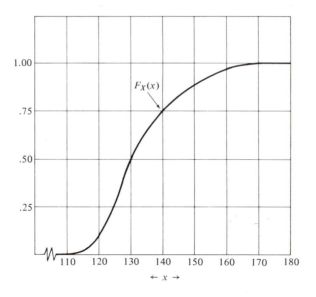

Figure 3.2: *A cumulative distribution function of a continuous random variable X.*

whenever $x_1 < x_2$. Hence, *the c.d.f. $F(x)$ of a random variable is a nondecreasing function*. Further, for any number x_0, it can be visually verified from Figures 3.1 and 3.2 that $F(x)$ tends to $F(x_0)$ as x decreases to x_0 ($x_0 \le x$), so that $F(x)$ is *continuous from the right*. In discussing Example 3.1, we gave an argument which always applies to show that if X has a largest possible value x_L, then $F(x) = 1$ for all $x \ge x_L$, while if X has a smallest value x_S, $F(x) = 0$ for all $x < x_S$. Even if X has neither a smallest nor a largest possible value, it is still the case that the graph of the c.d.f. $F(x)$ of a random variable X increases from 0 to 1 as we move from left to right with increasing x (see Figures 3.1 and 3.2).

Remark. It is worth noting that the properties (a) $F(x)$ is nondecreasing in x, (b) $F(x)$ is continuous from the right in x, and (c) $\lim_{x \to -\infty} F(x) = 0$, $\lim_{x \to \infty} F(x) = 1$, uniquely characterize c.d.f.'s. That is, any function $F(x)$ satisfying properties (a) through (c) is the c.d.f. of some random variable, and any function violating one or more of the three properties cannot be a c.d.f. of any random variable.

We often need to find the probabilities of such events as (i) $\{a < X\}$, (ii) $\{a < X \le b\}$, and (iii) $\{a \le X \le b\}$. Although we know from the previous discussion that all such probabilities are determined by the c.d.f. $F(x)$ of X, we still need formulas to tell us how to calculate such probabilities from knowledge of the values of the c.d.f. $F(x)$.

In case (i), note that we can apply the Law of Complementation (Rule 4.2 of Chapter 2) to obtain

$$P\{a < X\} = P\{\{X \le a\}^c\} = 1 - P\{X \le a\} = 1 - F(a).$$

The Law of Addition (Rule 4.3 of Chapter 2) implies that $P(A \cap B) = P(A) + P(B) - P(A \cup B)$. Letting $A = \{a < X\}$, $B = \{X \le b\}$, then $A \cup B = (-\infty, \infty)$, and

$$
\begin{aligned}
P\{a < X \le b\} = P(A \cap B) &= P(A) + P(B) - P(A \cup B) \\
&= P\{\{X \le a\}^c\} + P\{X \le b\} - 1 \\
&= (1 - F(a)) + F(b) - 1 = F(b) - F(a),
\end{aligned}
$$

solving case (ii). Finally, for case (iii),

$$
\begin{aligned}
P\{a \le X \le b\} &= P\{a < X \le b\} + P\{X = a\} \\
&= F(b) - F(a) + P\{X = a\},
\end{aligned}
$$

and $P\{X = a\}$ is the jump of $F(x)$ at $x = a$. Note that when X is a continuous random variable, $P\{X = a\} = 0$ and $P\{a \le X \le b\} = P\{a < X \le b\} = F(b) - F(a)$.

Table 3.2 gives formulas for calculating the probabilities of other events of interest when we know only the values of the c.d.f. $F(x)$ of the random variable X.

Example 3.1 (continued). In this example, the random variable X is the number of syllables in a randomly chosen German word. The c.d.f. $F(x)$ of X is given in Equation (3.4) and graphed in Figure 3.1. Using Table 3.2, the

Table 3.2: *Formulas for Calculating Probabilities of Certain Events Given the Values of the c.d.f. $F(x)$*

Event	Formula for probability of event
$\{X = a\}$	Height of jump of graph of $F(x)$ at $x = a$
$\{a < X\}$	$1 - F(a)$
$\{a \le X\}$	$1 - F(a) + P\{X = a\}$
$\{X \le b\}$	$F(b)$
$\{X < b\}$	$F(b) - P\{X = b\}$
$\{a < X \le b\}$	$F(b) - F(a)$
$\{a < X < b\}$	$F(b) - F(a) - P\{X = b\}$
$\{a \le X \le b\}$	$F(b) - F(a) + P\{X = a\}$
$\{a \le X < b\}$	$F(b) - F(a) + P\{X = a\} - P\{X = b\}$

probability that a randomly chosen German word has more than one syllable is $P\{1 < X\} = 1 - F(1) = 1 - 0.110 = 0.890$. Similarly,

$$P\{X = 5\} = F(5) - F(4) = 0.987 - 0.942 = 0.045,$$

$$P\{5 < X \le 7\} = F(7) - F(5) = 0.999 - 0.987 = 0.012,$$

$$P\{5 \le X \le 7\} = F(7) - F(5) + P\{X = 5\} = 0.999 - 0.987 + 0.045 = 0.057.$$

The problem of calculating probabilities of other events of the forms listed in Table 3.2 is handled in a similar fashion. Thus, Equation (3.4) can replace Table 3.1 for the purpose of summarizing the probability model of X in Example 3.1.

Estimating the Cumulative Distribution Function from Data

Suppose we make N independent trials of the random experiment that gives rise to a certain random variable X. We can estimate the value of the c.d.f. $F(x)$ of X from the relative frequencies of events of the form $X \le x$; that is,

$$\text{r.f.}\{X \le x\} = \frac{\text{number of observations that are } \le x}{N}.$$

For example, if we take $N = 10$ trials and obtain the numbers (arranged in order of size) -3.1, -1.9, -0.8, 0.1, 0.3, 0.3, 0.5, 1.3, 1.3, and 1.5, then

$$\text{r.f.}\{X \le 0.0\} = \tfrac{3}{10}, \qquad \text{r.f.}\{X \le 1.0\} = \tfrac{7}{10},$$

and so forth. We can graph r.f.$\{X \le x\}$ against x obtaining the graph in Figure 3.3.

The function r.f.$\{X \le x\}$ is called the *sample* (or *empirical*) *cumulative distribution function*. As $N \to \infty$, the frequency interpretation of probability

Figure 3.3: *Sample or empirical cumulative distribution function obtained from N = 10 trials.*

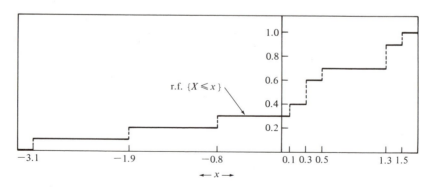

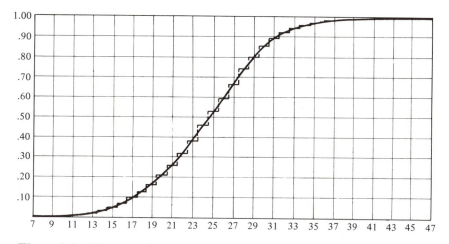

Figure 3.4: *The smooth curve is the theoretical cumulative distribution function. The step curve is the sample or empirical cumulative distribution function for large N.*

leads us to expect the graph of r.f.$\{X \le x\}$ to more and more closely resemble the graph of the true (theoretical) c.d.f. $F(x)$ (see Figure 3.4).

Mixed Discrete and Continuous Variables and Their Cumulative Distribution Functions

Although in most cases random variables are either continuous or discrete, occasionally we must deal with a random variable having the properties of both. For example, suppose that the random variable X denotes the amount of sales of a certain product. It is possible that with nonzero probability the demand for the product exceeds the maximum amount b that can be produced, so that all of b is sold, and $P\{X = b\}$ is positive. Similarly, there

Figure 3.5: *The mixed cumulative distribution function of the amount of sales.*

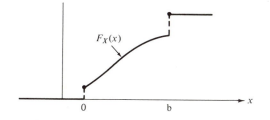

may be nonzero probability of zero demand, so that $P\{X = 0\}$ is also positive. On the other hand, all other numbers x between 0 and b (but not equal to 0 or to b) may be possible values for X, but over these values the probability model for X is that of a continuous variable. The graph of c.d.f. $F(x)$ of X is shown in Figure 3.5; $F(x)$ jumps twice, at $x = 0$ and at $x = b$, and otherwise rises smoothly.

4. PROBABILITY MASS FUNCTIONS OF DISCRETE RANDOM VARIABLES

Every random variable X has an associated cumulative distribution function $F(x)$ which contains the essential probabilistic information concerning X. However, the c.d.f. $F(x)$ is not the only function that contains such probabilistic information about X. Indeed, in certain cases other functional or graphical representations may be more directly interpretable. A useful function for describing the probability model of discrete random variables is the *probability mass function*.

The probability mass function of a discrete random variable X conveys the same information as does a table of probabilities of the simple events $\{X = x_j\}$, where $x_1, x_2, \ldots$ are the possible values (outcomes) of X. We denote the probability mass function of X by $p_X(x)$; when no confusion with probability mass functions of other variables is likely to arise, we drop the subscript X and write $p(x)$. The probability mass function $p_X(x)$ of X assigns to every number the probability of the event $\{X = x\}$; that is,

$$p_X(x) = P\{X = x\}.$$

Using Table 3.1, the probability mass function $p_X(x)$ of the number, X, of syllables in a randomly chosen German word is

$$p_X(x) = \begin{cases} 0.110, & \text{if } x = 1, \\ 0.313, & \text{if } x = 2, \\ 0.341, & \text{if } x = 3, \\ 0.178, & \text{if } x = 4, \\ 0.045, & \text{if } x = 5, \\ 0.010, & \text{if } x = 6, \\ 0.002, & \text{if } x = 7, \\ 0.001, & \text{if } x = 8, \\ 0.000, & \text{otherwise.} \end{cases}$$

This probability mass function is graphed in Figure 4.1.

The probability mass function and its graph may be thought of as a group of clumps, one clump placed at each of several points, $x_1, x_2, x_3, \ldots$, on a line. The weight of the clump placed at x_j corresponds to the probability

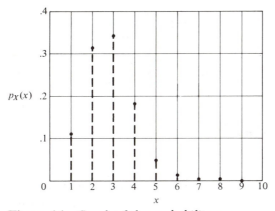

Figure 4.1: *Graph of the probability mass
function corresponding to the probability model
for the random variable X, the number of
syllables in a randomly chosen German word.*

that X equals x_j. We represent the magnitude of this weight by the height of
the line at x_j. Other examples of probability mass functions are shown in
Figures 4.2, 4.3, and 4.4.

    Remark. We note that a probability mass function $p(x)$ is characterized
by the properties (a) $p(x) \geq 0$ for all numbers x, (b) $p(x) \neq 0$ for at most a
countable[1] collection of x-values, and (c) $\sum_x p(x) = 1$. Also, recall that $p_X(x)$
can be obtained from the c.d.f. $F_X(x)$ of X by

$$p_X(x_0) = \text{height of jump of } F_X(x) \text{ at } x_0.$$

---

[1] A collection is countable if its elements can be put into a one-to-one correspondence with the
positive integers.

Figure 4.2: *Graph of the
probability mass function for
a distribution concentrated at
one point.*

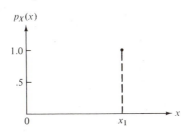

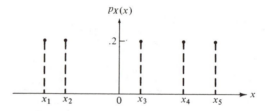

Figure 4.3: *Graph of the probability mass function for a uniform probability model.*

Since $p_X(x)$ gives the same information as a table of probabilities (e.g., Table 3.1), we can obtain $F_X(x)$ from $p_X(x)$ by the methods illustrated in Example 3.1.

Estimating a Probability Mass Function: Bar Graphs

If we make N independent trials of the random experiment that gives rise to the random X, and if we compute for every x the observed relative frequency r.f.$\{X = x\}$, then this *sample probability mass function* can also be represented by a bar graph. When N is large, the graph of the empirical probability mass function closely resembles the graph of $p_X(x)$.

Example 4.1. Williams (1956) made a statistical study of literary style and tabulated the number of letters in each of 2000 words in Thackeray's *Vanity Fair* (Table 4.1). If we let X equal the number of letters in a randomly chosen word and plot the graph of the relative frequency of that value (of X), we obtain Figure 4.5.

Figure 4.4: *Graph of the probability mass function for a symmetric and clustered model.*

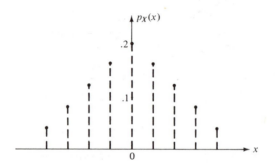

Table 4.1: *Letter Frequencies of 2000 Words in* Vanity Fair

Number of letters in the word	Frequency of words having that number of letters (out of 2000)	Relative frequency	Number of letters in the word	Frequency of words having that number of letters (out of 2000)	Relative frequency
1	58	0.029	8	100	0.050
2	315	0.1575	9	63	0.0315
3	480	0.240	10	43	0.0215
4	351	0.1755	11	16	0.008
5	244	0.122	12	15	0.0075
6	154	0.077	13	4	0.002
7	152	0.076	14	5	0.0025

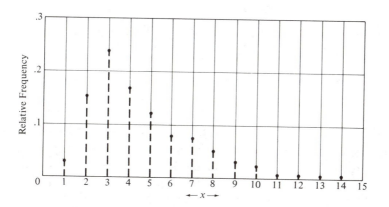

Figure 4.5: *Relative frequency distribution of the number of letters in words from Thackeray's* Vanity Fair.

5. DENSITY FUNCTIONS

For a continuous random variable X, the function analogous to the probability mass function is called the *density function*, denoted by $f_X(x)$ or, when no confusion is possible, simply $f(x)$. The density function of a random variable X usually may be given a graphical representation as in Figure 5.1. Here, the *area* under the graph between the numbers a and b represents $P\{a \le X \le b\}$. It thus follows that if $f_X(x)$ is smooth enough to permit integration,

(5.1)
$$P\{a \le X \le b\} = \int_a^b f_X(x)\, dx.$$

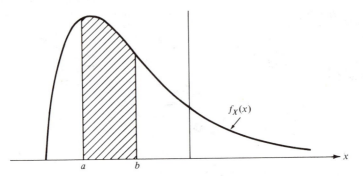

Figure 5.1: *A density function. The area of the shaded part is equal to the probability that X lies between a and b.*

Since the probability is 1 that X is equal to some real number, it follows that *the area under the graph of $f_X(x)$ over the entire horizontal axis is always equal to 1*, and

(5.2) $$\int_{-\infty}^{\infty} f_X(x)\, dx = 1.$$

We can think of the graph in Figure 5.1 as being a uniformly thick sheet of metal of total weight equal to one unit. The value of $f_X(x)$ at the number (point) x is then the *density* of the metal sheet at x. Notice that the *mass* of the sheet at any point x must be zero, but the mass over the line segment between a and b is not zero, and indeed represents $P\{a \le X \le b\}$.

From the density function, certain qualitative conclusions concerning the variation of X over repeated independent and identical trials are evident. If the density function is constant over the line segment from x_1 to x_2, then all subintervals of equal length are equally probable (see Figure 5.2).

On the other hand, if the density function is such that most of the area beneath it is concentrated within a very narrow range (see Figure 5.3), then repeated experiments on X tend to yield values of the random variable X mostly within the range of numbers where the area is concentrated.

The area between a and b under the graph of a density $f(x)$ equals $P\{a \le X \le b\}$. Note that if a approaches b, the area between a and b approaches 0; this is still another way of visualizing the fact that $P\{X = b\} = 0$ for any number b. It follows from this that when X is a continuous random variable

$$P\{a \le X \le b\} = P\{a < X \le b\} = P\{a \le X < b\} = P\{a < X < b\}.$$

This result is, in general, not true for discrete random variables.

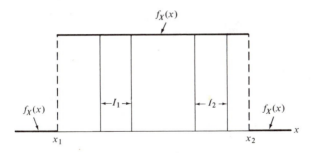

Figure 5.2: *Constant density between the values x_1 and x_2 implies equal probability for the intervals I_1 and I_2 of equal length L.*

We emphasize that $f_X(x)$ is *not* to be interpreted as being equal to the probability that X equals x. Instead, probabilities are computed by finding areas under the density function.

Example 5.1. A delicatessen has 400 pounds of luncheon meat delivered every day. A study has shown that if X is the weight of the luncheon meat sold in a morning (assumed measured *exactly*), then the probability model of the random variable X can be described by the density function $f(x)$ graphed in Figure 5.4.

The only possible weights for X are between 0 and 400 pounds. The equation for $f_X(x)$ is

$$f_X(x) = \begin{cases} (1.25)10^{-5}x, & \text{if } 0 \le x \le 400, \\ 0, & \text{if } x < 0 \text{ or } x > 400. \end{cases}$$

Figure 5.3: *A highly centralized density function.*

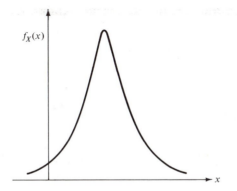

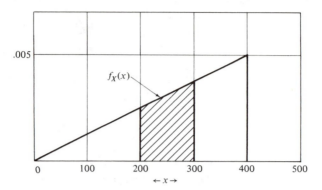

Figure 5.4: *Graph of the probability density function, $f_X(x)$, of the weight, X, of luncheon meat sold in a given morning. The shaded area gives the probability that between 200 and 300 pounds of luncheon meat is sold.*

Note that

$$\int_{-\infty}^{\infty} f_X(x)\, dx = \int_{-\infty}^{0} 0\, dx + \int_{0}^{400} (1.25)10^{-5}x\, dx + \int_{400}^{\infty} 0\, dx$$

$$= (1.25)10^{-5} \frac{x^2}{2} \Big]_0^{400} = 1.$$

If the delicatessen wants to know the probability that between 200 and 300 pounds of luncheon meat are sold on a given morning, this probability is

$$P\{200 \le X \le 300\} = \int_{200}^{300} (1.25)10^{-5}x\, dx = (1.25)10^{-5} \frac{x^2}{2} \Big|_{200}^{300} = 0.3125.$$

Example 5.2. The lifetime X (in playing hours) of a diamond phonograph needle is known to have the probability density function

$$f_X(x) = \begin{cases} (1/100)e^{-x/100}, & \text{if } x \ge 0, \\ 0, & \text{if } x < 0, \end{cases}$$

which is graphed in Figure 5.5.

The probability that a diamond needle will last more than 150 hours is given by

$$\int_{150}^{\infty} f_X(x)\, dx = \int_{150}^{\infty} \frac{1}{100} e^{-(x/100)}\, dx = -e^{-(x/100)} \Big|_{150}^{\infty} = e^{-1.5} = 0.223.$$

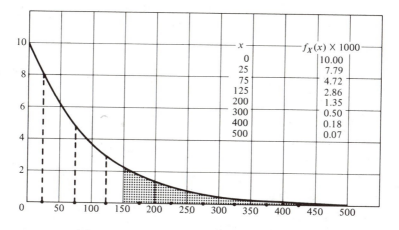

x	$f_X(x) \times 1000$
0	10.00
25	7.79
75	4.72
125	2.86
200	1.35
300	0.50
400	0.18
500	0.07

Figure 5.5: *The probability density function for the lifetime of a diamond phonograph needle. The shaded area gives the probability that the diamond needle lasts more than 150 hours.*

Relationship Between $f_X(x)$ and $F_X(x)$

The cumulative distribution function, $F_X(x)$, of a continuous random variable X is by definition $P\{X \le x\}$. If we know the probability density function, $f_X(x)$, of X, then

(5.3) $$F_X(x) = P\{X \le x\} = \int_{-\infty}^{x} f_X(u)\, du.$$

As a consequence,

(5.4) $$f_X(x) = \frac{d}{dx} F_X(x);$$

that is, $F_X(x)$ is obtained from $f_X(x)$ by integration, and $f_X(x)$ can be obtained from $F_X(x)$ by differentiation. Note that since we know from Section 3 that $F_X(x)$ is always nondecreasing in x, it follows that

$$0 \le \frac{d}{dx} F_X(x) = f_X(x),$$

so that the density function, $f_X(x)$ of a continuous random variable X is always nonnegative.

Example 5.1 (continued). The cumulative distribution function $F_X(x)$ of the random variable X described in Example 5.1 is

$$F_X(x) = \begin{cases} 0, & \text{if } x < 0, \\ \int_0^x (1.25)10^{-5}u \, du = (6.25)10^{-6}x^2, & \text{if } 0 \le x \le 400, \\ 1, & \text{if } x > 400. \end{cases}$$

For example, the probability that the delicatessen sells no more than 200 pounds of luncheon meat in a given day is $F_X(200) = (6.25)10^{-6}(200)^2 = 0.250$.

Example 5.3. An electronic control system provides a duplicate component in case a certain key component should fail. The entire system fails when both of the duplicate components fail. The combined lifetime Y of the two key components (in hours) has, under certain assumptions, a probability density function of the form

$$f_Y(y) = \begin{cases} 0, & \text{if } y < 0, \\ a^2 y e^{-ay}, & \text{if } y \ge 0, \end{cases}$$

where a is a given positive constant. The c.d.f. of Y is

$$F_Y(y) = \int_{-\infty}^y f_Y(u) \, du = \begin{cases} 0, & \text{if } y < 0, \\ \int_0^y a^2 z e^{-az} \, dz, & \text{if } y \ge 0. \end{cases}$$

Changing variables to $u = az$ and integrating by parts, we obtain

$$\int_0^y a^2 z e^{-az} \, dz = \int_0^{ay} u e^{-u} \, du = u(-e^{-u}) \Big|_0^{ay} - \int_0^{ay} (-e^{-u}) \, du$$

$$= -aye^{-ay} + (-e^{-u}) \Big|_0^{ay}$$

$$= 1 - (1 + ay)e^{-ay},$$

so that

$$F_Y(y) = \begin{cases} 0, & \text{if } y < 0, \\ 1 - (1 + ay)e^{-ay}, & \text{if } y \ge 0. \end{cases}$$

If $a = 0.01$, the probability that the system lasts no more than 150 hours is

$$F_Y(150) = 1 - [1 + (0.01)(150)]e^{-1.5} = 1 - (2.5)(0.223) = 0.4425.$$

Empirical Distributions: Histograms

Just as we found empirical c.d.f.'s and empirical probability mass functions for discrete random variables, so we can also find empirical density functions (called *histograms*) for continuous random variables.

One problem involved in constructing a histogram is that we are trying to show probabilistic properties of an infinite number of possible values with only a finite amount of data. Our data appear to be discrete, but the random variable observed is continuous. To resolve this problem, we try to construct representations of probabilities for certain intervals; these intervals are chosen before looking at the data in such a way that, hopefully, we get some idea of the way probability is spread over the line.

Consider the example in which we observe 417 forty-watt, internally frosted incandescent lamps that are kept lit continuously in a life test [Davis (1952)]. The lifetimes of the lamps vary between a minimum of 225 hours and a maximum of 1690 hours. If X is the random variable that equals the lifetime of such a lamp, we find the relative frequency of certain intervals of numbers. The intervals to be chosen should be (i) nonoverlapping, (ii) exhaustive of the data, and (iii) of equal length (if this is feasible). To have a representative picture of the data, we should have neither too few nor too many intervals. The rule of thumb frequently advocated is that we should use between 5 and 15 intervals.

In our life-testing problem, we might decide to use 15 intervals because we have many observations. (If there were fewer observations, fewer and broader intervals might be needed.) Then if we choose the left endpoint of our first interval to be 200 hours and the right endpoint of our last interval to be 1700 hours, all the data would be accounted for and the common length of our intervals is $(1700 - 200)/15 = 100$. The data of the lifetimes of the bulbs are now collected and summarized as shown in Table 5.1.[2]

From such a frequency distribution, it is a direct process to form the *histogram*. At the midpoint of each interval, a bar is constructed with height equal to the ratio of the relative frequency of the interval to the length of the interval. The relative frequency of the interval is now "spread" over the entire interval by constructing a rectangle over the interval with base equal to the length of the interval and height equal to the height of the bar (see Figure 5.6). The resulting construction is called a *modified relative frequency histogram*. The tops of the constructed rectangles form a curve; the area under this

---

[2] It would be better, in most problems, to construct the intervals in Table 5.1 in such a way that the end points of the intervals are not possible observations. Otherwise, if an observation falls on the boundary between two intervals, we do not know whether to assign it to the interval to the left or to the interval to the right. Thus, it is usually better to make the end points equal to values carried out to one more decimal place than are the data. Alternatively, as we have done here, we can always assign those observations which fall on the boundary between two intervals to the interval on the left.

Table 5.1: *Lifetimes of 417 Forty-Watt Bulbs*

Interval (hours)	Frequency	Interval (hours)	Frequency
200–300	1	1000–1100	85
300–400	0	1100–1200	80
400–500	0	1200–1300	44
500–600	3	1300–1400	23
600–700	10	1400–1500	9
700–800	21	1500–1600	3
800–900	45	1600–1700	2
900–1000	91	Total	417

curve over any interval of values is approximately equal to the probability of the event corresponding to that interval of values.

We use modified relative frequency histograms to estimate the probability density function $f_X(x)$ of X. For each number x, $f_X(x)$ can be estimated by the height, $f_X(x)$, of the modified relative frequency histogram over the point x on the horizontal axis. Thus, the probability density function $f_X(1150)$ of the lifetime X of a 40-watt bulb is estimated (see Figure 5.6) by $\hat{f}_X(1150) = 80/(100)(417) = 0.00192$. If it seems appropriate, we can obtain a more regular (continuous) estimate of $f_X(x)$ by drawing a smooth curve through the top of each rectangle of the histogram at the midpoint of the subinterval (see Figure 5.7). If this curve is carefully chosen, the area under the curve can be made to equal 1, and this curve can become our estimate of the probability density function.

Figure 5.6: *Modified relative frequency histogram of data on life testing.*

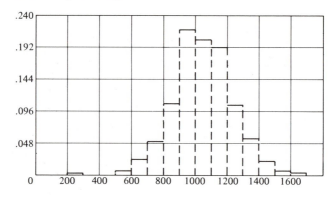

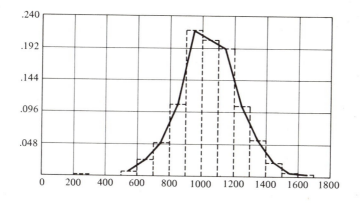

Figure 5.7: *Representation of life-testing data as a modified relative frequency curve.*

6. TRANSFORMATIONS OF A RANDOM VARIABLE

The random variables that we use to summarize the results of a random experiment are not always the obtained measurements themselves. Frequently, functions (or transformations) of these measurements are used. For example, if an angle θ is measured, the tangent, $\tan \theta$, of that angle may be reported. In some experiments, we may originally give our measurements in the units of one scale and later convert into the units of another scale. For example, measurements X of temperature in degrees Fahrenheit may have to be transformed into measurements $Y = (\frac{5}{9})X - \frac{160}{9}$ in degrees Celsius.

In the present section, we give some simple methods for obtaining distributions of transformed random variables $Y = g(X)$ from knowledge of the distribution of the original random variable X. Because both discrete and continuous distributions can be described by a cumulative distribution function (c.d.f.), our attention will be concentrated on formulas that allow us to obtain the c.d.f. $F_Y(y)$ of $Y = g(X)$ from knowledge of the c.d.f. $F_X(x)$ of X.

Linear Transformations

To illustrate our method of approach, consider the special case where Y is a *linear transformation*,

$$Y = aX + b, \quad a \neq 0,$$

of the random variable X.

We obtain the c.d.f. $F_Y(y)$ of $Y = aX + b$ from the c.d.f. $F_X(x)$ of X by

(6.1) $F_Y(y) = P\{Y \leq y\} = P\{aX + b \leq y\}$

$$= P\left\{X \leq \frac{y-b}{a}\right\} = F_X\left(\frac{y-b}{a}\right), \quad \text{for } a > 0,$$

(6.2) $F_Y(y) = P\{aX + b \leq y\} = P\left\{X \geq \frac{y-b}{a}\right\}$

$$= 1 - F_X\left(\frac{y-b}{a}\right) + P\left\{X = \frac{y-b}{a}\right\}, \quad \text{for } a < 0.$$

Suppose, now, that X is a discrete random variable. The probability mass function $p_Y(y)$ of $Y = aX + b$ can be obtained in terms of the probability mass function $p_X(x)$ of X through use of (6.1) or (6.2) and the relationship between the c.d.f. and probability mass function of a discrete random variable described in Sections 3 and 4. However, we can obtain $p_Y(y)$ from $p_X(x)$ more directly by noting that the events $\{Y = y\}$ and $\{X = (y-b)/a\}$ are the same event, so that

(6.3) $$p_Y(y) = P\{Y = y\} = P\left\{X = \frac{y-b}{a}\right\} = p_X\left(\frac{y-b}{a}\right).$$

When X is a continuous random variable, the probability density function $f_Y(y)$ of $Y = aX + b$ can be obtained from the probability density function $f_X(x)$ through the formula

(6.4) $$f_Y(y) = \frac{1}{|a|} f_X\left(\frac{y-b}{a}\right).$$

To verify (6.4), differentiate both sides of (6.1) and (6.2) with respect to y, using the chain rule. Thus, if $a > 0$,

$$f_Y(y) = \frac{d}{dy} F_Y(y) = \frac{d}{dy}\left(\frac{y-b}{a}\right) \frac{dF_X\left(\frac{y-b}{a}\right)}{d\left(\frac{y-b}{a}\right)} = \frac{1}{a} f_X\left(\frac{y-b}{a}\right),$$

while if $a < 0$, we obtain

$$f_Y(y) = -\frac{1}{a} f_X\left(\frac{y-b}{a}\right) = \frac{1}{|a|} f_X\left(\frac{y-b}{a}\right).$$

Example 6.1. Suppose that X has the probability mass function

$$p_X(x) = \begin{cases} \frac{64}{85}(\frac{1}{4})^x, & \text{if } x = 0, 1, 2, 3, \\ 0, & \text{otherwise.} \end{cases}$$

If $Y = 2X + 3$, then

$$p_Y(y) = p_X\left(\frac{y-3}{2}\right) = \begin{cases} \frac{64}{85}(\frac{1}{4})^{(y-3)/2}, & \text{if } \frac{y-3}{2} = 0, 1, 2, 3, \\ 0, & \text{otherwise,} \end{cases}$$

$$= \begin{cases} \frac{512}{85}(\frac{1}{2})^y, & \text{if } y = 3, 5, 7, 9, \\ 0, & \text{otherwise.} \end{cases}$$

Example 6.2. Suppose that X has the probability density function

$$f_X(x) = \begin{cases} e^{-x}, & \text{if } x > 0, \\ 0, & \text{otherwise.} \end{cases}$$

If $Y = (\frac{5}{9})X - \frac{160}{9}$, then

$$f_Y(y) = \begin{cases} \frac{9}{5}e^{-[(9y+160)/5]}, & \text{if } y > -\frac{160}{9}, \\ 0, & \text{otherwise.} \end{cases}$$

General Transformations: Discrete Case

When X is a discrete random variable and $Y = g(X)$ is a general transformation of X, the most direct method of obtaining the distribution of Y from the distribution of X is to determine, for each value of y, the probability of the event $\{X: g(X) = y\}$. That is, for each value of y, we list all values x of the random variable X that lead to that value of y, and then sum the probabilities of these values (as determined by the probability mass function of X). The resulting quantity is the value $p_Y(y)$ of the probability mass function of Y. Thus,

$$(6.5) \qquad p_Y(y) = \sum_{\{x:\ g(x)=y\}} p_X(x).$$

Note that this method is similar to the method by which we obtained the probability distribution of a random variable from the probabilities assigned to the simple events of an experiment possibly involving nonnumerical outcomes.

Example 6.3. Suppose that X has the probability mass function

$$p_X(x) = \begin{cases} \frac{4}{31}(\frac{1}{2})^x, & \text{if } x = -2, -1, 0, 1, 2, \\ 0, & \text{otherwise.} \end{cases}$$

If $Y = X^2$, then the possible values of Y are 0, 1, 4. Using (6.5), we obtain

$$p_Y(0) = \sum_{\{x:\ x^2 = 0\}} p_X(x) = p_X(0) = \tfrac{4}{31},$$

$$p_Y(1) = \sum_{\{x:\ x^2 = 1\}} p_X(x) = p_X(-1) + p_X(1) = \tfrac{8}{31} + \tfrac{2}{31} = \tfrac{10}{31},$$

$$p_Y(4) = \sum_{\{x:\ x^2 = 4\}} p_X(x) = p_X(-2) + p_X(2) = \tfrac{16}{31} + \tfrac{1}{31} = \tfrac{17}{31},$$

and $p_Y(y) = 0$ for all other values of y.

In special cases [e.g., when $g(x)$ is linear in x], we can determine the c.d.f. $F_Y(y)$ of Y from the c.d.f. $F_X(x)$ of X [see also Equations (6.6), (6.8), and (6.13)], and then from $F_Y(y)$ determine the probability mass function $p_Y(y)$ of Y. However, for discrete random variables the direct approach through Equation (6.5) is usually the most convenient approach.

Because individual values of continuous random variables have zero probability, a direct approach in the manner of Equation (6.5) does not aid us in determining the distribution of a transformation of such variables. For continuous random variables, an indirect approach through the cumulative distribution function is usually necessary. Since such an approach involves the need to consider special cases, and requires some care in application, we devote the remainder of this section to a consideration of methods for obtaining the distribution of a transformation $Y = g(X)$ of a continuous random variable X.

General Transformations: Continuous Case

The transformation $ax + b$, for $a > 0$, is a special case of a strictly increasing, differentiable transformation $g(x)$. A transformation $y = g(x)$ is strictly increasing if $x_1 < x_2$ implies that $g(x_1) < g(x_2)$, and differentiable if $(d/dx)g(x)$ exists at every x. The theory of implicit functions shows that if $y = g(x)$ and $g(x)$ is strictly increasing and differentiable, then there exists a differentiable increasing function $h(y)$ such that $x = h(y)$. In other words, $y = g(x)$ if and only if $x = h(y)$, so that there is a one-to-one correspondence between x and $y = g(x)$. Suppose that X is a continuous random variable. To find the density function of $Y = g(X)$, where X has probability density function $f_X(x) = (d/dx)F_X(x)$, first note that

(6.6) $F_Y(y) = P\{Y \le y\} = P\{g(X) \le y\} = P\{X \le h(y)\} = F_X(h(y)).$

Differentiating both sides of this equality with respect to y and using the chain rule [noting that $(d/dy)h(y) \ge 0$, since $h(y)$ is increasing] yields

(6.7) $f_Y(y) = \dfrac{dh(y)}{dy} f_X(h(y)).$

We may apply similar arguments to obtain the distribution of a strictly decreasing, differentiable transformation $Y = g(X)$. The transformation $y = g(x)$ is strictly decreasing if $x_1 < x_2$ implies that $g(x_1) > g(x_2)$. If $g(x)$ is strictly decreasing and differentiable, then there exists a differentiable decreasing function $h(y)$ such that $x = h(y)$. Thus, if $Y = g(X)$,

(6.8) $F_Y(y) = P\{Y \le y\} = P\{g(X) \le y\} = P\{X \ge h(y)\} = 1 - F_X(h(y)).$

Differentiating both sides of this equality with respect to y, we find that

(6.9) $$f_Y(y) = \frac{dh(y)}{dy}(-f_X(h(y))) = \left|\frac{dh(y)}{dy}\right| f_X(h(y)),$$

since $(d/dy)h(y) < 0$.

Putting (6.7) and (6.9) together, we can assert that when $Y = g(X)$ for a strictly monotone (increasing or decreasing) differentiable function $g(x)$ of x, then there exists a monotone differentiable function $h(y)$ of y such that $y = g(x)$ if and only if $x = h(y)$. The function $h(y)$ is called the *inverse function* of $g(x)$. In this case, if X is a continuous random variable, then the probability density function of $Y = g(X)$ is

(6.10) $$f_Y(y) = \left|\frac{dh(y)}{dy}\right| f_X(h(y)).$$

Example 6.4. Suppose that X has the probability density function

(6.11) $$f_X(x) = \begin{cases} \theta e^{-\theta x}, & \text{if } x > 0, \\ 0, & \text{otherwise,} \end{cases}$$

where $\theta > 0$ is a constant. Consider $Y = e^X$. Note that $g(x) = e^x$ is strictly increasing and differentiable and that $y = e^x$ if and only if $x = h(y) = \log y$. Hence, by (6.10), the probability density function of Y is

$$f_Y(y) = \left|\frac{d \log y}{dy}\right| f_X(\log y)$$

$$= \begin{cases} \dfrac{1}{y}\theta e^{-\theta \log y}, & \text{if } 0 \le \log y \le \infty, \\ 0, & \text{otherwise,} \end{cases}$$

$$= \begin{cases} \theta y^{-\theta-1}, & \text{if } 1 \le y \le \infty, \\ 0, & \text{otherwise.} \end{cases}$$

Example 6.5. An important application of the result (6.10) is the following: *If X is a continuous random variable with a strictly increasing cumulative distribution function $F_X(x) = F(x)$, then $Y = F(X)$ has the probability density function*

(6.12) $$f_Y(y) = \begin{cases} 1, & \text{if } 0 \le y \le 1, \\ 0, & \text{otherwise.} \end{cases}$$

The distribution defined by the probability density function (6.12) is called the *uniform distribution* on the interval [0, 1] (see Chapter 8, Section 4). The transformation $Y = F(X)$ is called the *probability integral transformation* of X, since

$$F(X) = \int_{-\infty}^{x} f_X(t)\, dt.$$

To prove the assertion that Y has a uniform distribution on the interval [0, 1], first note that since $F(x)$ is a cumulative distribution function, it must be the case that $0 \le F(x) \le 1$ for all values of x. Thus, $Y = F(X)$ can only take on values between 0 and 1. Next, note that by assumption, $F(x)$ is strictly increasing. Since also

$$\frac{d}{dx} F(x) = f_X(x),$$

$F(x)$ is differentiable. Thus, there exists an increasing differentiable function $h(y)$ for which $y = F(x)$ if and only if $x = h(y)$. By the Implicit Function Theorem,

$$\frac{dh(y)}{dy} = \frac{1}{f_X(h(y))},$$

and by (6.10),

$$f_Y(y) = \begin{cases} \left| \dfrac{dh(y)}{dy} \right| f_X(h(y)), & \text{if } 0 \le y \le 1, \\ 0, & \text{otherwise,} \end{cases}$$

$$= \begin{cases} 1, & \text{if } 0 \le y \le 1, \\ 0, & \text{otherwise.} \end{cases}$$

This proves our assertion.

If $F(x)$ is strictly increasing, then $1 - F(x)$ is strictly decreasing. Thus, we can repeat the arguments above and determine that if X is a continuous random variable with a strictly increasing c.d.f. $F_X(x) = F(x)$, then $Y = 1 - F(X)$ has a uniform distribution on the interval [0, 1].

Example 6.6. Suppose that X has a uniform distribution on the interval [0, 1]. Let $Y = -\log X$. Note that $g(x) = -\log x$ is strictly decreasing and differentiable, and that $y = -\log x$ if and only if $x = h(y) = e^{-y}$. Hence, by

(6.10), the probability density function of $Y = -\log X$ is

$$f_Y(y) = \left| \frac{de^{-y}}{dy} \right| f_X(e^{-y})$$

$$= \begin{cases} |(-e^{-y})| \cdot 1, & \text{if } 0 \le e^{-y} \le 1, \\ 0, & \text{otherwise}, \end{cases}$$

$$= \begin{cases} e^{-y}, & \text{if } 0 \le y \le \infty, \\ 0, & \text{otherwise}, \end{cases}$$

which we recognize as being a special case of the probability density function (6.11) with the constant $\theta = 1$.

When the transformation $g(x)$ is not strictly monotone, $g(x)$ may not be one-to-one, in that more than one value of x may yield the same value of $y = g(x)$. We can try to break the collection of x-values up into intervals such that over each interval $g(x)$ is strictly monotone, and then try to put together the pieces. For example, if $g(x) = x^2$, then $g(x)$ is not monotone over all x-values but is strictly decreasing for $-\infty < x < 0$ and strictly increasing for $0 \le x < \infty$. When $x < 0$, $h_1(y) = -\sqrt{y}$; when $x \ge 0$, $h_2(y) = \sqrt{y}$. Thus, if $Y = X^2$, then for $y \ge 0$,

$$F_Y(y) = P\{Y \le y\} = P\{X^2 \le y \text{ and } X < 0\} + P\{X^2 \le y \text{ and } X \ge 0\}$$

(6.13)
$$= P\{-\sqrt{y} \le X < 0\} + P\{0 \le X \le \sqrt{y}\}$$

$$= F_X(0) - F_X(-\sqrt{y}) + F_X(\sqrt{y}) - F_X(0)$$

$$= F_X(\sqrt{y}) - F_X(-\sqrt{y}).$$

Of course, if $y < 0$, then $F_Y(y) = P\{Y \le y\} = P\{X^2 \le y\} = 0$. If X is a continuous random variable, then differentiating both sides of (6.13) with respect to y, using the chain rule, we obtain

(6.14)
$$f_Y(y) = \begin{cases} \dfrac{1}{2\sqrt{y}} (f_X(\sqrt{y}) + f_X(-\sqrt{y})), & \text{if } y \ge 0, \\ 0, & \text{otherwise}. \end{cases}$$

Alternatively, we can obtain (6.13) and (6.14) by appealing to fundamentals. Let

(6.15)
$$A_y = \{x : g(x) \le y\};$$

then A_y is the event that $Y = g(X) \le y$, so that

(6.16)
$$F_Y(y) = P\{Y \le y\} = P(A_y).$$

Now in the case $g(x) = x^2$,

$$A_y = \{x: -\sqrt{y} \le x \le \sqrt{y}\},$$

so that

$$F_Y(y) = P\{A_y\} = P\{-\sqrt{y} \le X \le \sqrt{y}\} = F_X(\sqrt{y}) - F_X(-\sqrt{y}),$$

which is the result already obtained in (6.13).

Example 6.7. If X has probability density function

$$f_X(x) = \begin{cases} \frac{3}{8}(x-1)^2, & \text{if } -1 \le x \le 1, \\ 0, & \text{otherwise}, \end{cases}$$

and $Y = X^2$, then by (6.14),

$$f_Y(y) = \begin{cases} \dfrac{1}{2\sqrt{y}} \left[\dfrac{3}{8}(\sqrt{y}-1)^2 + \dfrac{3}{8}(-\sqrt{y}-1)^2 \right], & \text{if } y \ge 0, \\ 0, & \text{otherwise}, \end{cases}$$

$$= \begin{cases} \dfrac{3}{8} \dfrac{(y+1)}{\sqrt{y}}, & \text{if } y \ge 0, \\ 0, & \text{otherwise}. \end{cases}$$

Example 6.8. Let X have the probability density function

$$f_X(x) = \begin{cases} 140x^3(1-x)^3, & \text{if } 0 \le x \le 1, \\ 0, & \text{otherwise}. \end{cases}$$

We find the distribution of $Y = 4X(1-X)$ using piecewise monotone transformations. Note that $g(x) = 4x(1-x)$ is a quadratic that is strictly increasing for $x < \frac{1}{2}$ and strictly decreasing for $x \ge \frac{1}{2}$. For $x < \frac{1}{2}$, $y = g(x)$ if and only if $x = h_1(y) = \frac{1}{2}(1 - \sqrt{1-y})$; while for $x \ge \frac{1}{2}$, $y = g(x)$ if and only if $x = h_2(y) = \frac{1}{2}(1 + \sqrt{1-y})$. Also, for $0 \le x \le 1$, we have $0 \le g(x) \le 1$, so that $y = g(x)$ must lie between 0 and 1. Thus, for $0 \le y \le 1$,

$$\begin{aligned} F_Y(y) = P\{Y \le y\} &= P\{4X(1-X) \le y \text{ and } X < \tfrac{1}{2}\} \\ &\quad + P\{4X(1-X) \le y \text{ and } X \ge \tfrac{1}{2}\} \\ &= P\{X \le \tfrac{1}{2}(1 - \sqrt{1-y})\} + P\{X \ge \tfrac{1}{2}(1 + \sqrt{1-y})\} \\ &= F_X(\tfrac{1}{2}(1 - \sqrt{1-y})) + (1 - F_X(\tfrac{1}{2}(1 + \sqrt{1-y}))), \end{aligned}$$

whereas for $y < 0$, $F_Y(y) = 0$, and for $y > 1$, $F_Y(y) = 1$. Differentiating with respect to y, we see that $f_Y(y) = 0$ when $y < 0$ or $y > 1$, whereas when $0 \le y \le 1$,

$$f_Y(y) = \frac{1}{4\sqrt{1-y}} \left[f_X\left(\frac{1}{2}(1 - \sqrt{1-y})\right) + f_X\left(\frac{1}{2}(1 + \sqrt{1-y})\right) \right]$$

$$= \frac{140}{4\sqrt{1-y}} \left[\left(\frac{1 - \sqrt{1-y}}{2}\right)^3 \left(1 - \frac{1 - \sqrt{1-y}}{2}\right)^3 \right.$$

$$\left. + \left(\frac{1 + \sqrt{1-y}}{2}\right)^3 \left(1 - \frac{1 + \sqrt{1-y}}{2}\right)^3 \right]$$

$$= \frac{35y^3(1-y)^{-1/2}}{32}.$$

Thus,

$$f_Y(y) = \begin{cases} \frac{35}{32}y^3(1-y)^{-1/2}, & \text{if } 0 \le y \le 1, \\ 0, & \text{otherwise.} \end{cases}$$

The second approach [using (6.15) and (6.16)] yields similar results, since

$$A_y = \{x: g(x) \le y\} = \{x: 4(x)(1-x) \le y\}$$
$$= \{x: x \le \tfrac{1}{2}(1 - \sqrt{1-y}) \text{ or } x \ge \tfrac{1}{2}(1 + \sqrt{1-y})\},$$

and thus

$$F_Y(y) = P\{A_y\} = P\{X \le \tfrac{1}{2}(1 - \sqrt{1-y}) \text{ or } X \ge \tfrac{1}{2}(1 + \sqrt{1-y})\}$$
$$= P\{X \le \tfrac{1}{2}(1 - \sqrt{1-y})\} + P\{X \ge \tfrac{1}{2}(1 + \sqrt{1-y})\}$$
$$= F_X(\tfrac{1}{2}(1 - \sqrt{1-y})) + [1 - F_X(\tfrac{1}{2}(1 + \sqrt{1-y}))].$$

EXERCISES

1. At a certain hospital, the progress of 4 patients suffering from a certain disease is observed. After each patient has been in the hospital for a month, a prognosis is made as to whether the patient will recover (R) or will not recover (N).
    (a) What is the sample space for this experiment? List all possible outcomes.
    (b) The hospital assumes that the prognoses for the 4 patients are mutually independent, and that for each patient $P(R) = 0.75$. For each simple event of the random experiment, find the corresponding probability.
    (c) Find the probability model for the number, X, among the 4 patients who receive a favorable prognosis.
    (d) Find the c.d.f. $F_X(x)$ of X.

2. In the game "two-finger Morra," 2 players each show 1 or 2 fingers and simultaneously guess the number of fingers their opponent will show. If only 1 player guesses correctly, he wins an amount (in dollars) equal to the sum of the number of fingers shown by him and his opponent. If both players guess correctly, or if neither player guesses correctly, the game is a draw.
    (a) Write all the possible outcomes of one play of this game. Your outcomes should allow us to determine how many fingers each player holds up and how many fingers each player guesses that his opponent will hold up.
    (b) Let the random variable X equal the amount of money (in dollars) won by a specified player. What are the possible values of X?
    (c) Suppose that each player acts independently of the other player, and that each player makes his choice by a random mechanism that assigns equal probability to each of the possible alternatives. Find the resulting probability model for the random variable X.
    (d) Suppose that each player acts independently of the other player, but that both players hold up the *same* number of fingers that they guess their opponent will hold up. If each player guesses at random the number of fingers his opponent will hold up, find the resulting probability model for X.

3. There are 10 batteries in a bin at the grocery store. You sample $n = 2$ batteries randomly without replacement to use for your flashlight. Of the 10 batteries in the bin, 3 are defective. Let X be the number of defective batteries that you select in your sample. Find the probability model for X, and find the c.d.f. $F_X(x)$ of X.

4. The number X of floods in a wet season has the following probability model.

Simple event	$\{X = 0\}$	$\{X = 1\}$	$\{X = 2\}$	$\{X = 3\}$	$\{X = 4\}$	$\{X = 5\}$	$\{X = 6\}$
Probability	0.2466	0.3452	0.2417	0.1127	0.0395	0.0110	0.0033

    Find the c.d.f. $F_X(x)$ of X. Using the c.d.f., find $P\{X \geq 2\}$ and $P\{1 < X \leq 4\}$, and then check your answers by summing appropriate probabilities from the table.

5. The following is the cumulative distribution function $F_X(x)$ of a random variable X:

$$F_X(x) = \begin{cases} 0.0, & \text{if } x < 8.0, \\ 0.1, & \text{if } 8.0 \leq x < 9.0, \\ 0.4, & \text{if } 9.0 \leq x < 10.0, \\ 0.7, & \text{if } 10.0 \leq x < 11.0, \\ 0.9, & \text{if } 11.0 \leq x < 12.5, \\ 1.0, & \text{if } 12.5 \leq x. \end{cases}$$

    (a) Is X a continuous random variable or a discrete random variable?
    (b) What are the possible values of X?
    (c) What is the probability model for X?
    (d) Using the cumulative distribution function, find (i) $P\{X \le 9.5\}$, (ii) $P\{X > 10.8\}$, (iii) $P\{9.0 < X \le 11.5\}$, (iv) $P\{9.0 \le X \le 11.5\}$.

6. The cumulative distribution function $F_X(x)$ of a random variable X has the form

$$F_X(x) = \begin{cases} 0, & \text{if } x < 0, \\ x, & \text{if } 0 \le x \le 1, \\ 1, & \text{if } x > 1. \end{cases}$$

    (a) Is X a discrete random variable or a continuous random variable?
    (b) What are the possible values of X?
    (c) Using the values of $F_X(x)$, find $P\{\frac{1}{4} \le X \le \frac{3}{4}\}$.
    (d) Using the values of $F_X(x)$, find $P\{X > \frac{1}{2}\}$.

7. The cumulative distribution function of $F_Y(y)$ of a random variable Y has the form

$$F_Y(y) = \begin{cases} 0, & \text{if } y < 0, \\ y^2, & \text{if } 0 \le y \le 1, \\ 1, & \text{if } y > 1. \end{cases}$$

    (a) Is Y a discrete random variable or a continuous random variable?
    (b) What are the possible values of Y?
    (c) Using the values of $F_Y(y)$, find $P\{\frac{1}{4} \le Y \le \frac{3}{4}\}$.
    (d) Using the values of $F_Y(y)$, find $P\{Y > \frac{1}{2}\}$.

8. The cumulative distribution function $F_W(w)$ of a random variable W has the form

$$F_W(w) = \begin{cases} 0, & \text{if } w < 0, \\ w, & \text{if } 0 \le w < \frac{1}{2}, \\ \frac{1}{2}w + \frac{1}{2}, & \text{if } \frac{1}{2} \le w \le 1, \\ 1, & \text{if } w > 1. \end{cases}$$

    (a) What kind of random variable is W?
    (b) What are the possible values of W?
    (c) Find $P\{W = \frac{1}{2}\}$, $P\{W > \frac{1}{2}\}$, $P\{W < \frac{1}{4}\}$, and $P\{\frac{1}{3} \le W < \frac{2}{3}\}$.

9. Which of the following are c.d.f.'s? In each case, state your answer and give reasons for your answer.

    (a) $F(t) = \begin{cases} 0, & \text{if } t \le 1, \\ 1, & \text{if } 1 < t. \end{cases}$

    (b) $F(t) = \begin{cases} 0, & \text{if } t < 1, \\ 1, & \text{if } 1 \le t. \end{cases}$

(c) $F(t) = \begin{cases} 0, & \text{if } t < 0, \\ \log_{10}(1+t), & \text{if } 0 \le t. \end{cases}$

(d) $F(t) = \begin{cases} 0, & \text{if } t < 0, \\ \dfrac{t}{1+t}, & \text{if } 0 \le t. \end{cases}$

(e) $F(t) = \begin{cases} 0, & \text{if } t < 0, \\ \sin t, & \text{if } 0 \le t \le \dfrac{\pi}{2}, \\ 1, & \text{if } \dfrac{\pi}{2} < t. \end{cases}$

10. Consider the data given in Table 5.1 of Chapter 4 on the lifetimes X of incandescent lamps which we kept lit continuously in a life test.

(a) Find r.f.$\{X \le x\}$ at the right end points $x = 300, 400, 500, \ldots, 1700$ of the grouping intervals in the table. Use your results to construct and graph a sample cumulative distribution function for X. Such a function will change values (jump) only at the right end points of the grouping intervals.

(b) If we had the original ungrouped life test data (the *raw* data) rather than the grouped data in Table 5.1, and if we had used the raw data to construct a sample cumulative distribution function for X, would this function be the same as the one you obtained in part (a)? If the two functions are not the same, do they at least agree in value at certain x-points? Which ones?

11. Use the data in Example 4.1 to construct and graph the sample cumulative distribution function for X, the number of letters in a randomly chosen word from Thackeray's *Vanity Fair*.

12. In a large urban shopping mall, the stores are arranged as shown in the accompanying figure. Each store contains an underground garage. If a shopper wants to remain indoors, he or she can pass from store to store (and garage to garage) by the routes shown in the figure. The distances

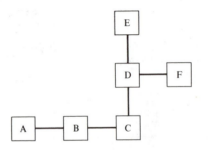

between adjacent stores are the same; let the distance between any two adjacent stores be the unit of distance. Thus, store A and store B are 1 unit apart.

(a) A shopper parks at the garage under a randomly chosen store and independently finishes shopping at another randomly chosen store. Let X equal the number of units of distance that the shopper has to walk to reach his or her car. Find the probability mass function of X.

(b) What is the probability that the shopper must walk more than 1 unit of distance to reach his or her car?

(c) Suppose that the parking garages C and D were closed for repair, so that the shopper could park only in the garages under stores A, B, E, and F. Under the same randomness assumptions as in part (a), find the probability mass function of the number Y of units of distance that the shopper has to walk to reach his or her car. Compare $p_X(x)$ and $p_Y(y)$.

13. Consider the discrete random variable X, which has possible values 0, 1, 2, ..., 9 and for which $P\{X = j\} = \frac{1}{10}, j = 0, 1, 2, ..., 9$.

(a) Let Y be the remainder obtained after dividing X^2 by 10 (e.g., 9^2 divided by 10 has remainder 1). Find the probability mass function of Y.

(b) Let Z be the remainder obtained after dividing Y^2 by 10. Find the probability mass function of Z.

(c) Let W be the remainder obtained after dividing Z^2 by 10. Find the probability mass function of W. Compare the probability mass functions of Z and of W.

(d) If, starting with the random variable W, we continue to define new random variables in terms of old random variables by the process described in parts (a), (b), and (c), do we ever obtain a probability mass function different from the probability mass function of W?

14. The following problem is a prototype of a set of problems known as "matches," "coincidences," or "rencontre." It has many variants, some of which date from the early eighteenth century. Suppose that we have N pairs of distinguishable tickets, divided into 2 decks of N tickets each. Each deck contains 1 ticket from each of N pairs. Both decks of N tickets are shuffled, and 1 ticket at a time is taken from the top of each deck and matched against a corresponding ticket from the top of the other deck. If the tickets agree, we say a *match* has occurred. Among the N tickets compared, we count the number X_N of matches. Assuming that both decks of tickets are well shuffled, find the probability mass function of X_N when (a) $N = 2$, (b) $N = 3$, (c) $N = 4$, and (d) $N = 5$.

15. The probability mass function $p_X(x)$ of a discrete random variable X has the form

$$p_X(x) = \begin{cases} (\frac{1}{2})^x(\frac{32}{31}), & \text{for } x = 1, 2, 3, 4, 5, \\ 0, & \text{otherwise.} \end{cases}$$

(a) Show that $\sum_{x=1}^{5} p_X(x) = 1$.
(b) Determine the cumulative distribution function $F_X(x)$ of X.
(c) Find (i) $P\{X \geq 3\}$, (ii) $P\{2 \leq X \leq 4\}$, and (iii) $P\{X < 4\}$.

16. A discrete random variable X has the following probability mass function

$$p_X(x) = \begin{cases} (\frac{1}{4})(\frac{3}{4})^x, & \text{for } x = 0, 1, 2, 3, \\ \frac{81}{256}, & \text{for } x = 4, \\ 0, & \text{otherwise.} \end{cases}$$

(a) Show that $\sum_{x=0}^{4} p_X(x) = 1$.
(b) Find the c.d.f. $F_X(x)$ of X.
(c) Find (i) $P\{X > 3\}$, (ii) $P\{1 \leq X \leq 5\}$, and (iii) $P\{X < 3\}$.

17. A discrete random variable X has probability mass function of the following form:

$$p_X(x) = \begin{cases} cx^2, & \text{for } x = 1, 2, 3, 4, \\ 0, & \text{otherwise.} \end{cases}$$

(a) What is the value of the constant c?
(b) Find the functional form of the c.d.f. $F_X(x)$ of X.
(c) Find $P\{1 < X \leq 3\}$ and $P\{X \text{ is an even integer}\}$.

18. (a) Write down the Taylor's expansion of e^λ about $\lambda = 0$.
(b) Use your result in part (a) to show that the following function is a probability mass function:

$$p(t) = \begin{cases} \dfrac{\lambda^t}{t!} e^{-\lambda}, & \text{if } t = 0, 1, 2, \ldots, \\ 0, & \text{otherwise,} \end{cases}$$

where λ is a given positive constant. [You must show that $p(t) \geq 0$, all t, and that $\sum_{t=0}^{\infty} p(t) = 1$.]
(c) If $\lambda < 0$, is $p(t)$ a probability mass function? Explain.

19. In a study of the incidence of dental caries in adults, a random sample of 100 adults was selected. These adults were offered free dental examinations at the beginning and end of a given year. A total of 80 adults

2	1	3	0	1	2	2	1	1	1	2	2	3	1	3	0	1	0	2	1
0	0	1	2	2	1	0	2	1	5	0	0	1	2	1	3	1	4	1	1
0	1	0	1	0	0	0	0	1	1	4	0	0	1	1	0	1	3	2	0
0	0	1	0	1	1	2	0	0	1	0	1	2	0	0	1	0	1	0	1

accepted the offer. For each adult studied, the number X of new caries that developed over the year were counted. The data obtained from this experiment are given above.

(a) Construct a sample probability mass function for X.
(b) Construct a bar graph for these data.
(c) Construct the sample cumulative distribution function for X.

20. The probability density function $f_X(x)$ of a continuous random variable X has the form

$$f_X(x) = \begin{cases} 0.4, & \text{if } 0 \le x < 1, \\ 0.2, & \text{if } 1 \le x < 2, \\ 0.3, & \text{if } 2 \le x < 3, \\ 0.1, & \text{if } 3 \le x < 4, \\ 0.0, & \text{if } x < 0 \text{ or } x \ge 4. \end{cases}$$

(a) Find the c.d.f. $F_X(x)$ of X.
(b) Find
   (i) $P\{0 < X \le 2\}$,   (ii) $P\{0 \le X \le 2\}$,
   (iii) $P\{2 \le X \le 3\}$,   (iv) $P\{\frac{1}{2} \le X < \frac{3}{2}\}$,
   (v) $P\{X \ge 1\}$,   (vi) $P\{X < 2\}$.
   [Hint: It will be helpful to graph $f_X(x)$ and compute probabilities by means of areas.]

21. Suppose that the continuous random variable Y has a probability density function $f_Y(y)$ of the form

$$f_Y(y) = \begin{cases} 0, & \text{if } y < 0, \\ 3y^2, & \text{if } 0 \le y \le 1, \\ 0, & \text{if } y > 1. \end{cases}$$

(a) Find $P\{0 \le Y \le \frac{1}{2}\}$ and $P\{\frac{1}{4} \le Y \le \frac{3}{4}\}$.
(b) Find the c.d.f. $F_Y(y)$ of Y.

22. The probability density function $f_Z(z)$ of a continuous random variable Z has the form

$$f_Z(z) = \begin{cases} 0, & \text{if } z < 1, \\ z - 1, & \text{if } 1 \le z < 2, \\ 3 - z, & \text{if } 2 \le z \le 3, \\ 0, & \text{if } z > 3. \end{cases}$$

(a) Find $P\{Z \le \frac{3}{2}\}$, $P\{\frac{3}{2} \le Z \le \frac{5}{2}\}$, and $P\{Z > \frac{5}{2}\}$.
(b) Find the c.d.f. $F_Z(z)$ of Z.

23. A continuous random variable W has a probability density function of the form

$$f_W(w) = \begin{cases} ce^{-2w}, & \text{if } w \ge 0, \\ 0, & \text{otherwise}. \end{cases}$$

(a) What is the value of the constant c?

(b) Find the functional form of the c.d.f. $F_W(w)$ of W.

(c) Find $P\{W \geq 1\}$ and $P\{2 \leq W < 5\}$.

(d) Find the value of d for which
$$P\{W \leq d\} = 0.5.$$

24. To show that a function $f(t)$ is a density function, you must show that $f(t) \geq 0$ for all t and that
$$\int_{-\infty}^{\infty} f(t)\, dt = 1.$$

Which of the following functions are density functions? Explain your answers.

(a) $f(t) = \begin{cases} 5t^4, & \text{if } 0 \leq t \leq 1, \\ 0, & \text{otherwise.} \end{cases}$

(b) $f(t) = \begin{cases} 2t, & \text{if } -1 \leq t \leq 2, \\ 0, & \text{otherwise.} \end{cases}$

(c) $f(t) = \begin{cases} \frac{1}{2}, & \text{if } -1 \leq t \leq 1, \\ 0, & \text{otherwise.} \end{cases}$

(d) $f(t) = \begin{cases} \dfrac{1}{\sqrt{t}}, & \text{if } \frac{1}{4} \leq t \leq 1, \\ 0, & \text{otherwise.} \end{cases}$

(e) $f(t) = \frac{1}{2}e^{-|t|}, \quad \text{if } -\infty < t < \infty.$

25. For each of the following c.d.f.'s find the corresponding density functions $f_X(t)$.

(a) $F_X(t) = \begin{cases} 0, & \text{if } t < 0, \\ 1 - e^{-10t}, & \text{if } t \geq 0. \end{cases}$

(b) $F_X(t) = \begin{cases} \frac{1}{2}e^t, & \text{if } t < 0, \\ 1 - \frac{1}{2}e^{-t}, & \text{if } t \geq 0. \end{cases}$

(c) $F_X(x) = \begin{cases} 0, & \text{if } X < 0 \\ 280\int_0^X u^4 (1-u)^3\, du, & \text{if } 0 \leq X \leq 1 \\ 1, & \text{if } X \leq 1 \end{cases}$

26. One measure of air pollution is the amount Y of beta radioactivity concentration in the air (measured in microcuries per cubic meter). Data for those states that have air sampling stations are collected by the federal government [*Statistical Abstracts of the United States* (1965)]. The results, listed alphabetically by states, are: 2.2, 6.5, 5.8, 3.9, 5.9, 4.0, 4.6, 4.9, 4.5, 2.6, 5.7, 4.7, 4.8, 3.5, 4.2, 5.0, 4.0, 4.1, 4.5, 5.0, 5.2, 3.9, 4.6, 4.1, 4.8, 3.9, 8.8, 5.9, 4.2, 4.8, 4.6, 5.5, 4.3, 5.0, 3.3, 5.2, 4.0, 4.6, 3.7, 5.7, 4.6, 5.6, 5.2, 4.4, 3.6, 4.9, 5.4, 6.2, 2.2. Prepare a modified relative frequency histogram from these data.

27. Let $W = (X - 2)^2$ be a transformation of the discrete random variable X. Find the probability mass function of W when X has the probability mass function $p_X(x)$ given in (a) Exercise 15, (b) Exercise 16, and (c) Exercise 17.

28. What is the probability density function of
    (a) $Y = Z - 2$
    (b) $X = (Z - 2)^2$
    when Z has the probability density function $f_Z(z)$ given in Exercise 22?

29. What is the probability density function of $Y = e^W$ when W has the probability density function $f_W(w)$ given in Exercise 23?

30. Let the random variable X have the probability density function

$$f_X(x) = \begin{cases} 1, & \text{if } 0 \le x \le 1, \\ 0, & \text{otherwise.} \end{cases}$$

Let $F(y)$ be a cumulative distribution function that is strictly increasing and differentiable. Let $F^{-1}(x) = y$ whenever $F(y) = x$ [i.e., $F^{-1}(x)$ is the inverse function to $F(y)$ and $F(F^{-1}(x)) = x$]. Show that the transformed random variable $Y = F^{-1}(X)$ has cumulative distribution function $F(y)$.

5

Descriptive Properties

of Distributions

1. INTRODUCTION

Cumulative distribution functions and probability mass functions or density functions are profiles that contain all the relevant information about the statistical properties of a random variable. We may refer to these profiles, whichever is used, as the *distribution* of the random variable.

For many purposes it is not necessary to use all the information contained in the distribution; rather, several descriptive properties may suffice. Two of the most frequently useful pieces of information about a distribution are its *location* and its *dispersion*.

To explain what we mean by the location of a distribution, consider Figure 1.1. Here we have two densities that have exactly the same graphical shape except that $f_X(x)$ concentrates about $x = x_0$, and $f_X^*(x)$ concentrates about $x = x_1$. Thus, we can see that $f_X^*(x)$ is, in some sense, $f_X(x)$ *relocated* from x_0 to x_1. If we had a descriptive measure that changed value by d whenever the distribution changed location by d, we would have a *measure of location* for the distribution.

On the other hand, a *measure of dispersion* (measure of variation) measures how strongly a distribution concentrates about a central value. A measure of dispersion is large when the spread of the distribution about a central value is large and is small when this spread is negligible, becoming zero when all the probability is concentrated at a single point. For example, the distribution graphed in Figure 1.2 has a smaller spread around the central point c than the distribution graphed in Figure 1.3, and thus its measure of dispersion should be smaller. Similarly, the distributions graphed in Figure 1.4 have different dispersions.

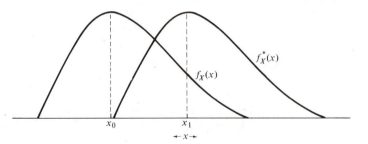

Figure 1.1: *Two densities differing in location.*

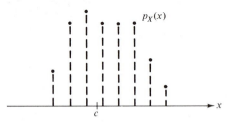
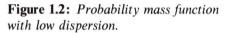

Figure 1.2: *Probability mass function with low dispersion.*

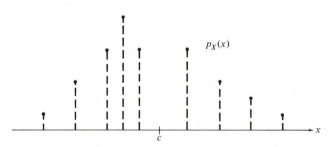

Figure 1.3: *Probability mass function with a larger dispersion than that of the mass function in Figure 1.2.*

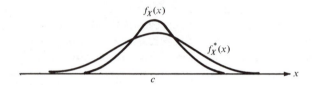

Figure 1.4: *Two density functions with different dispersions.*

In this chapter, we introduce several measures of location for a distribution (mean, median, quantile, mode), each of which can be used to answer questions about distributions. We also discuss measures of dispersion, one of which (the variance) plays a prominent role in statistical analysis. Other descriptive measures (the moments) and descriptive terminology (skewness, peakedness or kurtosis, and so on) are briefly mentioned. Finally, we show how a knowledge of a certain measure of location (the mean) and/or a certain measure of dispersion (the variance) for a random variable X in many cases allows us to state upper or lower limits for the values of the probabilities of certain events of interest concerning X, even when nothing else is known about the distribution of X.

2. MEASURES OF LOCATION FOR A DISTRIBUTION: THE MEAN

One of the most useful characteristics of a distribution is its *mean* (also called the *expected value* or the *expectation*). The mean of a distribution can be viewed as being the fulcrum, the center of gravity, or the balance point of the distribution of probability on the real line. Since in physics the center of gravity of a distribution is called the first moment of inertia of the distribution, the mean of a probability distribution is also called the *first moment* of that distribution.

The Mean of a Discrete Random Variable

In mechanics, the center of gravity of a distribution of discrete masses whose total weight is 1 is determined as the sum of the products of the magnitude of each weight times its distance from a given origin. If we regard the values $x_1, x_2, \ldots, x_k$ of a discrete random variable as being (signed) distances from the origin 0, and the probabilities $p_i = P\{X = x_i\}$, $i = 1, 2, \ldots,$ k, of these values as being masses, then the mean of X, denoted μ_X, is defined analogously as

$$(2.1) \qquad \mu_X = \sum_{i=1}^{k} x_i p_i .$$

If X has a countably infinite number of possible values $x_1, x_2, \ldots$, then k in Equation (2.1) is allowed to equal infinity; that is, $\mu_X = \sum_{i=1}^{\infty} x_i p_i$. Note that $p_i = P\{X = x_i\} = p_X(x_i)$, where $p_X(\cdot)$ is the probability mass function of X. Thus,

$$\mu_X = \sum_{x} x p_X(x) = \sum_{i=1}^{k} x_i p_X(x_i).$$

The probability mass function $p_X(x)$ and the mean μ_X of a discrete random variable X having five possible values are shown in Figure 2.1. Viewing the horizontal axis as a weightless seesaw supported by a fulcrum at μ_X, the probability masses p_1, p_2, p_3 on the left side of the fulcrum balance the probability masses p_4, p_5 on the right side.

Example 2.1. Let X be the number of children born to a randomly chosen lower-class urban family and Y be the number of children born to a randomly chosen family from an upper-class suburb. Experimental data are used to construct the following probability models for X and Y.

Number of children, x	0	1	2	3	4	5	6	7	8	9	10	11
$p_X(x)$	0.05	0.05	0.10	0.15	0.20	0.20	0.15	0.03	0.02	0.02	0.02	0.01
$p_Y(x)$	0.12	0.15	0.19	0.16	0.12	0.10	0.07	0.02	0.02	0.02	0.02	0.01

The means of X and Y are

$$\mu_X = (0)(0.05) + (1)(0.05) + \cdots + (11)(0.01) = 4.26,$$

$$\mu_Y = (0)(0.12) + (1)(0.15) + \cdots + (11)(0.01) = 3.20.$$

These means provide some indication as to how many children "on the average" are born in lower-class urban and upper-class suburban families,

Figure 2.1: *The probability mass function* $p_X(x)$ *of a discrete random variable,* X, *having possible values* x_1, x_2, x_3, x_4, x_5 *with* $p_X(x_j) = p_j, j = 1, 2, \ldots, 5$, *and with mean* μ_X.

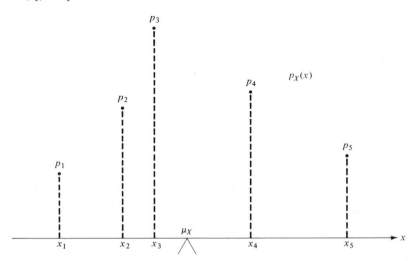

respectively. Although by themselves the two means do not allow us to compare the probability distributions of X and Y in detail, a comparison of $\mu_X = 4.26$ and $\mu_Y = 3.20$ does suggest that lower-class urban families tend to be larger than upper-class suburban families. The disadvantage of keeping only the information about X and Y provided by their means μ_X and μ_Y is that we do not obtain a complete picture of the differences between their distributions, and thus can miss important facts. It should be noted that similar advantages and disadvantages accrue to any other single index that is reported in place of the entire distribution of a random variable.

Example 2.2. In an investigation of the theory of Brownian motion in physics, the Swedish physicist T. Svedberg (1912) counted the number of particles X in an optically isolated volume of a colloidal solution of gold at a certain instance of time and obtained the following table of probabilities:

Number of particles, x	0	1	2	3	4	5	6	7
$P\{X = x\}$	0.182	0.310	0.264	0.150	0.064	0.022	0.006	0.002

for which the mean of the number of particles is $\mu_X = 1.704$.

We have remarked that μ_X is sometimes called the expected value of X. However, this terminology is not intended to imply that μ_X is the value expected as a result of performing the random experiment and observing X. Indeed, the mean of X need not even be a possible value of X. As examples, notice that in Example 2.1 neither $\mu_X = 4.26$ nor $\mu_Y = 3.20$ are possible values of the corresponding random variables X or Y, respectively.

The mean μ_X of X is, however, a number which in a large number of trials of the experiment that produces X would be very close to the "average" of the observed X values. The frequency interpretation of probability suggests that this should be true; for if $t_1, t_2, \ldots, t_N$ are the observed values of X in N independent trials of the random experiment that produces X, then the *arithmetic average* of the observed values of X is

(2.2)
$$\bar{t} = \frac{t_1 + t_2 + \cdots + t_N}{N} = \sum_t [t \times \mathrm{r.f.}\{X = t\}],$$

where the sum is over all distinct values t that are observed. As the number N of trials becomes very large, the frequency interpretation of probability tells us that $\mathrm{r.f.}\{X = t\}$ tends to be close to $P\{X = t\}$. Thus, when the number N of trials is large,

(2.3)
$$\bar{t} = \sum_t [t \times \mathrm{r.f.}\{X = t\}] \simeq \sum_t [t \times P\{X = t\}] = \mu_X,$$

where the symbol " $\simeq$ " means "approximately equal to." We conclude that *in a large number of trials in which X is observed, the arithmetic average* (also called *sample average* or *sample mean*) *of the observed values of X is approximately equal to* μ_X.

Example 2.3. The approximation given by Equation (2.3) is frequently used in practice to predict the magnitude of the sum $\sum_{i=1}^{N} t_i = N\bar{t}$ of the observed values of X before these values are actually observed. For example, suppose that in Example 2.1, a Board of Education wishes to predict the number of children who will use the schools. If a housing project consists of $N = 1000$ lower-class urban families, then the observations $t_1, t_2, \ldots, t_{1000}$ will be counts of the number of children in each of the 1000 families. Consequently,

$$\frac{t_1 + t_2 + \cdots + t_{1000}}{1000} \simeq \mu_X = 4.26,$$

which suggests that the Board of Education should plan schooling for approximately 4260 children.

Example 2.4. A man M who has just become 44 years of age and who is in good health applies to an insurance company, company A, for a one-year term insurance policy of $25,000 on his life. For this policy, Mr. M pays a premium of $100 to company A, and in turn company A will pay M's heirs $25,000 if he dies before he is 45 years of age. From a standard mortality table (*1967 Life Insurance Fact Book*, Institute of Life Insurance, New York), the probability that a man of age 44 will live to be age 45 is $p = 0.99681$. Let X equal the amount of profit (or loss) in dollars that Company A makes from the transaction. That is, if M lives to age 45, then company A obtains $X = \$100$, while if M dies before age 45, $X = \$100 - \$25,000$. The mean μ_X of the random variable X is then

$$\mu_X = (\$100)p + (\$100 - \$25,000)(1 - p)$$

$$= (\$100)(0.99681) + (-\$24,900)(0.00319)$$

$$= \$20.25,$$

indicating a profit for company A. Notice that $\mu_X = \$20.25$ does not really refer to the profit company A makes on the transaction with Mr. M (this profit is either $100 if M lives, or $-\$24,900$ if he dies). Instead, the value of μ_X refers to the *average* profit that company A can expect over a great many transactions of the type, and made under the same conditions of age and health, that company A made with Mr. M. The larger is the number N of such transactions that company A makes, the more accurately μ_X indicates what the average profit per transaction will be.

The Mean of a Discrete Random Variable Having an Infinite Number of Possible Values

Our discussion thus far has assumed that the sum in Equation (2.1) which defines μ_X has a well-defined finite value. When X has only a finite number $x_1, x_2, \ldots, x_k$ of possible values, this sum will always be well defined and finite. However, when X has an infinite number of possible values, then it is possible for either μ_X or $-\mu_X$ to be infinite. For example, suppose that X has possible values $x_i = i^2$, $i = 1, 2, \ldots$, and

$$P\{X = x_i\} = p_X(x_i) = \frac{6}{\pi^2} \frac{1}{i^2}, \quad i = 1, 2, \ldots .$$

This is a legitimate probability mass function, since

$$\sum_{i=1}^{\infty} p_X(x_i) = \frac{6}{\pi^2} \sum_{i=1}^{\infty} \frac{1}{i^2} = \left(\frac{6}{\pi^2}\right)\left(\frac{\pi^2}{6}\right) = 1.$$

Using Equation (2.1), we obtain

$$\mu_X = \sum_{i=1}^{\infty} x_i P\{X = x_i\} = \sum_{i=1}^{\infty} (i^2)\left(\frac{6}{\pi^2 i^2}\right) = \frac{6}{\pi^2} \sum_{i=1}^{\infty} 1 = \infty.$$

It is also possible for the sum in Equation (2.1) to not be well defined. This is the case, for example, when the possible values x_i and the probabilities $p_i = P\{X = x_i\}$ of X are such that $x_i p_i = (-1)^i$, so that $\mu_X = \sum_{i=1}^{\infty} x_i p_i = 1 - 1 + 1 - 1 + \cdots$. When μ_X or $-\mu_X$ equals ∞, or when μ_X is not well defined, the mean μ_X of X is said *not to exist* and the frequency interpretation of μ_X given earlier need not be valid.

The Mean of a Continuous Random Variable

We have mentioned the analogy existing between the center of gravity of the distribution of a discrete set of masses on a line and the mean μ_X of a discrete random variable. In mechanics, we can also speak of the center of gravity for a continuous distribution of mass on a line. Analogously, the mean μ_X of a continuous random variable X is defined to be

(2.4)
$$\mu_X = \int_{-\infty}^{\infty} x f_X(x) \, dx.$$

Example 2.5. Suppose that the continuous random variable X has density function

$$f_X(x) = \begin{cases} \frac{1}{2}, & \text{for } -1 \le x \le 1, \\ 0, & \text{for } |x| > 1. \end{cases}$$

This density is graphed in Figure 2.2. Then

$$\mu_X = \int_{-\infty}^{-1} (x)(0)\, dx + \int_{-1}^{1} (x)(\tfrac{1}{2})\, dx + \int_{1}^{\infty} (x)(0)\, dx = 0 + \frac{x^2}{4}\bigg|_{-1}^{1} + 0 = 0.$$

The result, $\mu_X = 0$, is intuitively reasonable because of the symmetry of the graph of $f_X(x)$ about the point $x = 0$.

Example 2.6. In Chapter 4, Section 5, we have given an example of a delicatessen that receives an order of 400 pounds of meat every day. The random variable X considered in this example is the total weight of meat sold in a day, and

$$f_X(x) = \begin{cases} (1.25)(10^{-5})x, & \text{for } 0 \le x \le 400, \\ 0, & \text{otherwise.} \end{cases}$$

The approximate total amount of meat that the delicatessen can expect to sell per day is

$$\mu_X = \int_{-\infty}^{0} (x)(0)\, dx + \int_{0}^{400} (x)(1.25)(10^{-5})x\, dx + \int_{400}^{\infty} (x)(0)\, dx$$

$$= \frac{(1.25)10^{-5}}{3} x^3 \bigg|_{0}^{400} = 266.67 \text{ pounds,}$$

and in a year the delicatessen can expect to sell approximately $365(\mu_X) = 97{,}333.33$ pounds of meat.

Figure 2.2: *Graph of the density function $f_X(x)$, which is flat from $x = -1$ to $x = 1$ and is zero elsewhere.*

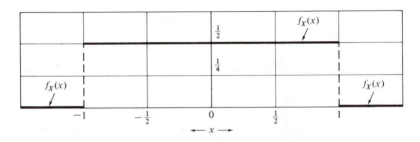

The Mean as a Measure of Location

The mean is a measure of location for a probability distribution. The interpretation of the mean as the center of gravity of a probability distribution gives some intuitive support to this assertion. Another way of seeing that μ_X is a measure of location is to consider what the mean $\mu_Z = \mu_{aX+c}$ of a linear transformation of X would be. If X is a discrete random variable with possible values $x_1, x_2, \ldots$ and $P\{X = x_i\} = p_i$, $i = 1, 2, \ldots$, then if $a \neq 0$,

$$p_i = P\{X = x_i\} = P\{Z = ax_i + c\} = P\{Z = z_i\},$$

where $z_i = ax_i + c$, $i = 1, 2, \ldots$. Thus,

$$\mu_Z = \sum_i z_i P\{Z = z_i\} = \sum_i (ax_i + c)P\{Z = ax_i + c\}$$

$$= \sum_i (ax_i + c)p_i = a \sum_i x_i p_i + c \sum_i p_i$$

$$= a\mu_X + c,$$

since $\sum_i p_i = 1$. Similarly, when X is a continuous random variable $f_Z(z) = (1/|a|)f_X((z - c)/a)$ and

$$\mu_Z = \int_{-\infty}^{\infty} z f_Z(z)\, dz = \int_{-\infty}^{\infty} \frac{z}{|a|} f_X\left(\frac{z - c}{a}\right) dz$$

$$= \int_{-\infty}^{\infty} (au + c)f_X(u)\, du$$

$$= a \int_{-\infty}^{\infty} u f_X(u)\, du + c \int_{-\infty}^{\infty} f_X(u)\, du$$

$$= a\mu_X + c,$$

where we changed the variable of integration from z to $u = (z - c)/a$ and used the fact that $\int_{-\infty}^{\infty} f_X(u)\, du = 1$. What we have thus shown is that when $a \neq 0$ and c is any number,

(2.5) $$\mu_{aX+c} = a\mu_X + c.$$

When $a = 1$, Equation (2.5) specializes to

(2.6) $$\mu_{X+c} = \mu_X + c,$$

showing that the mean of X shifts c units whenever the distribution of X changes location by c units, a property that we used to define a measure of location for the distribution of X.

3. OTHER MEASURES OF LOCATION FOR A DISTRIBUTION: MEDIAN, QUANTILE, MODE

The mean μ_X is not the only measure of location that we could use to describe the distribution of X. Other measures of location that are often used are medians, quantiles, and modes.

Quantiles and the Median

Let p be a fixed proportion, $0 \le p \le 1$. A pth *quantile* of the distribution of the random variable X is any number $Q_p(X)$ satisfying

(3.1) $P\{X < Q_p(X)\} \le p$ and $P\{X > Q_p(X)\} \le 1 - p$,

or, equivalently,

(3.2) $F_X(x) \le p \le F_X(Q_p(X))$,   for all $x < Q_p(X)$.

When $p = 0.5$ we obtain the median; that is,

(3.3) $Q_{0.5}(X) = \text{Med}(X)$.

In Figure 3.1, we illustrate in two cases ($p_1 = 0.50$ and $p_2 = 0.75$) how we can find a pth quantile of a discrete distribution from a graph of the c.d.f. Notice that the median or (0.50)th quantile is unique [$\text{Med}(X) = 30$], but that

Figure 3.1: *Finding a median,* $\text{Med}(X)$, *and* 0.75^{th} *quantile,* $Q_{0.75}(X)$, *of a discrete random variable,* X. *Here* $\text{Med}(X) = 30$, $Q_{0.75}(X) = 45$.

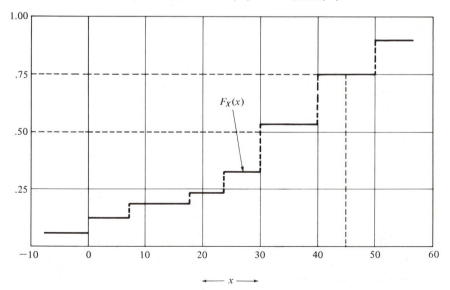

any number in the interval $40 \leq x < 50$ is a (0.75)th quantile of X. For the sake of specificity, we would report that $Q_{0.75}(X) = 45$, the midpoint of the interval of (0.75)th quantiles. In Figure 3.2, we show how the pth quantile of a continuous distribution can be found from a graph of the c.d.f.

We can also try to find $Q_p(X)$ analytically by solving the equation

(3.4)
$$F_X(x) = p$$

for x. If a unique solution to this equation exists ($p = 0.50$ in Figure 3.1 gives an example where a solution does not exist), then that solution x^* is the unique pth quantile of X. For example, if

$$F_X(x) = \begin{cases} 0, & \text{if } x < 0, \\ x^2, & \text{if } 0 \leq x \leq 1, \\ 1, & \text{if } 1 < x, \end{cases}$$

then solving

$$p = F_X(x) = x^2$$

for x, we find that $Q_p(X) = p^{1/2}$. If Equation (3.4) has no solution, this means that there exists a value x^* such that $F_X(x) < p$ for all $x < x^*$, and $F_X(x^*) > p$. In this case, x^* is the unique pth quantile of X. (This is the situation, for example, for $p = 0.50$ in Figure 3.1.) Finally, Equation (3.4) may be satisfied for an interval of x-values (e.g., in the case $p = 0.75$ in Figure 3.1), and in this case the midpoint of the interval of solutions is usually reported as the value of $Q_p(X)$.

If X is a discrete random variable, obtaining $Q_p(X)$ through a graph of $F_X(x)$, or analytically through the functional form of $F_X(x)$ by means of

Figure 3.2: *Finding a median,* Med(X), *and* 0.75^{th} *quantile,* $Q_{0.75}(X)$, *of a continuous random variable. Here* Med$(X) = 1$ *and* $Q_{0.75}(X) = 1.5$.

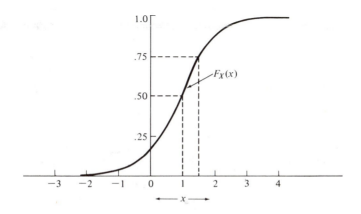

Equation (3.4), can become awkward. Instead, it is often easier to work with a *table of cumulated probabilities* of X [which, since the c.d.f. of $F_X(x)$ only changes at the possible values of X, is equivalent to $F_X(x)$]. Use of this method is illustrated by the following example.

Example 3.1. Suppose that the distribution of X is given by the following table of probabilities:

x	-2	-1	0	1	2
$P\{X = x\}$	0.12	0.13	0.21	0.30	0.24

We wish to find $Q_{0.90}(X)$, the (0.90)th quantile of X. Note that the values of X are listed in increasing order in the table. We now convert this table to a table of cumulative probabilities $P\{X \leq x\}$:

x	-2	-1	0	1	2
$P\{X \leq x\}$	0.12	0.25	0.46	0.76	1.00

where, for example, $P\{X \leq 0\}$ was computed by summing 0.12, 0.13, and 0.21. Since the jumps of $F_X(x)$ occur only at $x = -2, -1, 0, 1, 2$ and the graph of $F_X(x)$ is flat between jumps, we could use this table to graph $F_X(x)$. However, we need not do this; rather, we look at the entries $P\{X \leq x\}$ in the table and try to find the value of x at which $P\{X \leq x\}$ either equals 0.90 or jumps from under 0.90 to over 0.90. Here, this value is $x = 2$, since $P\{X \leq 1\} = 0.76 <$ 0.90, while $P\{X \leq 2\} = 1.00$. Hence, $Q_{0.90}(X) = 2$. Similarly, since $P\{X \leq -1\} = 0.25$, the number -1 is a (0.25)th quantile of X; and indeed all numbers $-1 \leq x < 0$ are (0.25)th quantiles of X. In this case, we would report $Q_{0.25}(X) = \frac{1}{2}(-1 + 0) = -0.5$, the midpoint of the interval of (0.25)th quantiles, as the value of $Q_{0.25}(X)$.

Example 2.1 (continued). Earlier we discussed probability models for the number X of children born to a randomly chosen lower-class urban family and the number Y of children born to a randomly chosen upper-class suburban family. From the table of probabilities given in Example 2.1, we can calculate the following table of cumulated probabilities:

Number of children, x	0	1	2	3	4	5	6	7	8	9	10	11
$P\{X \leq x\}$	0.05	0.10	0.20	0.35	0.55	0.75	0.90	0.93	0.95	0.97	0.99	1.00
$P\{Y \leq x\}$	0.12	0.27	0.46	0.62	0.74	0.84	0.91	0.93	0.95	0.97	0.99	1.00

From this table we can calculate various quantiles of the distributions of X and Y by the method illustrated in Example 3.1:

Quantile	$Q_{0.10}$	$Q_{0.25}$	$Q_{0.50} = $ Med	$Q_{0.75}$	$Q_{0.90}$
X	1.5	3.0	4.0	5.5	6.5
Y	0.0	1.0	3.0	5.0	6.0

Notice that all the quantiles calculated above satisfy the inequality $Q_p(X) \geq Q_p(Y)$, and these inequalities agree in direction with the inequality $\mu_X \geq \mu_Y$ that we earlier observed in connection with Example 2.1.

Approximating Quantiles from Tables

The c.d.f.'s of many continuous random variables cannot be given a convenient functional form, thus making it difficult to solve analytically for the pth quantile. Instead, tables of such c.d.f.'s $F_X(x)$, calculated at regularly spaced values of x, are often reported. From such a table, the pth quantile $Q_p(X)$ of X may be approximated by use of linear interpolation. The approximation comes from the fact that $F_X(x)$ is not tabled at all values and that we assume that $F_X(x)$ increases linearly between tabled x-values. The accuracy of approximation increases with the number of x-values used to tabulate $F_X(x)$.
We proceed as follows:

(1) If there is a unique value x^* of x for which $F_X(x^*) = p$, then $Q_p(X) = x^*$. If there is more than one value x for which $F_X(x) = p$, set $Q_p(X)$ equal to the average of the smallest and largest values of x for which $F_X(x) = p$.
(2) If there is no x^* for which $F_X(x^*) = p$, let x_1 be the largest value of x for which $F_X(x) < p$ and let x_2 be the smallest value of x for which $F_X(x) > p$. Then

$$Q_p(X) \simeq \frac{x_2(p - F_X(x_1)) + x_1(F_X(x_2) - p)}{[F_X(x_2) - F_X(x_1)]}.$$

For example, to obtain the (0.25)th quantile of X from Table 3.1, case (2) applies with $x_1 = -1.0$ and $x_2 = -0.5$, and

$$Q_{0.25}(X) \simeq \frac{(-0.5)(0.250 - 0.189) + (-1.0)(0.298 - 0.250)}{0.298 - 0.189} = -0.720.$$

Similarly, to approximate $Q_{0.5}(X) = \text{Med}(X)$, case (2) again applies with $x_1 = 0.0$ and $x_2 = 0.5$, and

$$\text{Med}(X) \simeq \frac{(0.5)(0.500 - 0.476) + (0.0)(0.583 - 0.500)}{0.583 - 0.476} = 0.112.$$

Table 3.1: *Values for a c.d.f.*
$F_X(x)$ Corresponding to Selected
Values of x

x	$F_X(x)$	x	$F_x(x)$
−3.0	0.001	0.5	0.583
−2.5	0.019	1.0	0.692
−2.0	0.057	1.5	0.783
−1.5	0.111	2.0	0.857
−1.0	0.189	2.5	0.954
−0.5	0.298	3.0	0.999
0.0	0.476		

Quantiles as Measures of Location

A pth quantile is a measure of location for a probability distribution in the sense that $Q_p(X)$ divides the distribution of probability mass into two parts, one having probability mass p and the other having mass $1 - p$ (see Figures 3.3, 3.4, and 3.5).

Since $F_{aX+b}(x) = F_X((x - b)/a)$ for any $a > 0$ and any constant b, it can be shown that

(3.5) $$Q_p(aX + b) = aQ_p(X) + b.$$

Because the median $\text{Med}(X)$ is a (0.50)th quantile, it follows that

(3.6) $$\text{Med}(aX + b) = a\,\text{Med}(X) + b.$$

Figure 3.3: *Graph of the probability mass function $p_X(x)$ of a discrete random variable, X, showing how the 0.25^{th} quantile $Q_{0.25}(X)$ divides the probability mass into two parts.*

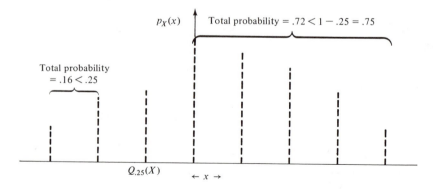

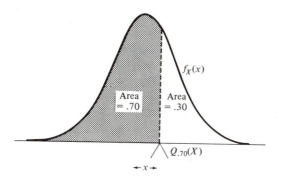

Figure 3.4: *Graph of the probability density function $f_X(x)$ of a continuous random variable, X, showing how the 0.70^{th} quantile $Q_{0.70}(X)$ divides the total area into two parts.*

Modes

A *mode* of a random variable is (one of) its "most probable" value(s). A mode of the distribution of a discrete random variable X is that possible value, Mode(X), of X at which the probability mass function $p_X(x)$ of X has its largest value. There can be one, or more than one, mode of the distribution of a discrete random variable (see Figures 3.6 and 3.7).

Since every possible value x of a continuous random variable X has

Figure 3.5: *Graph of the probability density function $f_X(x)$ of a continuous random variable, X, showing how the median Med(X) divides the total area into two parts.*

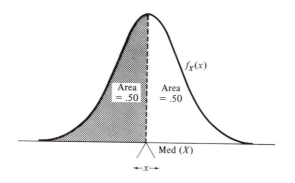

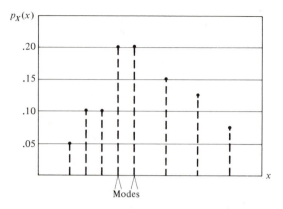

Figure 3.6: *A discrete unimodal distribution.*

zero probability of occurrence, the mode of a continuous random variable X with probability density function $f_X(x)$ is defined as that value of x at which $f_X(x)$ is a maximum. Again, there can be one, or more than one, mode of the distribution of a continuous random variable (see Figures 3.8 and 3.9).

Many books define a *relative* mode of a discrete (continuous) random variable X to be any possible value of X for which the probability mass function (probability density function) is *locally* the highest. Examples of this concept are shown in Figures 3.6 through 3.9. A mode, of course, is always a relative mode.

Remark. The descriptive terms "unimodal" and "bimodal" are often used in connection with distributions. A *unimodal distribution* has a single "hump" (see Figures 3.6 and 3.8). In contrast, a *bimodal distribution* has two "humps" (see Figures 3.7 and 3.9). A bimodal distribution has two different separated relative modes but may have a single unique mode.

Figure 3.7: *A discrete bimodal distribution.*

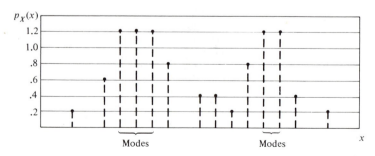

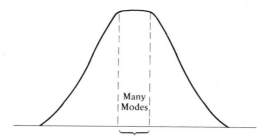

Figure 3.8: *A continuous unimodal distribution.*

Choice of a Measure of Location

There are many potential measures of location for a distribution: the mean μ_X, the median $\text{Med}(X)$, the quantiles $Q_p(X)$, and the mode. Each measure of location has both advantages and disadvantages so that there is no single best measure of location. The following example illustrates how the information that a measure of location gives about a distribution is influenced by the shape of the distribution, and why the choice of a measure of location is largely determined by the goals of the investigator.

Example 3.2. Two economists are studying the distribution of income in the United States. Economist A is concerned with poverty-oriented legislation; economist B is concerned with market research. The data available are the probability distribution of the total pretax income, X (in dollars), of a randomly chosen U.S. family. Although X is discrete, it is both mathematically and conceptually more convenient to treat X as if it were a continuous random variable with a probability density function $f_X(x)$. The shape of the

Figure 3.9: *A continuous bimodal distribution.*

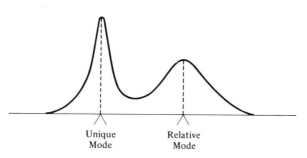

density function $f_X(x)$ of family income X in the United States resembles that of the function graphed in Figure 3.10. Important features of this graph include a long " tail " to the right, indicating incomes of the relatively few very wealthy families, and a single "hump" to the left, representing the low to moderate incomes of most U.S. families.

Economist A would not use the mean μ_X to describe the location of the distribution of family incomes X, because a few large possible value of X (corresponding to the incomes of the very wealthy) can inflate the value of μ_X to such an extent that this index does not reflect the incomes of the large majority of families. [If 99% of all families earned \$10,000 and the other 1% earned \$1,000,000, then $\mu_X = (10,000)(0.99) + (1,000,000)(0.01) = 19,900$ dollars. Thus, μ_X overstates (by nearly twice) the income of 99% of the population.] On the other hand, the value of Med(X) is not influenced by large incomes occurring with small probability. Indeed, we can calculate Med(X) without ever looking at the right-hand tail of the graph in Figure 3.10. Knowing that Med(X) = \$10,000, for example, tells us that 50% of all families have incomes less than or equal to \$10,000, and gives us some feel for the financial well-being of the majority of the population. Similarly, as long as p is not too close to 1, $Q_p(X)$ is not affected by extremely large incomes that have small probability. The values of $Q_{0.10}(X)$ and $Q_{0.25}(X)$, for example, might be used to give us some idea of the financial status of the poorest 10% and 25%, respectively, among U.S. families. Consequently, economist A would report one or more quantiles of the distribution of X.

Economist B is concerned with predicting family consumer purchases

Figure 3.10: *Probability density function of the income of a randomly chosen U.S. family.*

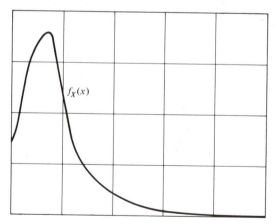

$\longleftarrow x \longrightarrow$

for any given year based on knowledge of family income in the immediately preceding year. If economist B reports $\text{Med}(X)$ instead of μ_X as the "average" or "typical" family income, this will ignore the incomes of the very rich families, and ultimately will lead to an underprediction of the amount of consumer spending for the coming year. Indeed, the mean μ_X allows us to estimate the total income for *all* families in the United States and thus permits prediction of total consumer purchases for the following year.

Neither economist A nor economist B would give much consideration to the mode of X as a measure of location. Economist A would not be interested because the mode tells little about how low the income of the lower income families actually are. Economist B is not interested in the mode because it does not help in calculating, even approximately, the total income of all U.S. families. In fact, the mode is rarely the preferred measure of location in practical problems. The mode is useful in situations where we want to predict a possible value (or interval of possible values) for a future observation of a random variable X and want to have the highest probability of making a correct prediction. In that case, the mode of X (or an interval containing the mode) is the proper predictor to use.

4. THE EXPECTED VALUE OF A FUNCTION OF A RANDOM VARIABLE

Given a probability model for a random variable X, we might also be interested in predicting (approximately) the mean of some function $g(X)$ of X. For example, if X represents measures of velocities of randomly selected atomic particles of known mass m, we might want to predict the average kinetic energy $g(X) = \frac{1}{2}mX^2$ of these particles. Let $Y = g(X)$. Then, as remarked in Chapter 4, Y is a random variable and thus has a probability distribution of its own. If X is a discrete random variable, then the probability that $Y = y$ can be obtained by summing the probabilities of all values x of X for which $g(x) = y$; that is,

$$P\{Y = y\} = \sum_{\substack{x \text{ for which} \\ g(x) = y}} P\{X = x\}.$$

In general, whether X is continuous or discrete, we can obtain the probability distribution of Y from the probability distribution for X (see Chapter 4, Section 6).

The average of the values of $g(X)$, in a large sample of values $t_1, t_2, \ldots,$ t_N observed on X can be approximated by μ_Y, the mean of the distribution of $Y = g(X)$. However, the formulas that we gave in Section 2 for calculating μ_Y were expressed in terms of the probability distribution of Y, so that it appears that we must first determine the probability distribution of Y. Fortunately, as suggested by the fact that

(4.1)   Average of $g(t_i)$-values $= \dfrac{1}{N} \sum_{i=1}^{N} g(t_i) = \sum_{t} g(t)$ r.f.$\{X = t\}$,

we can compute μ_Y directly from the distribution of X. To do so, it is convenient to define $E[g(X)]$, called the *expected value of* $g(X)$. In the case where X is a discrete random variable with probability mass function $p_X(x)$, we define

(4.2) $E[g(X)] = \sum_{x} g(x)p_X(x),$

and when X is a continuous random variable with probability density function $f_X(x)$, then we define

(4.3) $E[g(X)] = \int_{-\infty}^{\infty} g(x)f_X(x)\, dx.$

When the sum in (4.2) or the integral in (4.3) equal ∞ or $-\infty$, or are not well defined, then we say that the expected value $E[g(X)]$ *does not exist.*

Note from the definition of $E[g(X)]$ in the case $g(t) = t$ that

(4.4) $\mu_X = E[X].$

The notations μ_X and $E[X]$, however, are intended to play different roles in our vocabulary. The mean of X, μ_X, is an *index* of the distribution of X which helps locate this distribution on the horizontal axis (real line). On the other hand, $E[X]$ and, more generally, $E[g(X)]$ denote certain *computations* based on the probability distribution of X. We obtain the index μ_X by means of the computation $E[X]$.

We have asserted that if $Y = g(X)$, then μ_Y, the mean of Y, can be computed directly from the probability distribution of X, without the need to obtain the probability distribution of Y. Indeed,

(4.5) $\mu_Y = \mu_{g(X)} = E[g(X)].$

A general proof of this assertion would require us to exposit mathematical technicalities not needed elsewhere in the book. Thus, to illustrate the basic idea of the general proof, we use the example $Y = (X - 3)^2$, where X is a discrete random variable with probability distribution given by

x	0	1	2	3	4	5	6
$P\{X = x\}$	$\frac{1}{16}$	$\frac{2}{16}$	$\frac{3}{16}$	$\frac{3}{16}$	$\frac{5}{16}$	$\frac{1}{16}$	$\frac{1}{16}$

First, let us obtain μ_Y by calculating $p_Y(y)$ and using Equation (2.1). Now Y has possible values 0, 1, 4, and 9, and

$$P\{Y = 0\} = P\{X = 3\} = \tfrac{3}{16},$$

$$P\{Y = 1\} = P\{X = 2\} + P\{X = 4\} = \tfrac{8}{16},$$

$$P\{Y = 4\} = P\{X = 1\} + P\{X = 5\} = \tfrac{3}{16},$$

$$P\{Y = 9\} = P\{X = 0\} + P\{X = 6\} = \tfrac{2}{16},$$

so that from Equation (2.1),

$$\mu_Y = \sum_y yP\{Y = y\} = (0)(\tfrac{3}{16}) + (1)(\tfrac{8}{16}) + (4)(\tfrac{3}{16}) + (9)(\tfrac{2}{16}) = \tfrac{19}{8}.$$

On the other hand,

$$E[(X - 3)^2] = \sum_{x=0}^{6} (x - 3)^2 P\{X = x\} = (0 - 3)^2 (\tfrac{1}{16}) + \cdots + (6 - 3)^2 (\tfrac{1}{16})$$

$$= \tfrac{19}{8} = \mu_Y.$$

Thus, we have verified Equation (4.5) in this special case.

Example 4.1. Suppose that we have constructed a probability model for the velocity X of a randomly chosen atomic particle of unit mass $m = 1$. Since any system of such particles will contain a very large number of particles, it is convenient to regard X as a continuous variable. Suppose that all velocities between 0 and some number V have equal probability density; that is, X has density

$$f_X(x) = \begin{cases} \dfrac{1}{V}, & \text{if } 0 \le x \le V, \\ 0, & \text{otherwise.} \end{cases}$$

Then, if

$$Y = \tfrac{1}{2}mX^2 = \tfrac{1}{2}X^2$$

is the kinetic energy of a randomly selected atomic particle, the mean kinetic energy μ_Y of particles within the system can be computed from Equation (4.5) as

$$\mu_Y = E[\tfrac{1}{2}X^2] = \int_0^V (\tfrac{1}{2}x^2)\frac{1}{V}\,dx = \frac{V^2}{6}.$$

One of the advantages of working with the expected value notation is that for many discussions we do not have to distinguish between discrete and continuous random variables in order to derive important properties of indices of random variables, as long as these indices can be expressed in terms of an expected value of a function of a random variable. This lack of a need to distinguish between discrete and continuous random variables arises because of the following two important properties of the expected value.

Property 4.1 (Linearity of the Expected Value). If X is any random variable, $g(\cdot)$ and $h(\cdot)$ are any two functions, and a and b are any two constants, then

(4.6) $$E[ag(X) + bh(X)] = aE[g(X)] + bE[h(X)]$$

provided that all the expected values exist.

Property 4.2. If X is any random variable, and $g(X)$ is nonnegative with probability equal to 1, then $E[g(X)] \geq 0$.

The proof of Property 4.1 in the case when X is a discrete random variable is based on noting that

$$\sum_x [ag(x) + bh(x)]p_X(x) = a \sum_x g(x)p_X(x) + b \sum_x h(x)p_X(x),$$

while if X is a continuous random variable,

$$\int_{-\infty}^{\infty} [ag(x) + bh(x)]f_X(x)\, dx = a \int_{-\infty}^{\infty} g(x)f_X(x)\, dx + b \int_{-\infty}^{\infty} h(x)f_X(x)\, dx.$$

Property 4.2 follows by noting that $\sum_x g(x)p_X(x)$ and $\int_{-\infty}^{\infty} g(x)f_X(x)\, dx$ are a sum and an integral of nonnegative quantities, respectively. As an example of the use of Equation (4.6), notice that the conclusion $\mu_{aX+c} = a\mu_X + c$ of Equation (2.5) follows immediately from Equation (4.6) if we set $g(X) = X$, $a = a$, $h(X) = c$, and $b = 1$.

Remark. It is worth noting that the probability of any event A concerning X is the expected value of the indicator function

(4.7) $$I_A(X) = \begin{cases} 1, & \text{if } X \text{ satisfies the property defining event } A, \\ 0, & \text{otherwise.} \end{cases}$$

That is,

(4.8) $$P(A) = E[I_A(X)].$$

In deriving probabilities of events, use of Equation (4.8) permits us to take advantage of Properties 4.1 and 4.2 of the expected value (see Section 7).

5. MEASURES OF DISPERSION FOR A DISTRIBUTION: THE VARIANCE

The *variance* σ_X^2 of a random variable X is a measure of the dispersion of the distribution of X around the mean μ_X. We use the function $(X - \mu_X)^2$ to measure how close X is to its mean μ_X, and we define

(5.1) $$\sigma_X^2 = E[(X - \mu_X)^2].$$

In a large number of observations $t_1, t_2, \ldots, t_N$ upon X, σ_X^2 would be approximately equal to the *average squared distance* of the observations $t_1, t_2, \ldots, t_N$ from the mean μ_X. Thus, the variance σ_X^2 quantifies how extensively X varies "on the average" about its mean μ_X. Large values of the variance are associated with large dispersions of X about μ_X, and small values of the variance are associated with small dispersions. If the variance of X is equal to zero, then X does not vary at all about μ_X, and $P\{X = \mu_X\} = 1$.

The term *standard deviation* of X refers to the square root, σ_X, of the variance of X. The standard deviation is expressed in terms of the same units of measurement as X, while the variance is not. For this reason, the standard deviation is frequently used as a measure of dispersion in preference to the variance.

From Equation (5.1) and Equations (4.2) and (4.3), we conclude that if X is a discrete random variable, then σ_X^2 is calculated by

(5.2) $$\sigma_X^2 = \sum_x (x - \mu_X)^2 p_X(x),$$

while when X is a continuous random variable,

(5.3) $$\sigma_X^2 = \int_{-\infty}^{\infty} (x - \mu_X)^2 f_X(x)\, dx.$$

We now give some examples of distributions of discrete random variables X to show how the variance σ_X^2 of X is related to the spread, dispersion, or variation of X around its expected value. In Figure 5.1, the probability mass function is equally distributed at the numbers $x_1 = \mu - \Delta$, $x_2 = \mu + \Delta$, where Δ is some constant. That is, $P\{X = x_1\} = P\{X = x_2\} = \frac{1}{2}$. The mean of X is μ, and the variance of X is

$$\sigma_X^2 = (x_1 - \mu)^2 P\{X = x_1\} + (x_2 - \mu)^2 P\{X = x_2\}$$
$$= (-\Delta)^2 \tfrac{1}{2} + (\Delta)^2 \tfrac{1}{2} = \Delta^2.$$

The larger the value of Δ, the larger the spread of the distribution about its expected value becomes, and the larger the variance becomes.

The following example gives a less artificial indication of how the spread of a distribution about its mean is related to the variance.

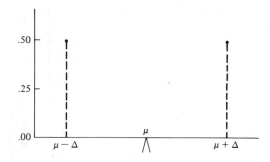

Figure 5.1: *Probability distributed equally between two values (of the discrete random variable X) equally spaced about the mean μ of X. The variance σ_X^2 of X is Δ^2.*

Example 5.1. A cereal company is required by law to fill their boxes of cereal with at least the amount of cereal (16 ounces) stated on the carton. If they underfill too many boxes, they may be in legal trouble, but if they overfill their boxes, they lose money. A study of the machines used to fill boxes indicates that the probability distribution of the amount X of creal placed in a randomly chosen box is given by

x	15.8	15.9	16.0	16.1	16.2
$P\{X = x\}$	0.15	0.20	0.30	0.20	0.15

The mean of X is

$$\mu_X = (15.8)(0.15) + (15.9)(0.20) + \cdots + (16.2)(0.15) = 16.0$$

so that the machines fill the correct amount "on the average." However, there is a high probability of under and overfilling the boxes. The variance of X is

$$\sigma_X^2 = (15.8 - 16.0)^2(0.15) + (15.9 - 16.0)^2(0.20) + (16.0 - 16.0)^2(0.30)$$
$$+ (16.1 - 16.0)^2(0.20) + (16.2 - 16.0)^2(0.15) = 0.016.$$

To decrease the dispersion, the machines are overhauled, yielding a new distribution of X:

x	15.8	15.9	16.0	16.1	16.2
$P\{X = x\}$	0.01	0.14	0.72	0.10	0.03

Now it is still the case that $\mu_X = 16.0$, so that the machines fill the correct amount "on the average," but the new variance is much smaller:

$$\sigma_X^2 = (15.8 - 16.0)^2(0.01) + (15.9 - 16.0)^2(0.14) + (16.0 - 16.0)^2(0.72)$$
$$+ (16.1 - 16.0)^2(0.10) + (16.2 - 16.0)^2(0.03) = 0.004.$$

This lower variance indicates less variation about the expected value, which is due to the reduction in the probabilities of under and overfilling under the new distribution of X.

Some caution should be noted with respect to the use of σ_X^2 as a measure of variation. The relationship of the degree of clustering of the distribution to the size of σ_X^2 holds in most practical cases; however, some mathematical examples can be given where the variance is not a good indicator of the shape of the distribution. For example, for the random variable X having the probability model

x	0	d
$P\{X = x\}$	$1 - \dfrac{1}{d}$	$\dfrac{1}{d}$

where $d > 1$ is a constant, the mean and variance of X are

$$\mu_X = (0)\left(1 - \frac{1}{d}\right) + (d)\left(\frac{1}{d}\right) = 1,$$

$$\sigma_X^2 = (0 - 1)^2\left(1 - \frac{1}{d}\right) + (d - 1)^2\left(\frac{1}{d}\right) = d - 1.$$

When $d = 10$, 100, and 1000, the probability models for X are

$d = 10$		
x	0	10
$P\{X = x\}$	0.9	0.1

$d = 100$		
x	0	100
$P\{X = x\}$	0.99	0.01

$d = 1000$		
x	0	1000
$P\{X = x\}$	0.999	0.001

In each case, $\mu_X = 1$, but $\sigma_X^2 = 9$, 99, and 999 for $d = 10$, 100, 1000, respectively. Thus, as d becomes larger, almost all of the probability mass of X becomes located at zero, with less and less probability being placed at d, and thus X is less and less variable. On the other hand, the variance σ_X^2 increases as d grows larger, suggesting that X is more and more variable.

The reason for this apparent contradiction is that the variance is unduly sensitive to a small amount of probability placed far out from the main body of the probability mass of the distribution. In Section 3, we noted that the mean μ_X had a similar sensitivity to extreme values assumed with small probability and that because of this sensitivity it is sometimes preferable to use one of the quantiles $Q_p(X)$ in place of μ_X as a measure of location. Similarly, one might use a measure of dispersion based on the quantiles in place of the variance in cases where extreme values may be assumed with small probability.

A measure of dispersion based on the quantiles is the *semi-interquartile range* $R = [Q_{0.75}(X) - Q_{0.25}(X)]/2$. Its justification is based on the fact that the interval of numbers between and including $Q_{0.25}(X)$ and $Q_{0.75}(X)$ includes at least 50% of the total probability mass of the distribution of X.

One very important property of the variance σ_X^2 of a random variable X is: If X is a random variable with variance σ_X^2, and if for any two numbers a and c we let $Z = aX + c$, then

(5.4) $$\sigma_Z^2 = \sigma_{aX+c}^2 = a^2\sigma_X^2.$$

Property (5.4) is most useful if we wish to change the units of measurement in terms of which we are measuring a given random variable X. For example, X may be a measurement of temperature in degrees Fahrenheit that we wish to convert to a measurement Z in degrees Celsius. Similarly, X may be a measurement of height in feet that we wish to convert to a measurement Z in inches. In the former case, $Z = \frac{5}{9}X - \frac{160}{9}$, so that $a = \frac{5}{9}$ and $c = -\frac{160}{9}$. In the latter case, $Z = 12X$, so that $a = 12$, $c = 0$. Suppose that we know the variance σ_X^2 of X. We could, of course, obtain the distribution of Z from knowledge of the distribution of X, and then from the distribution of Z compute the variance σ_Z^2 of Z by the methods described in this section. However, (5.4) provides a far easier way of obtaining σ_Z^2.

To verify (5.4), let $Z = aX + c$. It follows from Equation (2.5) that

$$\mu_Z = \mu_{aX+c} = a\mu_X + c,$$

and thus

$$\sigma_Z^2 = E(Z - \mu_Z)^2 = E[(aX + c) - (a\mu_X + c)]^2$$
$$= E[a(X - \mu_X)]^2 = E[a^2(X - \mu_X)^2] = a^2\sigma_X^2,$$

which proves (5.4).

As noted in Section 2, the first moment $\mu_X = E(X)$ of X about $c = 0$ is a measure of location for the distribution of X. The first moment of X around $c = \mu_X$ is $E(X - \mu_X) = 0$. The second moment of X around $c = \mu_X$ is the variance σ_X^2 of X and is a measure of dispersion for X. Part of the justification for choosing σ_X^2 comes from the physical interpretation of σ_X^2 as the second moment of inertia of the probability mass about its center of gravity; and part of the justification for σ_X^2 comes from the usefulness of μ_X as a center or measure of location for the distribution of X. However, a mathematical justification for choosing σ_X^2 is the following: *Among all quantities of the form*

$E(X - c)^2$, *choosing* $c = \mu_X$ *results in the smallest value*. That is, σ_X^2 is the smallest second moment about a central number c, and

(5.5) $$\sigma_X^2 \leq E(X - c)^2$$

for all constants c. Inequality (5.5) holds as a result of the very useful formula

(5.6) $$E(X - c)^2 = \sigma_X^2 + (\mu_X - c)^2,$$

which relates any second moment of X to the variance of X.

To verify (5.6), use Property 4.1:

$$\begin{aligned} E[(X - c)^2] &= E[(X - \mu_X + \mu_X - c)^2] \\ &= E[(X - \mu_X)^2 + 2(\mu_X - c)(X - \mu_X) + (\mu_X - c)^2] \\ &= E[(X - \mu_X)^2] + 2E[(\mu_X - c)(X - \mu_X)] + E(\mu_X - c)^2 \\ &= \sigma_X^2 + 2(\mu_X - c)E[(X - \mu_X)] + (\mu_X - c)^2. \end{aligned}$$

Since $E[X - \mu_X] = 0$, (5.6) follows.

Applying Equation (5.6) with $c = 0$, we obtain

(5.7) $$\sigma_X^2 = E[X^2] - (\mu_X)^2,$$

which often provides a convenient shortcut for computing the variance.

Example 5.1 (continued). In Example 5.1, we considered the discrete random variable X which was the amount of cereal placed by the loading machines of a cereal company in a randomly chosen box of cereal. After the machines were overhauled, X had the distribution

x	15.8	15.9	16.0	16.1	16.2
$P\{X = x\}$	0.01	0.14	0.72	0.10	0.03

For this distribution $\mu_X = 16.0$, and

$$\begin{aligned} E(X^2) &= (15.8)^2(0.01) + (15.9)^2(0.14) + (16.0)^2(0.72) \\ &\quad + (16.1)^2(0.10) + (16.2)^2(0.03) \\ &= 256.004, \end{aligned}$$

so that

$$\sigma_X^2 = E(X^2) - \mu_X^2 = 256.004 - (16.0)^2 = 0.004.$$

Example 5.2. Suppose that the continuous random variable X has density function

$$f_X(x) = \begin{cases} \dfrac{1.4430}{x}, & \text{if } 1 \leq x \leq 2, \\ 0, & \text{if } x < 1 \text{ or } x > 2. \end{cases}$$

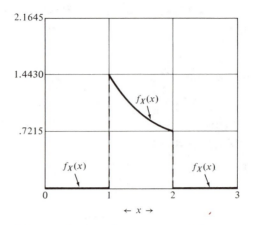

Figure 5.2: *Graph of the probability density function $f_X(x)$.*

A graph of $f_X(x)$ appears in Figure 5.2. We can compute

$$\mu_X = \int_{-\infty}^{\infty} x f_X(x)\, dx = 0 + \int_{1}^{2} x\left(\frac{1.4430}{x}\right) dx + 0 = 1.4430,$$

$$E[X^2] = \int_{-\infty}^{\infty} x^2 f_X(x)\, dx = 0 + \int_{1}^{2} x^2\left(\frac{1.4430}{x}\right) dx + 0$$

$$= \frac{1.4430}{2} x^2 \Big|_{1}^{2} = 2.1645,$$

so that, using Equation (5.7), we find that

$$\sigma_X^2 = E[X^2] - (\mu_X)^2 = 2.1645 - (1.4430)^2 = 0.08225.$$

6. HIGHER MOMENTS

As already noted, the mean μ_X of the random variable X is sometimes referred to as the *first moment* of the distribution of X. A generalization of this concept is $E[(X - c)^r]$, $r = 1, 2, \ldots$, which is referred to as the *rth moment about the point c* of the distribution of X. When $c = \mu_X$, we use the notation

(6.1) $$\mu_X^{(r)} = E[(X - \mu_X)^r], \quad r = 1, 2, \ldots$$

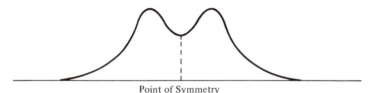

Point of Symmetry

Figure 6.1: *Graph of a symmetric probability density function.*

and call $\mu_X^{(r)}$ the rth *central moment*. Notice that the variance σ_X^2 of X is the second central moment; that is,

$$\sigma_X^2 = \mu_X^{(2)} = E[(X - \mu_X)^2].$$

Two more central moments, $\mu_X^{(3)}$ and $\mu_X^{(4)}$, are also often used as indices of certain properties of the distribution of X.

The third central moment $\mu_X^{(3)}$ is a measure of symmetry of the distribution of X about its mean. Notice that if X is measured in feet, then σ_X will also be measured in feet, while $\mu_X^{(3)}$ and σ_X^3 are expressible in units of (feet)^3. A measure of symmetry of a distribution whose magnitudes are *dimension-free* (do not depend on the units of measurement of X) is the *measure of skewness*

(6.2)
$$\gamma_1 = \frac{\mu_X^{(3)}}{\sigma_X^3}.$$

Notice that γ_1 and $\mu_X^{(3)}$ can be positive, zero, or negative, and that γ_1 and $\mu_X^{(3)}$ always agree in sign. The index γ_1 is zero when the distribution of X is symmetric about μ_X (for an example, see Figure 6.1). Negative values of γ_1 are usually found when the distribution of X is *skewed to the left* (or negatively

Figure 6.2: *Graph of a probability density function skewed to the left.*

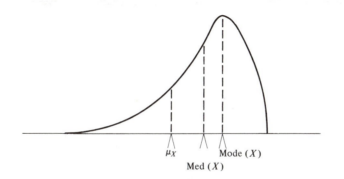

μ_X

Mode (X)

Med (X)

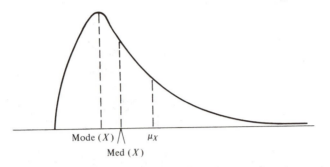

Mode (X) $\bigwedge$ μ_X
Med (X)

Figure 6.3: *Graph of a probability density function skewed to the right.*

skewed; see Figure 6.2), while γ_1 tends to be positive when the distribution of X is *skewed to the right* (positively skewed; see Figure 6.3).

The fourth central moment, $\mu_X^{(4)} = E(X - \mu_X)^4$, gives an indication of the "peakedness" or "kurtosis" of a distribution. As in the case of the measure γ_1 of symmetry, the dimension-free *measure of kurtosis* is

(6.3)
$$\gamma_2 = \frac{\mu_X^{(4)}}{\sigma_X^4}.$$

To determine the "peakedness" of a distribution using γ_2, we use the very famous *normal distribution* (Chapter 7) as a standard, since for any such distribution, $\gamma_2 = 3$. For any other kind of distribution for X, we say that this distribution is "more peaked" (or has greater kurtosis) than the normal distribution if $\gamma_2 > 3$ and that this distribution is "less peaked" (has less kurtosis)

Figure 6.4: *Graph of the probability density functions of the normal distribution* (a), *a more peaked distribution* (b), *and a less peaked distribution* (c).

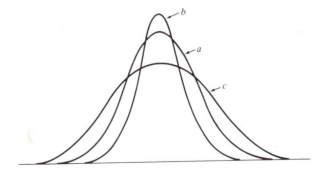

than the normal distribution if $\gamma_2 < 3$ (see Figure 6.4). Knowledge of the location, dispersion, skewness, and peakedness of a distribution is usually enough to give a coarse, but useful, picture of the distribution.

7. *INEQUALITIES FOR PROBABILITIES*

In the previous sections, various descriptive characteristics of distributions were discussed. Now suppose that we are given the values of some of these descriptive measures but are not informed as to the exact distribution which these values describe. How much of the probability distribution can we reproduce with only this information?

If only the mean $\mu_X = \mu$ of the distribution is given to us, we cannot in general make any nontrivial statements about the probability distribution of X. However, if we also know that $X \geq 0$ with probability equal to one (and that $P\{X = 0\} \neq 1$, so that $\mu_X > 0$), then we can obtain information about probabilities of certain events in the form of the following *Markov Inequality*, named after the Russian mathematician A. A. Markov (1856–1922), who first obtained the result.

Markov Inequality. *If X is a nonnegative random variable, and if $\mu_X = \mu$ is positive and finite, then for any positive constant c we have*

(7.1)
$$P\{X \geq c\mu\} \leq \frac{1}{c}.$$

To prove this result, let

$$I(x) = \begin{cases} 1, & \text{if } x \geq c\mu, \\ 0, & \text{otherwise} \end{cases}$$

be the indicator function of the event $\{X \geq c\mu\}$. From the remark at the end of Section 4, $E[I(X)] = P\{X \geq c\mu\}$. Note also that for $x \geq 0$,

$$h(x) = x - c\mu I(x) \geq 0.$$

Hence, since $P\{X \geq 0\} = 1$, we have $h(X) \geq 0$ with probability 1. Thus, by Properties 4.1 and 4.2 of expected values,

(7.2)
$$0 \leq E[h(X)] = E[X] - c\mu E[I(X)]$$
$$= \mu - c\mu P\{X \geq c\mu\}.$$

Since $c > 0$ and $\mu > 0$, the inequality (7.1) follows directly from (7.2).

Use of the Markov Inequality

Inequality (7.1) is of no value if $1/c$ exceeds 1, because $P\{X \geq c\mu\} \leq 1$ always. However, for c greater than 1, the information supplied by inequality (7.1) is useful. Table 7.1 shows the largest possible probabilities of the event $\{X \geq c\mu\}$ for various values of c (see Figure 7.1).

Example 7.1. A car rental company advertises that the mean time that a customer must wait to pick up a car is $\mu = 5$ minutes. Let X be the random variable that is the waiting time (in minutes) for a customer; then the company's announcement says that $\mu_X = \mu = 5$ minutes. A salesman in a hurry estimates that he can afford to wait no more than 60 minutes for a rental car. What is the probability that he will be delayed more than 60 minutes? Since $60 = 12(5) = 12\mu$, the value of c is 12. It follows from inequality (7.1) that regardless of the distribution of waiting times, the probability that the salesman will be delayed longer than 60 minutes waiting for a car cannot be larger than $1/c = \frac{1}{12} = 0.083$.

The Markov Inequality requires that X be nonnegative. However, not all random variables are nonnegative. Furthermore, we may not always be interested in events of the form $\{X \geq b\}$. Sometimes events of the form $\{X \leq a$ or $X \geq b\}$ are of greater interest to us. If we know both the mean μ and the

Table 7.1: *Largest Possible Values of the Probability $P\{X \geq c\mu\}$ in the Tail of the Distribution of a Nonnegative Random Variable X*

c	Largest possible value, $1/c$, of $P\{X \geq c\mu\}$
1.0	1.000
1.5	0.667
2.0	0.50
2.5	0.40
3.0	0.333
3.5	0.286
4.0	0.25
4.5	0.222
5.0	0.20
10.0	0.10
30.0	0.033
50.0	0.02
100.0	0.01

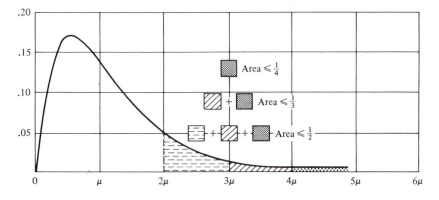

Figure 7.1: *Inequality on area based on the value of μ.*

variance σ^2 of a random variable X, then we can derive an inequality for the probability of certain events regardless of whether or not X takes on only nonnegative values.

Bienaymé–Chebyshev Inequality. *For any number $d > 0$,*

(7.3) $$P\{X \leq \mu - d\sigma \text{ or } X \geq \mu + d\sigma\} \leq \frac{1}{d^2}.$$

Inequality (7.3) is named after the French mathematician I. J. Bienaymé (1796–1878) and the Russian mathematician P. L. Chebyshev (Čebyšev) (1821–1894), both of whom independently discovered the result. We may prove this inequality in a manner similar to the proof of inequality (7.1). However, inequality (7.3) may also be obtained as a consequence of inequality (7.1) as follows. Note that $Y = (X - \mu)^2$ is a nonnegative random variable with mean

$$\mu_Y = E(X - \mu)^2 = \sigma^2$$

and that $Y \geq d^2\sigma^2$ if and only if $(X - \mu)^2 \geq d^2\sigma^2$, which in turn is true if and only if $X \leq \mu - d\sigma$ or $X \geq \mu + d\sigma$. Thus, applying inequality (7.1) to Y with $c = d^2$,

$$P\{X \leq \mu - d\sigma \text{ or } X \geq \mu + d\sigma\} = P\{Y \geq d^2\sigma^2\} \leq \frac{1}{d^2},$$

and this establishes inequality (7.3).

Use of the Bienaymé–Chebyshev Inequality

If $d \leq 1$, inequality (7.3) furnishes no nontrivial information about the distribution of X. If $d > 1$, then we gain some useful information about probabilities of events of the form $\{X \leq \mu - d\sigma \text{ or } X \geq \mu + d\sigma\}$. In Table 7.2, the

Table 7.2: Largest Possible Values of $P\{X \leq \mu - d\sigma$ or $X \geq \mu + d\sigma\}$ and Smallest Possible Values of $P\{\mu - d\sigma < X < \mu + d\sigma\}$

d	Largest possible probability that either $X \leq \mu - d\sigma$ or $X \geq \mu + d\sigma$	Smallest possible probability that $\mu - d\sigma < X < \mu + d\sigma$
1	1.000	0.000
1.5	0.444	0.556
2	0.250	0.750
2.5	0.160	0.840
3	0.111	0.889
3.5	0.082	0.918
4	0.063	0.937
4.5	0.049	0.951
5	0.040	0.960
10	0.010	0.990
30	0.0011	0.9989
50	0.0004	0.9996
100	0.0001	0.9999

largest value that the probability $P\{X \leq \mu - d\sigma$ or $X \geq \mu + d\sigma\}$ can possibly be is tabulated for various values of $d \geq 1$. Note that for any event E, $P(E) \leq p$ implies that $P(E^c) \geq 1 - p$. Hence, from inequality (7.3) we obtain the equivalent inequality

$$(7.4) \qquad\qquad P\{\mu - d\sigma < X < \mu + d\sigma\} \geq 1 - \frac{1}{d^2},$$

since $\{X \leq \mu - d\sigma$ or $X \geq \mu + d\sigma\}^c = \{\mu - d\sigma < X < \mu + d\sigma\}$. In Table 7.2, the smallest value that $P\{\mu - d\sigma < X < \mu + d\sigma\}$ can have is tabulated for various values of $d \geq 1$. Thus, we know almost certainly that X will be within 50 standard deviations of its mean. Indeed, the huge bulk of *any* distribution (over 95%) lies within $4\frac{1}{2}$ standard deviations of the mean.

Figure 7.2 illustrates this bound for $d = 1, 2,$ and 3 and X a continuous random variable.

Remark. Since the event $\{\mu - d\sigma \leq X \leq \mu + d\sigma\}$ includes the event $\{\mu - d\sigma < X < \mu + d\sigma\}$, the inequality

$$P\{\mu - d\sigma \leq X \leq \mu + d\sigma\} \geq 1 - \frac{1}{d^2}$$

holds for all random variables X that have mean μ and variance σ^2.

Example 7.2. A multiple-choice reading comprehension test (in which incorrect answers receive negative scores) has been constructed by a psychologist

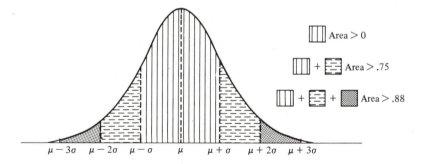

Figure 7.2: *Inequality on area based on the values of μ and σ.*

in such a way that the score X of a randomly chosen high school senior has mean $\mu_X = 0$ and variance $\sigma_X^2 = 100$. What can we say about a student who makes a score of 30? Using inequality (7.3) or Table 7.2,

$$P\{X \geq 30\} \leq P\{X \leq -30 \text{ or } X \geq 30\}$$
$$= P\{X \leq \mu_X - 3\sigma_X \text{ or } X \geq \mu_X + 3\sigma_X\}$$
$$\leq \frac{1}{(3)^2} = 0.111.$$

Thus, the student has a reading ability that is exceeded by at most 11.1% of the high school senior population.

Other Probability Inequalities

There are other inequalities similar to inequalities (7.3) and (7.4) in that they make use of the values of μ and σ^2. As an example, when X is a nonnegative random variable, and when $\mu_X = \mu$, $\sigma_X^2 = \sigma^2$, then for any $b > 0$,

(7.5)
$$P\{X \geq \mu + b\sigma\} \leq \frac{1}{1 + b^2}.$$

We can prove (7.5) in a manner similar to the proof of (7.1). Let

$$I(X) = \begin{cases} 1, & \text{if } X \geq \mu + b\sigma, \\ 0, & \text{otherwise,} \end{cases}$$

and

$$g(X) = \frac{[(X - \mu)b + \sigma]^2}{\sigma^2(1 + b^2)^2}.$$

Recall that $E[I(X)] = P\{X \geq \mu + b\sigma\}$, and note that $E[(X - \mu)b + \sigma]^2 = \sigma^2(b^2 + 1)$. Then

$$E[g(X)] = \frac{E[(X - \mu)b + \sigma]^2}{\sigma^2(1 + b^2)^2} = \frac{\sigma^2(b^2 + 1)}{\sigma^2(1 + b^2)^2} = \frac{1}{1 + b^2}.$$

Also, note that

$$h(X) = g(X) - I(X) \geq 0, \quad \text{if } X \geq 0,$$

so that $P\{h(X) \geq 0\} = 1$. From Properties 4.1 and 4.2,

$$0 \leq E[h(X)] = E[g(X)] - E[I(X)]$$

$$= \frac{1}{1 + b^2} - P\{X \geq \mu + b\sigma\},$$

which yields inequality (7.5). Table 7.3 gives values of the largest possible probability that X equals or exceeds $\mu + b\sigma$ for various values of b.

Since inequalities (7.1) and (7.5) are similar in form, it is of interest to compare Tables 7.1 and 7.3. To make this comparison, we need to set $c\mu = \mu + b\sigma$. Note that since $X \geq 0$ with probability 1, we have $\mu_X = \mu \geq 0$. Thus, let $k = \sigma/\mu$ and $c = 1 + bk$. For any value of k, this establishes a correspondence between the value of c in Table 7.1 and the value of b in Table 7.3.

Table 7.3: *Largest Possible Values of the Probability $P\{X \geq \mu + b\sigma\}$ in the Tail of the Distribution of a Nonnegative Random Variable X*

b	Largest possible value, $1/(1 + b^2)$, of $P\{X \geq \mu + b\sigma\}$
0.5	0.800
1	0.500
1.5	0.308
2	0.200
2.5	0.138
3	0.100
3.5	0.075
4	0.059
4.5	0.047
5	0.039
10	0.010
30	0.001
50	0.0004
100	0.0001

For example, if $k = 1$ and $b = 3$, then $c = 4$. From Table 7.3, we obtain

$$P\{X \geq \mu + 3\sigma\} = P\{X \geq 4\mu\} \leq 0.100,$$

while Table 7.1 gives the less precise result $P\{X \geq 4\mu\} \leq 0.250$. If $k = 3$, $b = 3$, then $c = 10$, and both tables yield $P\{X \geq 10\mu\} \leq 0.10$. Finally, suppose that $k = \frac{1}{2}$ and $b = 3$, so that $c = 1 + 3(\frac{1}{2}) = 2.5$. From Table 7.1, we obtain the bound $P\{X \geq 2.5\mu\} \leq 0.40$, while Table 7.3 yields the more precise bound

$$P\{X \geq \mu + 3(\tfrac{1}{2})\mu\} = P\{X \geq 2.5\mu\} \leq 0.10.$$

In general, Table 7.3 tells us more about probabilities of events of the form $\{X \geq 0\}$ than does Table 7.1. This fact is not surprising, since to use Table 7.3 we need to have more information (namely, the value of the variance σ^2 of X) than we do to use Table 7.1. *The more information that we have concerning a distribution, the better we are able to bound the probabilities of events concerning random variables which have that distribution.*

8. APPENDIX: ESTIMATION OF THE MEAN AND VARIANCE

In Chapters 6, 7, and 8, where we discuss various special probability distributions for random variables, we need a way of estimating the mean μ_X and variance σ_X^2 of a random variable X in situations where we do not know the exact distribution of X, but have available observations on X taken from N probabilistically independent trials of the random experiment. In the present appendix, we indicate how observations upon X can be used to estimate the values of μ_X and σ_X^2.

Suppose that we have N probabilistically independent observations t_1, $t_2, \ldots, t_N$ on X. In Section 2, we remarked that when N, the number of observations on X, is large, then μ_X should be approximately equal to the average

(8.1) $$\bar{t} = \frac{\sum_{i=1}^{N} t_i}{N} = \sum_t [t \times \text{r.f.}\{X = t\}]$$

of the observations $t_1, t_2, \ldots, t_N$. This suggests that we should estimate μ_X by the estimator

(8.2) $$\hat{\mu}_X = \bar{t}$$

regardless of how many observations on X we have available.

Similarly, in Section 5 we remarked that when N is large, the variance $\sigma_X^2 = E[(X - \mu_X)^2]$ will be approximately equal to the average $(1/N) \sum_{i=1}^{N} (t_i - \mu_X)^2$ of the squared distances of the observations from the mean μ_X. Since when N is large, $\hat{\mu}_X$ should be close to μ_X, this suggests estimating σ_X^2 by

(8.3) $$\hat{\sigma}_X^2 = \frac{1}{N} \sum_{i=1}^{N} (t_i - \hat{\mu}_X)^2 = \frac{1}{N} \sum_{i=1}^{N} (t_i - \bar{t})^2,$$

or equivalently by

(8.4) $$\hat{\sigma}_X^2 = \sum_t (t - \bar{t})^2 \; \text{r.f.}\{X = t\},$$

even when the number N of observations is not large.

Remark. A notational convention in the statistical literature, which we have adopted here, is to denote an estimate of an unknown quantity, say θ, by the symbol $\hat{\theta}$. Thus, $\hat{\mu}_X$ estimates μ_X, and $\hat{\sigma}_X^2$ estimates σ_X^2.

Example 8.1. If we obtain the observations $t_1 = 2$, $t_2 = 1$, $t_3 = 1$, $t_4 = 2$, $t_5 = 3$, $t_6 = 4$, $t_7 = 5$, $t_8 = 9$, $t_9 = 1$, $t_{10} = 1$, then

$$\hat{\mu}_X = \frac{2 + 1 + 1 + 2 + 3 + 4 + 5 + 9 + 1 + 1}{10} = \frac{29}{10} = 2.9.$$

Alternatively, note that by grouping similar values together in the sum above, we would obtain

$$\hat{\mu}_X = \frac{1 + 1 + 1 + 1 + 2 + 2 + 3 + 4 + 5 + 9}{10}$$

$$= (1)\left(\frac{4}{10}\right) + (2)\left(\frac{2}{10}\right) + (3)\left(\frac{1}{10}\right) + (4)\left(\frac{1}{10}\right) + (5)\left(\frac{1}{10}\right) + (9)\left(\frac{1}{10}\right)$$

$$= \sum_t [t \times \text{r.f.}\{X = t\}].$$

We remark that an argument of this sort will show that in general, for any function $g(\cdot)$,

(8.5) $$\frac{1}{N} \sum_{i=1}^{N} g(t_i) = \sum_t [g(t) \times \text{r.f.}\{X = t\}],$$

thus showing, for example, the equivalence of the formulas (8.3) and (8.4) in the case $g(x) = (x - \bar{t})^2$. Computing $\hat{\sigma}_X^2$ by (8.3), we obtain

$$\hat{\sigma}_X^2 = \tfrac{1}{10}[(2 - 2.9)^2 + (1 - 2.9)^2 + (1 - 2.9)^2 + \cdots + (1 - 2.9)^2] = 5.89.$$

Alternatively, if we had constructed an empirical probability mass function for our observations such as Table 8.1, we would use (8.4) to compute $\hat{\sigma}_X^2$, obtaining

$$\hat{\sigma}_X^2 = (1 - 2.9)^2(\tfrac{4}{10}) + (2 - 2.9)^2(\tfrac{2}{10}) + \cdots + (9 - 2.9)^2(\tfrac{1}{10}) = 5.89.$$

Table 8.1: *Empirical Probability Mass Function Based on the Observations $t_1 = 2$, $t_2 = 4$, etc.*

t	r.f.$\{X = t\}$
1	$\frac{4}{10}$
2	$\frac{2}{10}$
3	$\frac{1}{10}$
4	$\frac{1}{10}$
5	$\frac{1}{10}$
9	$\frac{1}{10}$

The original observations which we obtain on X are called the *raw data* for X, and Equations (8.1) through (8.4) give ways of computing $\hat{\mu}_X$ and $\hat{\sigma}_X^2$ from the raw data. However, it is often the case that the result of taking N observations upon X are reported to us in the form of Table 8.2, where the data have already been grouped into k intervals of equal length. (For example, such preliminary grouping would be done in constructing modified relative frequency histograms for the observations upon a continous random variable X; see Section 5 of Chapter 4.) In this case, the calculations for $\hat{\mu}_X$ and $\hat{\sigma}_X^2$ are done as if all observations in an interval were equal to the midpoint of that interval. That is,

$$(8.6) \quad \hat{\mu}_X = \sum_{i=1}^{k} (\text{midpoint of interval } i)(\text{relative frequency of interval } i)$$

and

$$(8.7) \quad \hat{\sigma}_X^2 = \sum_{i=1}^{k} (\text{midpoint} - \hat{\mu}_X)^2(\text{relative frequency of interval } i).$$

Table 8.2: *Relative Frequency Distribution Based on $N = 40$ Observations of the Random Variable X*

Midpoint x of interval of values	Observed relative frequency of the event $\{x - 0.5 \le X < x + 0.5)$
1.0	$\frac{1}{40}$
2.0	$\frac{2}{40}$
3.0	$\frac{4}{40}$
4.0	$\frac{7}{40}$
5.0	$\frac{12}{40}$
6.0	$\frac{8}{40}$
7.0	$\frac{5}{40}$
8.0	$\frac{1}{40}$

For example, the computations for $\hat{\mu}_X$ and $\hat{\sigma}_X^2$ based on the observations in Table 8.2 are

$$\hat{\mu}_X = (1.0)(\tfrac{1}{40}) + (2.0)(\tfrac{2}{40}) + (3.0)(\tfrac{4}{40}) + (4.0)(\tfrac{7}{40}) + (5.0)(\tfrac{12}{40})$$
$$+ (6.0)(\tfrac{8}{40}) + (7.0)(\tfrac{5}{40}) + (8.0)(\tfrac{1}{40})$$
$$= \tfrac{196}{40} = 4.90$$

and

$$\hat{\sigma}_X^2 = (1.0 - 4.90)^2(\tfrac{1}{40}) + (2.0 - 4.90)^2(\tfrac{2}{40}) + \cdots + (8.0 - 4.90)^2(\tfrac{1}{40})$$
$$= \frac{87.9192}{40} = 2.20.$$

Note that the rule (8.6) for calculating $\hat{\mu}_X$ from data grouped in intervals is almost certain to produce a different value for $\hat{\mu}_X$ than would have been obtained if $\hat{\mu}_X$ had been calculated from the raw data, since it is unlikely that every observation in an interval is equal to the midpoint of that interval. However, the error made by assuming that observations in an interval fall at the midpoint of the interval is, for each observation, never greater than half the common length l of each such interval. It follows that the discrepancy between $\hat{\mu}_X$, as calculated from the raw data, and $\hat{\mu}_X$ as calculated from the grouped data *can never be greater than* $l/2$. Hence, the smaller is the common length l of each interval used to construct the relative frequency histogram, the smaller is the discrepancy between the two estimates of μ_X.

EXERCISES

1. Let the random variable X have the following probability model:

x	0	1	2
Probability	$\tfrac{1}{3}$	$\tfrac{1}{2}$	$\tfrac{1}{6}$

(a) Find the mean, median, and mode of X. Which of these measures of location are possible values of X?

(b) Find the variance of X using both Equation (5.1) and also Equation (5.7). Verify that the results obtained by the two formulas are the same.

2. An integer is chosen at random from among the integers 1, 2, 3, 4, 5, 6, 7, 8, 9, and 10. Let X be the number of different positive integers that divide the chosen integer without remainder. (For example, if the integer 6 is chosen, then $X = 4$ since 6 can be divided without remainder by 1, 2, 3, and 6.)

(a) The possible values of X are 1, 2, 3, and 4. Find the probability mass function $p_X(x)$ of X.

(b) Find the mean, median, and mode of the distribution of X.

(c) If you were to play the "Guess the Number of Divisors Game," in which your opponent chooses an integer at random from among the integers 1, 2, ..., 9, 10, and you have to guess the number X of divisors that this integer has, which value would you guess for X? What is the probability that your guess would be exactly correct?

(d) Find the variance of X.

3. An insurance company wishes to establish the premium C for selling a 1-year term insurance policy of \$1000 to a man of age 45. Suppose that the probability that a 45-year-old man will die during the next year is 0.00363. Let the random variable X denote the gain (or loss) to the insurance company as a result of selling a 1-year term insurance policy of \$1000 to a 45-year-old man. Thus, the probability model for X is

x	C	$C - 1000$
Probability	0.99637	0.00363

(a) Determine C so that the company's expected gain is equal to 0.

(b) What premiums allow the company to make a profit "on the average"?

(c) If the company sets a premium of \$5 for every \$1000 of term insurance and sells \$500 worth of this term insurance to 45-year-old men (i.e., 100 \$1000 policies), approximately what total profit (or loss) will they receive?

4. A university loan office handles a special government loan program that awards emergency loans of \$200 to students. Past records show that 60% of all students completely repay the loan. Of those students who fail to repay all of their loans, 40% pay back $\frac{3}{4}$ of the loan, 30% pay back $\frac{1}{2}$ the loan, 20% pay back $\frac{1}{4}$ of the loan, and 10% fail to repay any of the loan.

(a) What is the mean loss (default) per loan for the loans under this special government program?

(b) The university has made loans to 1000 students under this program. Approximately what *total* amount of money can they expect to have defaulted on these loans?

5. In the problem on "matches" or "rencontres" (Exercise 14 of Chapter 4), we obtained the probability mass functions for the number X_N of matches among N pairs of identical tickets. For $N = 2$, 3, and 4, use these probability mass functions to find the mean and variance of X_N.

For convenience, the needed probability mass functions are as follows:

X_2:

x	0	1	2
Probability	$\frac{1}{2}$	0	$\frac{1}{2}$

X_3:

x	0	1	2	3
Probability	$\frac{2}{6}$	$\frac{3}{6}$	0	$\frac{1}{6}$

X_4:

x	0	1	2	3	4
Probability	$\frac{9}{24}$	$\frac{8}{24}$	$\frac{6}{24}$	0	$\frac{1}{24}$

6.   An experiment consists in selecting 2 balls (without replacement) from an urn that contains 2 white and 3 red balls, and then selecting 3 more balls (without replacement) from a second urn, which contains 3 white and 2 red balls. If X denotes the number of white balls among the 5 balls selected from the two urns, calculate the mean of X.

7.   Plates of cells are observed and the number of cells in the process of reproduction are counted. Suppose that each plate has 5 cells on it. Let X be the number of cells observed in reproduction on a given plate. The following is a possible probability model for this experiment.

x	0	1	2	3	4	5
Probability of x cells	$\frac{1}{243}$	$\frac{10}{243}$	$\frac{40}{243}$	$\frac{80}{243}$	$\frac{80}{243}$	$\frac{32}{243}$

Find the mean and variance of the random variable X.

8.   Let X be the yearly income (in thousands of dollars) of a person who earns more than K thousand dollars and who pays some federal income tax. The density function $f_X(x)$ of the random variable X is assumed to be of the form

$$f_X(x) = \begin{cases} aK^a/x^{a+1}, & \text{for } x \geq K, \\ 0, & \text{for } x < K. \end{cases}$$

Suppose that $K = 2$ and $a = 3$.
(a) Graph the probability density function $f_X(x)$.
(b) Find the mean, median, and mode of X.
(c) Find $Q_{0.25}(X)$ and $Q_{0.75}(X)$.
(d) What measure of location would you use to describe yearly income if you were interested in predicting tax revenues? What measure of location would you use if you wished to describe the economic

well-being of the typical American who earns more than $2000 a
year?

(e) Find the variance of X.

[Note: The distribution of X is a special case of a *Pareto distribution*. This application of Pareto distributions is discussed by Hagstroem (1960).]

9. Let X be a continuous random variable with a probability density function given by

$$f_X(x) = \begin{cases} \frac{3}{8}(2-x)^2, & \text{for } 0 \le x \le 2, \\ 0, & \text{for } x < 0 \text{ or } x > 2. \end{cases}$$

(a) Find the mean μ_X of X.

(b) Find the variance σ_X^2 of X.

(c) Show that the cumulative distribution function of X is given by

$$F_X(x) = \begin{cases} 0, & \text{if } x < 0, \\ 1 - (2-x)^3/8, & \text{if } 0 \le x \le 2, \\ 1, & \text{if } 2 < x. \end{cases}$$

Then find the median of X and $Q_{0.90}(X)$.

10. Suppose that the continuous random variable X satisfies
$P\{0 \le X \le a\} = 1$ for some number a, $0 < a < \infty$. (That is, X is a nonnegative and bounded random variable.) Let $F_X(x)$ be the cumulative
distribution function of X. Use integration by parts to show that

$$\mu_X = \int_0^a [1 - F_X(x)] \, dx.$$

Apply this result to obtain μ_X in Exercise 9.

11. (a) Extend the result of Exercise 10 to the case where
$P\{0 \le X < \infty\} = 1$ and $E(X) < \infty$. That is, prove that

$$\mu_X = E(X) = \int_0^\infty [1 - F_X(x)] \, dx$$

by making use of integration by parts. What use do you make of the
fact that $E(X) < \infty$?

(b) Apply the result of part (a) to obtain μ_X in Exercise 8.

(c) If X is any continuous random variable for which $E(|X|) < \infty$,
show that

$$\mu_X = E(X) = -\int_{-\infty}^0 F_X(x) \, dx + \int_0^\infty [1 - F_X(x)] \, dx.$$

[Hint: Note that $E(X) = \int_{-\infty}^0 x f_X(x) \, dx + \int_0^\infty x f_X(x) \, dx$. Handle each
integral separately using integration by parts.]

12. If $Y = 3X - 5$ is a transformation from the random variable X to the new random variable Y, and if $\mu_Y = 0$, $\sigma_Y^2 = 9$, find μ_X and σ_X^2.

13. If X is a random variable with $\mu_X = -1$, $\sigma_X^2 = 4$, find a and b so that $Y = aX + b$ has mean 0 and variance 1.

14. When $g(x) = ax + c$, $a > 0$, we have seen that

$$\mu_{g(X)} = \mu_{aX+c} = a\mu_X + c = g(\mu_X).$$

However, the result $\mu_{g(X)} = g(\mu_X)$ is not generally true. To see this, let X be a continuous random variable with probability density function

$$f_X(x) = \begin{cases} 1, & \text{if } 0 \leq x \leq 1, \\ 0, & \text{if } x < 0 \text{ or } x > 1, \end{cases}$$

and let $g(x) = e^x$. Find $\mu_{ex} = E[e^x]$ and show that it is not equal to $e^{\mu_X} = e^{1/2}$.

15. In contrast to what you showed in Exercise 14, show that for any random variable X with a unique pth quantile $Q_p(X)$, and for any strictly increasing function $g(x)$ it is the case that

$$Q_p(g(X)) = g(Q_p(X)).$$

[Hint: Use the fact that $g(x)$ has an inverse function $h(y)$ that is nondecreasing, $h(g(x)) = x$, and apply the definition of $Q_p(g(X))$.]

16. Let X be a random variable and let $g_1(x)$, $g_2(x)$ be two functions of x for which $g_1(x) \geq g_2(x)$, all x. Use the properties of the expected value computation to show that

$$E[g_1(X)] \geq E[g_2(X)].$$

17. Use the properties of the expected value computation to prove each of the following equalities:
(a) $E[X^2] = E[X(X-1)] + E(X) = \sigma_X^2 + \mu_X^2$.
(b) $E(X - \mu_X)^3 = E(X^3) - 3\sigma_X^2\mu_X - \mu_X^3$.
(c) $E[(X-2)(X-1)X] = E[X^3] - 3E[X(X-1)] - E(X)$.
(d) $E(X - \mu_X)^4 = E(X^4) - 4\mu_X E(X - \mu_X)^3 - 6\sigma_X^2\mu_X^2 - \mu_X^4$.
(e) $E[(X-3)(X-2)(X-1)X] = EX^4 - 6E[X(X-1)(X-2)]$
$\qquad\qquad\qquad\qquad\qquad - 7E[X(X-1)] - E(X)$.

18. For each of the following distributions find $\mu_X^{(3)}$, $\mu_X^{(4)}$, γ_1, and γ_2. Interpret your findings in terms of the skewness and kurtosis (peakedness) of the distribution, and compare your conclusions to a rough graph of the probability density function.

(a) $f_X(x) = \begin{cases} 2x, & \text{if } 0 \leq x \leq 1, \\ 0, & \text{if } x < 0 \text{ or } x > 1. \end{cases}$

(b) $f_X(x) = \begin{cases} 2(1-x), & \text{if } 0 \leq x \leq 1, \\ 0, & \text{if } x < 0 \text{ or } x > 1. \end{cases}$

(c) $f_X(x) = \begin{cases} 1, & \text{if } 0 \le x \le 1, \\ 0, & \text{if } x < 0 \text{ or } x > 1. \end{cases}$

(d) $f_X(x) = \begin{cases} 4x, & \text{if } 0 \le x \le \frac{1}{2}, \\ 4(1-x), & \text{if } \frac{1}{2} < x \le 1, \\ 0, & \text{otherwise.} \end{cases}$

19. Scores X on the verbal scale of the Scholastic Aptitude Test (SAT Verbal) are nonnegative and have a mean $\mu_X = 650$.

(a) Without making any other assumptions about the distribution of SAT Verbal scores X, find the largest possible values of the following probabilities: (i) $P\{X \ge 715\}$, and (ii) $P\{X \ge 780\}$.

(b) If someone claims that 30% of all SAT Verbal scores exceed 780, would you believe their claim?

(c) Suppose that we also know that the standard deviation σ_X of X is 64. Find the largest possible value of $P\{X \ge 780\}$. Do you now believe the claim that 30% of all SAT Verbal scores X exceed 780? Would you believe the claim that 80% of all SAT Verbal scores X exceed 715?

(d) Use the Bienaymé–Chebyshev Inequalities to find the smallest values for $P\{650 - 64d < X < 650 + 64d\}$ for $d = 1.1, 1.5, 1.8, 2.0$, and 2.2, and the largest values for $P\{X \ge 650 + 64b\}$ for $b = 0.5, 0.7, 1.0, 1.2$, and 1.5.

20. Suppose that X is a discrete random variable with the following probability mass function:

x	$\mu - k\sigma$	μ	$\mu + k\sigma$
Probability	$\dfrac{1}{2k^2}$	$1 - \dfrac{1}{k^2}$	$\dfrac{1}{2k^2}$

where μ is any number, σ is any positive number, and $k \ge 1$. Show that:
(a) the mean of X is μ;
(b) the variance of X is σ^2;
(c) $P\{X \ge \mu + k\sigma \text{ or } X \le \mu - k\sigma\} = 1/k^2$.

21. Studies have been made by quantitative linguists of the syllabication of words by various authors. For example, let X be the number of syllables in a word chosen at random from the works of the English author and lexicographer Samuel Johnson. In 2000 such words, the following sample probability mass function was obtained for X:

Number x of syllables	1	2	3	4	5	6
Observed relative frequency of the event $\{X = x\}$	$\frac{1268}{2000}$	$\frac{423}{2000}$	$\frac{195}{2000}$	$\frac{77}{2000}$	$\frac{29}{2000}$	$\frac{8}{2000}$

(a) Find the sample mean $\hat{\mu}_X$ and sample variance $\hat{\sigma}_X^2$ of X.
(b) The number Y of syllables used in a random word taken from the works of an American author, A., had a population mean $\mu_Y = 2.1$ and a population variance $\sigma_Y^2 = 0.75$. Assume that the sample mean and variance of X found in part (a) equal the population mean μ_X and variance σ_X^2 of X, respectively. Which author, Samuel Johnson or A., used more syllables on the average? Which author used a greater variety of word lengths (as measured in terms of the number of syllables)?

22. If we have observations $t_1, t_2, \ldots, t_N$ for X, show that

$$\hat{\sigma}_X^2 = \left(\frac{1}{N} \sum_{i=1}^{N} t_i^2\right) - (\hat{\mu}_X)^2$$

and also that

$$\hat{\sigma}_X^2 = \sum_t (t^2)(\text{r.f.}\{X = t\}) - (\hat{\mu}_X)^2.$$

Use these new formulas to recompute the estimates of $\hat{\sigma}_X^2$ based on Tables 8.1 and 8.2 of Section 8. [Hint: The new formulas for $\hat{\sigma}_X^2$ are analogs of Equation (5.7).]

6

Special Distributions:

Discrete Case

In any analysis of quantitative data, it is a major step when we can specify the form of the underlying probability distributions of those random variables that are of interest. Fortunately, there are certain basic probability distributions that are applicable in many diverse contexts and repeatedly arise in practice. In this and in the next two chapters, we describe some frequently encountered discrete and continuous distributions.

1. BERNOULLI AND BINOMIAL DISTRIBUTIONS

Whenever a random experiment has two possible outcomes, we have what is called a Bernoulli trial [after Jacob (James) Bernoulli, 1654–1705]. For example, we have a Bernoulli trial when we are interested in whether a student's grade is pass or fail, whether a person has or has not contracted a given disease, whether a stock price rises or does not rise, and so on. Because there are two possible outcomes in a Bernoulli trial, we might, somewhat arbitrarily, call one outcome a "success" and the other outcome a "failure." On the other hand, we might perhaps call one outcome a "head" and the other outcome a "tail." The choice, as we have said, is arbitrary. Since the terminology "success" and "failure" has meaning in many contexts where Bernoulli trials occur, we adopt it in the present section.

Bernoulli Distribution

We can define a random variable associated with every Bernoulli trial as follows. If the outcome ω is a success, set $X(\omega) = 1$: whereas if the outcome ω is a failure, set $X(\omega) = 0$. Consequently, we have the probability model

(1.1)

x	0	1
$p_X(x)$	$1 - p$	p

where p is the probability of success. Any random variable having this probability mass function for some probability p is said to be a *Bernoulli random variable*, and the distribution (1.1) is called a *Bernoulli distribution*. The form of the probability mass function (1.1) depends upon the constant p, which is called the *parameter* of the Bernoulli distribution.

Remark. The choice of values for the random variable X is as arbitrary as our choice of names for the two outcomes of a Bernoulli trial. Any two unequal numbers a and b could be assigned to the outcomes success and failure, respectively, so as to provide a numerical indicator of when a success (or failure) occurs. The advantage of the choice $a = 1$, $b = 0$ is that in repeated observations of X the sample average $\hat{\mu}_X$ of the observations is equal to the relative frequency of a success.

If X has a Bernoulli distribution with parameter p, then the mean and variance of X are

$$\mu_X = (0)(1 - p) + (1)(p) = p,$$
$$\sigma_X^2 = (0 - p)^2(1 - p) + (1 - p)^2 p = p(1 - p).$$

Binomial Distribution

If we make n repeated trials of the random experiment that gives rise to a Bernoulli distribution, it is natural to consider the random variable Z that records the *number of successes* in the n trials. This random variable Z can then assume any of the possible values $0, 1, 2, \ldots, n$.

For example, if $n = 3$, we may observe any one of the following eight outcomes: (S, S, S), (S, S, F), (S, F, S), (F, S, S), (F, F, S), (F, S, F), (S, F, F), (F, F, F), where S denotes occurrence of a success and F denotes occurrence of a failure. Corresponding respectively to these outcomes are the following values of Z: 3, 2, 2, 2, 1, 1, 1, 0. To find the appropriate probability model for Z, we first find probabilities for the simple results. Let E_j be the event that a success occurs on the jth trial, $j = 1, 2, 3$. Since each trial is a Bernoulli trial with probability p of success and $1 - p$ of failure, we have

$$P(E_j) = p, \qquad P(E_j^c) = 1 - p.$$

Further, since the $n = 3$ trials are independent, probabilities of the simple results of the experiment can be calculated as shown in Table 1.1. From Table 1.1, we obtain the following probability model for Z:

z	0	1	2	3
$P\{Z = z\}$	$(1 - p)^3$	$3p(1 - p)^2$	$3p^2(1 - p)$	p^3

Notice from Table 1.1 that all simple events that are associated with the same value of Z have the same probability. For example, the simple events $\{(S, S, F)\}$, $\{(S, F, S)\}$, and $\{(F, S, S)\}$ all correspond to the value $Z = 2$ and all have the same probability, $p^2(1 - p)$. Hence, we could obtain $P\{Z = 2\}$ by multiplying the common probability $p^2(1 - p)$ of the simple events associated with the value $Z = 2$ by the number $C(3, 2) = 3$ of such events.

In general, if we take n independent Bernoulli trials and count the number Z of successes, then we can find the appropriate probability model for Z as follows. First, we note that the event $\{Z = k\}$ is the same as the event that there are k successes and $(n - k)$ failures. Any simple event corresponding to outcomes in which k S's and $n - k$ F's appear has, because of independence, a probability of the form $p^k(1 - p)^{n-k}$. The probability of the event $\{Z = k\}$ is now found by multiplying this common probability $p^k(1 - p)^{n-k}$ of all such simple events by the number $C(n, k)$ of such simple events. However, the number $C(n, k)$ is equal to the number of distinguishable ways of ordering n items consisting of k S's and $n - k$ F's; that is,

$$C(n, k) = \binom{n}{k}.$$

Consequently, the probability mass function, $p_Z(k)$, for Z is given by the equation

(1.2) $\quad p_Z(k) = P\{Z = k\} = \binom{n}{k} p^k(1 - p)^{n-k}, \quad k = 0, 1, 2, \ldots, n.$

Table 1.1: *Calculating Probabilities of Simple Events for an Experiment consisting of $n = 3$ Independent Bernoulli Trails*

Simple event	Probability	Value of Z
$\{(S, S, S)\} = E_1 \cap E_2 \cap E_3$	$P(E_1)P(E_2)P(E_3) = p^3$	3
$\{(S, S, F)\} = E_1 \cap E_2 \cap E_3^c$	$P(E_1)P(E_2)P(E_3^c) = p^2(1 - p)$	2
$\{(S, F, S)\} = E_1 \cap E_2^c \cap E_3$	$P(E_1)P(E_2^c)P(E_3) = p^2(1 - p)$	2
$\{(F, S, S)\} = E_1^c \cap E_2 \cap E_3$	$P(E_1^c)P(E_2)P(E_3) = p^2(1 - p)$	2
$\{(F, F, S)\} = E_1^c \cap E_2^c \cap E_3$	$P(E_1^c)P(E_2^c)P(E_3) = p(1 - p)^2$	1
$\{(F, S, F)\} = E_1^c \cap E_2 \cap E_3^c$	$P(E_1^c)P(E_2)P(E_3^c) = p(1 - p)^2$	1
$\{(S, F, F)\} = E_1 \cap E_2^c \cap E_3^c$	$P(E_1)P(E_2^c)P(E_3^c) = p(1 - p)^2$	1
$\{(F, F, F)\} = E_1^c \cap E_2^c \cap E_3^c$	$P(E_1^c)P(E_2^c)P(E_3^c) = (1 - p)^3$	0

For example, if $n = 6$, we obtain the following probability model.

k	0	1	2	3	4	5	6
$p_Z(k)$	$(1-p)^6$	$6(1-p)^5p$	$15(1-p)^4p^2$	$20(1-p)^3p^3$	$15(1-p)^2p^4$	$6(1-p)p^5$	p^6

The distribution given by (1.2) is called the *binomial distribution*. The functional form of $p_Z(k)$ is determined by two *parameters n* and *p*, where *n* is the number of repeated Bernoulli trials and *p* is the probability of success on each one of these trials.

In Table 1.2, we illustrate how values of the parameter *p* affect probabilities for the binomial distribution in the particular case when the parameter *n* equals 10. Note that for any fixed value of *p*, the probabilities $p_Z(k)$ increase, reach a maximum, and then decrease as *k* increases from $k = 0$ to $k = 10$.

To show the dependence of the probability mass function $p_Z(k)$ on both *p* and *n*, we write $p_Z(k; n, p)$ for $p_Z(k)$.

Tables of Binomial Probabilities

There is a basic symmetry in the function $p_Z(k; n, p)$ that permits us to calculate probabilities $P\{Z = k\}$ for $p > 0.5$ from knowledge of the probabilities $P\{Z = n - k\}$ for $p < 0.5$. Since

$$\binom{n}{n-k} = \frac{n!}{(n-k)!\,k!} = \frac{n!}{k!\,(n-k)!} = \binom{n}{k},$$

Table 1.2: *Values of $p_Z(k)$ for the Binomial Distribution When $n = 10$ and p Varies*

					p				
k	0.1	0.2	0.3	0.4	0.5	0.6	0.7	0.8	0.9
0	0.3487	0.1074	0.0282	0.0060	0.0010	0.0001	0.0000	0.0000	0.0000
1	0.3874	0.2684	0.1211	0.0403	0.0098	0.0016	0.0001	0.0000	0.0000
2	0.1937	0.3020	0.2335	0.1209	0.0439	0.0106	0.0014	0.0001	0.0000
3	0.0574	0.2013	0.2668	0.2150	0.1172	0.0425	0.0090	0.0008	0.0000
4	0.0112	0.0881	0.2001	0.2508	0.2051	0.1115	0.0368	0.0055	0.0001
5	0.0015	0.0264	0.1029	0.2007	0.2461	0.2007	0.1029	0.0264	0.0015
6	0.0001	0.0055	0.0368	0.1115	0.2051	0.2508	0.2001	0.0881	0.0112
7	0.0000	0.0008	0.0090	0.0425	0.1172	0.2150	0.2668	0.2013	0.0574
8	0.0000	0.0001	0.0014	0.0106	0.0439	0.1209	0.2335	0.3020	0.1937
9	0.0000	0.0000	0.0001	0.0016	0.0098	0.0403	0.1211	0.2684	0.3874
10	0.0000	0.0000	0.0000	0.0001	0.0010	0.0060	0.0282	0.1074	0.3487

we have

(1.3) $p_Z(k; n, p) = \binom{n}{k} p^k (1 - p)^{n-k}$

$= \binom{n}{n-k}(1 - p)^{n-k} p^k = p_Z(n - k; n, 1 - p)$.

That is, *the probability of k successes in n trials when $p = p_0$ equals the probability of $n - k$ successes in n trials when $p = 1 - p_0$.* For example, the probability of 3 successes in 10 trials when $p = 0.8$ is 0.0008, while the probability of $10 - 3 = 7$ successes in 10 trials when $p = 1 - 0.8 = 0.2$ is also 0.0008, as can be seen from Table 1.2. Because of the relationship (1.3), tables of binomial probabilities need only be tabulated for values of $p \leq 0.5$. Table T.1 gives the probability mass functions $p_Z(k)$ of binomially distributed random variables Z for $p = 0.01, 0.05(0.05)0.50$, $n = 1(1)10$.

The Mean and Variance of a Binomial Distribution

The recursive relationship

(1.4) $kp_Z(k; n, p) = (np)p_Z(k - 1; n - 1, p)$, $k = 0, 1, 2, \ldots, n$,

between binomial probabilities is useful both for tabulating probabilities and also for obtaining the moments of the binomial distribution. To verify (1.4), note that for $k = 1, 2, \ldots, n$,

$$kp_Z(k; n, p) = k\binom{n}{k} p^k (1 - p)^{n-k} = k \frac{n!}{k!\,(n - k)!} p^k (1 - p)^{n-k}$$

$$= np \frac{(n - 1)!}{(k - 1)!\,(n - 1 - (k - 1))!} p^{k-1}(1 - p)^{n-1-(k-1)}$$

$$= npp_Z(k - 1; n - 1, p),$$

since $k! = k(k - 1)!$ and $n! = n(n - 1)!$. For $k = 0$, both the left and right sides of (1.4) are equal to 0.

Using (1.4), the mean μ_Z of a binomial distribution with parameters n and p is

$$\mu_Z = E(Z) = \sum_{k=0}^{n} kp_Z(k; n, p) = 0 + \sum_{k=1}^{n} kp_Z(k; n, p)$$

$$= np \sum_{k=1}^{n} p_Z(k - 1; n - 1, p) = np \sum_{j=0}^{n-1} p_Z(j; n - 1, p)$$

$$= np,$$

where we have made the change of index from k to $j = k - 1$ and used the fact that the probabilities of a binomial distribution sum to 1.

Using (1.4) twice,

$$k(k-1)p_Z(k; n, p) = (k-1)np p_Z(k-1; n-1, p)$$
$$= n(n-1)p^2 p_Z(k-2; n-2, p),$$

from which, by an argument similar to that used to obtain μ_Z,

$$E[Z(Z-1)] = \sum_{k=0}^{n} k(k-1)p_Z(k; n, p) = n(n-1)p^2.$$

Thus,

$$\sigma_Z^2 = E[Z^2] - (\mu_Z)^2 = E[Z(Z-1)] + \mu_Z - (\mu_Z)^2$$
$$= n(n-1)p^2 + np - (np)^2 = np(1-p).$$

To summarize, if Z has a binomial distribution with parameters n and p, then

(1.5) $$\mu_Z = np, \qquad \sigma_Z^2 = np(1-p).$$

The Practical Significance of the Binomial Distribution

The binomial distribution can arise whenever we select a random sample of n units *with replacement*. Each unit in the population is classified into one of two categories according to whether it does or does not possess a certain property. For example, the unit may be a person and the property may be whether he intends to vote "yes" on a certain bond issue or whether he has ever been arrested. If the unit is a machine part, this property may be whether the part is defective; if the unit is a leaf, the property may be whether the leaf has worm damage; and so on. Suppose that the proportion of units in the population possessing the property of interest is p. Then, if Z denotes the number of units in the sample (of size n) that possess the given property, Z has a binomial distribution with parameters n and p.

The binomial distribution also arises in contexts other than in random sampling experiments.

Example 1.1 (Genetics). The inheritance of biological characteristics depends on special carriers, called genes. These genes appear in pairs. In the simplest genetic model, each gene of a pair may assume one of two forms: a or A. Consequently, three combinations, called genotypes, may be formed: aa, Aa, AA. (Note that aA and Aa are indistinguishable.)

Fisher and Mather (1936) describe an experiment (called a "backcross") in which the gene controlling straight (W) or wavy (w) hair in mice was segregated. The female parents were all wavy-haired of genotype (ww), and the male parents were all straight-haired of genotype (wW). If the Mendelian laws of inheritance are true, each mouse in the offspring of these parents has a probability 0.5 of being straight-haired (wW) and a probability of 0.5 of being wavy-haired (ww). Assume that the hair character of one offspring is independent of the hair character of any other offspring. Thus, in litters of 8 mice, the theoretical distribution of Z, the number of straight-haired offspring, is the binomial distribution with $n = 8$ and $p = 0.5$. This distribution has the following probability mass function:

k	0	1	2	3	4	5	6	7	8
$p_Z(k)$	0.0039	0.0312	0.1094	0.2188	0.2734	0.2188	0.1094	0.0312	0.0039

In the genetic experiment described by Fisher and Mather, 32 such litters of 8 mice were observed. These litters represent $N = 32$ trials of the random experiment in which Z is observed. If the genetic theory given above truly describes these litters, and if the number $N = 32$ of trials is large enough for the stability of relative frequencies to start to hold, then we would expect that

$$\text{r.f.}\{k \text{ straight-haired mice in litter}\} \simeq p_Z(k)$$

for $k = 0, 1, 2, \ldots, 8$. Equivalently, we would anticipate that the frequency of litters in which k straight-haired mice are observed would be equal to $N p_Z(k) = 32 p_Z(k)$. The numbers $32 p_Z(k)$ are the *expected frequencies*, or *theoretical frequencies*, of litters with k straight-haired mice among the 32 litters observed. A comparison of the theoretical frequencies with the frequencies actually observed is given in Table 1.3. Note that there are differences be-

Table 1.3: Comparison of Observed and Theoretical Frequency Distributions for the Number of Straight-Haired Mice in a Litter of 8 Mice

Number of straight-haired mice in a litter of 8 mice	Observed frequency	Theoretical frequency
0	0	0.1
1	1	1.0
2	2	3.5
3	4	7.0
4	12	8.7
5	6	7.0
6	5	3.5
7	2	1.0
8	0	0.1
	32	31.9

Source: R. A. Fisher and K. Mather (1936). A linkage test with mice. *Annals of Eugenics*, vol. 7, pp. 265–80. Reprinted with permission of Cambridge University Press.

tween the observed and theoretical frequencies. Whether these differences arise due to failure of the theory to describe the situation ("lack of fit" of the model) or are merely due to the fact that the relative frequencies have not completely stabilized ("chance variation") is a question of statistical inference. Ways of answering this question are discussed in statistics textbooks under the topic "goodness-of-fit tests."

Supplementary Tables of Binomial Probabilities

Table T.1 in the Appendix gives only a limited tabulation of binomial probabilities. In particular, the range of values of n covered is only from 1 to 10. More detailed tabulations can be found in National Bureau of Standards [1950; $n = 2(1)49$, $p = 0.01(0.01)0.50$] and in Romig [1953; $n = 50(5)100$, $p = 0.01(0.01)0.50$].

2. HYPERGEOMETRIC DISTRIBUTION

The binomial distribution quite frequently arises from a random experiment in which there is sampling with replacement. In contrast, the hypergeometric distribution typically results from a random experiment in which we select a random sample of units without replacement from a population of N units. In such experiments, each unit in the population can be classified into one of two categories (which we call, arbitrarily, "success" and "failure") according to whether the unit does or does not possess a certain property of interest. We count the number X of "successes" in our sample; it is this random variable that has the hypergeometric distribution.

As an example, consider the problem in which we choose 2 individuals from a population of $N = 5$ individuals by means of random sampling without replacement. Of the 5 individuals in the population, 3 of them, H_1, H_2, H_3, rate high on a certain achievement scale, and 2 of them, L_4, L_5, rate low on that scale. In Table 2.1, we list the 20 possible outcomes of the experiment.

Table 2.1: Outcomes of a Simple Random Sample of Size 2 Without Replacement from a Population Consisting of the Units H_1, H_2, H_3, L_4, L_5

$\omega_1 = (H_1, H_2)$	$\omega_6 = (H_2, H_3)$	$\omega_{11} = (H_3, L_4)$	$\omega_{16} = (L_4, L_5)$
$\omega_2 = (H_1, H_3)$	$\omega_7 = (H_2, L_4)$	$\omega_{12} = (H_3, L_5)$	$\omega_{17} = (L_5, H_1)$
$\omega_3 = (H_1, L_4)$	$\omega_8 = (H_2, L_5)$	$\omega_{13} = (L_4, H_1)$	$\omega_{18} = (L_5, H_2)$
$\omega_4 = (H_1, L_5)$	$\omega_9 = (H_3, H_1)$	$\omega_{14} = (L_4, H_2)$	$\omega_{19} = (L_5, H_3)$
$\omega_5 = (H_2, H_1)$	$\omega_{10} = (H_3, H_2)$	$\omega_{15} = (L_4, H_3)$	$\omega_{20} = (L_5, L_4)$

Let X denote the number of persons who have a high achievement rating. From Table 2.1, we have the following assignment of the possible values of X to the outcomes ω_i:

$X(\omega_1) = 2$	$X(\omega_6) = 2$	$X(\omega_{11}) = 1$	$X(\omega_{16}) = 0$
$X(\omega_2) = 2$	$X(\omega_7) = 1$	$X(\omega_{12}) = 1$	$X(\omega_{17}) = 1$
$X(\omega_3) = 1$	$X(\omega_8) = 1$	$X(\omega_{13}) = 1$	$X(\omega_{18}) = 1$
$X(\omega_4) = 1$	$X(\omega_9) = 2$	$X(\omega_{14}) = 1$	$X(\omega_{19}) = 1$
$X(\omega_5) = 2$	$X(\omega_{10}) = 2$	$X(\omega_{15}) = 1$	$X(\omega_{20}) = 0.$

Since the probability $P\{\omega_i\}$ of each simple event $\{\omega_i\}$, $i = 1, 2, \ldots, 20$, is equal to $\frac{1}{20}$,

$$P\{X = 0\} = P\{\omega_{16}, \omega_{20}\} = \tfrac{2}{20},$$

$$P\{X = 1\} = P\{\omega_3, \omega_4, \omega_7, \omega_8, \omega_{11}, \omega_{12}, \omega_{13}, \omega_{14}, \omega_{15}, \omega_{17}, \omega_{18}, \omega_{19}\} = \tfrac{12}{20},$$

$$P\{X = 2\} = P\{\omega_1, \omega_2, \omega_5, \omega_6, \omega_9, \omega_{10}\} = \tfrac{6}{20},$$

so that the probability distribution of X is given by

x	0	1	2
$P\{X = x\}$	$\tfrac{2}{20}$	$\tfrac{12}{20}$	$\tfrac{6}{20}$

In the foregoing example, we found the distribution of X through an explicit enumeration of the outcomes ω resulting from random sampling without replacement. This is not a difficult task when the size N of the population and the sample size n are both small. However, when either N or n are large, enumeration of the outcomes ω_i becomes quite cumbersome. (For example, when $N = 100$, $n = 3$, there are 970,200 outcomes to list.) Fortunately, the distribution of X can be obtained without the need to enumerate the outcomes.

Derivation of the Hypergeometric Distribution

Assume that we draw a random sample of size n without replacement from a population of N units of which Np are successes and Nq are failures, $q = 1 - p$. (Note that Np and Nq must be integers.) In the sample of size n that we have drawn, let X denote the number of successes that appear. From the definition of random sampling without replacement, every ordered sample of size n in which no unit appears more than once has the same probability $(N - n)!/N!$ of being drawn. Thus,

$$(2.1) \qquad P\{X = k\} = A_k \frac{(N - n)!}{N!},$$

where A_k denotes the number of ordered samples in which no unit appears more than once, and in which k units are successes and $n - k$ units are failures. We can think of forming each such sample in the following three steps: (i) First, select without replacement a collection of k units from among the Np units in the population that are successes; (ii) second, select without replacement a collection of $n - k$ units from among the Nq failures in the population; and (iii) finally, combine the two collections chosen into a single collection of n units and choose the order (permutation) in which to list these n units in our ordered sample.

If $k > Np$ or $n - k > Nq$, either there are not enough successes or there are not enough failures in the population to complete our sample, and $A_k = 0$. Thus, suppose that $k \le Np$ and $n - k \le Nq$. From the counting rules in Chapter 3, Section 6, step (i) can be performed in one of $\binom{Np}{k}$ distinguishable ways. For each such way, there are $\binom{Nq}{n-k}$ ways in which we can complete step (ii). Finally, for every collection of n units that we draw in steps (i) and (ii), there are $n!$ distinguishable ways of ordering these units so as to complete step (iii). Consequently, when $k \le Np$ and $n - k \le Nq$,

$$(2.2) \qquad\qquad A_k = \binom{Np}{k}\binom{Nq}{n-k}n!.$$

From Equations (2.1) and (2.2), it follows that for $k = 0, 1, \ldots, n$,

$$(2.3) \qquad P\{X = k\} = \begin{cases} 0, & \text{when } k > Np \text{ or } n - k > Nq, \\[2mm] \dfrac{\binom{Np}{k}\binom{Nq}{n-k}}{\binom{N}{n}}, & \text{when } k \le Np \text{ and } n - k \le Nq. \end{cases}$$

Of course, $P\{X = k\} = 0$ when $k < 0$, $k > n$, or k is not an integer.

Example 2.1. A radio supply house has $N = 200$ transistor radios, of which $Np = 3$ are improperly soldered (i.e., $p = 0.015$) and $Nq = 197$ are properly soldered ($q = 0.985$). The supply house randomly draws $n = 4$ radios without replacement and sends them to a customer. The probability that the supply house sends $X = 2$ improperly soldered radios to their customer is

$$P\{X = 2\} = \frac{\binom{3}{2}\binom{197}{2}}{\binom{200}{4}} = \frac{57,918}{64,684,950} = 0.000895.$$

On the other hand, $P\{X = 4\} = 0$ since the supply house only has $Np = 3$ improperly soldered radios in their inventory.

Any random variable X possessing a distribution of the form (2.3) is said to have a *hypergeometric distribution* with parameters n, N, and p. The parameters n and N are frequently known to us, but the parameter p may have to be determined either by theory or by observing repeated trials of X.

The Mean and Variance of the Hypergeometric Distribution

To explicitly show the dependency of the hypergeometric probabilities on the parameters n, N, and p, write

$$p_X(k) = p_X(k; n, N, p).$$

Using cancellation of factorials, in a manner similar to the way in which we derived Equation (1.4), we can obtain the following useful recurrence relation for hypergeometric probabilities:

$$(2.4) \qquad k p_X(k; n, N, p) = (np) p_X\left(k - 1; n - 1, N - 1, \frac{Np - 1}{N - 1}\right),$$

for $k = 0, 1, \ldots, n$. Using this relation, it is then straightforward to show that

$$(2.5) \qquad\qquad\qquad \mu_X = np.$$

Iterating (2.4) yields

$$k(k - 1) p_X(k; n, N, p) = np(k - 1) p_X\left(k - 1; n - 1, N - 1, \frac{Np - 1}{N - 1}\right)$$

$$= np(n - 1)\left(\frac{Np - 1}{N - 1}\right) p_X\left(k - 2; n - 2, N - 2, \frac{Np - 2}{N - 2}\right),$$

from which we obtain

$$E[X(X - 1)] = n(n - 1) p\left(\frac{Np - 1}{N - 1}\right)$$

and thus

$$(2.6) \quad \sigma_X^2 = E[X(X - 1)] + \mu_X - (\mu_X)^2$$

$$= n(n - 1) p\left(\frac{Np - 1}{N - 1}\right) + np - (np)^2 = \left(\frac{N - n}{N - 1}\right) np(1 - p).$$

Note that for given values of N, n, and p, the mean of the hypergeometric distribution agrees with the mean of the binomial distribution with parameters n and p. On the other hand, the variances of the hypergeometric and binomial distributions differ. The ratio of the variance of the hypergeometric distribution to the variance $np(1 - p)$ of the binomial distribution is equal to $(N - n)/(N - 1)$. This number is sometimes called the *finite population factor*. As the population size N becomes infinitely large, the factor $(N - n)/(N - 1)$

tends to 1, so that we might say that in infinite populations there is no difference in location and dispersion between binomial and hypergeometric distributions having the same values of the parameters n and p.

Approximation of the Hypergeometric Distribution by the Binomial Distribution

We have remarked that both the binomial and hypergeometric distributions can arise from sampling situations where at each draw of the sample either a success or a failure occurs. The basic difference between the binomial and hypergeometric distributions in such contexts arises from the differing nature of the dependencies among the draws of the sample. For the binomial distribution, sampling is done with replacement and the draws are independent of one another. For the hypergeometric distribution, sampling is without replacement and the draws are dependent.

We can derive the hypergeometric distribution in a manner similar to that in which we derived the binomial distribution. To illustrate, let us use the case of the radio supply house in Example 2.1. To obtain $P\{X = 2\}$, we let E_j be the event that the jth radio drawn is improperly soldered (is a "success"), $j = 1, 2, 3, 4$, and, in terms of the E_j's, we list all ways that we could obtain 2 successes in $n = 4$ draws. These are the following:

$$E_1 \cap E_2 \cap E_3^c \cap E_4^c, \qquad E_1 \cap E_2^c \cap E_3^c \cap E_4,$$

$$E_1 \cap E_2^c \cap E_3 \cap E_4^c, \qquad E_1^c \cap E_2 \cap E_3^c \cap E_4,$$

$$E_1^c \cap E_2 \cap E_3 \cap E_4^c, \qquad E_1^c \cap E_2^c \cap E_3 \cap E_4.$$

As argued when we derived the binomial distribution, there are always $\binom{n}{k}$ distinguishable ways to assign k successes and $n - k$ failures to positions in the sample; this is true regardless of how we draw units from the population. Here, we have explicitly listed the $\binom{4}{2} = 6$ possibilities. If the radios had been drawn with replacement (binomial distribution), each of these 6 events would have the same probability. However, the radios are drawn without replacement (hypergeometric distribution). Nevertheless, the 6 events still have a common probability. For example,

$$P(E_1 \cap E_2 \cap E_3^c \cap E_4^c) = P(E_1)P(E_2 \mid E_1)P(E_3^c \mid E_1 \cap E_2)P(E_4^c \mid E_1 \cap E_2 \cap E_3^c)$$

$$= \tfrac{3}{200}(\tfrac{2}{199})(\tfrac{197}{198})(\tfrac{196}{197})$$

and

$$P(E_1^c \cap E_2 \cap E_3^c \cap E_4) = P(E_1^c)P(E_2 \mid E_1^c)P(E_3^c \mid E_1^c \cap E_2)P(E_4 \mid E_1^c \cap E_2 \cap E_3^c)$$

$$= \tfrac{97}{200}(\tfrac{3}{199})(\tfrac{196}{198})(\tfrac{2}{197})$$

$$= \tfrac{3}{200}(\tfrac{2}{199})(\tfrac{197}{198})(\tfrac{196}{197}).$$

Thus,

$$P\{X = 2\} = 6P(E_1 \cap E_2 \cap E_3^c \cap E_4^c)$$

$$= 6(\tfrac{3}{200})(\tfrac{2}{199})(\tfrac{197}{198})(\tfrac{196}{197}) = 0.000895,$$

which is the answer already obtained using Equation (2.3).

If we had assumed that the radios were drawn independently (sampling with replacement), then we would use the binomial distribution with parameters $n = 4$ and $p = 0.015$ to calculate the probability that the customer receives 2 improperly soldered radios. We would obtain

$$P\{Z = 2\} = 6P(E_1 \cap E_2 \cap E_3^c \cap E_4^c)$$

$$= 6(\tfrac{3}{200})(\tfrac{3}{200})(\tfrac{197}{200})(\tfrac{197}{200}) = 0.001310,$$

which differs from the correct answer, 0.000895, by only 0.000415. Note that the finite population factor in this case is

$$\frac{N - n}{N - 1} = \frac{200 - 4}{200 - 1} = \frac{196}{199} = 0.984925,$$

which is very close to 1. We have already seen that when the finite population factor is approximately equal to 1, then the binomial and hypergeometric distributions with the same values of the parameters n and p do not differ in location or dispersion. Our calculations for the example of the radio supply house also suggest that when the finite population factor is close to 1, corresponding probabilities for the binomial and hypergeometric distributions (with the same values of the parameters n and p) do not differ appreciably. Noting that the finite population factor $(N - n)/(N - 1)$ is close to 1 when N is large with respect to n, it seems reasonable to expect that the binomial distribution will serve as a good approximation to the hypergeometric distribution when N is large relative to n (say, $N \geq 10n$).

Thus, suppose that X has a hypergeometric distribution with parameters n, N, and p. We would like to show that when n and p are fixed, and N is large enough, then

(2.7)
$$P\{X = k\} \simeq \binom{n}{k} p^k (1 - p)^{n-k}.$$

Using either the conditional probability arguments used above to obtain $P\{X = 2\}$ in the case of Example 2.1, or by direct simplification of (2.3), we can show that for $k \leq Np$, $n - k \leq Nq$,

(2.8)
$$P\{X = k\} = \binom{n}{k}\left[\left(\frac{Np}{N}\right)\left(\frac{Np - 1}{N - 1}\right) \cdots \left(\frac{Np - k - 1}{N - k - 1}\right)\right]$$

$$\times \left[\left(\frac{Nq}{N - k}\right)\left(\frac{Nq - 1}{N - k - 1}\right) \cdots \left(\frac{Nq - n + k + 1}{N - n + 1}\right)\right].$$

Since, as $N \to \infty$, each of the terms $(Np - i)/(N - i)$ tends to p, $i = 0, 2, \ldots,$ $k - 1$, while each of the terms $(Nq - j)/(N - k - j)$ tends to q, $j = 0, 1, \ldots,$ $n - k - 1$, it follows from (2.8) that (2.7) must hold. An inspection of (2.8) shows that N does not have to be very large with respect to n for the approximation (2.7) to be reasonably exact.

For example, suppose that $N = 100$, $p = 0.1$, $n = 10$, so that the finite population factor is $(N - n)/(N - 1) = \frac{90}{99} = 0.909$. Then, a comparison of the exact hypergeometric probabilities with the probabilities given by the approximation using the binomial distribution with $n = 10$, $p = 0.1$, yields

k	0	1	2	3	4	5	6	7	8	9	10
Hypergeometric probability	0.330	0.408	0.202	0.052	0.008	0.001	0.000	0.000	0.000	0.000	0.000
Binomial probability	0.349	0.387	0.194	0.057	0.011	0.002	0.000	0.000	0.000	0.000	0.000

As can be seen, the comparison of exact with approximate probabilities is quite favorable, the maximum difference being 0.021. Since the binomial probabilities are simpler to compute and more readily available in tabular form, they are frequently used to approximate the hypergeometric probabilities.

Uses of the Hypergeometric Distribution

The hypergeometric distribution is very important in the analysis of opinion surveys. Most opinion surveys involve drawing a random sample of individuals from a population of individuals (voters, taxpayers, school children, etc.) and asking them one or more questions. The questions asked are usually a kind admitting only one of two possible answers, and the concern is with the number X of people in the sample who answer positively.

Example 2.2. A random sample of $n = 5$ students is drawn without replacement from among $N = 300$ seniors, and each of these 5 seniors is asked if he or she has tried a certain drug. Two of the students interviewed have tried the drug. How probable is this result if 50% of the total population of seniors actually have tried the drug? From Equation (2.3), noting that $p = 0.5$, so that $Np = Nq = 150$, we find that

$$P\{X = 2\} = \frac{\binom{150}{2}\binom{150}{3}}{\binom{300}{5}} = 0.3146.$$

Note that the probability given for the event $\{X = 2\}$ by the binomial distribution with parameters $n = 5$, $p = 0.5$ is 0.3125. The finite population factor $(N - n)/(N - 1)$ in this example is $(300 - 5)/(300 - 1) = 0.986622$.

Another application of the hypergeometric distribution arises in the area of quality control. From an inventory of N items, a random sample of n items without replacement is sampled and the number X of defective items is noted. From knowledge of the value $X = x$, and from knowledge of the form of the distribution (2.3) for X for various values of the parameter p, statistical inferences may be made concerning the proportion p of defective items in the inventory.

There are many other examples of applications of the hypergeometric distribution.

Example 2.3 (Card Playing). Pearson (1924) reported on the records of 25,000 actual deals of the card game whist. His concern was in the determination of whether the shuffling was "perfect." This was done by comparing actual and theoretical results. Table 2.2 gives observed and theoretical frequencies for the event that X trump cards appear in the first hand of a deal of whist. The data are taken from a sample of 3400 deals of the cards. The theoretical observed frequencies are obtained by multiplying the theoretical probability $P\{X = x\}$ by 3400.

Table 2.2: Comparison of Observed and Theoretical Frequencies for the Event That X Trump Cards Appear in the First Hand of a Deal of the Card Game Whist

Number x of trump cards in the hand	Observed frequency	Theoretical frequency computed from the hyper-geometric distribution with $N = 52$, $n = 13$, $p = \frac{1}{4}$
0	35	43.4
1	290	272.3
2	697	700.2
3	937	973.4
4	851	811.2
5	444	424.0
6	115	141.4
7	21	29.9
8	10	3.7
≥ 9	0	0.5
	3400	3400.0

Source: K. Pearson (1924). On a certain double hypergeometrical series and its representation by continuous frequency surfaces. *Biometrika*, vol. 16, pp. 172–88. Reprinted with permission.

Description of Tables

Table T.2 in the Appendix provides values for the probability mass function of the hypergeometric distribution for $N = 2(1)9$ and selected values of n and p. A more detailed tabulation for $N = 2(1)50(10)100$ can be found in Lieberman and Owen (1961).

3. POISSON DISTRIBUTION

If we study data collected on accidents by insurance companies, an interesting fact emerges. Suppose, for example, that we are interested in the number X of individuals injured when getting out of an automobile in a given year. For any particular individual, the probability of such an accident in a given year is quite small; in fact, it is almost zero. However, data reveal that a number of such accidents occur every year, and that the number X of such accidents varies from year to year. Over many years, the relative frequencies of the events $\{X = k\}$, for k an integer, exhibit stability. This type of statistically regular behavior is closely described by the *Poisson distribution* (named after the French mathematician Simeon D. Poisson, 1781–1840). This discrete distribution has a probability mass function of the form

(3.1) $$p_X(k) = P\{X = k\} = \frac{e^{-\lambda}\lambda^k}{k!}, \quad k = 0, 1, 2, \ldots,$$

where the constant λ can be any positive number. The effect of the parameter λ on the behavior of the probability mass function $p_X(k)$ of the Poisson distribution can be observed from Table 3.1.

In Table 3.1, note that when $\lambda > 1$, the probabilities $p_X(k)$ increase until k is approximately equal to λ, and then decrease. Indeed, for integer values of λ, both λ and $\lambda - 1$ are modes of the Poisson distribution, while when λ is not an integer, the mode is the integer k which lies between $\lambda - 1$ and λ. In general, regardless of whether or not λ is an integer, when $\lambda > 1$, the values of $p_X(k)$ increase, reach a maximum, and then decrease as k increases from 0 to infinity.

Mean and Variance of the Poisson Distribution

The recursive relation

(3.2) $$k\,p_X(k) = \lambda\,p_X(k - 1), \quad k = 0, 1, \ldots,$$

can be used (as in Sections 1 and 2) to show that if X has a Poisson distribution with parameter λ, then

(3.3) $$\mu_X = \lambda, \qquad \sigma_X^2 = \lambda.$$

Table 3.1: *Values of $p_X(k)$ for the Poisson Distribution When $\lambda = 0.5, 1, 2, 3, 5$*

	λ				
k	0.5	1	2	3	5
0	0.6065	0.3679	0.1353	0.0498	0.0067
1	0.3033	0.3679	0.2707	0.1494	0.0337
2	0.0758	0.1839	0.2707	0.2240	0.0842
3	0.0126	0.0613	0.1804	0.2240	0.1404
4	0.0016	0.0153	0.0902	0.1680	0.1755
5	0.0002	0.0031	0.0361	0.1008	0.1755
6		0.0005	0.0120	0.0504	0.1462
7		0.0001	0.0034	0.0216	0.1044
8			0.0009	0.0081	0.0653
9			0.0002	0.0027	0.0363
10				0.0008	0.0181
11				0.0002	0.0082
12				0.0001	0.0034
13					0.0013
14					0.0005
15					0.0002
16					0.0000

The fact that $\sigma_X^2 = \lambda$ follows by using (3.2) to obtain

$$k(k-1)p_X(k) = \lambda^2 p_X(k-2), \quad k = 0, 1, 2, \ldots,$$

from which $E[X(X-1)] = \lambda^2$ and

$$\sigma_X^2 = E[X(X-1)] + \mu_X - (\mu_X)^2 = \lambda^2 + \lambda - \lambda^2 = \lambda.$$

Applications of the Poisson Distribution

The wide variety of random phenomena that give rise to random variables X having a Poisson distribution is truly astonishing. Some examples for which the Poisson distribution provides a reasonable model are given in this section.

In many biological examples a surface is subdivided into equal sections, from which counts of specimens are made. This occurs, for example, in studies of blood samples, yeast cultures, and so on.

Example 3.1 (Zoology). A horizontal quarry surface was divided into 30 squares about 1 meter on a side. In each square the number X of specimens of the extinct mammal *Litolestes notissimus* was counted. The results are shown in Table 3.2.

Unlike previous examples which we have given in which a distribution is tested against data, in the present example there is no background theory that enables us to specify the value of the parameter λ. We must thus try to let the data suggest a value for λ. Recall that for a Poisson distribution with parameter λ, $\mu_X = \lambda$. Thus, the sample average $\hat{\mu}_x$ provides us with an estimate

$$\hat{\lambda} = 0\left(\tfrac{8}{15}\right) + 1\left(\tfrac{3}{10}\right) + 2\left(\tfrac{1}{10}\right) + 3\left(\tfrac{1}{30}\right) + 4\left(\tfrac{1}{30}\right) + 0 = \tfrac{11}{15} = 0.73$$

of λ. Using (3.1) with $\lambda = 0.73$, we can compute values of $p_X(k)$ for $k = 0, 1, 2, \cdots$ For example,

$$p_X(2) = \frac{e^{-0.73}(0.73)^2}{2!} = 0.13.$$

From such calculations, the following theoretical probability mass function is obtained for X:

k	0	1	2	3	4	5 and over
$p_X(k)$	0.48	0.35	0.13	0.03	0.01	0.00

The theoretical frequencies, $Np_X(k) = 30\,p_X(k)$, are then computed for comparison with the observed frequencies, as in Table 3.3.

Judging by Table 3.3, a Poisson distribution with parameter $\lambda = 0.73$

Table 3.2: *Observed Frequencies of the Event* $\{X = k\}$ *in 30 Trials (Squares) of the Random Experiment in Which We Count the Number* X *of Specimens of* Litolestes notissimus *in a Square of Horizontal Quarry Space*

Number of specimens per square, k	Observed frequency of squares with k specimens	Observed relative frequency of squares with k specimens
0	16	8/15
1	9	3/10
2	3	1/10
3	1	1/30
4	1	1/30
≥ 5	0	0
	30	1

Source of data: *Quantitative Zoology* by G. G. Simpson, A. Roe, and R. C. Lewontin, by permission of the authors, holders of copyright.

Table 3.3: *Comparison of Observed and Theoretical Frequencies of the Event* $\{X = k\}$ *in 30 Trials* (*Squares*) *of the Random Experiment in Which We Count the Number X of Specimens of* Litolestes notissimus

Number of specimens per square, k	Observed frequency	Theoretical frequency
0	16	14.4
1	9	10.5
2	3	3.9
3	1	0.9
4	1	0.3
≥ 5	0	0.0

Source of data: *Quantitative Zoology* by G. G. Simpson, A. Roe, and R. C. Lewontin, by permission of the authors, holders of copyright.

agrees closely with the observed frequencies. If the number N of squares observed had been larger than 30, we would expect the fit to have been better.

Remark. Although here we estimated λ using the sample average $\hat{\mu}_X$, alternative techniques for estimating λ are available in the statistical literature. However, these techniques are peripheral to our main interest, and we have selected a method that appears to be both useful and easily understood.

We have already remarked upon the versatility of application of the Poisson distribution. In industry, engineers have applied the Poisson distribution to the distribution of the number of flaws in capacitors, to the number of defects per linear unit of wire and of rope, and to the number of strands in a cross section of thread. In agriculture and biology, applications have been made to the number of beetle larvae, to the number of fish caught in a day, to the number of photons reaching the retina, and to bacteria counts. In medicine, we find the Poisson distribution applied to the number of defective teeth per individual, and to the number of victims suffering from various specific diseases. In sociology, the Poisson distribution has been applied to the number of vacancies in the Supreme Court and to the number of labor strikes. Applications of the Poisson distribution have also been made to the number of words misread in a text, to the number of misprints, and to the frequency of earthquakes.

The following are two additional examples where the Poisson distribution is tested for fit on data. In each case, the appropriate value of the parameter λ is estimated by taking the sample average of the observations.

Because in these examples the sample average is computed from the data *before* the data are condensed for the tables, there may be discrepancies between the stated value of the average and the average that can be obtained from the results shown in the tables.

Example 3.2 (Telephone Connections). Groups of coin-box telephones in a large transportation terminal were observed. The calls made on several such telephones in each of about 20 five-minute intervals between noon and 2:00 P.M. were monitored on each of 7 days (excluding Saturday and Sunday), resulting in a total of 267 trials. A count was made of the number X of calls to a wrong number, and a Poisson distribution with $\lambda = 8.74$ was fitted to the resulting data (see Table 3.4).

Table 3.4: *Observed Frequencies of k Wrong Connections in 267 Trials of the Random Experiment in Which Calls Were Monitored on Groups of Coin-Box Telephones over a 5-Minute Interval*

Number of wrong connections, k	Observed frequency	Theoretical frequency, $\lambda = 8.74$
≤ 2	1	1.87
3	5	4.81
4	11	10.41
5	14	18.16
6	22	26.43
7	43	33.11
8	31	36.05
9	40	34.98
10	35	30.71
11	20	24.30
12	18	17.62
13	12	12.02
14	7	7.48
15	6	4.27
16	2	2.40
≥ 17	0	2.38
	267	267.00

Source: F. Thorndike (1926). Applications of Poisson's probability summation. *Bell System Technical Journal*, vol. 5, pp. 604–24. Copyright 1926, American Telephone and Telegraph Company, reprinted by permission.

Example 3.3 (Radioactive Emissions). Rutherford and Geiger (1910) observed the collision of an α particle emitted from a small bar of polonium with a small screen placed at a short distance from the bar. The number of such collisions in each of 2608 eight-minute intervals was recorded; the distance between the bar and screen was gradually decreased so as to compensate for the decay of the radioactive substance. A Poisson distribution with $\lambda = 3.870$ was fitted to the data (see Table 3.5).

The Poisson Distribution as an Approximation to the Binomial Distribution

In almost all cases, random variables possessing a Poisson distribution can be conceptualized as having arisen from a random experiment consisting of a large number of trials. On each trial a certain rare occurrence may or may not take place. Because the occurrence is rare, the probability of its

Table 3.5: *Observed Frequencies of k Alpha Particles in 2608 Trials of the Random Experiment in Which Alpha Particles Emitted from a Small Bar of Polonium Are Counted Over an 8-Minute Interval of Time*

Number of Alpha particles observed in the interval, k	Observed frequency	Theoretical frequency, $\lambda = 3.870$
0	57	54.77
1	203	211.25
2	383	406.85
3	525	524.21
4	532	508.56
5	408	393.81
6	273	252.98
7	139	140.83
8	45	67.81
9	27	28.69
10	10	10.43
11	4	5.22
≥ 12	2	2.59
	2608	2608.00

Sources: Sir Ernest Rutherford, James Chadwick, and C. D. Ellis (1930). *Radiations from Radioactive Substances.* New York: Macmillan Publishing Co., Inc., pp. 171–72. Ernest Rutherford and Hans Geiger (1910). The probability variations in the distribution of α particles. *Philosophical Magazine*, vol. 20, pp. 698–707.

taking place is small. However, the rarity of the occurrence is offset by the fact that a large number of trials are performed. The resultant of these two effects is the Poisson distribution (3.1). This fact can be demonstrated by obtaining the Poisson distribution as an approximation to the binomial distribution when the two effects are present.

The derivation is as follows. Under a binomial distribution with parameters n and p, $P\{X = k\} = \binom{n}{k} p^k (1 - p)^{n-k}$. Let $p = \lambda/n$, so that a large sample size n will be offset by the diminution of p to produce a constant mean number of successes $\mu_X = np = n(\lambda/n) = \lambda$ for all values of n. Then as $n \to \infty$,

$$P\{X = k\} = \binom{n}{k} \left(\frac{\lambda}{n}\right)^k \left(1 - \frac{\lambda}{n}\right)^{n-k}$$

$$= \frac{\lambda^k}{k!} \left(1 - \frac{\lambda}{n}\right)^n \frac{n!}{(n-k)! \, n^k} \left(1 - \frac{\lambda}{n}\right)^{-k}$$

(3.4)
$$= \frac{\lambda^k}{k!} \left(1 - \frac{\lambda}{n}\right)^n \left[\left(\frac{n}{n}\right)\left(\frac{n-1}{n}\right) \cdots \left(\frac{n-k+1}{n}\right)\right]\left(1 - \frac{\lambda}{n}\right)^{-k}$$

$$\cong \frac{\lambda^k}{k!} e^{-\lambda}.$$

since $(1 - (\lambda/n))^n$ approaches $e^{-\lambda}$ and both $(1 - (\lambda/n))^{-k}$ and the term in square brackets approach 1. Comparing our result with Equation (3.1) verifies our assertion that the binomial probability that a rare occurrence will take place exactly k times is closely approximated by the probability (3.1). For this reason binomial probabilities for small values of p and large values of n are often approximated using the Poisson distribution, with $\lambda = np$.

For example, suppose that $n = 10$ and $p = 0.10$. The comparison between the exact probability mass function for the binomial distribution and the Poisson approximation with $\lambda = np = 1$ is

k	0	1	2	3	4	5	6	7 or more
Binomial	0.349	0.387	0.194	0.057	0.011	0.002	0.000	0.000
Poisson	0.3679	0.3679	0.1839	0.0613	0.0153	0.0031	0.0005	0.0001

If n becomes larger, say 20, and p smaller, say $p = 0.05$, then it is still the case that $\lambda = np = 1$, and we obtain the comparison

k	0	1	2	3	4	5	6	7 or more
Binomial	0.3585	0.3774	0.1887	0.0596	0.0133	0.0022	0.0003	0.0000
Poisson	0.3679	0.3679	0.1839	0.0613	0.0153	0.0031	0.0005	0.0001

As can be seen, the comparison in the first case is good even for n as small as 10. However, it is even closer when $n = 20$. Finally, if $n = 100$ and $p = 0.01$, the comparison becomes

k	0	1	2	3	4	5	6	7 or more
Binomial	0.3660	0.3697	0.1849	0.0610	0.0149	0.0029	0.0005	0.0001
Poisson	0.3679	0.3679	0.1839	0.0613	0.0153	0.0031	0.0005	0.0001

In general, the Poisson distribution with parameter λ provides a good approximation to the binomial distribution with parameters n and $p = \lambda/n$ in cases when n is large and p is small, and when $\lambda = np$ is of moderate size (say, $\lambda \le 20$).

Tables of the Poisson Distribution

Table T.3 in the Appendix gives values of the probability mass function $p_X(k)$ of the Poisson distribution for $\lambda = 0.1(0.1)10(1)20$. For more detailed Tables with $\lambda = 0.001(0.001)1.000(0.01)10.00$ see Kitagawa (1952), and for $\lambda = 0.001(0.001)0.010(0.010)0.30(0.1)15.0(1)100$ see Molina (1942).

4. GEOMETRIC DISTRIBUTION

If we make a fixed number n of independent Bernoulli trials, then we know that the random number Z of successes in those n trials has a binomial distribution. Suppose, however, that instead of fixing the number of trials that we make and observing the resulting random number of successes, we decide to fix the number of successes that we require and then observe the random number of trials needed to obtain this number of successes. For example, let us continue to make independent Bernoulli trials until we obtain the first success. If the probability of success on any single Bernoulli trial is p, then the random number X of trials required to obtain the first success has a geometric distribution with parameter p.

Example 4.1 (Medicine). A blood bank needs type B negative blood, and will continue purchasing blood from individuals until the first time a sample of this kind of blood is purchased. If blood purchases are independently made, the random number X of purchases that must be made until a type B negative blood is purchased has a geometric probability distribution.

Example 4.2 (Gambling). A gambler plays roulette at Monte Carlo (see Figure 4.1) and continues gambling, wagering the same amount c each time on the same eventuality, until he wins for the first time. Suppose that spins of

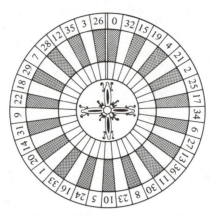

Figure 4.1: *Roulette wheel used at Monte Carlo.*

the roulette wheel are independent. Then the number X of gambles until (and including) the gamble on which the gambler wins for the first time has a geometric distribution. If the gambler always bets on " red," then the parameter p of the distribution of X is equal to $\frac{18}{37}$.

In the examples given above, the random variable X, which is the number of trials required to obtain the first success, is identified with the cost of running the experiment. If the length of time between trials is uniform, the number X of trials that must be made until a success is observed can also be equated with the *length of time* required until a success occurs. For example, if in Example 4.1 the blood bank makes one blood purchase an hour, then X is the number of hours they must wait until they purchase their first sample of type B negative blood. Thus, X frequently refers to a waiting time until a certain event occurs.

Derivation of the Geometric Distribution

We now obtain the probability mass function of the geometric distribution with parameter p. Assume that we observe independent Bernoulli trials with probability p of success on each trial, and let X be the random number of trials required to obtain the first success. Let E_j be the event that we obtain a success on the jth trial, $j = 1, 2, \ldots$. Note that since the trials are independent and E_j depends only on the jth trial, the events $E_1, E_2, \ldots$ are independent. Also, $P(E_j) = p = 1 - P(E_j^c)$ for $j = 1, 2, \ldots$. We stop at the kth trial only if success has not occurred on trials $1, 2, \ldots, k - 1$, and the kth trial results in a success. Thus, $P\{X = 1\} = P(E_1) = p$,

$$P\{X = 2\} = P(E_1^c \cap E_2) = P(E_1^c)P(E_2) = (1 - p)p,$$

and in general,

$$P_X(k) = P\{X = k\} = P(E_1^c \cap E_2^c \cap \cdots \cap E_{k-1}^c \cap E_k)$$
$$= P(E_1^c)P(E_2^c) \cdots P(E_{k-1}^c)P(E_k) = (1 - p)^{k-1}p,$$

for $k = 1, 2, \ldots$. Thus, the probability mass function of the geometric distribution with parameter p is

$$(4.1) \qquad p_X(k) = (1 - p)^{k-1}p, \quad k = 1, 2, \ldots,$$

with $p_X(x) = 0$ when x is not a positive integer. Note that the probabilities $P\{X = k\} = p_X(k)$ decrease rapidly to 0 as k increases from 1 to infinity. Thus, for any p, the mode of the geometric distribution is always 1.

Unlike the other discrete distributions which we consider in this chapter, the geometric distribution has a cumulative distribution function $F_X(x)$ with a simple functional form. For any positive integer x,

$$F_X(x) = \sum_{k=1}^{x}(1 - p)^{k-1}p = p \sum_{k=1}^{x}(1 - p)^{k-1}$$

$$(4.2) \qquad = \sum_{k=1}^{x}(1 - p)^{k-1} - (1 - p)\sum_{k=1}^{x}(1 - p)^{k-1}$$

$$= \sum_{k=1}^{x}(1 - p)^{k-1} - \sum_{k=2}^{x+1}(1 - p)^{k-1} = 1 - (1 - p)^x,$$

while $F_X(x) = 0$ for $x < 1$, and $F_X(x) = F_X(k)$ when $k \le x < k + 1$, k any positive integer.

Values of $p_X(k)$ can easily be computed for any p and all but the largest values of k. For this reason, tables of the probability mass function of the geometric distribution are omitted.

Example 4.2 (*continued*). The gambler continues to bet on "red" at Monte Carlo until he wins for the first time. Thus, the number X of trials required to win for the first time has a geometric distribution with parameter $p = \frac{18}{37}$. Suppose that the gambler has only enough money for 5 trials. Then the probability that he will win before he exhausts his funds is

$$P\{X \le 5\} = F_X(5) = 1 - (1 - \frac{18}{37})^5 = 0.964.$$

The probability that he wins on the second trial is

$$P\{X = 2\} = p_X(2) = (1 - \frac{18}{37})(\frac{18}{37}) = 0.250.$$

Note that from (4.2),

$$\sum_{k=1}^{\infty}p_X(k) = \lim_{x \to \infty}F_X(x) = 1 - \lim_{x \to \infty}(1 - p)^x = 1,$$

since $(1 - p)^x$ goes to 0 as $x \to \infty$. This result verifies that the function (4.1) is indeed a probability mass function, since $p_X(x) \geq 0$ for all numbers x and the probabilities $p_X(k)$ sum to 1.

The Mean and Variance of the Geometric Distribution

When the random variable X has a geometric distribution with parameter p, the mean μ_X and the variance σ_X^2 of X are given by

(4.3) $$\mu_X = \frac{1}{p}, \qquad \sigma_X^2 = \frac{1-p}{p^2}.$$

To verify (4.3), we make use of a result from the theory of infinite series. For any positive integer r,

(4.4) $$\frac{1}{p^r} = \sum_{k=r-1}^{\infty} \binom{k}{r-1}(1-p)^{k-r+1}.$$

When $r = 2$, (4.4) becomes

$$\frac{1}{p^2} = \sum_{k=1}^{\infty} \binom{k}{1}(1-p)^{k-1} = \sum_{k=1}^{\infty} k(1-p)^{k-1},$$

so that

$$\mu_X = \sum_{k=1}^{\infty} kp(1-p)^{k-1} = p \sum_{k=1}^{\infty} k(1-p)^{k-1} = \frac{p}{p^2} = \frac{1}{p}.$$

When $r = 3$, (4.4) becomes

$$\frac{1}{p^3} = \sum_{p=2}^{\infty} \binom{k}{2}(1-p)^{k-2} = \sum_{k=2}^{\infty} \frac{k(k-1)}{2}(1-p)^{k-2}$$

$$= \frac{1}{2(1-p)} \sum_{k=1}^{\infty} k(k-1)(1-p)^{k-1},$$

from which we see that

$$E[X(X-1)] = \sum_{k=1}^{\infty} k(k-1)p(1-p)^{k-1} = \frac{2(1-p)}{p^2},$$

$$\sigma_X^2 = E[X(X-1)] + \mu_X - (\mu_X)^2 = \frac{2(1-p)}{p^2} + \frac{1}{p} - \left(\frac{1}{p}\right)^2 = \frac{1-p}{p^2}.$$

It is worth noting that when $r = 1$, (4.4) asserts that

$$\frac{1}{p} = \sum_{k=0}^{\infty} \binom{k}{0}(1-p)^k = 1 + (1-p)\sum_{k=1}^{\infty}(1-p)^{k-1},$$

so that

$$\sum_{k=1}^{\infty} p(1 - p)^{k-1} = \frac{p}{1 - p}\left(\frac{1}{p} - 1\right) = 1,$$

once again demonstrating that the probabilities (4.1) of the geometric distribution sum to 1.

Applications of the Geometric Distribution

When the parameter p of the geometric distribution must be estimated from data, use of (4.3) suggests that we estimate p by $1/\hat{\mu}_X$.

Example 4.3 (Transportation). Studies have been made in an attempt to estimate the number of passengers carried per car in urban vehicular traffic. Data were collected Tuesday, March 24, 1959, as to the number X of passengers carried in each of 1011 passenger cars traveling through the intersection of Wilshire and Bundy boulevards in Los Angeles between 10:00 A.M. and 10:40 A.M. These data are presented (in condensed form) in Table 4.1. Theoretically, we might or might not expect a geometric distribution to explain the variation shown in Table 4.1. However, let us try to fit a geometric distribution to these data.

From the original (raw) data,

$$\hat{\mu}_X = 1.52,$$

Table 4.1: *Frequency Distribution for the Number of Occupants in a Passenger Car*

Number of occupants, k	Frequency of k occupants
1	678
2	227
3	56
4	28
5	8
≥ 6	14
	1011

Source: F. A. Haight (1970). Group size distributions, with applications to vehicle occupancy, in G. P. Patil (ed.), *Random Counts in Physical Science, Geo Science, and Business,* vol. 3, pp. 95–105. Reprinted by permission of the Pennsylvania State University Press.

so that

$$\hat{p} = \frac{1}{\hat{\mu}_X} = \frac{1}{1.52} = 0.658.$$

Alternatively, we could obtain $\hat{\mu}_X$ from the grouped data in Table 4.1, provided that we assume that no observations larger than 6 were observed:

$$\hat{\mu}_X = (1)(\tfrac{678}{1011}) + (2)(\tfrac{227}{1011}) + \cdots + (6)(\tfrac{14}{1011}) = \tfrac{1536}{1011} = 1.519.$$

This yields the same estimate of p that we obtain from the raw data. (However, it will not always be the case that estimates of $\hat{p}$ based on the raw data and the grouped data agree.) Using $p = 0.658$, we can now find the theoretical probabilities for the geometric distribution. The theoretical frequencies for the events $\{X = k\}$, $k = 1, 2, 3, 4, 5$, are given by

$$(1011)p_X(k) = (1011)(0.342)^{k-1}(0.658), \quad k = 1, 2, 3, 4, 5,$$

while the theoretical frequency for the event $\{X \geq 6\}$ is given by

$$(1011)P\{X \geq 6\} = (1011)P\{X > 5\} = (1011)[1 - F_X(5)] = (1011)(0.342)^5.$$

The results appear in Table 4.2.

The agreement between observed and expected frequencies is reasonably close, except for the last entry, which may be due to the fact that some exceptionally large cars (limousines) can hold up to 9 passengers. These exceptional cars should perhaps be treated separately.

The System of Doubling the Bet

The geometric distribution, being of a comparatively simple form, enables us to make some interesting calculations. In particular, the following example provides an interesting illustration of the fact that the mean of a discrete random variable can be infinitely large.

Table 4.2: *Comparison of Observed and Theoretical Frequencies for the Number of Occupants in a Passenger Car*

Number of occupants, k	Observed frequencies	Theoretical frequencies
1	678	665
2	227	227
3	56	78
4	28	27
5	8	9
≥ 6	14	5

Example 4.4 (Doubling the Bet). Suppose that a gambler has a probability $p = \frac{1}{2}$ of winning on any play of a given game of chance. Then, from (4.1) with $p = \frac{1}{2}$, his probability of winning the game for the first time at the kth play is $(\frac{1}{2})^k$. Suppose that the gambler follows a betting system of *doubling* his bets until he wins, and that his initial bet is 1 dollar. Then, if he loses on the first play of the game, his second bet is 2 dollars; if he again loses, his third bet is 4 dollars; the fourth bet 8 dollars, and so on. This system is advocated by many people as a sure way to win since all losses are recaptured and a dollar is won as soon as the gambler wins for the first time.

Let Z denote the amount the gambler needs to be able to continue betting until he wins for the first time. That is, if he were to win on the first toss he would only need 1 dollar (to make the first bet). If he does not win until the fifth trial, he needs $1 + 2 + 4 + 8 + 16 = 2^5 - 1$ dollars. In general, if he does not win until the kth play, he needs $2^k - 1$ dollars. Thus, Z is a random variable that takes on the values $2^k - 1$ with probability $(\frac{1}{2})^k$ for $k = 1, 2, 3, \ldots$, that is, Z has the probability mass function given as follows:

z	1	3	7	15	$\cdots$	$2^k - 1$	$\cdots$
$p_Z(z)$	$\dfrac{1}{2}$	$\dfrac{1}{4}$	$\dfrac{1}{8}$	$\dfrac{1}{16}$	$\cdots$	$\dfrac{1}{2^k}$	$\cdots$

The expected amount (mean amount) of money needed to sustain the betting system is

$$\mu_Z = \sum_{k=1}^{\infty} (2^k - 1)\left(\frac{1}{2^k}\right) = \tfrac{1}{2} + \tfrac{3}{4} + \tfrac{7}{8} + \tfrac{15}{16} + \cdots .$$

We see that the size of the terms in the sum above is increasing; consequently, $\mu_Z = \infty$. Thus, *on the average, no finite amount of money is sufficient to sustain this betting system.*

In some books, the geometric distribution is said to refer to the number of *failures* observed before the first success is obtained. In this case, we do not count the trial on which the first success is obtained. If Y is the number of failures observed before the first success and X is the number of trials needed to obtain the first success, then $Y = X - 1$. Thus, the events $\{X = k + 1\}$ and $\{Y = k\}$ are the same event, and from (4.1) we find that

$$p_Y(k) = p_X(k + 1) = p(1 - p)^k, \quad k = 0, 1, 2, \ldots,$$

which is the probability mass function that some texts call the geometric distribution. However, since $Y = X - 1$, either variable gives the same information about the random phenomenon being studied, so that the choice of variable (X or Y) which is said to have a geometric distribution is strictly a matter of taste and convenience.

5. NEGATIVE BINOMIAL DISTRIBUTION

The geometric distribution is the distribution of the number of independent Bernoulli trials required to obtain a single success. Now let us generalize this distribution by finding the distribution of the random variable X that equals the number of trials required until (and including) the trial upon which the rth success occurs, $r = 1, 2, \ldots$. For $r > 1$, it is sometimes more convenient to study the random variable $Y = X - r$, which is equal to the number of failures that are observed before the rth success is achieved. We use X or Y, whichever is more appropriate, and say that the random variables X and Y have the *negative binomial distribution* with parameters r and p.

Example 5.1 (Fishery Management). To estimate the relative number of fish of a certain species in a given pond, fish are caught one at a time, and for every fish that is caught, the species of that fish is noted before the fish is returned to the pond. When r fish of the species of interest have been caught, the total number Y of fish that were caught and were not of the species of interest is recorded. Since fish are returned to the pond after being caught, sampling is with replacement, and thus the catches act as independent Bernoulli trials. If there are a total of N fish in the pond, of which M belong to the species of interest, then the probability that a fish caught at any catch is of the species of interest is $p = M/N$. The random variable Y thus has a negative binomial distribution with parameters r and $p = M/N$.

Example 5.2 (Politics). A political canvasser requires $r = 100$ signatures for a petition. The probability that a person will agree to sign the petition is $\frac{1}{10}$, and each person approached acts independently of any other person. The canvasser can ask one person to sign every 5 minutes. Let X be the number of persons that must be approached before the 100 needed signatures have been gathered. Then $5X$ is the number of minutes that the canvasser must spend trying to obtain the needed signatures. The random variable X has a negative binomial distribution with parameters $r = 100$ and $p = \frac{1}{10}$.

As can be seen from Example 5.2, random variables (X or Y) having a negative binomial distribution are often amenable to a waiting-time interpretation. Suppose that we observe a given process at equally spaced intervals of time. During the time between observations, some random event may or may not occur. If the event occurs, at the end of that time period we record a success; if not, at the end of the time period we record a failure. Let X be the total time elapsed until r successes have been recorded; $Y = X - r$ is the part of the total elapsed time that is truly "random" (r units of time must occur before r successes are observed, but the amount of time beyond that is variable).

We now obtain the probability mass function of the random number Y of failures that must be observed before r successes are observed in repeated independent Bernoulli trials with probability p of success on any trial. Let E_j be the event that success occurs on the jth trial, $j = 1, 2, \ldots$. The events E_j, $j = 1, 2, \ldots$, are independent since the trials are independent, and $P(E_j) = p = 1 - P(E_j^c)$ for all j. For any integer u, $u \geq 0$, the event $\{Y = u\}$ means that u failures are observed before the rth success is observed. This event occurs when any pattern of $r - 1$ successes and u failures is observed in the first $(u + r - 1)$ trials, and then the $(u + r)$th trial is a success. Thus, let A_u be the event that $(r - 1)$ successes and u failures occur in the first $(u + r - 1)$ Bernoulli trials; thus,

$$P(A_u) = \binom{u + r - 1}{r - 1} p^{r-1}(1 - p)^u.$$

Since the first $(u + r - 1)$ trials and the $(u + r)$th trial are independent, and $\{Y = u\} = A_u \cap E_{u+r}$, we have

$$p_Y(u) = P\{Y = u\} = P(A_u)P(E_{u+r}) = \left[\binom{u + r - 1}{r - 1} p^{r-1}(1 - p)^u\right]p$$

or

$$(5.1) \quad p_Y(u) = \binom{u + r - 1}{r - 1} p^r(1 - p)^u, \quad u = 0, 1, 2, \ldots,$$

and $p_Y(u) = 0$ for all other values of u. The probability mass function (5.1) defines the *negative binomial distribution* with parameters r and p. This distribution is sometimes called the Pascal distribution (after the French mathematician Blaise Pascal).

Table T.4 in the Appendix gives the probability mass functions (evaluated for integers $k = 0, 1, 2, \ldots$) associated with the negative binomial distribution for $r = 2, 3, 4, 5$ and $p = 0.2, 0.4, 0.5, 0.6, 0.8$.

Generalization of the Negative Binomial Distribution

The negative binomial distribution occurs in a wider range of practical situations than just those in which the random variable Y is a result of sequential trials. In these other applications, however, it has frequently proved useful to generalize the form of the probability mass function (5.1) so as to permit noninteger values of the parameter r. To do this, let us write the probability mass function (5.1) in the form

$$(5.2) \qquad p_Y(u) = H(u, r)p^r(1 - p)^u, \quad u = 0, 1, 2, \ldots.$$

where

$$(5.3) \qquad H(u, r) = \frac{(u + r - 1)!}{(r - 1)!\,u!} = \frac{\Gamma(u + r)}{\Gamma(r)\Gamma(u + 1)},$$

and $\Gamma(a)$ is the gamma function. The gamma function has played an important role in models of physical phenomena, and consequently this function has been extensively investigated and tabled. Some of the principal properties of the gamma function (and related functions) are discussed in the Appendix. We have used the gamma function here because (as shown in the Appendix) it generalizes the factorial computation $k!$ to noninteger values of k and thus permits us to formally define the negative binomial distribution for any positive value of the parameter r.

Computation of the values of $p_Y(u)$ in (5.2) for values of $u = 0, 1, 2, \ldots$ can be simplified by noting that

$$(5.4) \qquad\qquad \Gamma(a + 1) = a\Gamma(a),$$

for any $a > 0$. Repeated application of (5.4) to $\Gamma(u + r)$ yields

$$\frac{\Gamma(1 + r)}{\Gamma(r)} = \frac{r\Gamma(r)}{\Gamma(r)} = r, \qquad \frac{\Gamma(2 + r)}{\Gamma(r)} = \frac{(1 + r)(r)\Gamma(r)}{\Gamma(r)} = (1 + r)(r),$$

and in general

$$(5.5) \qquad \frac{\Gamma(u + r)}{\Gamma(r)} = (u + r - 1)(u + r - 2) \cdots (r), \quad u = 2, 3, \ldots.$$

Thus,

$$(5.6) \qquad H(u, r) = \frac{\Gamma(u + r)}{\Gamma(r)\Gamma(u + 1)} = \frac{(u + r - 1)(u + r - 2) \cdots (r)}{u!},$$

since $\Gamma(u + 1) = u!$. For example,

$$H(3, 1.2) = \frac{(3.2)(2.2)(1.2)}{3!} = \frac{8.448}{6} = 1.408.$$

If Y has a generalized negative binomial distribution with parameters $r = 1.2$ and $p = 0.6$, then

$$p_Y(3) = H(3, 1.2)(0.6)^{1.2}(0.4)^3 = (1.408)(0.542)(0.064) = 0.0488.$$

The Mean and Variance of the Generalized Negative Binomial Distribution

To make explicit the dependence of the generalized negative binomial distribution on the parameters r and p, write

$$p_Y(u; r, p) = p_Y(u) = H(u, r)p^r(1 - p)^u.$$

Making use of (5.6), it is straightforward to show that $p_Y(u; r, p)$ satisfies the following recurrence relationship:

$$(5.7) \qquad u\, p_Y(u; r, p) = \frac{r(1 - p)}{p} p_Y(u - 1; r + 1, p), \quad u = 0, 1, \ldots.$$

This recurrence relationship can be helpful in the computation of the values of $p_Y(u)$.

It follows directly from (5.7) that if Y has a generalized negative binomial distribution with parameters r and p, then

(5.8)
$$\mu_Y = \frac{r(1-p)}{p}.$$

Further, applying (5.7) twice,

$$u(u-1)p_Y(u; r, p) = \frac{r(r+1)(1-p)^2}{p^2} p_Y(u-2; r+2, p),$$

for $u = 0, 1, 2, \ldots$, and thus

$$E[Y(Y-1)] = \frac{r(r+1)(1-p)^2}{p^2}$$

and

$$\sigma_Y^2 = E[Y(Y-1)] + \mu_Y - (\mu_Y)^2$$

(5.9)
$$= \frac{r(r+1)(1-p)^2}{p^2} + \frac{r(1-p)}{p} - \frac{r^2(1-p)^2}{p^2}$$

$$= \frac{r(1-p)}{p^2}.$$

Solving (5.8) and (5.9) for r and p in terms of μ_Y and σ_Y^2, we obtain

(5.10)
$$r = \frac{(\mu_Y)^2}{\sigma_Y^2 - \mu_Y}, \qquad p = \frac{\mu_Y}{\sigma_Y^2}.$$

Hence, μ_Y and σ_Y^2 could be used in place of r and p as parameters of the generalized negative binomial distribution (of Y).

Applications of the Generalized Negative Binomial Distribution

The generalized negative binomial distribution has been applied in a variety of contexts. A theoretical derivation of this distribution was given by Greenwood and Yule (1920) in an attempt to justify the close fit of this distribution to the probabilities associated with the number Y of accidents suffered by any given woman working on high explosive shells. The number of accidents suffered by a specific woman is assumed to have a Poisson distribution with a parameter λ; however, different women have different values of λ, and λ is assumed to be itself a random variable (defined over the population of all women) with a certain continuous distribution. The resulting distribution for Y is a generalized negative binomial distribution.

Greenwood and Yule used data gathered on 647 women observed over a period of 5 weeks in order to determine the parameters r and p and to demonstrate the fit of the negative binomial distribution to the data. Their data are shown in Table 5.1.

From Table 5.1,

$$\hat{\mu}_Y = (0)\left(\tfrac{447}{647}\right) + (1)\left(\tfrac{132}{647}\right) + \cdots + (5)\left(\tfrac{2}{647}\right) = \tfrac{301}{647} = 0.465$$

and

$$\hat{\sigma}_Y^2 = (0 - 0.465)^2\left(\tfrac{447}{647}\right) + (1 - 0.465)^2\left(\tfrac{132}{647}\right)$$

$$+ \cdots + (5 - 0.465)^2\left(\tfrac{2}{647}\right) = 0.691.$$

We can now use Equation (5.10) and the values of $\hat{\mu}_Y$ and $\hat{\sigma}_Y^2$ to estimate r and p:

$$\hat{r} = \frac{(0.465)^2}{0.691 - 0.465} = 0.96, \qquad \hat{p} = \frac{0.465}{0.691} = 0.67.$$

Remark. If $\hat{\mu}_Y > \hat{\sigma}_Y^2$, then our method for estimating r and p will yield inadmissible values for r and p, namely $r < 0$ and $p > 1$. Although $\hat{\mu}_Y$ can exceed $\hat{\sigma}_Y^2$ even when the data are observations on a random variable Y that has a generalized negative binomial distribution, this happening has a low probability. Thus, if we observe that $\hat{\mu}_Y$ exceeds $\hat{\sigma}_Y^2$, some distribution other than a generalized negative binomial distribution is likely to provide a better fit to the data.

Table 5.1: *Observed Frequency Distribution for the Number of Accidents Suffered by Women Working on 6-Inch Shells for 5 Weeks*

Number of accidents u	Observed frequency of u accidents
0	447
1	132
2	42
3	21
4	3
5	2
	647

Source: M. Greenwood and G. U. Yule (1920). An inquiry into the nature of frequency distributions representative of multiple happenings with particular reference to the occurrence of multiple attacks of disease or of repeated accidents. *Journal of the Royal Statistical Society.* Series A, vol. 83, pp. 255–79. Reprinted with permission.

To determine the probability mass function (5.2) that results from the values $r = 0.96$, $p = 0.67$ for the parameters r and p, we require values of $H(u, 0.96)$ and $(0.33)^u$ for $u = 0, 1, 2, \ldots$. Using Equation (5.6) to obtain $H(u, 0.96)$ and direct multiplication, we obtain the following results:

u	0	1	2	3	4	5	6
$H(u, 0.96)$	1	0.96	0.941	0.928	0.919	0.912	0.905
$(0.33)^u$	1	0.33	0.109	0.0359	0.0119	0.0039	0.0013

We also need the value of $(0.067)^{0.96}$, which equals 0.681. Substituting these results into Equation (5.2), we obtain the following table of theoretical probabilities for Y:

u	0	1	2	3	4	5	6
$p_Y(u)$	0.681	0.216	0.070	0.023	0.007	0.002	0.001

We note that for $u > 6$, $p_Y(u) = 0.000$ to three-decimal-place accuracy. Because there are a total of 647 observations, we must multiply each theoretical probability by 647 to obtain the theoretical frequency of observations yielding u accidents, $u = 0, 1, 2, 3, 4, 5, 6$. From Table 5.2 it appears that the negative binomial distribution with parameters $r = 0.96$ and $p = 0.67$ provides a rather good fit to the data.

Three additional applications of the generalized negative binomial distribution are summarized in the next examples.

Table 5.2: *Comparison of Actual Data and Theoretical Frequency Distribution Based on a Negative Binomial Distribution with* $r = 0.96$, $p = 0.67$

Number of accidents u	Observed frequency	Theoretical frequency
0	447	440.6
1	132	139.8
2	42	45.3
3	21	14.9
4	3	4.5
5	2	1.3
6	—	0.6
	647	647.0

Example 5.3 (Geology). Griffiths (1960) obtained zircon counts in rock samples taken from several counties in Pennsylvania. The zircon counts in his studies followed a probability distribution closely approximated by a generalized negative binomial distribution with parameters $r = 0.239$ and $p = 0.186$.

Example 5.4 (Market Research). A number of studies of buying habits of consumers are summarized by Chatfield (1970). In one of these studies a random sample of consumers kept a record of their own purchases over an extended time period. Analyses were made every 4 weeks, and the number Y of units purchased of a particular package size of a particular brand was recorded (see Table 5.3). From Table 5.3, we can calculate

$$\hat{\mu}_Y = 0.08857, \qquad \hat{\sigma}_Y^2 = 0.28073$$

and use these estimates and Equation (5.10) to obtain

$$\hat{r} = \frac{(0.08857)^2}{0.28073 - 0.08857} = 0.041, \qquad \hat{p} = \frac{0.08857}{0.28073} = 0.315.$$

These estimates of r and p are used to obtain the theoretical probabilities $p_Y(u)$, $u = 0, 1, \ldots, 9$, and $P\{Y \geq 10\} = 1 - \sum_{u=0}^{9} p_Y(u)$. The theoretical frequencies are obtained by multiplying each of the theoretical probabilities by 1750. As shown in Table 5.3, the observed and theoretical frequencies agree fairly closely.

Table 5.3: *Comparison of Observed and Theoretical Frequency Distributions for Number of Units Purchased in 4 Weeks of a Particular Brand of Item*

Number of units purchased, u	Observed frequency	Theoretical frequency: $r = 0.041$, $p = 0.315$
0	1671	1669.49
1	43	46.65
2	19	16.62
3	9	7.74
4	2	4.03
5	3	2.23
6	1	1.28
7	0	0.76
8	0	0.46
9	2	0.28
≥ 10	0	0.46
	1750	1750.00

Example 5.5 (Transportation). In Example 4.3 of Section 4 we noted that the geometric distribution served as a reasonably good model for the probability distribution for the number X of occupants in a car. Since X cannot be 0 (there must be at least one occupant, the driver, in a car), it does not seem profitable to try to fit a negative binomial distribution to the distribution of X. We can, however, try to fit a negative binomial ditribution to the distribution of the number, $Y = X - 1$, of occupants of the car who are passengers (i.e., not driving). Note that if X has a geometrical distribution, $Y = X - 1$ has a negative binomial distribution with $r = 1$. Thus, if we find that a negative binomial distribution with $r \neq 1$ gives a good fit to the data, closer investigation should be made as to which model (geometrical or negative binomial with $r \neq 1$) is more appropriate. The data are given in Table 5.4. In comparing these data with the data given in Section 4, Example 4.3, remember that the number of passengers in a car is equal to the number of occupants minus 1 (for the driver).

From the raw data,

$$\hat{\mu}_Y = 0.52, \qquad \hat{\sigma}_Y^2 = 0.722,$$

and thus the estimated values of r and p are

$$\hat{r} = \frac{(0.52)^2}{0.722 - 0.52} = 1.337, \qquad \hat{p} = \frac{0.52}{0.722} = 0.720.$$

Although the value of r obtained in this manner does not differ radically from the value $r = 1$ expected if the geometric distribution holds, the values of p obtained from the generalized negative binomial and geometric distributional assumptions differ considerably. The determination of which kind of distribution provides a better fit to the data requires more sensitive techniques of analysis.

Table 5.4: Comparison of Observed and Theoretical Frequency Distributions for the Number of Passengers in a Car

Number of passengers u	Observed frequency	Theoretical frequency: $r = 1.337$, $p = 0.720$
0	678	685
1	227	206
2	56	74
3	28	28
4	8	11
≥ 5	14	7
	1011	1011

Tables of the Generalized Negative Binomial Distribution

Extensive tables of the probability mass function of the generalized negative binomial distribution are given in Williamson and Bretherton (1963). In using the tables of Williamson and Bretherton, the reader should be aware that in their notation for the probability mass function (5.2), their k is our r, their n is our p, and their p is our u.

6. DISCRETE UNIFORM DISTRIBUTION

Suppose that K slips of paper labeled $1, 2, \ldots, K$ are placed in a hat. One slip is drawn from the hat in such a way that each slip in the hat has the same chance of being selected. Let the random variable X denote the number selected. Then the probability mass function of X has the constant value $1/K$ for all possible values k of X, $k = 1, 2, 3, \ldots, K$. That is, the probability mass function, $p_X(k)$, of X has the form

(6.1) $$p_X(k) = \frac{1}{K}, \quad k = 1, 2, \ldots, K,$$

or equivalently,

x	1	2	$\cdots$	K
$p_X(x)$	$\dfrac{1}{K}$	$\dfrac{1}{K}$	$\cdots$	$\dfrac{1}{K}$

The distribution given by (6.1) is called the *discrete uniform distribution.*

The constant K in (6.1) is the parameter of the discrete uniform distribution; once K is known, the uniform distribution is completely determined. Unlike the other discrete distributions considered in Sections 1 through 5, in most applications of the discrete uniform distribution, the parameter K is known. However, there are some exceptions.

Example 6.1 (Estimating the Size of Enemy Production). During World War II, it was of considerable interest to learn the number of new tanks that the enemy was producing. An observer sent in behind enemy lines might obtain an estimate of this number by the following argument: "Every tank made by the enemy is given a consecutive serial number according to the order in which it is manufactured. Assuming that I know the smallest such serial number (this might be the serial number of the last tank observed of the enemy's previous production), I can subtract this number from the observed serial number of any new tank that I see. The resulting possible numbers are then $1, 2, 3, \ldots, K$, where K is the total number of new tanks produced by the enemy. Assuming that I have equal probability of seeing every new enemy tank while I am behind the lines, the random variable X, which equals the serial number of the new tank that I observe minus the serial number of the last tank observed of the enemy's previous production, has the discrete uniform distribution with parameter K."

The Mean and Variance of the Uniform Distribution

The following are two well-known summation identities:

$$\sum_{i=1}^{K} i = \frac{K(K+1)}{2}, \qquad \sum_{i=1}^{K} i^2 = \frac{K(K+1)(2K+1)}{6}.$$

Using these identities, the mean and variance of a random variable X having the discrete uniform distribution with parameter K is given by

(6.2) $$\mu_X = \sum_{i=1}^{K} (i)\left(\frac{1}{K}\right) = \frac{K+1}{2},$$

and

(6.3) $$\sigma_X^2 = E(X^2) - (\mu_X)^2 = \sum_{i=1}^{K} (i^2)\left(\frac{1}{K}\right) - \left(\frac{K+1}{2}\right)^2$$

$$= \frac{(K+1)(2K+1)}{6} - \left(\frac{K+1}{2}\right)^2 = \frac{K^2 - 1}{12}.$$

Given several observations on a random variable which is thought to have a discrete uniform distribution with parameter K, we could use the sample average $\hat{\mu}_X$ to estimate μ_X and from that estimate obtain an estimate of K (i.e., $K = 2\hat{\mu}_X - 1$). However, the method usually used in practice is to set K equal to the largest value of X that is observed.

Example 6.1 (continued). The observer behind enemy lines records the following values of X (the difference between an observed tank serial number and the highest previously observed serial number): 53, 146, 397, 601, 747, 1048, 1219, 1428, 1621, 1825. The average of these 10 numbers is $\hat{\mu}_X = 908.5$, so that if we estimated K by $2\hat{\mu}_X - 1 = 1816$, we would obtain a value of K under which the probability model (6.1) would assign the *observed* value 1825 a probability of zero. On the other hand, the estimate of K that sets K equal to the largest observed value of X (i.e., $K = 1825$) cannot result in a discrete uniform probability distribution whose assignment of probabilities conflicts with what is observed. Thus, the observer should report that the enemy has produced (at least) 1825 new tanks.

Lotteries

The discrete uniform distribution is of particular importance in lotteries. The principle that each value of a random variable X has the same chance of occurring is an essential feature of lotteries. Although lotteries can take place without any use of numbers at all (e.g., by drawing slips of paper with names written on them from some container), it is frequently useful to index participants in a lottery by some identifying number (e.g., social security number, number of the raffle ticket, birth date, license number), and then select a number from among all such numbers according to the probability model (6.1).

The use of a lottery as a fair method of choosing among rival candidates or alternatives is quite old. The Old Testament contains a number of examples in which lotteries were used to determine the division of land, the choice of a scapegoat as a sacrifice, the allocation of duties in the temple, as well as in the division of an estate between brothers. [For details concerning these examples, see Hasofer (1970) and Rabinovitch (1973).]

More recently, the lottery has been used in a variety of circumstances. Its use in gambling is traditional. Lotteries are also often used to select random samples of units, individuals, and so on, for various purposes in scientific research. Jury lists are often selected by a lottery from among all eligible voters in a given community. In the state of Arizona there is an annual hunt of buffalo. The U.S. Department of the Interior determines how thin the herd should be and designates certain animals for the hunt. Hunters register for the hunt and are then chosen by lottery (*New York Times*, October 12, 1971, p. 45).

Perhaps the most controversial lotteries in the United States in this century have been the draft lotteries of 1917, 1940, and 1970. Although there has been controversy concerning the draft lotteries, the issue lies not in whether a lottery in itself is a fair procedure, but rather in the question whether equal probabilities were guaranteed to all the possible outcomes. [For details, see Fienberg (1971).]

Recently, various states have used lotteries as a source of income. For example, New Jersey has a lottery in which a single winner of $50,000 is chosen from a total of 1,000,000 ticket holders. This is done by requiring that all 6 digits on the winning ticket match the number chosen by a certain process. This process acts as if it generates a random variable X having a discrete uniform distribution with parameter $K = 1,000,000$, and then subtracts 1 from the observed value of X (i.e., the process chooses one of the numbers 0 through 999,999, giving each such number equal probability of being chosen).

Generalized Discrete Uniform Distribution

In general, a random variable Y of the form $Y = c + hX$, where X has a discrete uniform distribution (6.1) with some parameter K, h is a given positive constant, and c is a given number, is frequently of interest. The random variable Y then has possible values $c + h, c + 2h, c + 3h, \ldots, c + Kh$, and the probability mass function $p_Y(X)$ is given by

$$(6.4) \qquad p_Y(X) = \begin{cases} \dfrac{1}{K}, & \text{if } X = c + h, c + 2h, c + 3h, \ldots, c + Kh, \\ 0, & \text{otherwise.} \end{cases}$$

A probability distribution defined by (6.4) generalizes the discrete uniform distribution (6.1). Such a probability distribution is appropriate whenever the possible values of a random variable Y are equally spaced and equally probable. When Y has the generalized discrete uniform distribution defined by the probability mass function (6.4), the mean μ_Y and the variance σ_Y^2 of Y are given by

$$(6.5) \qquad \mu_Y = c + h\left(\frac{K+1}{2}\right), \qquad \sigma_Y^2 = h^2\left(\frac{K^2-1}{12}\right),$$

as can be seen by noting that $\mu_{c+hX} = c + h\mu_X$, $\sigma_{c+hX}^2 = h^2\sigma_X^2$.

7. OTHER DISCRETE DISTRIBUTIONS

Because researchers are interested in a wide variety of phenomena, we cannot expect any short list of distributions to serve as a model for all data. Very often special classes of distributions are developed in order to deal with a particular study, only to find that these distributions are useful in other contexts. In this section, we give an example of two such special classes of distributions.

The Zeta Distribution

Many discrete quantitative phenomena in the behavioral sciences appear to behave according to a "the-higher-the-fewer" rule. For example, the higher the value x of an executive's salary X, the fewer executives there are who will have that salary; the higher the frequency x of a word in prose, the fewer are the words which have that frequency; the larger the size x of a claim against an insurance company, the fewer claims of that size there are; and so on. We might try to approximate the probability mass function of a random variable X that obeys such a the-higher-the-fewer rule by a function of the form

$$(7.1) \qquad p_X(x) = P\{X = x\} = c\,\frac{1}{x^{\alpha+1}}, \qquad x = 1, 2, \ldots,$$

with $p_X(x)$ equal to 0 otherwise. Here the constant α, $\alpha > 0$, measures the rate of decrease of the probability of the value x with increase in x. Since

$$(7.2) \qquad 1 = \sum_{x=1}^{\infty} p_X(x) = c \sum_{x=1}^{\infty} \frac{1}{x^{\alpha+1}}$$

if (7.1) is to be a probability mass function, the constant c is determined by α, and α is the sole parameter of the distribution (7.1). The function

$$(7.3) \qquad \zeta(s) = \frac{1}{1^s} + \frac{1}{2^s} + \frac{1}{3^s} + \cdots = \sum_{k=1}^{\infty} \frac{1}{k^s}, \quad s > 1,$$

is well known and arises in many mathematical disciplines. It is called the *Riemann zeta function* [after the German mathematician Georg Friedrich Bernhard Riemann (1826–1866)] and is extensively tabulated [e.g., see Davis (1935)]. From (7.2), we see that $c = 1/\zeta(\alpha + 1)$, and hence

$$(7.4) \qquad p_X(x) = \begin{cases} \dfrac{1}{\zeta(\alpha + 1)} \dfrac{1}{x^{\alpha+1}}, & \text{if } x = 1, 2, 3, \ldots, \\ 0, & \text{otherwise.} \end{cases}$$

A probability distribution of the form (7.4) was used by the Italian economist and sociologist Vilfredo Pareto (1848–1923) to describe the distribution of family incomes. In Pareto's studies, the law fairly accurately described income distribution in almost every country that was developed enough to keep income statistics. Even after the progressive tax legislation adopted in almost every industrialized country in the present century, the probability law (7.4) provides a reasonably good fit to personal income data.

Despite Pareto's contribution to the development of probability laws of the form of (7.4), such probability laws are usually associated with the name of G. K. Zipf, since it was Zipf who applied such distributions to a variety of contexts [Zipf (1949)] and, through the success of such applications, succeeded in popularizing distributions of this form.

We have already mentioned the relationship of the probability mass function (7.4) to the Riemann zeta function. Because of this relationship, the probability distribution (7.4) is also known to some probabilists and statisticians as the *zeta distribution*. If X is a random variable having a zeta distribution with parameter α, then

$$(7.5) \qquad \mu_X = \frac{\zeta(\alpha)}{\zeta(\alpha + 1)}, \qquad \sigma_X^2 = \frac{\zeta(\alpha - 1)\zeta(\alpha + 1) - [\zeta(\alpha)]^2}{[\zeta(\alpha + 1)]^2}.$$

since

$$EX^r = \sum_{\tau=1}^{\infty} \tau^r \frac{1}{\zeta(\alpha + 1)\tau^{\alpha+1}} = \frac{1}{\zeta(\alpha + 1)} \sum_{\tau=1}^{\infty} \frac{1}{\tau^{\alpha-r+1}} = \frac{\zeta(\alpha - r + 1)}{\zeta(\alpha + 1)}.$$

In applying (7.5), it should be noted that $\zeta(w) = \infty$ for $w \leq 1$. Thus, when X has a zeta distribution with parameter α, the variance of X is infinite when $\alpha \leq 1$.

To determine the value of the parameter α from the data, we estimate μ_X by means of the sample average $\hat{\mu}_X$, and then solve

$$\hat{\mu}_X = \frac{\zeta(\alpha)}{\zeta(\alpha + 1)}$$

for α. Because fitting the probability distribution (7.4) to the data requires some laborious interpolations in tables of $\zeta(s)$, we will not attempt here to give an example showing how to fit this distribution. However, the following example of the application of the probability distribution (7.4) illustrates the usefulness of the zeta distribution in describing the variation of random variables observed in many contexts of human behavior.

Example 7.1 (Steel Production). Simon and Bonini (1958) ranked steel producers according to size. Out of the total capacity of the top 10 steel producers, they asked how well the distribution (7.4) with $\alpha = 1$ described the proportion $p_X(k)$ of this total capacity produced by the steel producer whose size rank X was equal to k. That is, suppose that we mark every ton of steel produced by the kth largest steel producer with the rank k of that producer, $k = 1, 2, \ldots, 10$. From the steel potentially produced by the 10 leading steel companies in a given year, we can conceive of drawing 1 ton of steel through a random sample. The number k appearing on that ton of steel is an observed value of the random variable X which gives the rank (in size) of the steel producer that produced that ton of steel. The ratio of the capacity of the steel producer having size rank k to the total capacity of all 10 steel firms then has the properties of a relative frequency (from among 99,400,000 trials of the random experiment described above), and the capacity of the kth largest steel producer is like the frequency of the event $\{X = k\}$ in those 99,400,000 trials. This discussion provides motivation for Table 7.1, which is taken from the article of Simon and Bonini and shows how well the actual capacities of the steel companies agreed with the predicted theoretical capacities:

$$\text{Theoretical capacity of }k\text{th largest firm} = (99,400,000)\frac{1}{\zeta(2)}\frac{1}{k^2}$$

obtained from a distribution of the form (7.4) with $\alpha = 1$.

The Truncated Poisson Distribution

In some experiments, we may believe that we are observing a random variable X which has a Poisson distribution. Because of the nature of the phenomenon, however, the event $\{X = 0\}$ cannot occur. For example, we may

Table 7.1: *Comparison of Actual and Theoretical Ingot Capacities of 10 Leading Steel Producers in the United States*

Producer	Rank order	Actual capacity	Theoretical capacity: zeta distribution with $\alpha = 1$
U.S. Steel	1	38.7	34.3
Bethlehem	2	18.5	17.1
Republic	3	10.3	11.3
Jones and Laughlin	4	6.2	8.5
National	5	6.0	6.8
Youngstown	6	5.5	5.2
Armeo	7	4.9	4.8
Inland	8	4.7	4.2
Colorado Fuel and Iron	9	2.5	3.8
Wheeling	10	2.1	3.4
		99.4	99.4

believe that the number X of people in a certain kind of social group has a Poisson distribution. The event "the group has 0 people" cannot happen, however, because a group by definition has to have at least 1 member. The resulting range of possible values of X is thus "truncated" to exclude the value 0.

If X is assumed to have a Poisson distribution with some parameter λ, the fact that the event $\{X = 0\}$ cannot occur, or, equivalently, that the event $\{X > 0\}$ has occurred, means that we must redefine the probability model *conditional* upon this fact. Thus, the probability of the event $\{X = k\}$ is 0 if $k = 0$ and equals

$$P(\{X = k\} \mid \{X > 0\}) = \frac{P\{X = k \text{ and } X > 0\}}{P\{X > 0\}}$$

(7.6)
$$= \frac{P\{X = k\}}{1 - P\{X = 0\}} = \frac{(\lambda^k/k!)e^{-\lambda}}{1 - e^{-\lambda}}$$

$$= \frac{\lambda^k}{k!\,(e^\lambda - 1)}$$

for $k = 1, 2, \ldots$.

A random variable Y having a probability mass function

(7.7)
$$p_Y(y) = \begin{cases} \dfrac{\lambda^y}{y!\,(e^\lambda - 1)}, & \text{if } y = 1, 2, \ldots, \\ 0, & \text{otherwise,} \end{cases}$$

has the same probabilities as the (conditional) probabilities of a Poisson random variable X with parameter λ truncated at 0, and thus Y is said to have a *truncated Poisson distribution* with parameter λ. The mean and variance of such a distribution are

(7.8)
$$\mu_Y = \frac{\lambda e^\lambda}{e^\lambda - 1}, \qquad \sigma_Y^2 = \frac{\lambda e^{2\lambda} - \lambda(\lambda + 1)e^\lambda}{(e^\lambda - 1)^2}.$$

These formulas follow directly from the identity

$$
\begin{aligned}
E[Y(Y-1) \cdots (Y-r+1)] &= \sum_{y=1}^\infty (y)(y-1) \cdots (y-r+1)\frac{\lambda^y}{y!\,(e^\lambda - 1)} \\
&= \sum_{y=r}^\infty \frac{\lambda^r \lambda^{y-r}}{(y-r)!\,(e^\lambda - 1)} = \frac{\lambda^r}{e^\lambda - 1} \sum_{i=0}^\infty \frac{\lambda^i}{i!} \\
&= \frac{\lambda^r e^\lambda}{e^\lambda - 1}
\end{aligned}
$$

for $r = 1, 2, \ldots$.

The parameter λ of the truncated Poisson distribution can be estimated from data by calculating the sample average $\hat{\mu}_Y$ and solving

(7.9)
$$\hat{\mu}_Y = \frac{\lambda e^\lambda}{e^\lambda - 1}$$

for λ. Table 7.2 provides solutions to (7.9) corresponding to given values μ of $\hat{\mu}_Y$. For example, if $\hat{\mu}_Y = 2.500$, then from Table 7.2, the solution to (7.9) is $\lambda = 2.231$.

Table 7.2: *Values of λ Corresponding to Given Values of μ*

Between 1.00 and 1.99

μ	0.00	0.01	0.02	0.03	0.04	0.05	0.06	0.07	0.08	0.09
1.0	0.000	0.020	0.040	0.060	0.078	0.098	0.118	0.137	0.155	0.175
1.1	0.193	0.212	0.232	0.250	0.268	0.287	0.305	0.323	0.340	0.358
1.2	0.377	0.395	0.412	0.430	0.447	0.465	0.482	0.498	0.515	0.533
1.3	0.550	0.567	0.583	0.600	0.617	0.633	0.650	0.667	0.683	0.699
1.4	0.715	0.732	0.748	0.763	0.780	0.795	0.812	0.827	0.843	0.858
1.5	0.874	0.890	0.905	0.920	0.936	0.952	0.967	0.982	0.997	1.012
1.6	1.027	1.042	1.057	1.072	1.087	1.102	1.117	1.131	1.146	1.160
1.7	1.175	1.190	1.204	1.218	1.233	1.247	1.262	1.276	1.290	1.304
1.8	1.318	1.332	1.347	1.360	1.375	1.389	1.402	1.416	1.430	1.444
1.9	1.458	1.471	1.485	1.499	1.513	1.526	1.540	1.553	1.567	1.580

(continued)

Table 7.2: (continued)

Between 2.0 and 9.9

μ	0.0	0.1	0.2	0.3	0.4	0.5	0.6	0.7	0.8	0.9
2.0	1.594	1.726	1.856	1.984	2.109	2.231	2.353	2.472	2.590	2.706
3.0	2.821	2.935	3.048	3.160	3.271	3.381	3.490	3.599	3.707	3.814
4.0	3.921	4.027	4.133	4.238	4.343	4.447	4.551	4.655	4.759	4.862
5.0	4.965	5.068	5.170	5.273	5.375	5.477	5.579	5.681	5.782	5.884
6.0	5.985	6.086	6.187	6.288	6.389	6.490	6.591	6.692	6.792	6.893
7.0	6.994	7.094	7.195	7.295	7.395	7.496	7.596	7.696	7.797	7.897
8.0	7.997	8.098	8.198	8.298	8.398	8.498	8.598	8.699	8.799	8.899
9.0	8.999	9.099	9.199	9.299	9.399	9.499	9.599	9.699	9.799	9.900

Example 7.2 (Social Groups). James (1953) studied the size Y of small groups that form in a variety of social contexts. Tables 7.3 through 7.5 show the fit of truncated Poisson distributions with parameters $\lambda = 0.892$, $\lambda = 0.889$, and $\lambda = 1.362$ to the distribution of the number of pedestrians grouped together on a spring morning in Eugene, Oregon, the distribution of the size of shopping groups in department stores and public markets, and the distribution of the size of playgroups in the playgrounds of 14 elementary schools. The theoretical frequencies shown in Tables 7.3, 7.4, and 7.5 in each case were obtained through the following steps:

(1) Compute the sample average $\hat{\mu}_Y$ computed from the tabled data.
(2) Solve for λ in (7.9) using interpolation in Table 7.2. (The values for λ shown in Tables 7.3, 7.4, and 7.5 actually come from more detailed tables than Table 7.2)
(3) Using this value of λ, compute $p_Y(y)$ in (7.7) for $y = 1, 2, \ldots$, and then find the theoretical frequencies by multiplying each theoretical probability $p_Y(y)$ by the number N of groups observed.

Table 7.3: *Size of Groups of Pedestrians in Eugene, Oregon, on a Spring Morning*

Size of group, k	Observed frequency	Theoretical frequency (truncated Poisson, $\lambda = 0.892$)
1	1486	1501
2	694	670
3	195	199
4	37	44
5	10	8
6	1	1
	$N = \overline{2423}$	$\overline{2423}$

Table 7.4: *Size of Shopping Groups*

Size of group, k	Observed frequency	Theoretical frequency (truncated Poisson, $\lambda = 0.889$)
1	316	316
2	141	141
3	44	42
4	5	9
5	4	2
	$N = 510$	510

Table 7.5: *Size of Playgroups*

Size of group, k	Observed frequency	Theoretical frequency (truncated Poisson, $\lambda = 1.362$)
1	570	599
2	435	408
3	203	185
4	57	63
5	11	17
6	1	4
7	0	1
	$N = 1277$	1277

Source: Tables 7.3, 7.4, and 7.5 from C. Chatfield (1970). Distributions in market research, in G. P. Patil, ed., *Random Counts in Physical Science, Geo Science, and Business*, pp. 163–81. Reprinted by permission of the Pennsylvania State University Press.

In each of Tables 7.3, 7.4, and 7.5, the resulting theoretical frequencies are reasonably close to the actual frequencies, indicating that a truncated Poisson distribution provides a reasonable approximation to the actual distribution of the group sizes.

SUMMARY OF DISCRETE DISTRIBUTIONS

The various discrete distributions discussed in Sections 1 through 6 of this chapter are listed below, together with their parameters, means, and variances. In this summary, p is always a number in the interval $[0, 1]$ and $q = 1 - p$. The values of the probability mass function are given only for possible values of the associated random variable.

Distribution	Probability mass function	Parameters	Mean	Variance
Bernoulli	$p^k q^{1-k}$ for $k = 0, 1$	p	p	pq
Binomial	$\binom{n}{k} p^k q^{n-k}$ for $k = 0, 1, \ldots, n$	p $n = 1, 2, \ldots$	np	npq
Hypergeometric	$\dfrac{\binom{Np}{k}\binom{Nq}{n-k}}{\binom{N}{n}}$ for $k = 0, 1, \ldots, n,$ $k \le Np, n - k \le Nq$	p $N = 1, 2, \ldots$ $n = 1, 2, \ldots, N$ (Np must be an integer)	np	$\left(\dfrac{N-n}{N-1}\right)npq$
Poisson	$\dfrac{e^{-\lambda}\lambda^k}{k!}$ for $k = 0, 1, 2, \ldots$	$\lambda > 0$	λ	λ
Geometric (X = number of trials)	$q^{k-1}p$ for $k = 1, 2, \ldots$	p	$\dfrac{1}{p}$	$\dfrac{1-p}{p^2}$
Negative binomial (Y = number of failures)	$\binom{r+k-1}{r-1} p^r q^k$ for $k = 0, 1, 2, \ldots$	p $r = 1, 2, \ldots$	$\dfrac{r(1-p)}{p}$	$\dfrac{r(1-p)}{p^2}$
Generalized negative binomial	$H(k, r)p^r q^k$ for $k = 0, 1, 2, \ldots$	p $r > 0$	$\dfrac{r(1-p)}{p}$	$\dfrac{r(1-p)}{p^2}$
Discrete uniform	$\dfrac{1}{K}$ for $k = 1, 2, \ldots, K$	$K = 1, 2, \ldots$	$\dfrac{K+1}{2}$	$\dfrac{K^2-1}{12}$
Generalized discrete uniform	$\dfrac{1}{K}$ for $k = c + h, c + 2h, \ldots,$ $c + Kh$	$K = 1, 2, \ldots$ $-\infty < c < \infty$ $h > 0$	$c + h\left(\dfrac{K+1}{2}\right)$	$h^2\left(\dfrac{K^2-1}{12}\right)$

NOTES AND REFERENCES

The books by Patil (1965) and Johnson and Kotz (1969) discuss a number of discrete distributions that have been used in theory and practice, and provide an extensive list of references.

EXERCISES

1. An examination consists of 10 multiple-choice questions. Each question has 5 possible answers, of which only one is correct. Suppose that a student takes this examination and independently guesses the answer to each question.
   (a) If Z denotes the number of correct answers obtained on the examination, what is the probability distribution of Z?
   (b) Find the mean, μ_Z; median, Med(Z); and variance, σ_Z^2, of Z.
   (c) Suppose that 5 or more correct answers constitutes a passing grade. What is the probability that the student passes?

2. Another student takes the examination described in Exercise 1, but this student has studied, and consequently has a probability of 0.8 of correctly answering a question. Complete parts (a) through (c) of Exercise 1 for this student.

3. Five individuals are chosen from among the registered voters of a community and are asked if they favor a certain bond issue. Assume that the choice of these 5 individuals can be regarded as resulting from a simple random sample with replacement. If only 30% of the voters in the community favor the bond issue, what is the probability that the majority (more than half) of the 5 individuals chosen will favor the proposal?

4. The Internal Revenue Service has a special letter-opening machine that opens and removes the contents from an envelope. If the envelope is fed improperly into the machine, the contents of the envelope may not be removed or may be damaged. In this case, we say that the machine has "failed." Assume that the fate of each envelope entering the machine is independent of the fate of all other envelopes handled by that machine.
   (a) If the machine has a probability of failure of 0.1, what is the probability of more than 3 failures occurring in a batch of 10 envelopes?
   (b) Suppose that the probability of failure of the machine is 0.02. What is the probability that more than 3 failures of the machine will occur in a batch of 100 envelopes?
   (c) If the probability of failure of the machine is 0.01, what is the probability that more than 3 failures of the machine will occur in a batch of 100 envelopes?

5. In some military courts, 9 judges are appointed. Both the prosecution and the defense attorneys are entitled to a peremptory challenge of any judge, in which case that judge is removed from the case and is *not* replaced. Suppose that the prosecution attorney does not exercise his right to challenge the selection of judges, and the defense is limited to

two challenges. A defendant is declared guilty only if there are a majority of votes cast in favor of "guilty," and is declared innocent otherwise. Suppose that each judge has a probability of 0.6 of voting "guilty," and that judges make their decisions independently.

If Z denotes the *number* of votes for a verdict of "guilty," find the values of the probability mass function for Z when there are

(a) 9 judges,   (b) 8 judges,   (c) 7 judges.

What is the probability that the defendant is declared guilty when there are

(d) 9 judges,   (e) 8 judges,   (f) 7 judges?

6. The following question is, in essence, the question which the Chevalier de Méré asked Pascal (see Chapter 1): Which of the following probabilities is the larger?

(i) The probability that a "6" comes up at least *once* in 4 tosses of a balanced die.

(ii) The probability that *at least once* in 24 tosses of 2 balanced dice, both dice simultaneously show a "6."

(The Chevalier de Méré thought that he had given a theoretical argument which proved that these probabilities are equal, yet his observations suggested otherwise. He asked Pascal to explain the apparent contradiction.)

7. There is currently much discussion concerning minority representation in a venire (panel from which a jury is chosen). Complete the following table assuming that choice of one individual for the panel is statistically independent of the choice of any other individual for the panel. In a recent study, 30 venires were sampled and all had 5 or fewer nonwhites. Is this result surprising? [Note: For further discussion of such problems, see Finkelstein (1966).]

Probability p of selecting a single nonwhite venireman	Probability P of selecting a single venire of 30 with 5 or fewer nonwhites	Probability P^{30} of selecting 30 such venires
0.20		
0.10		
0.05		

8. Suppose that Z has a binomial distribution with parameters $n = 8$ and p. For $p = 0.3$ and $p = 0.5$,

(a) calculate the values of $p_Z(k)$, $k = 0, 1, \ldots, 8$;

(b) find the c.d.f. $F_Z(z)$ for all values of z;

(c) find the mode of Z;

(d) find the median of Z.

9. Suppose that the random variable Z has a binomial distribution with parameters n and p.
   (a) If $p = 0.6$ and $\mu_Z = 6$, find n and σ_Z^2.
   (b) If $\mu_Z = 6.0$ and $\sigma_Z^2 = 4.2$, find n and p.
   (c) If $n = 25$ and $\mu_Z = 10$, find p and σ_Z^2.
   (d) Is it possible for $\mu_Z = 3$ and $\sigma_Z^2 = 5$ if Z has a binomial distribution?

10. Show that if Z has a binomial distribution with parameters n and p, then for $r = 3, 4$,
$$E[Z(Z - 1) \cdots (Z - r + 1)] = (n)(n - 1) \cdots (n - r + 1)p^r.$$
   Use this result to find $\mu_Z^{(3)}$ and $\mu_Z^{(4)}$ and state your conclusions concerning the skewness and kurtosis of binomial distributions for various values of p.

11. A large number (53,680) of German families with 8 children were contacted and, for each family, the number X of male children was noted. The results are given in Table E.1. Fit a binomial distribution to these data. [Remark. These data are taken from a study by Geissler and are given by R. A. Fisher (1950). *Statistical Methods for Research Workers* 11th ed. Edinburgh: Oliver & Boyd Ltd. Geissler's data are also discussed in G. G. Simpson, A. Roe, and R. C. Lewontin (1960). *Quantitative Zoology*. New York: Harcourt, Brace & World, Inc.]

Table E.1: *Results of Observing the Number X of Male Children in Each of 53,680 German Families*

Number x of boys in German families of 8 children	Observed frequency
0	215
1	1485
2	5331
3	10,649
4	14,959
5	11,929
6	6678
7	2092
8	342
	53,680

12. In the study of Mendelian laws of inheritance, litters of 5 mice were studied and the number of "dominant" mice in each litter noted. The

actual mating was a "backcross," that is, a mating of a parent having
one dominant and one recessive allele with a parent having two recessive alleles. See Table E.2. If the theoretical probability of a "dominant"
character is 0.5 for each mouse, what are the theoretical frequencies that
you would obtain using the binomial distribution? [Remark: Data from
J. A. Detlefsen (1918). Fluctuations of sampling in a Mendelian population. *Genetics*, vol. 3, pp. 599–607. A discussion of these data can also
be found in S. Wright (1968). *Genetic and Biometric Foundations*, vol. 1.
Chicago: University of Chicago Press.]

Table E.2: *Results of Observing the
Number X of Mice with the Dominant
Character in 330 Litters of 5 Mice*

Number x of mice in each litter that show the dominant character	Observed frequency
0	9
1	47
2	106
3	103
4	51
5	14
	330

13. Suggest how to estimate p from data if you do not have theoretical
reasons to specify the value of p. Using the data of Exercise 12, estimate
the value of p and find theoretical frequencies using this estimated value
of p. Which value of p seems to fit the data more closely, the theoretical
value $p = 0.5$, or your estimated value?

14. Suppose that 20% of all patients who come to a clinic and exhibit
certain symptoms have a certain disease. Final diagnosis of this disease
depends on a blood test. Sixty individuals with symptoms of the disease
come to the clinic. Individual blood tests are expensive, so the hematol-
ogist uses the following screening device. The blood of a number n of
individuals is combined and tested. If none of the n persons has the
disease, then the composite blood test is negative. However, if at least
one person has the disease, the composite test will be positive. Assuming

that the n persons whose blood is tested are chosen from among the 60 individuals by a simple random sample *without replacement*, what is the probability that the composite blood test will be negative if
(a) $n = 2$, (b) $n = 4$, (c) $n = 6$, (d) $n = 10$?

15. Consider a group of 10 individuals consisting of 7 males and 3 females. We choose a committee of n people from among these 10 individuals by means of a simple random sample without replacement. We want the committee size n to be as small as possible but sufficiently large that the probability that the committee will have at least 1 male and at least 1 female is no less than 0.9. How large should the committee size n be?

16. A radio manufacturer intends to purchase 100 radio tubes from a supply house. He expects that some of these tubes will be defective but is willing to tolerate only 4 defective tubes in the batch of 100 tubes. He checks the quality of the batch by drawing a random sample of 3 tubes without replacement from among the 100 tubes, and testing these tubes.
    (a) Let X be the number of defective tubes in the sample of 3 tubes. Determine the values of the probability mass function $p_X(k)$ of X if there are exactly 4 defective tubes in the batch of 100 tubes.
    (b) What is the mode of the distribution of X? What is the median of X? What is the mean of X?
    (c) The radio manufacturer decides to reject the batch of radio tubes if X exceeds 1. What is the probability that he will reject the batch of tubes?

17. Let Z be a random variable that has a hypergeometric distribution with parameters $N = 10$, $n = 8$, and p. Find the probability mass function $p_Z(k)$ and the c.d.f. $F_Z(z)$ of Z when $p = 0.3$ and $p = 0.5$. Compare these mass functions and c.d.f.'s with those that you obtained for the binomial distribution for similar values of p.

18. Repeat Exercise 17 for $N = 100$.

19. Let A_k be the number of ways that we can choose an ordered sample of n items from a population of size N containing M successes so that exactly k successes appear in the sample. Prove that $\sum_{k=0}^{n} A_k = N!/(N - n)!$.

20. Let X have a hypergeometric distribution with parameters N, $n = 5$, and p.
    (a) If $\mu_X = 1$, $\sigma_X^2 = 0.40$, find N and p.
    (b) Can you find the value of N if you only know the values of p and μ_X? Can you find the value of N if you know p and σ_X^2? Can you find the value of N if you know Np and μ_X?

21. Using (2.4) of Section 2 show that if X has a hypergeometric distribution with parameters n, N, and p, then for $r = 3, 4$,

$$E[X(X-1) \cdots (X-r+1)]$$
$$= \frac{(n)(n-1) \cdots (n-r+1)(Np)(Np-1) \cdots (Np-r+1)}{(N)(N-1) \cdots (N-r+1)}.$$

Use this result to find $\mu_X^{(3)}$ and $\mu_X^{(4)}$.

22. There are an unknown number N of fish in a pond. However, we do know that $M = 30$ of these fish have tags placed on them. We catch $n = 5$ fish in a net, and assume that this process is equivalent to drawing the fish randomly without replacement. Let X be the number of tagged fish among the 5 fish caught; then X has a hypergeometric distribution with parameters $n = 5$, N, and $p = 30N^{-1}$. If we observe $X = 2$, what estimate would you give for N? What estimate would you give for N if $X = 0$?

23. There has been much concern with the problem of abandoned cars on highways. Suppose that the number X of cars abandoned in a week on a particular highway has a Poisson distribution with parameter $\lambda = 2.0$.
    (a) Find the probability mass function $p_X(k)$ of X.
    (b) It costs the state $100 per car to tow away and dispose of an abandoned car. What is the expected cost per week to the state to dispose of cars abandoned on the given highway?
    (c) What is the most probable amount of money that the state must pay to dispose of cars abandoned on the highway in a given week?
    (d) How probable is it that the state will have to spend more than $400 in a given week to dispose of cars abandoned on this highway?
    (e) At present the state pays private tow truck operators $25 a car to tow away abandoned automobiles. The cost of purchasing and running its own tow truck is $50 a week. Would you recommend that the state purchase its own tow truck?

24. An archeologist has come to the conclusion that along a certain stretch of coastal shelf, every acre of ocean bottom has a probability 0.01 of containing a wrecked ancient ship. He randomly chooses 100 acres of ocean bottom on this shelf and applies to a foundation for money to search for ancient ships. Assume that the presence or lack of presence of ancient ships on these acres of shelf bottom are mutually independent, and that at most one ship can be found on each acre.
    (a) What is the exact probability distribution of X?
    (b) We can use the Poisson distribution to approximate the probabilities of the events $\{X = k\}$. Why? What value of λ should we use?
    (c) What is the probability that the archeologist will find at least 1 ancient ship among the 100 acres surveyed? Compute both the exact and approximate probabilities of this event.

25. In Exercise 4, use the Poisson approximation to the binomial distribution to find the probabilities asked for in parts (a), (b), and (c). Compare your results with the exact probabilities.

26. Let X have a hypergeometric distribution with parameters n, N, and $p = \lambda/n$. Show that when N and n go to infinity, with n/N going to 0, the distribution of X has probabilities that can be approximated by a Poisson distribution with parameter λ.

27. Show that if X has a Poisson distribution with parameter λ, then for $r = 3, 4$,

$$E[X(X - 1) \cdots (X - r + 1)] = \lambda^r.$$

Using this result, find $\mu_X^{(3)}$ and $\mu_X^{(4)}$ and discuss the skewness and kurtosis of the Poisson distribution.

28. In studying the frequency of wars, Richardson (1944) tallied the number of wars X which began in each of the calendar years from A.D. 1500 to A.D. 1931. The number of wars X in a calendar year was hypothesized to have a Poisson distribution. The 432 observed values of X are summarized in Table E.3. Fit a Poisson distribution to these data, assuming that $\hat{\mu}_X = 0.70$ (instead of the observed value of $\hat{\mu}_X = 0.692$).

Table E.3: *Results of Observing the Number X of Wars in Each of the Calendar Years* A.D. *1500–1931*

Number x of outbreaks of war in a particular year	Observed frequency
0	223
1	142
2	48
3	15
4	4
≥ 5	0
	$\overline{432}$

29. The number of lost articles found in a large municipal office building and turned into the lost-and-found office was noted. Records for each day (except Sundays and holidays) were kept for the period November 1, 1923, to September 30, 1925 (excluding the summer months June, July, and August, when there might be variations in the population of the building). The data so obtained are given in Table E.4. [Thorndike (1926)]. Fit a Poisson distribution to these data.

Table E.4: Results of Observing the Number X of Lost Articles Turned into the Lost-and-Found Office of a Large Municipal Office Building for Each Working Day from November 1, 1923, to September 30, 1925

Number x of lost articles in a day	Observed frequency of x lost articles in a day
0	169
1	134
2	74
3	32
4	11
5	2
6	0
7	1
	423

30. A doctor wants to do a case study of a certain genetic disease appearing in 5% of all individuals. Patients are assumed to come to her genetic clinic as if they were randomly sampled *with* replacement from the population. Let X be the number of patients the doctor must see until the individual with the genetic disease is discovered.
(a) What is the distribution of X?
(b) What is the mean number μ_X of patients that must be seen?
(c) What is the probability that no more than 10 patients must be seen?

31. If X has a geometric distribution with parameter p, find the median, $\text{Med}(X)$, of X when $p = 0.4$ and when $p = 0.8$. What is the mode of X in both these cases?

32. If X has a geometric distribution with parameter p, find $P(\{X \geq k + 1\} | \{X \geq k\})$ for $k = 1, 2, \dots$.

33. Show that if X has a geometric distribution with parameter p, then for $r = 3, 4,$

$$E[X(X - 1) \cdots (X - r + 1)] = r! \frac{(1 - p)^{r-1}}{p^r}.$$

Use this result to find $\mu_X^{(3)}$ and $\mu_X^{(4)}$.

34. The number X of occupants carried in each of 1469 passenger cars was tabulated by Haight (1970). The cars were both eastbound and westbound, and were observed at the corner of Wilshire and Bundy (in Los

Angeles) between 11:10 and 12:00 on the morning of March 23. The data obtained by Haight are given in Table E.5.

Table E.5: *Observations of the Number X of Occupants Carried in Each of 1469 Passenger Cars*

Number x of occupants	Observed frequency
1	902
2	403
3	106
4	38
5	16
≥ 6	4
	1469

Assuming that the 4 observations corresponding to "6 or more occupants" were actually observations of "exactly 6 occupants," fit a geometric distribution to these data.

35. A sociologist studying the publication behavior of researchers takes a sample of 363 active sociologists and asks these individuals to state the number of journals to which their most recent published paper was submitted before it was finally published. The data are summarized in Table E.6. Fit a geometric distribution to these data.

Table E.6: *Observations of the Number X of Journals to Which a Sociologist's Most Recent Paper Was Submitted Before Publication*

Number x of journals	Observed frequency
1	261
2	75
3	24
4	3
≥ 5	0
	363

36. Two baseball teams, A and B, meet in the World Series. Assume that every game is statistically independent of the other games between these teams and that the teams are evenly matched. In the World Series, the first team to win 4 games wins the series.
  (a) For $k = 4, 5, 6, 7$, what is the probability that A wins the series in k games? What is the probability that the series lasts k games? What is the conditional probability that A wins the series given that the series lasts k games?
  (b) What is the expected number of games that will be played in the World Series?

37. Suppose, in Example 5.1, that there are $N = 1000$ fish in the pond and $M = 200$ of these are of the desired species. What is the probability that 10 or more fish that are not of the desired species are caught before 3 fish of the desired species are observed?

38. An organization has two open positions but has very stringent requirements for filling the positions that are open, so stringent that only 10% of all possible applicants are likely to be qualified. Applications come in one at a time, and the qualifications for any applicant are independent of the qualifications for all other applicants. The organization will stop accepting applications when it finds 2 suitable candidates. Let X be the number of applications that the university must process until it finds 2 suitable candidates, and let $Y = X - 2$.
  (a) Find the probability mass functions $p_X(x)$ and $p_Y(y)$ for X and Y, respectively.
  (b) What is the mean number μ_X of applications that must be processed? What is the mean number of unqualified applications that must be processed?
  (c) What is the median number, $\text{Med}(X)$, of applications that must be processed?
  (d) It costs the university 0.5 labor-hour to process each application. What is the probability that the cost will exceed 10 labor-hours?

39. Let s be any positive number. For any $x \neq 1$, and for $k = 1, 2, \ldots$, show that

$$\left(\frac{d}{dx}\right)^k \frac{1}{(1-x)^s} = [s(s+1) \cdots (s+k-1)] \frac{1}{(1-x)^{s+k}}$$

$$= \frac{\Gamma(s+k)}{\Gamma(s)} \frac{1}{(1-x)^{s+k}}.$$

Note that for $-1 \leq x < 1$, and for each $k = 1, 2, \ldots$, these derivatives are continuous in x. Thus, by Taylor's theorem [see, e.g., Schwartz (1967), p. 793], for each $n \geq 1$ there exists a number ω between x and 0 such that

$$\frac{1}{(1-x)^s} = 1 + \sum_{k=1}^{n} \frac{\Gamma(s+k)}{\Gamma(s)} \frac{x^k}{k!} + \frac{\Gamma(s+n+1)}{\Gamma(s)(n+1)!} \frac{x^{n+1}}{(1-\omega)^{s+n+1}}.$$

Now, show that if $0 < x < 1$, then

$$\lim_{n \to \infty} \frac{\Gamma(s+n+1)}{\Gamma(s)(n+1)!} \frac{x^{n+1}}{(1-\omega)^{s+n+1}} = 0.$$

Hence (e.g., see Schwartz (1967), p. 799), conclude that

$$\frac{1}{(1-x)^s} = \sum_{k=1}^{\infty} \frac{\Gamma(s+k)}{\Gamma(s)k!} x^k.$$

Use this result to show that if $0 < p < 1$, then

$$\frac{1}{p^r} = \sum_{k=r+1}^{\infty} \binom{k}{r-1}(1-p)^{k-r+1}.$$

Then prove that for $r > 0$,

$$\sum_{u=0}^{\infty} \frac{\Gamma(u+r)}{\Gamma(r)\Gamma(u+1)} p^r (1-p)^u = 1,$$

which shows that (5.2) is indeed a probability mass function.

40. Let Y have a generalized negative binomial distribution with parameters r and p. Show that for $s = 3, 4$,

$$E[Y(Y-1)\cdots(Y-s+1)] = \frac{r(r+1)\cdots(r+s)(1-p)^s}{p^s}.$$

Use this result to find $\mu_Y^{(3)}$ and $\mu_Y^{(4)}$.

41. The data summarized in Table E.7 list the number X of 1-day industrial

Table E.7: *Observations of the Number X of 1-Day Industrial Absences During 1957 for Each of 195 Inspectors*

Number x of 1-day absences	Observed frequency of x 1-day absences
0	27
1	35
2	35
3	37
4	19
5	20
6	10
7	6
8	3
9	3
≥ 10	0
	195

absences (all causes) during 1957 of each of 195 nonsupervisory married male inspectors employed in one company throughout the year. Fit a negative binomial distribution to these data.

42. The data summarized in Table E.8 list the number X of traffic accidents incurred during the period 1952–1955 by each of 708 public corporation bus drivers who drove regularly throughout the period. Fit a negative binomial distribution to these data.

Table E.8: *Observations of Number X of Traffic Accidents Incurred During 1952–1955 by Each of 708 Bus Drivers*

Number x of traffic accidents during the period	Observed frequency of x accidents
0	117
1	157
2	158
3	115
4	78
5	44
6	21
7	7
8	6
9	1
10	3
11	1
≥ 12	0
	708

Source: Data from P. Froggatt (1970). Application of discrete distribution theory to the study of noncommunicable events in medical epidemiology, in G. P. Patil, ed., *Random Counts in Biomedical and Social Sciences*, vol. 2. University Park, Pa.: Pennsylvania State University Press.

43. When an electric light bulb is purchased, the bulb can be tested to determine whether it is operating properly. If it is defective, the bulb is put aside to be returned to the manufacturer, and another bulb is tested. Bulbs are tested until a good one is found. Suppose that we start with a box of 5 bulbs of which 2 are defective. The experiment consists of testing bulbs (without replacement) until 2 good ones are found. Let X denote the number of bulbs tested. Find the probability mass function of X.

44. It can be shown that if $1 \le r \le K$,

$$\sum_{i=1}^{K} [i(i-1) \cdots (i-r+1)] = \sum_{i=r}^{K} [i(i-1) \cdots (i-r+1)]$$
$$= \frac{(K+1)(K) \cdots (K-r+1)}{r+1}.$$

Suppose that X has a discrete uniform distribution with parameter K. Use the foregoing result to find $\mu_X^{(3)}$ and $\mu_X^{(4)}$.

45. Suppose that X has a discrete uniform distribution with parameter K.
(a) Find the c.d.f. $F_X(x)$ of X.
(b) Find the median of X. [Hint: There are two cases: K odd and K even.]
(c) Show that every possible value of X is a mode.

46. Fit a truncated Poisson distribution to the data in Table 7.1.

47. In a geological study, Krumbein (1954) noted the number X of granite pebbles in each of 100 samples of 10 beach pebbles of moderate size: 16–32 mm in diameter. The resulting observations are summarized in Table E.9.

Table E.9: *Observations of the Number X of Granite Pebbles in 100 Samples of Beach Pebbles*

Number x of granite pebbles in a sample	Observed frequency of samples with x granite pebbles
0	58
1	33
2	7
3	2
	100

One could give an argument for any one of the following distributions as providing a possible good fit to these data: (i) binomial with $n = 10$, p unknown, (ii) hypergeometric with $n = 10$, N and p unknown, (iii) Poisson, (iv) geometric, and (v) generalized negative binomial. Calculate the theoretical frequencies predicted by these distributions.

7

The Normal Distribution

1. INTRODUCTION

Of all distributions of continuous random variables, the most widely studied and frequently used is the *normal distribution*. The form of the normal distribution was discovered early in the history of probability as a convenient approximation to binomial probabilities (Section 5), and the properties of this distribution were of great theoretical interest to de Moivre, Laplace, and other early probabilists. One of the very early applications of the normal distribution in modeling quantitative phenomena was in describing the variability of errors of measurement in observations of the motions of planetoids. The success of the normal distribution in describing the variability of measurement errors in the physical sciences led to its acceptance as the "Law of Errors." Because the German mathematician and physicist Gauss (1777–1855) played such a prominent role in demonstrating the usefulness of the normal distribution in error analysis, this distribution is also known as the *Gaussian* distribution.

Among the first scientists to apply the normal distribution to physiological and behavioral phenomena were L. A. J. Quetelet (1796–1874), a Belgian statistician, and Sir Francis Galton (1822–1911), a British pioneer in many scientific fields. Quetelet used the concept of the "average" or "normal" man (*l'homme moyen*) in summarizing the central tendency of anthropometric data, and used the Law of Errors to describe anthropomorphic variation. This may provide an explanation of how the term "normal distribution" originated.

The pioneering work of Quetelet and Galton was followed in the early 1900s by numerous studies of anthropometric data in which the normal distribution was asserted to describe (fit) the variability in the data.

Example 1.1. In a study of the inheritance of physical characteristics in human beings, Pearson and Lee (1903) examined family data and went into considerable detail to try to justify their assertion that physical measurements

Table 1.1: *Distribution of Mothers'*
Stature

(1052 cases; estimated μ = 62.484 in., estimated
σ^2 = 5.7140 in.^{2})

Stature (in inches)	Observed frequency	Normal frequency
52–53	1.5	}0.9
53–54	0.5	
54–55	1	
55–56	2	2.6
56–57	6.5	7.9
57–58	18	20.9
58–59	34.5	44.5
59–60	79.5	80.8
60–61	135.5	124.1
61–62	163	160.3
62–63	183	174.3
63–64	163	159.4
64–65	114.5	122.8
65–66	78.5	79.5
66–67	41	43.2
67–68	16	20.1
68–69	7.5	7.7
69–70	4.5	2.5
70–71	2	0.8

such as stature, arm span, and length of forearm have normal distributions. In Table 1.1 and Figure 1.1 we show the distribution of the mothers' stature in inches.

Example 1.2. In a study by Pearl (1905), variations in brain weight (in grams) in 416 Swedish males between the ages of 20 and 80 were studied (see Table 1.2 and Figure 1.2).

Many other examples of these kinds of studies could be given: measurements on birds' eggs [Latter (1902)], measurements taken on criminals and noncriminals to see if criminality could be predicted from physiological measurements [Macdonell (1902)], weights of human viscera taken to see if healthy and diseased hearts could be distinguished by weight [Greenwood (1904)], and so on.

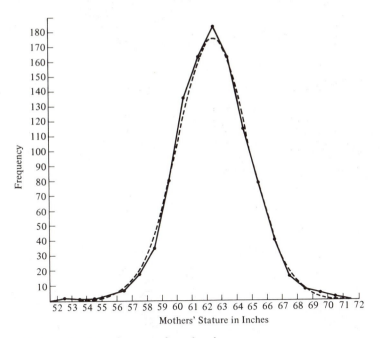

Figure 1.1: *Distribution of mothers' stature.*

Table 1.2: *Distribution of Observed and Theoretical Frequencies of Brain Weight of 416 Adult Swedish Males*

(Estimated $\mu = 1400.481$, estimated $\sigma^2 = 4.5224$)

Grams of brain weight	Observed	Theoretical
Under 1100	0	0.981
1100–1150	1	2.9
1150–1200	10	8.5
1200–1250	21	20.3
1250–1300	44	39.0
1300–1350	53	60.4
1350–1400	86	75.2
1400–1450	72	75.3
1450–1500	60	60.8
1500–1550	28	39.4
1550–1600	25	20.6
1600–1650	12	8.7
1650–1700	3	2.9
1700–1750	1	0.8
1750 and over	0	0.036
Totals	416	415.817

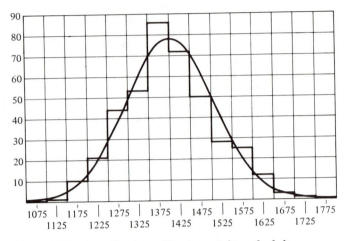

Figure 1.2: *Distribution of brain weights of adult Swedish males.*

2. THE NORMAL DENSITY AND ITS PROPERTIES

The probability density function $f_X(x)$ of a normal distribution is defined by the equation

$$(2.1) \qquad f_X(x) = \frac{1}{\sqrt{2\pi\sigma^2}} \exp\left[-\frac{(x-\mu)^2}{2\sigma^2}\right], \quad -\infty < x < \infty.$$

Here, exp (s) is a convenient way of writing e^s. The constants μ, $-\infty < \mu < \infty$, and σ^2, $\sigma^2 > 0$ are the parameters of the normal distribution. The graph of $f_X(x)$, which is a bell-shaped curve, is shown in Figure 2.1. Note that this graph is symmetric about the point $x = \mu$ on the horizontal axis and that it extends over the entire horizontal axis.

As suggested by our notation, the parameter μ gives the value of the mean μ_X of the normal density (2.1), while the parameter σ^2 gives the value of the variance σ_X^2. As a shorthand way of saying that the random variable X has a normal distribution with mean μ and variance σ^2, we write $X \sim \mathcal{N}(\mu, \sigma^2)$. Thus, $X \sim \mathcal{N}(80, 16)$ means that the random variable X has a normal distribution with mean $\mu = 80$ and variance $\sigma^2 = 16$. The symbol "$\sim$," which means "is distributed as," is widely utilized in the literature of probability and statistics.

Since $f_X(x)$ is symmetric about $x = \mu$ and achieves its maximum value at $x = \mu$ (see Figure 2.1), the median and the mode of the normal distribution both have the value μ.

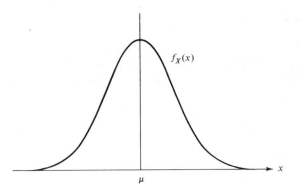

Figure 2.1: *Graph of a normal density function,* $f_X(x)$. *The graph is symmetric about the point* μ.

Although we have asserted that (2.1) is a density function, and it is clear from (2.1) that $f_X(x)$ is always nonnegative, it is far from obvious that $\int_{-\infty}^{\infty} f_X(x)\,dx = 1$. However, making the change of variables from x to $z = (x - \mu)/\sigma$, we see that

$$(2.2) \qquad \int_{-\infty}^{\infty} f_X(x)\,dx = \int_{-\infty}^{\infty} \frac{1}{\sqrt{2\pi}} \exp\left(-\tfrac{1}{2}z^2\right) dz = \frac{2}{\sqrt{2\pi}} \int_{0}^{\infty} \exp\left(-\tfrac{1}{2}z^2\right) dz,$$

since $\exp\left(-\tfrac{1}{2}z^2\right)$ is an even function of z. From (A.7) in the Appendix,

$$(2.3) \qquad \int_{0}^{\infty} z^{2s-1} \exp\left(-\tfrac{1}{2}z^2\right) dz = 2^{s-1}\Gamma(s).$$

Hence, applying (2.3) with $s = \tfrac{1}{2}$ to (2.2), we conclude that

$$\int_{-\infty}^{\infty} f_X(x)\,dx = \frac{2}{\sqrt{2\pi}} \frac{\Gamma(\tfrac{1}{2})}{\sqrt{2}} = 1,$$

since $\Gamma(\tfrac{1}{2}) = \sqrt{\pi}$. Thus, (2.1) is indeed a density function.

Using the same change of variables used in (2.2),

$$(2.4) \qquad E[(X - \mu)^r] = \int_{-\infty}^{\infty} (x - \mu)^r f_X(x)\,dx = \sigma^r \int_{-\infty}^{\infty} \frac{z^r \exp\left(-\tfrac{1}{2}z^2\right)}{\sqrt{2\pi}}\,dz.$$

When r is an odd integer, the integrand of the last integral in (2.4) is an odd function of z, and thus the integral equals 0. Consequently, $E[(X - \mu)^r] = 0$

for $r = 1, 3, 5, \ldots$. In particular, $E[X - \mu] = 0$, so that, in agreement with our earlier assertion,

$$\mu_X = E[X - \mu] + \mu = \mu.$$

When r is an even integer, the integrand of the last integral in (2.4) is an even function of z, and an application of (2.3) yields

$$E[(X - \mu)^r] = 2\sigma^r \int_0^\infty \frac{z^r \exp\left(-\tfrac{1}{2}z^2\right) dz}{\sqrt{2\pi}} = \frac{\sigma^r 2^{r/2} \Gamma(\tfrac{1}{2}(r + 1))}{\sqrt{\pi}}.$$

When $r = 2$,

$$\sigma_X^2 = E[(X - \mu_X)^2] = E[(X - \mu)^2] = \frac{2\sigma^2 \Gamma(\tfrac{3}{2})}{\sqrt{\pi}} = \sigma^2,$$

as earlier asserted. In general, we have shown that

$$(2.5) \qquad E[(X - \mu)^r] = \begin{cases} \dfrac{\sigma^r 2^{r/2} \Gamma(\tfrac{1}{2}(r + 1))}{\Gamma(\tfrac{1}{2})}, & \text{if } r = 2, 4, 6, \ldots . \\[2ex] 0, & \text{if } r = 1, 3, 5, \ldots, \end{cases}$$

In particular,

$$(2.6) \qquad \mu_X^{(3)} = E[(X - \mu_X)^3] = 0, \qquad \mu_X^{(4)} = E[(X - \mu_X)^4] = 3\sigma^4.$$

3. PROBABILITY CALCULATIONS: THE STANDARD NORMAL DISTRIBUTION

The change of variables from x to $z = (x - \mu)/\sigma$ corresponds to a linear transformation from the random variable X to the random variable

$$(3.1) \qquad Z = \frac{X - \mu}{\sigma} = \frac{1}{\sigma} X + \left(-\frac{\mu}{\sigma}\right).$$

Since $\mu_X = \mu$ and $\sigma_X^2 = \sigma^2$, it is easily shown that Z has mean 0 and variance 1. In general, the *standardization* Z of any random variable X, whether normally distributed or not, is defined by $Z = (X - \mu_X)/\sigma_X$ and is constructed to have mean 0, variance 1, and to take values on a scale independent of the units of measurement in which X was originally measured. The standardization (or *standardized random variable*) Z expresses distances of X from its mean μ_X in standard deviation units. Thus, $Z = 2.3$ means that the value of X is 2.3 standard deviation units larger than the mean μ_X of the distribution (population) of X-values.

Example 3.1. Two exams were given in a certain class. Scores on the first exam had a mean $\mu_X = 80$ and a standard deviation $\sigma_X = 4$. The second exam was harder, and $\mu_X = 65$ with $\sigma_X = 5$. A student had a score of 82 on the first exam and a score of 75 on the second exam. Although the student's score on the first exam was larger, the standardized scores of $Z = (82 - 80)/4 = 0.5$ and $Z = (75 - 65)/5 = 2.0$ show that the student's performance relative to the class was superior on the second exam.

When X has a normal distribution, then the standardization Z of X has a normal distribution with parameters 0 and 1. Indeed,

(3.2) $$f_Z(z) = \frac{1}{|1/\sigma|} f_X(\sigma(z + \mu)) = \frac{\exp\left(-\frac{1}{2}z^2\right)}{\sqrt{2\pi}}$$

for $-\infty < z < \infty$. The density (3.2) is said to be the density of a *standard normal distribution*.

When $X \sim \mathcal{N}(\mu, \sigma^2)$,

(3.3) $$F_X(x) = F_Z\left(\frac{x - \mu}{\sigma}\right).$$

Hence, the c.d.f. of any normally distributed random variable X can be evaluated if we have available a table of the c.d.f. $F_Z(z)$ of the standard normal distribution. Table T.5 in the Appendix gives values of $F_Z(k)$ for $k = 0.00(0.01)4.00$. Because of symmetry (see Figure 3.1), $P\{Z \leq -k\} =$

Figure 3.1: *Graph of the density of the $\mathcal{N}(0, 1)$ distribution. The area, $P\{Z \leq -k\}$, under the graph over the horizontal axis to the left of $Z = -k$ equals the area, $P\{Z > k\}$, under the graph to the right of $Z = k$.*

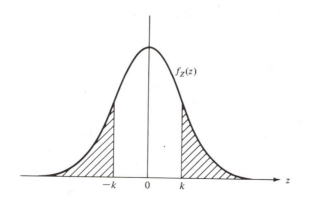

$P\{Z > k\}$. Since $P\{Z \le -k\} = F_Z(-k)$ and $P\{Z > k\} = 1 - F_Z(k)$, we conclude that

(3.4) $$F_Z(-k) = 1 - F_Z(k).$$

For example,

$$F_Z(-1.96) = 1 - F_Z(1.96) = 1 - 0.97500 = 0.02500.$$

Thus, only the values of $F_Z(k)$ for $k > 0$ need to be given in Table T.5.

Example 3.2. Two students take a reading comprehension test for which the scores X are known to have a normal distribution with mean $\mu = 80$ and variance $\sigma^2 = 16$. One student obtains a score of 85, and the other student obtains a score of 70. Since $X \sim \mathcal{N}(80, 16)$, the quantile rank of the first student among the total population of individuals taking the test is

$$P\{X \le 85\} = F_X(85) = F_Z\left(\frac{85 - 80}{\sqrt{16}}\right) = F_Z(1.25) = 0.89435.$$

Using (3.4), the quantile rank of the second student is

$$F_X(70) = F_Z\left(\frac{78 - 80}{\sqrt{16}}\right)$$
$$= F_Z(-2.5) = 1 - F_Z(2.5) = 1 - 0.99379 = 0.00621.$$

The graph of the cumulative distribution function $F_X(x)$ of a normally distributed random variable X is shown in Figure 3.2. Notice that the graph

Figure 3.2: *Graph of the c.d.f. of a $\mathcal{N}(\mu, \sigma^2)$ distribution.*

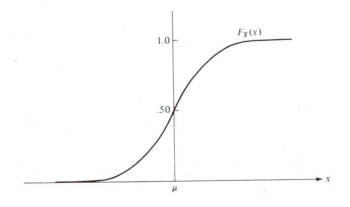

has an S-shaped form. The graph of the c.d.f. $F_X(x)$ of a normally distributed random variable is often called a *normal ogive*.

Recall that for any continuous random variable X,

$$(3.5) \qquad P\{a \leq X \leq b\} = F_X(b) - F_X(a),$$

and that the probabilities of the events $\{a < X < b\}$, $\{a \leq X < b\}$, and $\{a < X \leq b\}$ are equal to (3.5) since $P\{X = x\} = 0$ for any x. If $X \sim \mathcal{N}(\mu, \sigma)$, then X is a continuous random variable and

$$(3.6) \qquad P\{a \leq X \leq b\} = F_X(b) - F_X(a) = F_Z\left(\frac{b - \mu}{\sigma}\right) - F_Z\left(\frac{a - \mu}{\sigma}\right).$$

Example 3.2 (continued). The fraction of all scores on the reading comprehension test that are between 78 and 84 is, since $X \sim \mathcal{N}(80, 16)$,

$$P\{78 \leq X \leq 84\} = F_Z\left(\frac{84 - 80}{\sqrt{16}}\right) - F_Z\left(\frac{78 - 80}{\sqrt{16}}\right)$$

$$= F_Z(1.00) - F_Z(-0.50)$$

$$= F_Z(1.00) - [1 - F_Z(0.50)]$$

$$= 0.84134 + 0.69146 - 1 = 0.53280.$$

When $b = \infty$, (3.6) reduces to

$$(3.7) \qquad P\{a \leq X\} = 1 - F_X(a) = 1 - F_Z\left(\frac{a - \mu}{\sigma}\right)$$

and gives the *tail probability* of the distribution of X.

Example 3.3. At an archaeological site, suppose that skeletal heights, X, have a $\mathcal{N}(170, 100)$ distribution when measured in centimeters. The skeletal height of a discovered skeleton is 188 cm. Since

$$P\{188 \leq X\} = 1 - F_X(188) = 1 - F_Z\left(\frac{188 - 170}{\sqrt{100}}\right) = 1 - 0.96407 = 0.03593,$$

the individual whose skeleton was found would have been unusually tall compared to others at that time.

Using (3.4) and (3.6), we can compute the probability that X is within $\pm k$ standard deviations of its mean μ:

$$(3.8) \qquad P\{-k\sigma \leq X - \mu \leq k\sigma\} = P\{\mu - k\sigma \leq X \leq \mu + k\sigma\}$$

$$= F_Z(k) - F_Z(-k) = 2F_Z(k) - 1.$$

From (3.8) and Table T.5, we obtain Table 3.1, which gives values for $P\{-k\sigma \leq X - \mu \leq k\sigma\}$ for $k = 0.5(0.5)3.0$. (A pictorial representation of these

Table 3.1: *Probability That a Normally Distributed Random Variable Is Within ±k Standard Deviations of Its Mean*

k	$P\{-k\sigma \leq X - \mu \leq k\sigma\}$	Bienaymé–Chebychev bounds
0.5	0.38292	0.00000
1.0	0.68268	0.00000
1.5	0.86638	0.55556
2.0	0.95450	0.75000
2.5	0.98758	0.84000
3.0	0.99730	0.88889

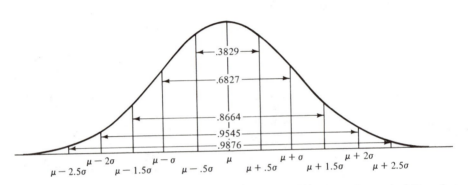

Figure 3.3: *The areas indicated by the arrows are equal to the probability that a normally distributed random variable is within ±k standard deviations σ of its mean μ.*

probabilities appears in Figure 3.3.) It is of interest to compare these probabilities with the lower bounds for such probabilities given by the Bienaymé–Chebychev Inequality (Chapter 5).

In computing probabilities for the normal distribution, a rough sketch of the graph of $f_X(x)$ showing the area under the graph required to obtain a desired probability is often helpful. For example, suppose that $X \sim \mathcal{N}(15, 49)$ and that we want to calculate the probability that $X < 8$ *or* $X > 29$. A sketch such as Figure 3.4 tells us that the desired probability is the sum of $F_X(8) = 0.15866$ and $1 - F_X(29) = 0.00275$.

4. FITTING A NORMAL DISTRIBUTION

We now consider a classical example of the applicability of the normal distribution.

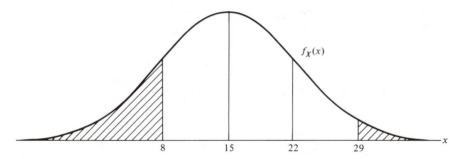

Figure 3.4: *Sketch of areas required (shaded areas) to obtain the probability* $P\{X < 8 \text{ or } X > 29\}$ *for a* $\mathcal{N}(15, 49)$ *distribution.*

Example 4.1 (Intelligence Quotients). In his well-known book, *The Intelligence of School Children*, L. M. Terman describes several studies on school children which utilized the Stanford Revision of the Binet–Simon Intelligence Scale. In one of these studies, 112 children (65 boys and 47 girls) attending five kindergarten classes in San Jose and San Mateo, California, were measured on this intelligence scale. In Table 4.1, the original ungrouped (raw) data are given. The data are grouped in Table 4.2 to form a frequency distribution and are graphed as a frequency histogram in Figure 4.1. It appears from Figure 4.1 that a normal distribution might serve as an approximation to the actual probability distribution of IQs.

Figure 4.1: *Histogram corresponding to grouped data (Table 4.2) in the Terman study.*

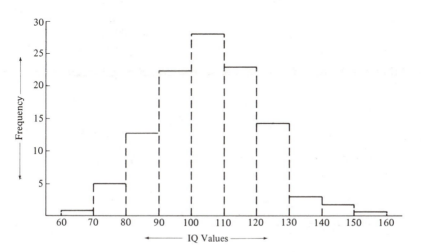

Table 4.1: *Original IQ Data in the Terman Study of 112 Children Attending Kindergarten Classes in San Jose and San Mateo, California*

(If a value appeared more than once in the data, the number of times that it appeared appears in parentheses after the value.)

152	123	112(2)	102(5)	92	80(4)
146	122	111(2)	101(2)	91(2)	79
142	121(5)	110(4)	100(3)	90(4)	77
136	120	109(5)	99	88	76
130(2)	119(2)	108(2)	98(3)	86(2)	75
129	118	107(4)	97(2)	85(3)	72
126(2)	117(2)	106(3)	96(4)	84	61
125	114(6)	105	94(2)	82	
124(2)	113(4)	103(3)	93(3)	81	

Table 4.2: *Observed Frequency Distribution of IQ of 112 Children in Terman Study*

Interval of IQ values	Midpoint of IQ intervals	Frequency
60–69	65	1
70–79	75	5
80–89	85	13
90–99	95	22
100–109	105	28
110–119	115	23
120–129	125	14
130–139	135	3
140–149	145	2
150–159	155	1
		112

As a first step in fitting a normal distribution to the observed data, we estimate values for μ and σ^2 from the raw data of Table 4.1:

$$\hat{\mu} = \hat{\mu}_X = 104.5, \qquad \hat{\sigma}^2 = \hat{\sigma}_X^2 = 263.66.$$

Thus, we wish to fit a $\mathcal{N}(104.5, 263.66)$ distribution to the grouped data in Table 4.2.

Remark. Values of $\hat{\mu} = \hat{\mu}_X$ and $\hat{\sigma}^2 = \hat{\sigma}_X^2$ obtained from the grouped data of Table 4.2 would differ slightly from the estimates obtained above.

To find theoretical frequencies to compare to the observed frequencies in Table 4.2 (or to the histogram in Figure 4.1), we need theoretical probabilities for the intervals $[-\infty, 60]$, $[60, 70]$, $[70, 80]$, ..., $[150, 160]$, $[160, \infty]$. To obtain these probabilities, we begin by calculating $F_X(x)$ for $x = 60, 70, \ldots,$ 160 using (3.3) and (3.4). For example, since $\sqrt{263.66} = 16.23761 \simeq 16.2$,

$$F_X(60) = F_Z\left(\frac{60 - 104.5}{16.2}\right) = 1 - F_Z(2.74) = 0.00347,$$

while

$$F_X(160) = F_Z\left(\frac{160 - 104.5}{16.2}\right) = F_Z(3.43) = 0.99970.$$

Our results appear in Table 4.3.

Now, theoretical probabilities of the intervals in Figure 4.1 are found using (3.5): $P\{-\infty < X \le 60\} = F_X(60) = 0.00347$, $P\{60 \le X \le 70\} = F_X(70) - F_X(60) = 0.01312$, ..., $P\{150 \le X \le 160\} = F_X(160) - F_X(150) = 0.00218$, $P\{160 \le X < \infty\} = 1 - F_X(160) = 0.00030$. Each such theoretical probability then is multiplied by 112 to yield the theoretical frequencies in Table 4.4.

The normal distribution with parameters $\mu = 104.5$ and $\sigma^2 = 263.66$ appears to provide a reasonably close approximation to the distribution of IQ scores.

Remark. It is a more common practice to construct the histogram in Figure 4.1 using intervals of the form $[59.5, 69.5]$, $[69.5, 79.5]$, and so on. Although this way of constructing the histogram would require us to compute

Table 4.3: *Theoretical Probabilities for the Distributions in Table 4.2*

x	$F_X(x)$	x	$F_X(x)$
60	0.00347	120	0.83147
70	0.01659	130	0.94179
80	0.06552	140	0.98574
90	0.18406	150	0.99752
100	0.38974	160	0.99970
110	0.63307		

Table 4.4: *Comparison of Observed and Theoretical Frequency Distributions for the IQs of 112 Children in the Terman Study*

Interval of IQ scores	Observed frequency	Theoretical frequency if $X \sim \mathcal{N}(104.5, 263.66)$
< 60	0	0.39
60–69	1	1.46
70–79	5	5.48
80–89	13	13.28
90–99	22	23.04
100–109	28	27.25
110–119	23	22.22
120–129	14	12.36
130–139	3	4.92
140–149	2	1.32
150–159	1	0.24
> 160	0	0.03
	112	111.99

probabilities of events such as $\{59.5 \le X \le 69.5\}$, $\{69.5 \le X \le 79.5\}$, and so on, the resulting theoretical frequencies would differ only slightly from those given in Table 4.4.

5. THE CENTRAL LIMIT THEOREM

We have seen that the normal distribution can be used to describe the variation in many quantitative phenomena. However, the normal distribution is also useful in other ways. In the present section, we describe other ways in which the normal distribution is used in experimental investigations.

Suppose that we are interested in estimating the mean μ_X of a random variable X, and that we have made n independent observations $X_1, X_2, \ldots, X_n$ on X; that is, observations are taken from n independent trials of the random experiment that generated the random variable X. We may estimate μ_X by the sample average

$$(5.1) \qquad \hat{\mu}_X = \frac{1}{n} \sum_{i=1}^{n} X_i,$$

which, because each observation X_i is a random variable, is itself a random variable. We cannot expect $\hat{\mu}_X$ always to be equal to μ_X, since every time we take n independent observations on X, a new value of $\hat{\mu}_X$ will be obtained. We can, however, measure how closely $\hat{\mu}_X$ varies about μ_X by calculating the probability

(5.2) $$P\{-\varepsilon \leq \hat{\mu}_X - \mu_X \leq \varepsilon\}$$

that a specified accuracy $\varepsilon > 0$ is obtained. The distribution of $\hat{\mu}_X$, and thus the probability (5.2), can be difficult to obtain, especially when the distribution of X is not simple. It is a remarkable fact that, regardless of the distribution of X, as the number n of independent trials increases, the probability (5.2) can be approximated by a probability computed from the standard normal distribution:

(5.3) $$P\{-\varepsilon \leq \hat{\mu}_X - \mu_X \leq \varepsilon\} \simeq P\left\{\frac{-\varepsilon\sqrt{n}}{\sigma_X} \leq Z \leq \frac{\varepsilon\sqrt{n}}{\sigma_X}\right\} = 2F_Z\left(\frac{\varepsilon\sqrt{n}}{\sigma_X}\right) - 1,$$

where $Z \sim \mathcal{N}(0, 1)$, σ_X^2 is the variance of X, and " $\simeq$ " means "approximately equal to." More generally,

(5.4) $$P\{a \leq \hat{\mu}_X - \mu_X \leq b\} \simeq F_Z\left(\frac{\sqrt{n}\,b}{\sigma_X}\right) - F_Z\left(\frac{\sqrt{n}\,a}{\sigma_X}\right).$$

Example 5.1. An astronomer plans to take independent measurements X_1, $X_2, \ldots, X_n$ of the true distance D in light-years between his observatory and a distant star. These measurements are known to have a common distribution with mean $\mu_X = D$ and variance $\sigma_X^2 = 4$. To estimate D with an accuracy of ± 0.25 light-year, the astronomer might use the sample average $\hat{\mu}_X = (100)^{-1} \sum_{i=1}^{100} X_i$ of $n = 100$ observations. Using (5.3) with $\varepsilon = 0.25$, $n = 100$, the probability of obtaining the desired accuracy is

$$P\{-0.25 \leq \hat{\mu}_X - D \leq 0.25\} \simeq 2F_Z\left(\frac{\sqrt{100}(0.25)}{\sqrt{4}}\right) - 1$$

$$= 2(0.89435) - 1 = 0.78870.$$

When $n = 400$ observations are used, this probability increases to

$$P\{-0.25 \leq \hat{\mu}_X - D \leq 0.25\} = 2F_Z\left(\frac{\sqrt{400}(0.25)}{\sqrt{4}}\right) - 1 = 0.98758.$$

Equations (5.3) and (5.4) are consequences of the famous Central Limit Theorem.

Statements of the Central Limit Theorem

An informal statement of the Central Limit Theorem is that, for large enough values of n, the sample average $\hat{\mu}_X = \sum_1^n X_i/n$ of n independent observations $X_1, \ldots, X_n$ on a random variable X has approximately a normal distribution with mean $\mu = \mu_X$ and variance $\sigma^2 = \sigma_X^2/n$. Furthermore, this result holds regardless of the underlying distribution of X. That is, for any number u, and for large enough values of n,

(5.5)
$$P\{\hat{\mu}_X \le u\} \simeq F_Z\left(\frac{u - \mu_X}{\sigma_X/\sqrt{n}}\right),$$

where $F_Z(z)$ is the c.d.f. of $Z \sim \mathcal{N}(0, 1)$. We remark that in Chapter 10 it is shown that for all values of n, the mean and variance of $\hat{\mu}_X$ are exactly μ_X and σ_X^2/n, respectively. Thus, the approximation referred to in this informal statement of the Central Limit Theorem concerns only the distribution of $\hat{\mu}_X$, not its mean or variance.

A formal statement of the Central Limit Theorem is the following.

Central Limit Theorem. *Let* $X_1, X_2, \ldots, X_n$ *be mutually independent observations on a random variable* X *having a well-defined mean* μ_X *and variance* σ_X^2. *Let*

(5.6)
$$Z_n = \frac{(1/n)\sum_1^n X_i - \mu_X}{\sigma_X/\sqrt{n}} = \frac{\hat{\mu}_X - \mu_X}{\sigma_X/\sqrt{n}},$$

and let $F_{Z_n}(z)$ *be the cumulative distribution function of the random variable* Z_n. *Then for all* z, $-\infty < z < \infty$,

(5.7)
$$\lim_{n \to \infty} F_{Z_n}(z) = F_Z(z),$$

where $F_Z(z)$ *is the c.d.f. of* $Z \sim \mathcal{N}(0, 1)$.

A proof of the Central Limit Theorem is given in Chapter 11, Section 2.

It follows directly from the Central Limit Theorem that for any fixed interval $[a, b]$,

(5.8)
$$\lim_{n \to \infty} P\{a \le Z_n \le b\} = \lim_{n \to \infty} [F_{Z_n}(b) - F_{Z_n}(a)] = F_Z(b) - F_Z(a).$$

However, our earlier applications (5.3), (5.4), and (5.5) of the Central Limit Theorem involved intervals that varied with the sample size n. Since it can be shown that the limit in (5.7) is achieved uniformly in z, $-\infty < z < \infty$, it then follows that for any sequence of numbers $\{z_n\}$, we have

(5.9)
$$\lim_{n \to \infty} [F_{Z_n}(z_n) - F_Z(z_n)] = 0,$$

where $Z \sim \mathcal{N}(0, 1)$. That is, for large enough n, $F_{Z_n}(z_n) \simeq F_Z(z_n)$. It follows from (5.6) and (5.9) that for large enough n,

$$P\{\hat{\mu}_X \le u\} = P\left\{Z_n \le \frac{u - \mu_X}{\sigma_X/\sqrt{n}}\right\} \simeq F_Z\left(\frac{u - \mu_X}{\sigma_X/\sqrt{n}}\right),$$

verifying (5.5). Since

$$P\{a \leq \hat{\mu}_X - \mu_X \leq b\} = P\{\hat{\mu}_X \leq b + \mu_X\} - P\{\hat{\mu}_X \leq a + \mu_X\},$$

(5.3) and (5.4) follow directly from (5.5).

Let $S_n = \sum_{i=1}^{n} X_i$ be the sample total of the observations $X_1, X_2, \ldots, X_n$. From (5.6),

$$Z_n = \frac{(1/n)S_n - \mu_X}{\sigma_X/\sqrt{n}} = \frac{S_n - n\mu_X}{\sigma_X\sqrt{n}},$$

and thus (5.9) yields

(5.10) $$P\{S_n \leq s\} = P\left\{Z_n \leq \frac{s - n\mu_X}{\sqrt{n}\,\sigma_X}\right\} \simeq F_Z\left(\frac{s - n\mu_X}{\sqrt{n}\,\sigma_X}\right),$$

when n is large enough. That is: *For large enough values of n, the sample total* S_n *of n probabilistically independent observations* $X_1, X_2, \ldots, X_n$ *on a random variable X has approximately a normal distribution with mean* $\mu = n\mu_X$ *and variance* $\sigma^2 = n\sigma_X^2$. As in the case of $\hat{\mu}_X$, the approximation concerns only the distribution of S_n, since $\mu_{S_n} = n\mu_X$, $\sigma_{S_n}^2 = n\sigma_X^2$, exactly.

A Justification of the Normal Distribution by the Central Limit Theorem

The Central Limit Theorem applied to the sample total helps to explain why so many quantitative phenomena have (at least approximately) a normal distribution. Suppose, for example, that we are interested in the height Y of an individual. We can think of this height as being the result of the sum $Y = \sum_{i=1}^{n} X_i$ of a large number of individual growth spurts X_i, each spurt taking place over some short interval of time. Over very short time spans, it is reasonable to assume that the X_i's are mutually probabilistically independent and have a common distribution. In this case, application of the Central Limit Theorem allows us to conclude that the height $Y = S_n = \sum_{i=1}^{n} X_i$ has, at least approximately, a normal distribution.

The Accuracy of the Central Limit Theorem

A concrete indication of the accuracy of the Central Limit Theorem as an approximation to the distribution of Z_n can be obtained by considering probabilistically independent observations $X_1, X_2, \ldots, X_n$ of a Bernoulli random variable X having parameter $p = 0.4$. In this case, X has only two possible values, 0 and 1, and from Chapter 6, Section 1, we know that

$$\mu_X = p = 0.4, \qquad \sigma_X^2 = p(1 - p) = 0.24.$$

Thus,

$$Z_n = \frac{S_n - n\mu_X}{\sqrt{n}\,\sigma_X} = \frac{S_n - n(0.4)}{\sqrt{n}\,\sqrt{0.24}}.$$

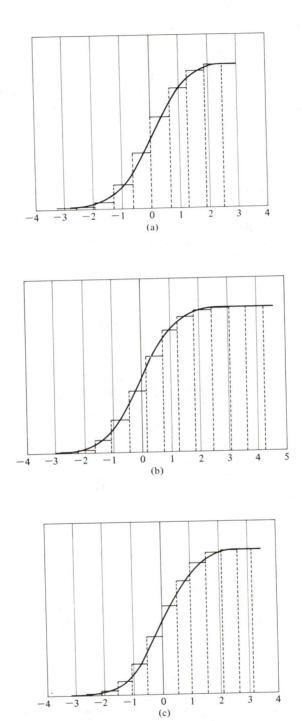

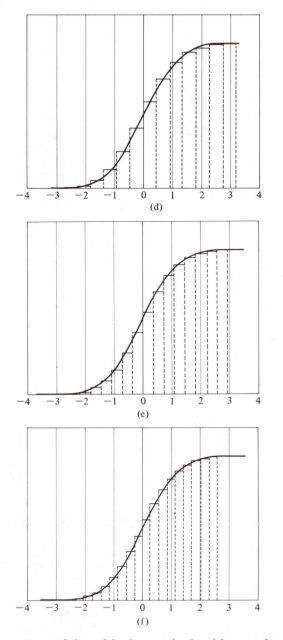

Figure 5.1: *Comparison of the c.d.f. of a standardized binomial random variable and the c.d.f. of a standard normal random variable when the parameters of the distribution of the binomial variable are* $p = 0.4$ *and* (a) $n = 10$, (b) $n = 12$, (c) $n = 15$, (d) $n = 20$, (e) $n = 30$, (f) $n = 50$.

Note that

$$
F_{Z_1}(z) = \begin{cases} 1, & \text{if } \dfrac{0.6}{\sqrt{0.24}} \le z < \infty, \\[2ex] 0.4, & \text{if } -\dfrac{0.4}{\sqrt{0.24}} \le z < \dfrac{0.6}{\sqrt{0.24}}, \\[2ex] 0, & \text{if } -\infty < z < -\dfrac{0.4}{\sqrt{0.24}}, \end{cases}
$$

and that the graph of this c.d.f. bears little similarity to the graph of the c.d.f. of a $\mathcal{N}(0, 1)$ random variable. We have calculated and graphed $F_{Z_n}(z)$ for $n = 10, 12, 15, 20, 30, 50$, making use of the fact that

$$
F_{Z_n}(z) = P\{Z_n \le z\} = P\{S_n \le (0.4)n + z\sqrt{n}\sqrt{0.24}\}
$$

and that $S_n = \sum_{i=1}^{n} X_i$ has a binomial distribution with parameters $p = 0.4$ and n. For comparison to the graph of $F_{Z_n}(z)$ in Figure 5.1, in each case we have superimposed a graph of the c.d.f. $F_Z(z)$ of the standard normal distribution. From Figure 5.1, it can be seen that the graphs of $F_{Z_n}(z)$ and $F_Z(z)$ become more and more alike as n increases, becoming virtually identical when $n = 50$.

In this example, the value of the sample size n needed so that the $\mathcal{N}(0, 1)$ distribution would serve as a good approximation to the distribution of Z_n was somewhere between $n = 20$ and $n = 50$, depending on how close an approximation was desired. If we had used Bernoulli observations $X_1, X_2, \ldots$ with parameter $p = 0.2$, a larger value of n would have been required, while if $p = 0.5$, even a sample size of $n = 12$ would do. Drawing graphs of the probability mass functions of X in the cases $p = 0.2$, $p = 0.4$, $p = 0.5$ shows that the distribution of X is quite skewed (asymmetric) when $p = 0.2$, less skewed when $p = 0.4$, and symmetric about its mean when $p = 0.5$. Since the normal distribution is also symmetric about its mean, our discussion in this special case suggests that in general the more skewed the original distribution of X is, the larger the sample size n must be before $F_Z(z)$ provides a good approximation to $F_{Z_n}(z)$.

There is one case where Z_n (and also $\hat{\mu}_X$ and S_n) has *exactly* a normal distribution *regardless* of how many observations are taken on X. This is the case when X has a $\mathcal{N}(\mu_X, \sigma_X^2)$ distribution. When $X_1, X_2, \ldots, X_n$ are mutually independent with a common $\mathcal{N}(\mu_X, \sigma_X^2)$ distribution, it is shown in Chapter 10 that $\hat{\mu}_X \sim \mathcal{N}(\mu_X, \sigma_X^2/n)$, $S_n \sim \mathcal{N}(n\mu_X, n\sigma_X^2)$, and $Z_n \sim \mathcal{N}(0, 1)$, regardless of how small the sample size n may be. [Indeed, when $n = 1$, $\hat{\mu}_X = S_1 = X_1 \sim \mathcal{N}(\mu_X, \sigma_X^2)$, and Z_1 is the standardization of X_1.]

Example 5.2. A psychologist studying disruptive schoolchildren wants to estimate the mean IQ μ of these children within an accuracy of ± 5, and assumes on the basis of Terman's study (Example 4.1) that IQ measurements X for these children have a normal distribution with mean $\mu_X = \mu$ and variance $\sigma_X^2 = 263.66$. The sample average $\hat{\mu}_X = (20)^{-1} \sum_{i=1}^{20} X_i$ based on IQ data X_1, $X_2, \ldots, X_n$ obtained independently from $n = 20$ of the children is used to estimate μ. Since the observations $X_1, X_2, \ldots, X_{20}$ are independent and normally distributed, $\hat{\mu}_X$ has an exact $\mathcal{N}(\mu, 263.66/20)$ distribution, even though the number $n = 20$ of observations taken is not very large. Thus,

$$P\{-5 \le \hat{\mu}_X - \mu \le 5\} = 2F_Z\left(\frac{\sqrt{20}(5)}{\sqrt{263.66}}\right) - 1 = 2F_Z(1.38) - 1 = 0.83242.$$

Notice from Table 3.1 that $2F_Z(2.00) - 1 = 0.95450$. Thus, if the estimate of μ is to be accurate within ± 5 with probability 0.9545, then we must choose the sample size n to satisfy

$$\frac{\sqrt{n}(5)}{\sqrt{263.66}} = 2.00,$$

or $n = 42.19$. Thus, taking data on $n = 43$ children should assure the desired accuracy.

The Continuity Correction

Suppose that $S_n = \sum_1^n X_i$ or $\hat{\mu}_X = S_n/n$ is computed from random variables $X_1, X_2, \ldots, X_n$ having discrete distributions whose possible values are equally spaced at intervals of length 2Δ. For moderate sample sizes n, a *continuity correction* often helps correct for the error incurred in approximating a discrete distribution by the continuous normal distribution. If s and u are possible values of the random variable S_n and $\hat{\mu}_X$, then we use the approximations

$$P\{S_n \le s\} \simeq F_Z\left(\frac{s + \Delta - n\mu}{\sqrt{n}\,\sigma}\right),$$

$$P\{\hat{\mu}_X \le u\} \simeq F_Z\left(\frac{u + \Delta/n - \mu}{\sigma/\sqrt{n}}\right).$$

Note that if X has possible values equally spaced in steps of 2Δ, then so does $S_n = \sum_1^n X_i$. To justify the approximation for S_n, we think of the probability mass $P\{S_n = s\}$ located at the possible values s of S_n as being spread uniformly over the interval $[s - \Delta, s + \Delta]$, thus making the distribution of S_n into a continuous distribution whose probability density function has the form of a histogram (see Figure 5.2). The probability $P\{S_n = s\}$ is now given by the area under the histogram over the interval $[s - \Delta, s + \Delta]$; thus, to compute $P\{S_n \le s\}$, we must actually find the area under the histogram over the interval $(-\infty, s + \Delta)$. Areas under the density function of the normal distribution with mean $n\mu$ and variance $n\sigma^2$ are used to approximate areas under the

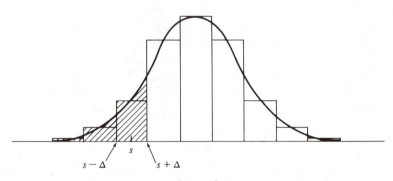

Figure 5.2: *Approximating $P\{S_n \leq s\}$ by the normal curve.*

histogram (Figure 5.2), leading to the approximation given by the foregoing continuity correction. A similar explanation can be given to justify the continuity correction for $\hat{\mu}_X$.

Example 5.3 (Discrete Case). A gambler playing blackjack decides to play 100 times and to bet \$10 each time. Assume that the hands are independently played, and that on each play, the gambler's probability of winning \$10 is 0.49; the probability of losing \$10 is 0.51. Let $X_i (i = 1, 2, \ldots, 100)$ denote the gambler's winnings on the ith play. Then

$$\mu_{X_i} = \$10(0.49) + (-\$10)(0.51) = -\$0.20,$$

$$\sigma^2_{X_i} = (\$10 + \$0.20)^2(0.49) + (-\$10 + \$0.20)^2(0.51)$$

$$= 99.96 \text{ dollars squared},$$

$$\sigma_{X_i} = \sqrt{99.96} = \$10.00.$$

The total winnings of the gambler are $S_{100} = \sum_{i=1}^{100} X_i$. Note that each X_i is a discrete random variable whose values are spaced \$20 apart. Thus, to use the continuity correction, we let $\Delta = 10$. The probability, $P\{S_{100} > 0\}$, that the gambler finishes the evening with a profit is $1 - P\{S_{100} \leq 0\}$, which by the Central Limit Theorem (using the continuity correction) is approximately equal to

$$1 - F_Z\left(\frac{S + \Delta - n\mu}{\sqrt{n}\sigma}\right) = 1 - F_Z\left(\frac{0 + \$10 - 100(-\$0.20)}{\sqrt{100}(\$10.00)}\right) = 1 - F_Z(0.30).$$

From Table T.5, we find that this probability is 0.38209. It is interesting to calculate the probability that the gambler makes a profit if 1000 hands are played instead of 100. In this case, the gambler's total winnings, S_{1000}, has a probability approximately $1 - F_Z(0.66)$ or 0.25463 of being greater than 0. We see that the probability that the gambler finishes the evening with a profit decreases with the number of plays.

One of the important special uses of the Central Limit Theorem is to provide an approximation for the probabilities of a binomial distribution. If S_n has the binomial distribution with parameters n and p, then we have already remarked that $S_n = \sum_{i=1}^{100} X_i$, where the X_i are independent Bernoulli trials each having parameter p. For n even moderately large [say such that $np \geq 5$ and $n(1 - p) \geq 5$], the standard normal c.d.f. gives a good approximation to the c.d.f. of $(S_n - np)/\sqrt{np(1 - p)}$. (We have already demonstrated this fact when $p = 0.4$; see Figure 5.1.) This approximation is often improved by a continuity correction with $\Delta = \frac{1}{2}$ because each X_i is a discrete random variable with values a distance of 1 apart. Thus, when S_n has a binomial distribution with parameters n and p,

$$P\{S_n \leq s\} \simeq F_Z\left(\frac{s + \frac{1}{2} - np}{\sqrt{np(1 - p)}}\right).$$

Example 5.4. A genetic theory asserts that 25% of the offspring of a certain crossbreeding experiment will have the characteristic C. A total of $n = 50$ offspring of such an experiment are observed and only six are found to have characteristic C. If the genetic theory is true, the probability of observing six or fewer offspring with characteristic C is

$$P\{S_{25} \leq 6\} \simeq F_Z\left(\frac{6 + \frac{1}{2} - (50)(0.25)}{\sqrt{(50)(0.25)(0.75)}}\right) = F_Z(-1.96) = 1 - F_Z(1.96) = 0.02500.$$

Thus, we have observed an outcome belonging to an event that is very unlikely if the genetic theory is correct.

EXERCISES

1. Suppose $Z \sim \mathcal{N}(0, 1)$; let $A = \{-0.3 \leq Z \leq 0.7\}$ and $B = \{0.1 \leq Z \leq 3.0\}$. Find
   (a) $P(A)$,   (c) $P(A \cup B)$,   (e) $P(A \mid B)$,
   (b) $P(B)$,   (d) $P(A \cap B)$,   (f) $P(B \mid A)$.

2. Suppose that $X \sim \mathcal{N}(3, 4)$. Find
   (a) $P\{X > 5\}$,   (d) $P\{3 < X \leq 5\}$,   (g) $P\{X < 2 \text{ or } X \geq 5\}$,
   (b) $P\{X \geq 5\}$,   (e) $P\{2 \leq X \leq 5\}$,   (h) $P\{X \leq 3 \text{ or } X > 5\}$,
   (c) $P\{X \leq 5\}$,   (f) $P\{2 < X \leq 5\}$,   (i) $P\{X > 3 \text{ or } X \leq 5\}$.

3. Let $X \sim \mathcal{N}(\mu, \sigma^2)$. Show that $Y = aX + b \sim \mathcal{N}(a\mu + b, a^2\sigma^2)$ by finding $f_Y(y)$. Thus, show that if $Z \sim \mathcal{N}(0, 1)$, then $\sigma Z + \mu \sim \mathcal{N}(\mu, \sigma^2)$.

4. Two IQ tests are commonly used. One test has scores $X \sim \mathcal{N}(100, 324)$, while the other test has scores $Y \sim \mathcal{N}(50, 100)$. Under the assumption that both tests measure the same phenomenon ("general intelligence"), what Y-score is comparable to an X-score of 127?

5. If $X \sim \mathcal{N}(100, 225)$, find
   (a) the (0.25)th quantile $Q_{0.25}(X)$;
   (b) the (0.95)th quantile $Q_{0.95}(X)$;
   (c) the median, $\text{Med}(X)$;
   (d) the interval $[a, b]$ of length $b - a = 10$ having the highest probability $P\{a \leq X \leq b\}$.

6. The scores X of individuals on a certain test of personality are normally distributed, but the publisher of the test keeps the mean μ and variance σ^2 of the scores on the test confidential. However, the publisher does announce that the (0.90)th quantile is $Q_{0.90}(X) = 87$ and that the (0.95)th quantile is $Q_{0.95}(X) = 96$. Are the values of the mean μ and variance σ^2 of the test scores really confidential?

7. Men's shirts are classified according to size as S, M, L, XL, corresponding respectively to neck circumferences of under 15 inches (for S), between 15 and 16 inches (for M), between 16 and 17 inches (for L), and over 17 inches (for XL). Suppose that neck circumferences X for adult males have a $\mathcal{N}(15.75, 0.49)$ distribution.
   (a) How many shirts should be manufactured in each category of shirt size if 1000 shirts are to be manufactured?
   (b) If you wanted to define categories S, M, L, XL so that each category contained 25% of the total population of adult males, what neck sizes would you assign to each of these categories?

8. An airfreight company has 5000 cubic feet of cargo space available for a given flight. The company accepts packages for shipment on the basis of weight alone. However, there is the question of storing the cargo. Previous experience has shown that the cargo space X required (in cubic feet) to store a shipment of Y pounds in weight is a random variable having approximately a $\mathcal{N}((0.34)Y, (0.01)Y^2)$ distribution. The company figures that their plane can fly safely with a cargo of up to 16,000 pounds. They want to determine a weight limit for acceptance of packages that with probability no less than 0.99 will allow them to store all packages accepted for shipment. Can the company continue to accept packages for shipment up to the weight limit of 16,000 pounds? If not, what is the largest weight of shipment that they can accept?

9. The College Entrance Examination Board Advanced Placement Examination in English was administered on May 18, 1964, to 11,329 secondary school students seeking advanced placement in college. The test consists of three parts: (i) analysis of a poem, (ii) literature, and (iii) composition. A detailed distributional analysis of 370 of the composition scores appears in Table E.1.

Table E.1: *Distribution of Scores X on the Composition Portion of the College Entrance Examination Board Advanced Placement Examination in English*

Intervals of scores	Observed frequency
68–71	2
64–67	6
60–63	13
56–59	21
52–55	35
48–51	41
44–47	58
40–43	63
36–39	46
32–35	34
28–31	28
24–27	17
20–23	3
16–19	1
12–15	1
8–11	1
	370

For these data, the sample average $\hat{\mu}_X$ is 42.96, the sample variance is $\hat{\sigma}_X^2 = 101.2036$, and the sample standard deviation is $\hat{\sigma}_X = 10.06$. To simplify the numerical computations, use the approximate values $\hat{\mu}_X = 43.0$ and $\hat{\sigma}_X = 10.0$. Fit a theoretical normal distribution to the data by calculating theoretical frequences for the intervals of scores given. [Remark: It would be currently more acceptable to use intervals of the form [7.5, 11.5], [11.5, 15.5], and so on, rather than the kinds of intervals used in Table E.1.]

10. In data collected at the Lick Observatory on 80 bright stars in a certain celestial area, the radial velocity X is measured. The data [reported by Trumpler and Weaver (1953)] are shown in Table E.2. Fit a theoretical normal distribution to these data. (See the remark after Exercise 9.)

11. The reaction time of cats to a certain stimulus is normally distributed with expected value 0.1 second and variance 0.000169 (second)². Three cats are chosen independently and subjected to this stimulus.
    (a) What is the probability that all three cats respond in less than 0.126 second to the stimulus?

Table E.2: *Observed Values of Radial Velocity X for 80 Bright Stars*

Interval of velocities	Midpoint of interval	Observed frequency
-80 to -70	-75	1
-70 to -60	-65	2
-60 to -50	-55	2
-50 to -40	-45	2
-40 to -30	-35	8
-30 to -20	-25	24
-20 to -10	-15	26
-10 to 0	-5	11
0 to 10	5	2
10 to 20	15	1
20 to 30	25	$\dfrac{1}{80}$

(b) What is the probability that at least one of the cats takes more than 0.113 second to respond?

(c) What is the probability that the average of the response times of the three cats is strictly greater than 0.110 second?

(d) If the stimulus is given to 100 cats, what is the probability that the average of the response times of these 100 cats is strictly greater than 0.110 second?

12. Simple flashlights generally contain two or three batteries. Suppose that the life of each battery is normally distributed with a mean of 30 hours and a variance of 25 (hours)^2. The flashlight will cease to function if one or more of its batteries fails. Assuming that the lifetimes of batteries in a flashlight are probabilistically independent, find the probability that a flashlight will operate for at least 30 hours if the flashlight has (a) 2 batteries, and (b) 3 batteries. If you take a 2-battery flashlight and a 3-battery flashlight on a trip, what is the probability that at least one of these flashlights will last the trip if the trip requires use of each of the flashlights for 40 hours?

13. An actuary fits the Poisson distribution to accident data, using $n = 900$ observations $X_1, X_2, \ldots, X_{900}$, where each X_i is an independent observation of the number X of accidents in a given time period. To estimate the parameter λ of the Poisson distribution, we use the value of $\hat{\mu}_X$. The actuary wants the estimate of λ not to be in error by more than ± 0.1. Using the Central Limit Theorem, find $P\{-0.1 \leq \hat{\mu}_X - \lambda \leq 0.1\}$, the probability that the desired accuracy is achieved when the value of λ is

actually equal to 9. [Warning: What is the variance σ_X^2 of the Poisson distribution?]

14. An educational administrator is planning for future elementary schools in a given community. Several new subdivisions are being built in the community, and 1000 new families are expected. The administrator assumes that each new family is an independent trial of a random experiment in which X, the number of children of elementary school age in the family, is observed. The probability model for X is given by

X	0	1	2	3
$P\{X = x\}$	0.35	0.45	0.15	0.05

(a) Find μ_X and σ_X^2.
(b) Additional classroom space will be needed for $Y = \sum_{i=1}^{1000} X_i$ new children. The administrator plans to add classrooms for 1000 new students. Use the Central Limit Theorem with continuity correction to find the probability $P\{Y > 1000\}$ that there will be insufficient room for the new children.

15. The probability of a male infant on a given birth is $p = 0.51$. Use the Central Limit Theorem, both with and without the continuity correction, to answer the following:
(a) In $n = 100$ probabilistically independent births, what is the approximate probability of obtaining 50 or fewer male births?
(b) In $n = 400$ births, what is the probability of obtaining 200 or fewer male births?
(c) In $n = 1000$ births, what is the probability of obtaining 500 or fewer male births?
Reach conclusions about (i) the importance of the continuity correction as n increases, and (ii) the probability of 50% or fewer male births as n increases.

16. Let Y have a binomial distribution with $p = 0.80$ and $n = 100$. Use the Central Limit Theorem with the continuity correction to find
(a) $P\{Y \le 80\}$, (b) $P\{60 < Y \le 90\}$, (c) $P\{70 \le Y\}$, (d) $P\{Y = 80\}$.

17. The Internal Revenue Service uses a letter-opening machine which removes checks from the envelopes. Suppose that the probability that the machine fails to remove a check from any given envelope is $p = 0.03$, and assume that the actions of the machine on the various envelopes are mutually probabilistically independent trials. On a certain day, $n = 500$ envelopes are opened. Let Y be the number of envelopes for which the machine fails to remove a check.
(a) What is the exact distribution of Y?

(b) Use the Poisson distribution to approximate $P\{Y \le 5\}$.

(c) Use the Central Limit Theorem with continuity correction to approximate $P\{Y \le 5\}$.

18. In a shipment of 30,000 cans of peaches, a total of 300 cans are damaged. In a quality control inspection, a sample of $n = 100$ cans are randomly sampled *without* replacement. Let Y be the number of damaged cans observed in the sample.

(a) What is the exact distribution of Y?

(b) If the cans had been sampled *with* replacement, what would be the exact distribution of Y?

(c) Use the Central Limit Theorem with continuity correction to approximate $P\{Y \ge 3\}$.

(d) Use the Poisson distribution to approximate $P\{Y \ge 3\}$.

19. Let $X \sim \mathcal{N}(0, \sigma^2)$. If (a) $Y = |X|$, and (b) $Y = X^2$, find the cumulative distribution function $F_Y(y)$ and then the probability density function $f_Y(y)$ of Y.

8

Special Distributions:

Continuous Case

In Chapter 6, we presented some discrete distributions that frequently arise in applications. In the present chapter we discuss some continuous distributions other than the normal distribution that are frequently used. The continuous distributions discussed are the exponential, gamma, Weibull, uniform, beta, lognormal, and Cauchy distributions.

1. THE EXPONENTIAL DISTRIBUTION

A light bulb is turned on and left to burn until it burns out. How long will it last? A student is given a learning task to master. How long will it take to complete the task? An alpha particle is emitted from a sample of radioactive material. How long will it be before the next alpha particle is emitted?

Many scientific experiments involve the measurement of the duration of time X between an initial point of time and the occurrence of some phenomenon of interest. For example, psychologists have studied the length of time X between successive presses of a bar by a rat placed in a Skinner box [see Mueller (1950)]. Engineers have investigated the period of time X between successive failures of components in complex machines. Telephone companies are interested in the duration X of a telephone call. Finally, both psychologists and biophysicists have been interested in the time X between successive electrical impulses in the spinal cords of various mammals.

If we make repeated measurements of X in such experiments, the histogram frequently resembles Figure 1.1. One probability density function whose

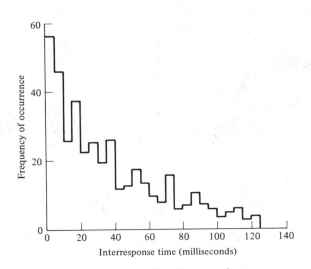

Figure 1.1: *Frequency distribution of interresponse times in the spontaneous activity of a single spinal interneuron. The distribution consists of 391 intervals recorded by a micropipet in the spinal cord of a cat.* [*From W. J. McGill (1963) in Luce, Bush, and Galanter,* Handbook of Mathematical Psychology. *New York: John Wiley & Sons, Inc.*]

graph resembles the histogram shown in Figure 1.1 is the exponential density function

(1.1) $$f_X(x) = \begin{cases} \theta e^{-\theta x}, & \text{if } x \geq 0, \\ 0, & \text{if } x < 0, \end{cases}$$

where $\theta > 0$. A random variable X having a density function (1.1) is said to have an *exponential distribution*. Figure 1.2 provides graphs of the exponential density function for $\theta = 0.1$, 0.5, 1.0, and 2.0. Notice that the graph of the exponential density function $f_X(x)$ falls off rapidly as x increases from 0 to infinity, with the steepness of the fall increasing with increasing values of the parameter θ.

The Mean and Variance of the Exponential Distribution

If X has an exponential distribution with parameter θ, then for any $r \geq 0$, using (A.1) of the Appendix,

(1.2) $$E(X^r) = \int_0^\infty x^r (\theta e^{-\theta x}) \, dx = \frac{1}{\theta^r} \int_0^\infty v^r e^{-v} \, dv = \frac{\Gamma(r+1)}{\theta^r}.$$

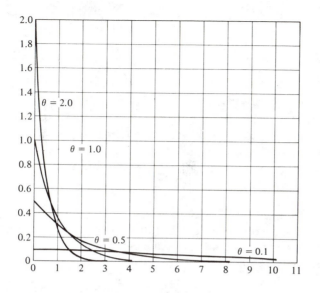

Figure 1.2: *Graphs of the exponential density function for* $\theta = 0.1, 0.5, 1.0,$ *and* 2.0.

In particular,

(1.3)
$$\mu_X = \frac{\Gamma(2)}{\theta} = \frac{1}{\theta}, \qquad \sigma_X^2 = \frac{\Gamma(3)}{\theta^2} - \mu_X^2 = \frac{1}{\theta^2}.$$

Incidentally, an application of (1.2) with $r = 0$ shows that $\int_0^\infty f_X(x)\, dx = 1$.

Probability Calculations

The cumulative distribution function $F_X(x)$ of a random variable X which has an exponential distribution with parameter θ is given by the expression

(1.4)
$$F_X(x) = \begin{cases} 1 - e^{-\theta x}, & \text{if } x \geq 0, \\ 0, & \text{if } x < 0. \end{cases}$$

To see this, note that $F_X(x) = 0$ when $x < 0$. When $x \geq 0$,

$$F_X(x) = \int_0^x f_X(t)\, dt = \int_0^x \theta e^{-\theta t}\, dt = -e^{-\theta t}\Big|_0^x = -e^{-\theta x} + 1.$$

We can obtain the probability of any interval using $F_X(x)$. For example, if $a \geq 0$,

$$(1.5) \qquad P\{X \geq a\} = P\{X > a\} = 1 - F_X(a) = e^{-\theta a},$$

$$(1.6) \quad P\{a \leq X \leq b\} = P\{a \leq X < b\} = P\{a < X < b\} = P\{a < X \leq b\}$$
$$= F_X(b) - F_X(a) = e^{-\theta a} - e^{-\theta b},$$

since X is a continuous random variable. To obtain the probabilities (1.5), (1.6), and $F_X(x) = P\{X \leq x\}$, it is helpful to have a table of values of e^{-t}. Table T.6 provides values of e^{-t} for $t = 0.00(0.01)7.99$; values of e^{-t} for larger values of t are smaller than 0.0004.

Example 1.1. Mueller (1950) describes an experiment in which white rats were periodically conditioned. The duration of time X in seconds between presses of the rat on a bar has an exponential distribution with parameter $\theta = 0.20$. Using (1.6) and Table T.6,

$$P\{1 \leq X \leq 3\} = e^{-(0.20)(1)} - e^{-(0.20)(3)} = 0.819 - 0.549 = 0.270.$$

Similarly, using (1.5) and Table T.6,

$$P\{X > 3\} = e^{-(0.20)(3)} = 0.549.$$

Fitting the Exponential Distribution to Data

Proschan (1963) provides records giving the durations of time between successive failures of the air-conditioning systems of each member of a fleet of 13 Boeing 720 jet airplanes. These durations of time are listed in Table 1.1. Thus, plane number 7907 had a failure of its air-conditioning system after 194 hours of service, a second failure 15 hours of service after the first failure, a third failure 41 hours of service after the second failure, and so on. After roughly 2000 hours of service, each plane received a major overhaul; if a time interval between two failures of the air-conditioning system included the major overhaul, the length of this time interval is not listed in Table 1.1, since its magnitude may have been affected by the fact that repairs were made on the plane.

Equation (1.3) suggests that we estimate θ from the data in Table 1.1 by calculating the sample average

$$\hat{\mu}_X = \tfrac{1}{213}(194 + 15 + 41 + \cdots + 66 + 61 + 34) = 93.14 \text{ hours,}$$

and then letting $\hat{\theta} = 1/\hat{\mu}_X = 1/93.14 = 0.0107$. The density (1.1) with $\theta = 0.0107$, as well as a modified relative frequency histogram obtained from Table 1.1, are shown in Figure 1.3.

Table 1.1: *Table of the Durations of Time Between Successive Failures of the Air-Conditioning Systems of Each Member of a Fleet of 13 Boeing 720 Jet Airplanes (213 Observations in All)*

Plane identification number												
7907	7908	7909	7910	7911	7912	7913	7914	7915	7916	7917	8044	8045
194	413	90	74	55	23	97	50	359	50	130	487	102
15	14	10	57	320	261	51	44	9	254	493	18	209
41	58	60	48	56	87	11	102	12	5		100	14
29	37	186	29	104	7	4	72	270	283		7	57
33	100	61	502	220	120	141	22	603	35		98	54
181	65	49	12	239	14	18	39	3	12		5	32
	9	14	70	47	62	142	3	104			85	67
	169	24	21	246	47	68	15	2			91	59
	447	56	29	176	225	77	197	438			43	134
	184	20	386	182	71	80	188				230	152
	36	79	59	33	246	1	79				3	27
	201	84	27	a	21	16	88				130	14
	118	44	a	15	42	106	46					230
	a	59	153	104	20	206	5					66
	34	29	26	35	5	82	5					61
	31	118	326		12	54	36					34
	18	25			120	31	22					
	18	156			11	216	139					
	67	310			3	46	210					
	57	76			14	111	97					
	62	26			71	39	30					
	7	44			11	63	23					
	22	23			14	18	13					
	34	62			11	191	14					
		a			16	18						
		130			90	163						
		208			1	24						
		70			16							
		101			52							
		208			95							

ᵃ Indicates major overhaul.

The Survival Function

Suppose that the lifetime X of a certain system is observed. The c.d.f. $F_X(x) = P\{X \leq x\}$ of X gives the probability that the system will "die" before x units of time have passed. Thus, the quantity $1 - F_X(x) = P\{X > x\}$, called

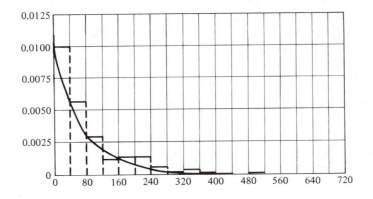

Figure 1.3: *Modified relative frequency histogram and fitted exponential density function for the data of Table 1.1 representing durations of time between successive failures of the air conditioning systems of each member of a fleet of 13 Boeing 720 jet airplanes.*

the *survival function*, gives the probability that the "system" *survives* more than x units of time.

For the exponential distribution with parameter θ, the survival function is given by the equation

$$1 - F_X(x) = \begin{cases} 1, & \text{if } x < 0, \\ e^{-\theta x}, & \text{if } x \geq 0. \end{cases}$$

Figure 1.4 provides graphs of the survival functions $1 - F_X(x)$ corresponding to exponentially distributed random variables with parameter values $\theta = 0.5$, 1.0, 2.0, 5.0, and 10.0.

In the case of the data of Table 1.1 (where the lifetime X of interest is the length of time during which the air-conditioning system of the jet plane is operational), a comparison (see Figure 1.5) of the graphs of the theoretical survival function and the sample survival function

$$1 - F_n(x) = \frac{\text{number of observations} > x}{n}$$

bears out our earlier finding that there is a close fit between the observed data and the exponential distribution with parameter $\theta = 0.0107$.

An Important Property of the Exponential Distribution

Suppose that the random variable X measures the duration of time until the occurrence of a given phenomenon and that it is known that X has an

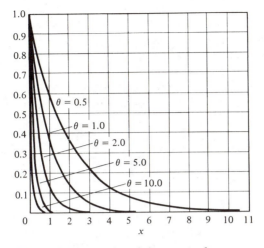

Figure 1.4: *Graphs of the survival function* $1 - F_X(x)$ *for the exponential distribution with values of the parameter* $\theta = 0.5,\ 1.0,\ 2.0,\ 5.0,\ and\ 10.0.$

Figure 1.5: *Comparison between the theoretical survival function* $1 - F_X(x) = \exp - (0.0107x)$ *and the sample survival function* $1 - F_n(x)$ *calculated from the data of Table 1.1.*

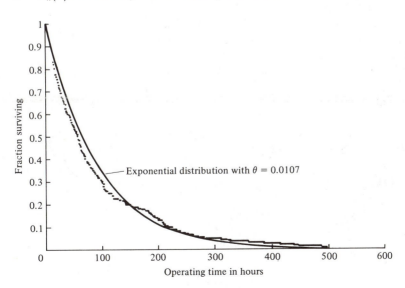

exponential distribution with parameter θ. The time is now x^* and the phenomenon of interest has not yet occurred. Thus, we know that $X > x^*$. Let $Y = X - x^*$ equal the additional time that we must wait for the phenomenon to occur. What is the *conditional distribution* of Y given that we have already waited x^* units of time?

The conditional distribution of $Y = X - x^*$ given that the event $A = \{X > x^*\}$ has occurred can be found from the conditional c.d.f.:

$$(1.7) \qquad F_Y(y \,|\, A) = P(\{Y = X - x^* \leq y\} \,|\, A) = \frac{P(\{X - x^* \leq y\} \cap A)}{P\{X > x^*\}}.$$

If $y < 0$, then $F_Y(y \,|\, A) = 0$. If $y \geq 0$, we can use (1.5), (1.6), and (1.7) to conclude that

$$F_Y(y \,|\, A) = \frac{e^{-\theta x^*} - e^{-\theta(x^* + y)}}{e^{-\theta x^*}} = 1 - e^{-\theta y}.$$

Thus,

$$(1.8) \qquad F_Y(y \,|\, A) = \begin{vmatrix} 1 - e^{-\theta y}, & \text{if } y \geq 0, \\ 0, & \text{if } y < 0, \end{vmatrix} = F_X(y).$$

Since the cumulative distribution function of a random variable determines its distribution, it follows that the conditional distribution of the additional waiting time $Y = X - x^*$ *given* that we have already waited x^* units of time (i.e., $\{X > x^*\}$ has occurred) is the *same* as the unconditional distribution of the original waiting time X. In this sense, the exponential distribution "has no memory" of the past.

In terms of the length, X, of a telephone conversation, the foregoing property implies that if you have been waiting at a phone booth for 10 minutes for a conversation to end so that you can use the phone, then the remaining time that you have to wait has the *same* distribution as if the conversation had just started. Consequently, if one phone booth has just been occupied as you arrive, and another phone booth has been occupied for some time, it does not matter which phone booth you decide to wait for, since the probability distribution of the length of time you must wait will be the same for either phone booth!

Among all distributions of nonnegative continuous variables, only the exponential distributions "have no memory." To see this, let

$$h(s) = P\{X > s\}, \quad s > 0.$$

The memoryless property asserts that

$$(1.9) \qquad P\{X > s + t \,|\, X > s\} = P\{X > t\}, \quad s > 0, t > 0.$$

But

$$(1.10) \quad P\{X > s + t \,|\, X > s\} = \frac{P(\{X > s + t\} \cap \{X > s\})}{P\{X > s\}} = \frac{P\{X > s + t\}}{P\{X > s\}}.$$

Thus, (1.9) and (1.10) together yield the famous functional equation

$$(1.11) \qquad\qquad h(s + t) = h(s)h(t), \quad s > 0, t > 0,$$

which is known as the *Cauchy equation* [e.g., see Aczél (1966)]. Under mild regularity conditions on the function h, the only solution of (1.11) is

$$h(x) = e^{-cx},$$

where c is any constant. However, $F_X(x) = 1 - h(x)$ is a c.d.f. and thus $\lim_{x \to \infty} F_X(x) = 1$. This forces c to be positive, and we recognize $h(x)$ as being the survival function of the exponential distribution with parameter $\theta = c$.

2. THE GAMMA DISTRIBUTION

Visualize a system (such as a radio, an assembly line, an airplane, and so on) in which the proper functioning of a certain component is essential for the proper functioning of the system as a whole. In order to increase the reliability of the system, the system may be designed to carry $(r - 1)$ spare parts. When the original component fails, one of the $(r - 1)$ spare components is activated to take its place. If this component fails, one of the $(r - 2)$ remaining spare components takes over. This process continues until the original component and the $(r - 1)$ spares have failed, at which point the entire system suffers failure. Assuming that all other components of the system except for the essential component have infinite lifetimes, the lifetime (time until failure) of the entire system is the sum $Y = \sum_{i=1}^{r} X_i$ of the lifetimes $X_1, X_2, \ldots, X_r$ of the r duplicates of the essential component. If each of the lifetimes $X_1, X_2, \ldots, X_r$ has the same exponential distribution with parameter θ, and if the lifetimes $X_1, X_2, \ldots, X_r$ are probabilistically independent, it is shown in Chapter 10 that the lifetime $Y = \sum_{i=1}^{r} X_i$ of the system has the *gamma distribution* with density function

$$(2.1) \qquad f_Y(y) = \begin{cases} \dfrac{\theta^r y^{r-1} e^{-\theta y}}{\Gamma(r)}, & \text{if } y \geq 0, \\[2mm] 0, & \text{if } y < 0, \end{cases}$$

where θ and r are positive numbers and $\Gamma(r)$ is the gamma function. The constants θ and r in Equation (2.1) are the parameters of the gamma distribution. Although in the example of the system having spare components, the parameter r of the gamma distribution is an integer, this need not always be the case. In many applications of the gamma distribution, r is not an integer.

Example 2.1 (Radioactivity). Slack and Krumbein (1955) describe an experiment in which a representative measure Y of radioactivity (alphas per minute) within a sample of Pennsylvania black shale is obtained. A gamma distribution with parameters $\theta = 0.9083$ and $r = 2.493$ is found to provide a reasonably close fit to the observed frequency distribution of Y.

Probability Calculations: The Standard Gamma Density

When Y has a gamma distribution with parameters θ and r, probabilities can be obtained from the cumulative distribution function $F_Y(y)$. Except in special cases, $F_Y(y)$ does not have a simple functional form, and thus must be tabulated. Since the distribution of Y has two parameters θ and r, the problem of tabulation appears formidable. Fortunately, as in the case of the normal distribution, some standardization is possible. In Chapter 4, Section 6, we showed that if $V = aY + b$ is a linear transformation of Y, then $f_V(v) = (|a|)^{-1}f_Y((v-b)/a)$. Let $a = \theta$, $b = 0$. Then $V = \theta Y$ and

$$(2.2) \qquad f_V(v) = \frac{1}{\theta} f_Y\left(\frac{v}{\theta}\right) = \begin{cases} \dfrac{v^{r-1}e^{-v}}{\Gamma(r)}, & \text{if } v \geq 0, \\[2mm] 0, & \text{if } v < 0. \end{cases}$$

A random variable V with the density (2.2) is said to have the *standard gamma distribution* with parameter r. Since $V = \theta Y$, it follows that

$$(2.3) \qquad F_Y(y) = P\{Y \leq y\} = P\left\{\frac{V}{\theta} \leq y\right\} = F_V(\theta y),$$

so that the c.d.f. of the gamma distribution with parameters θ and r can be determined from tables of the c.d.f. of the standard gamma distribution with parameter r.

Graphs of the standardized gamma density function are given in Figure 2.1(a) for $r = 0.5$, 2, 4, and 5. Note that when $r \leq 1$, the graph of the corresponding standardized gamma density function drops off steeply, just as in the case of the exponential density function. Indeed, when $r = 1$ the density function of the standardized gamma function is identical to the density function of the exponential distribution with parameter $\theta = 1$. When $r > 1$, the graph of the standardized gamma density function takes on a "humped" form. The density function has a long right-hand "tail," but this skewness becomes less pronounced as r increases. For r large enough (say, $r \geq 50$), the standard gamma distribution closely resembles a normal distribution with mean and variance both approximately equal to r [see Figure 2.1(b)].

Figure 2.1 shows the influence of the parameter r on the shape of the gamma density. The parameter θ in the (nonstandardized) gamma density (2.1) influences the shape of the density in the same manner as θ influences the shape of the exponential density (1.1); that is, as θ increases, the density is

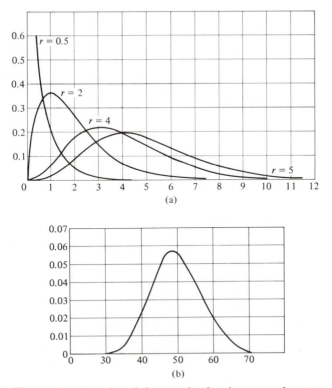

Figure 2.1: *Graphs of the standardized gamma density function for* (a) $r = 0.5, 2, 4, 5$; (b) $r = 50$.

"squeezed" in such a manner that the right-hand tail of the density function moves closer to the vertical axis (see Figure 2.2).
     The function

(2.4)
$$F_V(t) = \int_0^t \frac{v^{r-1}e^{-v}}{\Gamma(r)}\, dv = I_r(t), \quad t \geq 0,$$

is extensively tabulated and is known as the *incomplete gamma function*. Tables of $I_r(t) = F_V(t)$ for $r = 1(1)5$, $t = 0.2(0.2)8.0(0.5)15.0$ and for $r = 6(1)10$, $t = 1.0(0.2)8.0(0.5)17.0$, are given in Table T.7 in the Appendix. Combining (2.3) and (2.4), we see that when Y has a gamma distribution with parameters θ and r, then the cumulative distribution function of Y is

(2.5)
$$F_Y(y) = I_r(\theta y),$$

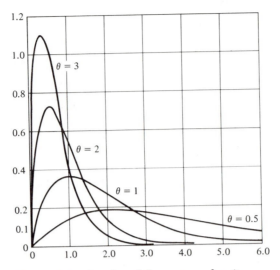

Figure 2.2: *Graphs of the gamma density function for r = 2, θ = 0.5, 1, 2, 3.*

where we adopt the convention that $I_r(t) = 0$ for $t < 0$. It follows from (2.5) that

(2.6) $P\{Y \geq a\} = P\{Y > a\} = 1 - F_Y(a) = 1 - I_r(\theta a)$

and

(2.7) $P\{a \leq Y \leq b\} = F_Y(b) - F_Y(a) = I_r(\theta b) - I_r(\theta a).$

Example 2.2. In studying the task-solving ability of children, the length of time Y (in minutes) that it takes the child to complete a complex learning task is measured. One theory of learning is that a child solves a complex task by independently and sequentially solving r simple tasks, each of approximately the same difficulty, and that the time, X_i, taken to solve the ith simple task is exponentially distributed with parameter θ for $i = 1, 2, \ldots, r$. In consequence, the total length of time $Y = \sum_{i=1}^{r} X_i$ required to finish the entire learning task has a gamma distribution with parameters θ and r. A large number of children are given a certain complex learning task, and it is found that the distribution of Y is approximately a gamma distribution with parameters $\theta = 0.40 = (2.50)^{-1}$ and $r = 5.01$. Because the theory says that r must be an integer, the gamma distribution with $\theta = (2.50)^{-1}$ and $r = 5$ is used to model the variation of Y. Then, to calculate the probability that a child cannot finish the learning task in, say, 30 minutes, (2.6) and Table T.7 yield

$$P\{Y > 30\} = 1 - I_5\left(\frac{30}{2.5}\right) = 1 - I_5(12) = 1 - 0.9924 = 0.0076.$$

Similarly, from (2.7) and Table T.7,

$$P\{15 \leq Y \leq 20\} = I_5\left(\frac{20}{2.5}\right) - I_5\left(\frac{15}{2.5}\right) = I_5(8) - I_5(6)$$

$$= 0.90037 - 0.71494 = 0.18543.$$

Equation (2.5) and Table T.7 allow us to graph either the cumulative distribution function $F_Y(y)$ or the survival function $1 - F_Y(y)$ of random variables Y having a gamma distribution. Graphs of the cumulative distribution functions (Figure 2.3) and survival functions (Figure 2.4) for random variables having gamma distributions with various pairs of parameters (θ, r) are given in Figures 2.3 and 2.4.

Figure 2.3: *Graphs of the gamma cumulative distribution functions with* (a) $r = 2$, $\theta = 0.5$, 1.0, 2.0; (b) $r = 5$, $\theta = 1.0$, 2.0, 4.0.

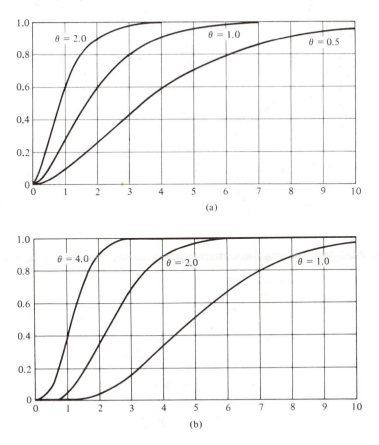

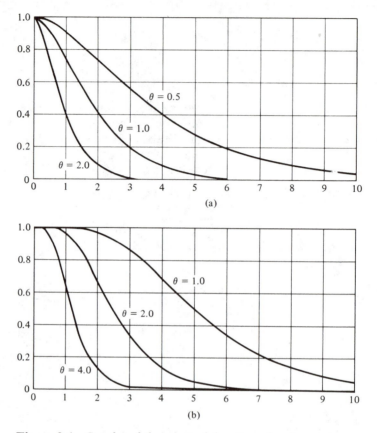

Figure 2.4: *Graphs of the survival functions for the gamma distribution with parameters* (a) $r = 2$, $\theta = 0.5$, 1.0, 2.0; (b) $r = 5$, $\theta = 1.0$, 2.0, 4.0.

Moments of the Gamma Distribution

When V has the standard gamma distribution with parameter r, then using (A.1) of the Appendix,

(2.8)
$$E(V^k) = \int_0^\infty v^k \left(\frac{v^{r-1} e^{-v}}{\Gamma(r)} \right) dv = \frac{\Gamma(r+k)}{\Gamma(r)},$$

so that if Y has a gamma distribution with parameters θ and r,

(2.9)
$$E(Y^k) = \frac{E(V^k)}{\theta^k} = \frac{\Gamma(r+k)}{\Gamma(r)\theta^k}.$$

Letting $k = 0$ in (2.9) shows that $\int_0^\infty f_Y(y) \, dy = 1$. Since $f_Y(y) \geq 0$ for all y, this shows that (2.1) is a probability density function. Further applying (2.9) with $k = 1, 2$ yields the mean and variance of Y:

$$(2.10) \qquad \mu_Y = \frac{r}{\theta}, \qquad \sigma_Y^2 = E(Y^2) - (\mu_Y)^2 = \frac{\Gamma(r+1)}{\theta^2} - \frac{r^2}{\theta^2} = \frac{r}{\theta^2}.$$

Note that the parameters θ and r can be determined from (2.10) by

$$(2.11) \qquad \theta = \frac{\mu_Y}{\sigma_Y^2}, \qquad r = \frac{(\mu_Y)^2}{\sigma_Y^2}.$$

Fitting the Gamma Distribution to Data

To illustrate fitting the gamma distribution to data, consider the following example.

Example 2.3. The data given in Table 2.1 are the failure times Y (in weeks) of 31 transistors (out of 34 under test) that failed to survive an accelerated life test. These observations are reported by Wilk, Gnanadesikan, and Huyett (1962), who indicate that from past experience there is reason to believe that the gamma distribution might reasonably approximate the distribution of failure times.

To find values of θ and r from these data, we first find

$$\hat{\mu}_Y = \tfrac{1}{31}(3 + 4 + 5 + \cdots + 42 + 42 + 52) = 15.71$$

and

$$\hat{\sigma}_Y^2 = \tfrac{1}{31}[(3 - \hat{\mu}_Y)^2 + (4 - \hat{\mu}_Y)^2 + \cdots + (52 - \hat{\mu}_Y)^2] = 141.50.$$

From (2.11),

$$\hat{\theta} = \frac{15.71}{141.50} = 0.11, \qquad \hat{r} = \frac{(15.71)^2}{141.50} = 1.74.$$

Using these values of θ and r and (2.1), we can prepare a graph of the (fitted) probability density function of the failure times Y of the transistors; this graph appears in Figure 2.5.

Table 2.1: *Times Until Failure of 31 Transistors That Failed an Accelerated Life Test*

3	4	5	6	6	7	8	8
9	9	9	10	10	11	11	11
13	13	13	13	13	17	17	19
19	25	29	33	42	42	52	

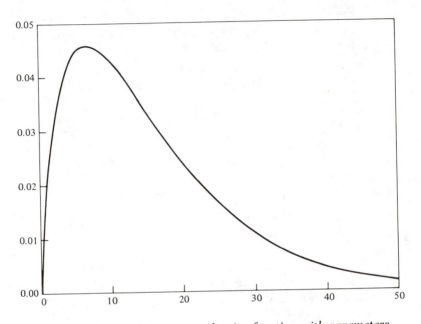

Figure 2.5: *Graph of the gamma density function with parameters* $\theta = 0.11$, $r = 1.74$.

Remark. In the article by Wilk, Gnanadesikan, and Huyett from which the data in Table 2.1 are taken, different estimates of the parameters θ and r are obtained, because these authors use a different method to estimate these parameters.

A Connection Among the Exponential, Gamma, and Poisson Distributions

The exponential, gamma, and Poisson distributions are related in an interesting way. Suppose that we have a batch of light bulbs whose lifetimes are each exponentially distributed with parameter θ. We start at a given time (say $t = 0$) and burn 1 bulb until extinction. We replace that bulb instantly with another bulb, wait until that bulb burns out, replace it with another, wait until this third bulb burns out, and so on. At a given time t, we stop this process and count the number L of light bulbs that we have burned out. It then follows that L is a random variable having the Poisson distribution with parameter $\lambda = \theta t$.

To see this, we make use of the gamma distribution. Let us find the probability of the event $\{L \geq k\}$ that at least k bulbs have been burned when we stop the experiment at time t. Let $X_1, X_2, \ldots, X_k$ be the lifetimes of the first k bulbs that we (might) have burned. Then, $X_1, X_2, \ldots, X_k$ are, by assumption, probabilistically independent of one another, and each lifetime X_i has an exponential distribution with parameter θ. We are able to burn at least these k bulbs if $X_1 + X_2 + \cdots + X_k \le t$, since in that case the k bulbs will have all burned out (one after the other) before our time is up. Recall that $Y = \sum_{i=1}^{k} X_i$ has a gamma distribution with parameters θ and k. Hence,

$$P\{L \ge k\} = P\{Y \le t\} = I_k(\theta t) = \int_0^{\theta t} \frac{v^{k-1} e^{-v} \, dv}{\Gamma(k)}$$

$$= -\frac{v^{k-1} e^{-v}}{\Gamma(k)} \bigg|_0^{\theta t} + \int_0^{\theta t} \frac{(k-1) v^{k-2} e^{-v} \, dv}{\Gamma(k)}$$

$$= -\frac{(\theta t)^{k-1} e^{-\theta t}}{(k-1)!} + \int_0^{\theta t} \frac{v^{k-2} e^{-v} \, dv}{\Gamma(k-1)}$$

where we have integrated by parts and used the fact that $\Gamma(k) = (k-1)\Gamma(k-1) = (k-1)!$ (see the Appendix). We continue integrating by parts in this fashion $(k-1)$ more times, to obtain

$$P\{L \ge k\} = -\sum_{i=1}^{k-1} \frac{(\theta t)^i e^{-\theta t}}{i!} + \int_0^{\theta t} \frac{e^{-v}}{\Gamma(1)} \, dv$$

$$= 1 - \sum_{i=0}^{k-1} \frac{(\theta t)^i e^{-\theta t}}{i!},$$

since $\int_0^{\theta t} e^{-v} \, dv = 1 - e^{-\theta t}$. However,

$$P\{L = k\} = P\{L \ge k\} - P\{L \ge k + 1\}$$

$$= \left[1 - \sum_{i=0}^{k-1} \frac{(\theta t)^i e^{-\theta t}}{i!} \right] - \left[1 - \sum_{i=0}^{k} \frac{(\theta t)^i e^{-\theta t}}{i!} \right]$$

$$= \frac{(\theta t)^k e^{-\theta t}}{k!},$$

so that the probability mass function $p_L(k) = P\{L = k\}$ of the number, L, of bulbs burned is that of a Poisson distribution with parameter $\lambda = \theta t$.

We have shown by this example that L has a Poisson distribution with parameter θT. However, in the process, we have also shown that if Y has a gamma distribution with parameters θ and r, and r is an integer, then

(2.12) $$F_Y(t) = P\{Y \le t\} = 1 - \sum_{i=0}^{r-1} \frac{(\theta t)^i}{i!} e^{-\theta t},$$

which gives a convenient formula by which to compute the cumulative distribution function of Y using a table of Poisson probabilities (Table T.3).

3. THE WEIBULL DISTRIBUTION

Important examples of nonnegative random variables occurring in applications are lifetimes, waiting times, learning times, durations of epidemics, and traveling times. Nontemporal examples of nonnegative random variables include material strengths, particle dimensions, radioactive intensities, rainfall amounts, and costs of industrial accidents. Although exponential or gamma distributions provide reasonable fits to the frequency distributions of some of these random variables, in some cases the fit is not as close as is desired, and in other cases the fit is unsatisfactory. Hence, other classes of distributions have been introduced to explain the variability of some of these phenomena. One such family of distributions are the *Weibull distributions*, named after the Swedish physicist Waloddi Weibull, who in 1939 suggested this family of distributions.

If a random variable X has a Weibull distribution, then the probability density function $f_X(x)$ of X has the form

(3.1) $$f_X(x) = \begin{cases} \dfrac{\beta}{\alpha}\left(\dfrac{x-v}{\alpha}\right)^{\beta-1} \exp\left[-\left(\dfrac{x-v}{\alpha}\right)^{\beta}\right], & \text{if } x \geq v, \\ 0, & \text{if } x < v. \end{cases}$$

The three constants $\beta > 0$, $\alpha > 0$, and $v \geq 0$ are parameters of the distribution. Note that the parameter v tells us the smallest possible value of the random variable X. The parameter β determines the shape of the density (3.1), while $1/\alpha$ plays the same role that the parameter θ did for the family of gamma densities. The roles of β and α are illustrated in Figures 3.1 and 3.2; in Figure 3.1, $f_X(x)$ is graphed for $\beta = 2.0$, $v = 0.0$, $\alpha = 0.5, 1.0, 2.0$, and in Figure 3.2 $f_X(x)$ is graphed for $\beta = 10.0$, $v = 0.0$, $\alpha = 0.5, 1.0, 2.0$. Finally, we see that v plays the role of a location parameter for the density (3.1) by graphing $f_X(x)$ for $\beta = 0.5$, $\alpha = 1.0$, $v = 0.0, 0.5$ in Figure 3.3(a) and for $\beta = 2.0$, $\alpha = 1.0$, $v = 0.0, 0.5$ in Figure 3.3(b).

Relations Between the Weibull and Exponential Distributions

From (3.1) we see that *when $\beta = 1$ and $v = 0$, the Weibull distribution is the same as the exponential distribution with parameter $\theta = 1/\alpha$.*

On the other hand, *if X has a Weibull distribution with parameters α, β, and v, then $Y = [(X - v)/\alpha]^{\beta}$ has an exponential distribution with parameter $\theta = 1$.* To verify this assertion, we find the probability density function of Y:

$$f_Y(y) = \left|\frac{d}{dy}(\alpha y^{1/\beta} + v)\right| f_X(\alpha y^{1/\beta} + v) = \begin{cases} e^{-y}, & \text{if } y \geq 0, \\ 0, & \text{if } y < 0. \end{cases}$$

We can use the fact that $Y = [(X - v)/\alpha]^{\beta}$ has an exponential distribution with parameter $\theta = 1$ to obtain the cumulative distribution function $F_X(x)$ of

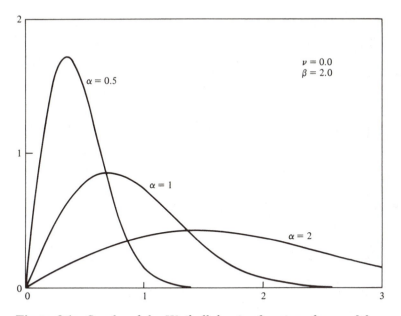

Figure 3.1: *Graphs of the Weibull density functions for* $v = 0.0$, $\beta = 2.0$, $\alpha = 0.5, 1.0, 2.0$.

Figure 3.2: *Graphs of the Weibull density functions for* $v = 0.0$, $\beta = 10.0$, $\alpha = 0.5, 1.0, 2.0$.

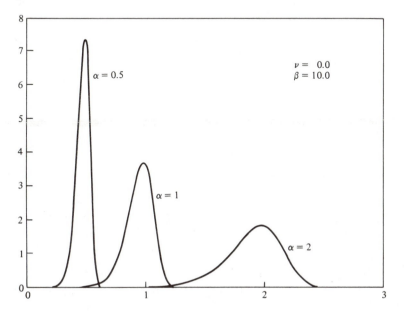

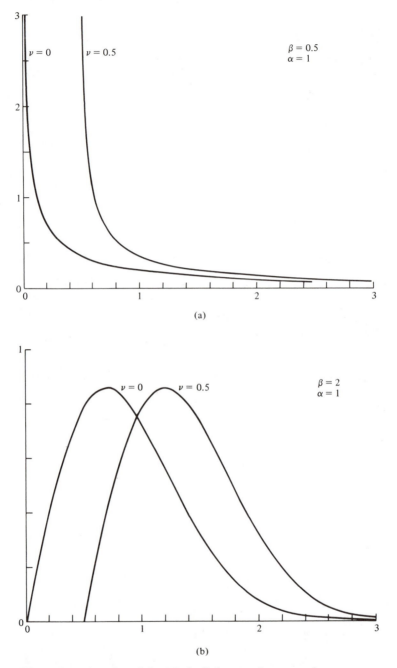

Figure 3.3: *Graphs of the Weibull density functions for*
(a) $\beta = 0.5$, $\alpha = 1.0$, $v = 0.0$, 0.5; (b) $\beta = 2.0$, $\alpha = 1.0$, $v = 0.0$, 0.5.

X. As already noted, $X - v$ is nonnegative with probability 1, so that $F_X(x) = 0$ when $x < v$. When $x \geq v$,

$$F_X(x) = P\{X \leq x\} = P\left\{Y = \left(\frac{X - v}{\alpha}\right)^\beta \leq \left(\frac{x - v}{\alpha}\right)^\beta\right\}$$

$$= F_Y\left(\left(\frac{x - v}{\alpha}\right)^\beta\right) = 1 - \exp\left[-\left(\frac{x - v}{\alpha}\right)^\beta\right].$$

Hence, if X has a Weibull distribution with parameters β, α, and v,

(3.2)
$$F_X(x) = \begin{cases} 1 - \exp\left[-\left(\frac{x - v}{\alpha}\right)^\beta\right], & \text{if } x \geq v, \\ 0, & \text{if } x < v. \end{cases}$$

From $F_X(x)$, probabilities such as $P\{a \leq X \leq b\}$ are easily determined.

Finally, let Y have an exponential distribution with parameter $\theta = 1$ and let $W = \alpha Y^{1/\beta} + v$. Since $Y \geq 0$ with probability equal to 1, then W must always be $\geq v$ and it follows that $F_W(w) = 0$ for $w < v$. For $w \geq v$,

$$F_W(w) = P\{\alpha Y^{1/\beta} + v \leq w\} = P\left\{Y \leq \left(\frac{w - v}{\alpha}\right)^\beta\right\}$$

$$= 1 - \exp\left[-\left(\frac{w - v}{\alpha}\right)^\beta\right].$$

Thus, W has the same c.d.f. as a Weibull distribution with parameters β, α, and v. We have thus established the following useful distributional representation for the Weibull distribution: *If Y has an exponential distribution with parameter $\theta = 1$, then $X = \alpha Y^{1/\beta} + v$ has a Weibull distribution with parameters β, α, and v.*

The Mean and Variance of a Weibull Distribution

We have seen that if X has a Weibull distribution with parameters β, α, and v, and Y has an exponential distribution with parameter $\theta = 1$, then X and $\alpha Y^{1/\beta} + v$ have the same distribution, and hence have the same mean, variance, and other moments. Thus,

(3.3)
$$\mu_X = E(\alpha Y^{1/\beta} + v) = \alpha E(Y^{1/\beta}) + v,$$
$$E(X^2) = E(\alpha Y^{1/\beta} + v)^2 = \alpha^2 E(Y^{2/\beta}) + 2v\alpha E(Y^{1/\beta}) + v^2.$$

Applying (1.2) of Section 1 with $r = 1/\beta$, $r = 2/\beta$, and $\theta = 1$, we have

$$E(Y^{1/\beta}) = \Gamma\left(1 + \frac{1}{\beta}\right), \qquad E(Y^{2/\beta}) = \Gamma\left(1 + \frac{2}{\beta}\right),$$

so that from (3.3),

(3.4)
$$\mu_X = \alpha \Gamma\left(1 + \frac{1}{\beta}\right) + v$$

and

(3.5)
$$\sigma_X^2 = E(X^2) - (\mu_X)^2 = \alpha^2 \left\{ \Gamma\left(1 + \frac{2}{\beta}\right) - \left[\Gamma\left(1 + \frac{1}{\beta}\right)\right]^2 \right\}.$$

From (3.5), note that the variance of a random variable X having a Weibull distribution with parameters β, α, and v depends upon β and α, but not on v. This should not be surprising since we have already seen (Figure 3.3) that v acts like a measure of location, and does not affect the *shape* of the graph of the density function $f_X(x)$.

A Theoretical Explanation

The experience of many investigators has shown that the Weibull distributions provide good probability models for describing "length of life" and other endurance data. One explanation for this success is related to the following "weakest link" interpretation of endurance. Suppose that an object is put under stress. Think of the object as being composed of a large number of separate parts, each of which has its own probabilistically independent random endurance time (lifetime). If any one of these parts fails under stress, the whole object experiences failure. For example, the object could be a metal chain composed of a large number of links. If any link breaks, so does the chain. Thus, the lifetime of the object is equal to the minimum lifetime of any of its parts. If the lifetimes of any class of objects have this property, then it can be shown that a Weibull distribution provides a close approximation to the distribution of these lifetimes.

Fitting the Weibull Distribution to Data

Suppose that we have n probabilistically independent observations X_1, $X_2, \ldots, X_n$ of a random variable X which we believe has a Weibull distribution, and we want to find the values of the parameters α, β, and v. Since v is the smallest possible value of X, it seems reasonable to estimate v by the smallest observed value $X_{\min}$ of X. That is,

(3.6)
$$\hat{v} = \text{minimum of } X_1, X_2, \ldots, X_n = X_{\min}.$$

Remark. In many cases, we know for theoretical reasons that v has some value v_0. In this case, we do not need to estimate v from the data.

The remaining two parameters, α and β, can now be obtained through the use of Equations (3.4) and (3.5). First note that

$$\mu_X - v = \alpha\Gamma\left(1 + \frac{1}{\beta}\right), \qquad \sigma_X^2 + (\mu_X - v)^2 = \alpha^2\Gamma\left(1 + \frac{2}{\beta}\right).$$

Thus, β can be obtained from

(3.7)
$$\frac{(\mu_X - v)^2}{\sigma_X^2 + (\mu_X - v)^2} = \frac{\Gamma[1 + (1/\beta)]\Gamma[1 + (1/\beta)]}{\Gamma[1 + (2/\beta)]}.$$

Once we have estimated μ_X and σ_X^2 by $\hat{\mu}_X$ and $\hat{\sigma}_X^2$, we can use the estimated value $\hat{v} = X_{\min}$ for v and then solve (3.7) for β. In order to facilitate this computation, Table 3.1 provides values of $\Gamma(1 + z)\Gamma(1 + z)/\Gamma(1 + 2z)$ for values of z between 0 and 1.

Remark. If z exceeds 1, Table 3.1 can still be used by recalling that $\Gamma(z + 1) = z\Gamma(z)$.

Once an estimate of β has been determined from Equation (3.7), we can obtain an estimate of α through the formula

(3.8)
$$\alpha = \frac{\mu_X - v}{\Gamma[1 + (1/\beta)]},$$

which is a consequence of (3.4).

In a paper published in 1951, Weibull provided three examples in which a distribution of the form (3.1) was fit to data. Since that time, there have been many other applications of the Weibull distribution.

Table 3.1: Values of $\Gamma(1 + z)\Gamma(1 + z)/\Gamma(1 + 2z)$

	0.00	0.01	0.02	0.03	0.04	0.05	0.06	0.07	0.08	0.09
0.00	1.0000	0.9998	0.9993	0.9985	0.9975	0.9961	0.9945	0.9928	0.9906	0.9884
0.10	0.9858	0.9830	0.9801	0.9768	0.9735	0.9699	0.9664	0.9625	0.9585	0.9545
0.20	0.9502	0.9458	0.9412	0.9367	0.9319	0.9271	0.9221	0.9170	0.9119	0.9067
0.30	0.9015	0.8961	0.8906	0.8852	0.8796	0.8741	0.8683	0.8626	0.8568	0.8512
0.40	0.8453	0.8395	0.8336	0.8274	0.8215	0.8156	0.8095	0.8035	0.7975	0.7914
0.50	0.7854	0.7793	0.7732	0.7672	0.7611	0.7550	0.7489	0.7428	0.7367	0.7307
0.60	0.7246	0.7186	0.7125	0.7064	0.7004	0.6944	0.6885	0.6825	0.6765	0.6706
0.70	0.6647	0.6588	0.6529	0.6470	0.6412	0.6354	0.6296	0.6239	0.6182	0.6125
0.80	0.6068	0.6012	0.5955	0.5899	0.5844	0.5789	0.5734	0.5679	0.5625	0.5571
0.90	0.5518	0.5464	0.5411	0.5359	0.5306	0.5254	0.5203	0.5152	0.5101	0.5050

Example 3.1 (Size Distribution of Particles). The term "fly ash" refers to fine solid particles of noncombustible ash that are carried out of a bed of solid fuel by the draft of combustion. Table 3.2 shows a frequency distribution of the size (particle diameter) of fly ash, based on $n = 211$ observations.

For these data, Weibull notes that the smallest observed value of X is equal to 30 microns or 1.5 20-micron units. Thus, the estimated value for the parameter v based on these data is $v = 1.5$ (in 20-micron units). Notice that we do not obtain this estimate from Table 3.2, since from that table we can only determine the *interval* of numbers in which the smallest observed value falls.

From Table 3.2,

$$\hat{\mu}_X = 2\left(\tfrac{3}{211}\right) + 3\left(\tfrac{11}{211}\right) + \cdots + 13\left(\tfrac{6}{211}\right) + 14\left(\tfrac{3}{211}\right) = 7.18,$$

$$\hat{\sigma}_X^2 = (2.00 - 7.18)^2\left(\tfrac{3}{211}\right) + \cdots + (14.00 - 7.18)^2\left(\tfrac{3}{211}\right) = 6.70.$$

Thus,

$$\frac{(\hat{\mu}_X - v)^2}{\hat{\sigma}_X^2 + (\hat{\mu}_X - v)^2} = \frac{(7.18 - 1.50)^2}{6.70 + (7.18 - 1.50)^2} = 0.828.$$

Solving the equation

$$0.828 = \frac{\Gamma[1 + (1/\beta)]\Gamma[1 + (1/\beta)]}{\Gamma[1 + (2/\beta)]}$$

Table 3.2: *Observed Frequency Distribution of the Size X of Fly Ash (in 20-Micron Units)*

Midpoint x of interval of sizes	Observed frequency of the event $\{x - 0.5 \leq X \leq x + 0.5\}$
2	3
3	11
4	20
5	22
6	29
7	41
8	24
9	25
10	13
11	9
12	5
13	6
14	3
	211

for β, we find from Table 3.1 that $z = 1/\beta$ lies between 0.42 and 0.43. By linear interpolation we obtain $z = 0.429$, so that $\hat{\beta} = 1/0.429 = 2.331$. Finally, from Equation (3.8) and Table A.1 of the Appendix,

$$\hat{\alpha} = \frac{\hat{\mu}_X - \hat{v}}{\Gamma(1 + 1/\hat{\beta})} = \frac{7.18 - 1.5}{\Gamma(1.429)} = \frac{5.68}{0.8860} = 6.41.$$

Thus, based on the data in Table 3.2, we would fit a Weibull distribution with parameters $\beta = 2.331$, $\alpha = 6.41$, and $v = 1.50$.

Example 3.2 (Strength of Steel). Weibull (1951) also discusses an example in which the yield strength X of a Bofors steel was measured. The resulting frequency distribution is given in Table 3.3.

The minimum value of the yield strength observed was 38.57 kilograms per millimeter squared (kg/mm^2) or 30.25 in units of 1.275 kg/mm^2. Thus, to fit a Weibull distribution to the data in Table 3.3, we take $v = 30.25$.
From Table 3.3, we find that

$$\hat{\mu}_X = 36.152, \qquad \hat{\sigma}_X^2 = 3.995.$$

Thus,

$$\frac{(\hat{\mu}_X - v)^2}{\hat{\sigma}_X^2 + (\hat{\mu}_X - v)^2} = \frac{(36.152 - 30.250)^2}{3.995 + (36.152 - 30.250)^2} = 0.8971,$$

Table 3.3: *Observed Frequency Distribution of the Yield Strength X in 1.275-kg/mm^2 Units of a Bofors Steel*

Midpoint x of interval of yield strengths	Observed frequency of the event $\{x - 0.5 \leq X \leq x + 0.5\}$
32	10
33	23
34	48
35	80
36	63
37	65
38	47
39	33
40	14
41[a]	6
	389

[a] In the published version of Weibull's paper, this number is given as 42, not 41. However, we suspect that this is a typographical error.

so that from Equation (3.7) and by interpolation in Table 3.1,

$$\hat{\beta} = \frac{1}{0.308} = 3.247.$$

Finally, from (3.8) and Table A.1 of the Appendix,

$$\hat{\alpha} = \frac{36.152 - 30.250}{\Gamma(1.308)} = \frac{5.902}{0.8963} = 6.58.$$

Therefore, we would fit a Weibull distribution with parameters $\beta = 3.247$, $\alpha = 6.58$, and $v = 30.25$ to these data.

The Survival Function

As in the case of the exponential distribution, the survival function of a Weibull distribution can be expressed in a convenient form. If X has a Weibull distribution with parameters α, β, and v, then the survival function $1 - F_X(x)$ of X is given by the equation

$$(3.9) \qquad 1 - F_X(x) = \begin{cases} 1, & \text{if } x < v, \\ \exp\left[-\left(\dfrac{x - v}{\alpha}\right)^{\beta} \right], & \text{if } x \geq v. \end{cases}$$

We can use the survival function (3.9) to evaluate the fit of the Weibull distribution to data. In Figure 3.4, we compare the theoretical survival function and sample survival function based on the data in Example 3.1. This comparison suggests that the variability of the random variable in this example is adequately described by a Weibull distribution.

Figure 3.4: *Comparison of the observed and theoretical survival functions for the data of Example 3.1. The theoretical survival function is based on the Weibull distribution with parameters $\beta = 2.331$, $\alpha = 6.41$, $v = 1.5$.*

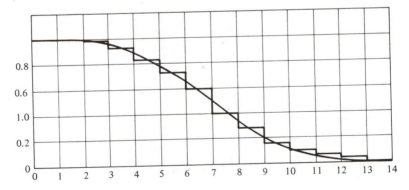

Remark. Both the gamma family of distributions and the Weibull family of distributions generalize the exponential family of distributions in the sense that an exponentially distributed random variable X with parameter θ can also be said to have a gamma distribution with parameters θ and $r = 1$, and a Weibull distribution with parameters $\alpha = 1/\theta$, $\beta = 1$, and $v = 0$. However, there are gamma distributions that are not also Weibull distributions (e.g., any gamma distribution with parameter $r \neq 1$) and Weibull distributions that are not gamma distributions (e.g., any Weibull distributions with $\beta \neq 1$ or $v \neq 0$).

4. THE UNIFORM DISTRIBUTION

A continuous random variable X has a *uniform distribution* over the interval of numbers from a to b, and we write $X \sim U[a, b]$, if intervals of possible values that have equal length have equal probability of containing an observed value of X.

The probability density function $f_X(x)$ of a random variable X having a uniform distribution over the interval of numbers from a to b is given by the equation

(4.1)
$$f_X(x) = \begin{cases} \dfrac{1}{b-a}, & \text{if } a \leq x \leq b, \\ 0, & \text{if } x < a \text{ or } x > b. \end{cases}$$

The interval end points a and b, $-\infty < a < b < \infty$, are the parameters of the uniform distribution. Note that a is the smallest possible value of X and that b is the largest possible value of X. The density (4.1) is uniform in value over the interval $[a, b]$.

Graphs of the density functions $f_X(x)$ for a $U[0, 1]$ and a $U[1, 3]$ distribution are shown in Figure 4.1. From Figure 4.1, we can see why the uniform distribution is also called the *rectangular distribution*.

If $X \sim U[a, b]$, then for $k \geq 0$,

$$E(X^k) = \int_a^b x^k \left(\frac{1}{b-a}\right) dx = \left(\frac{1}{b-a}\right) \frac{x^{k+1}}{k+1}\Big|_a^b = \frac{b^{k+1} - a^{k+1}}{(k+1)(b-a)},$$

from which

(4.2)
$$\mu_X = E(X) = \frac{b^2 - a^2}{2(b-a)} = \frac{b+a}{2},$$

$$\sigma_X^2 = E(X^2) - (\mu_X)^2 = \frac{b^3 - a^3}{3(b-a)} - \left(\frac{b+a}{2}\right)^2 = \frac{(b-a)^2}{12}.$$

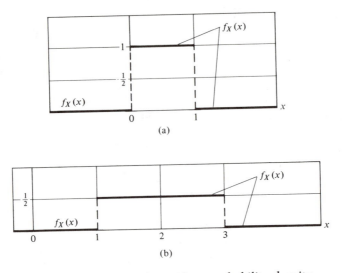

Figure 4.1: *Graphs of the uniform probability density function $f_X(x)$ for (a) $a = 0$, $b = 1$; (b) $a = 1$, $b = 3$.*

It follows from (4.2) that $a = \mu_X - \sqrt{3}\,\sigma_X$, $b = \mu_X + \sqrt{3}\,\sigma_X$. Thus, knowledge of μ_X and σ_X^2 is equivalent to knowledge of the parameters a and b. However, in most applications of the uniform distribution, the parameters a and b are known and do not need to be estimated. Further, using estimates of μ_X and σ_X^2 to obtain a and b can lead to unreasonable values of a and b. When probabilistically independent observations $X_1, X_2, \ldots, X_n$ are taken on $X \sim U[a, b]$, then since a is the smallest possible value of X and b is the largest possible value of X, we use the estimates $\hat{a} = \min\{X_1, X_2, \ldots, X_n\}$, $\hat{b} = \max\{X_1, X_2, \ldots, X_n\}$ for a and b, respectively.

Example 4.1. Let X denote the length of gaps (in feet) between following cars on a two-lane road. We assume that $X \sim U[a, b]$, where the parameter b is the largest gap between successive cars and a is the smallest such gap. In an experiment, gaps between 11 cars were observed. The 10 observed gaps are 15.1, 31.3, 64.2, 51.6, 87.4, 25.0, 103.3, 74.9, 98.5, and 118.7. Since the smallest observed gap is 15.1 feet and the largest observed gap is 118.7 feet, $\hat{a} = 15.1$ and $\hat{b} = 118.7$.

Probability Calculations

Probability calculations for the uniform distribution are quite easy to perform. Suppose, for example, that $X \sim U[a, b]$; then,

$$(4.3) \qquad F_X(x) = \begin{cases} 0, & \text{if } x < a, \\ \dfrac{x-a}{b-a}, & \text{if } a \le x \le b, \\ 1, & \text{if } x > b. \end{cases}$$

The graph of $F_X(x)$ for $a = 1$, $b = 3$, is shown in Figure 4.2.

Example 4.2. A train is scheduled to depart at 6:00 P.M. A person who wants to catch that train takes a cab at 5:43 P.M. from an intersection which under ideal driving conditions is 10 minutes from the train station but which can be as much as 50 minutes driving time away from the station if traffic is heavy. Assume that $X \sim U[10, 50]$, where X is the driving time from the intersection to the station, and assume that upon reaching the station, the person needs 2 minutes to board the train. Thus, if the individual is to catch the train, X must be no greater than $6:00 - 5:43 - 0:02 = 15$ minutes. The probability that the person catches the train is thus

$$P\{X \le 15\} = F_X(15) = \frac{15 - 10}{50 - 10} = \frac{1}{8} = 0.125.$$

Example 4.3. A biologist takes homing pigeons from their cote and keeps them for several days in a dark room. If the pigeons orient themselves by light, then when they are brought back into the open air, they may be confused. The assumption is that once released, a confused pigeon will pick a direction in which to fly according to a uniform distribution. More precisely, if the bird is released at a point A (see Figure 4.3), then its line of flight will make an angle of X degrees with line L, where $X \sim U[0, 360]$. It has been observed that correct lines of flight, taken by pigeons who were not confused, vary from an angle of 210 degrees from line L to an angle of 220 degrees from

Figure 4.2: *Graph of the c.d.f. $F_X(x)$ of the uniform distribution with parameters $a = 1$, $b = 3$.*

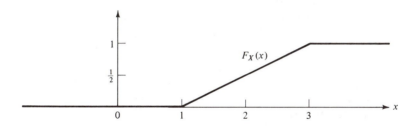

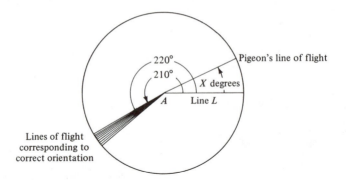

Figure 4.3: *Diagram illustrating the experiment described in Example 4.3. The pigeon is released at point A. Its line of flight makes an angle of X degrees with the fixed line L. Lines of flight making angles of 210–220 degrees correspond to orientations ordinarily taken by pigeons returning home.*

line L. Under the assumption of a uniform distribution, the probability that a confused pigeon chooses a correct line of flight is

$$P\{210 \leq X \leq 220\} = F_X(220) - F_X(210) = \frac{220 - 0}{360 - 0} - \frac{210 - 0}{360 - 0} = \frac{1}{36}.$$

Thus, under the assumption of a uniform distribution, it is rather improbable that a confused pigeon will choose a correct line of flight.

Other Applications

Rounding Errors. In measurement theory, the uniform distribution with $a = 0$, $b = (0.5)10^{-k}$ is often used to represent the distribution of "rounding-off" errors in values tabulated to the nearest k decimal places. That is, the unsigned difference, $X = |Y - Y_R|$, between the actual observed value Y of a random variable and the rounded-off value Y_R is assumed to have a uniform distribution with parameters $a = 0$ and $b = (0.5)10^{-k}$.

Computer Simulations. The $U[0, 1]$ distribution is widely used in what are called *Monte Carlo simulation methods.* Computer programs called *random number generators* provide a succession of values that resemble, in their variation, observations obtained from the $U[0, 1]$ distribution. Using the output of such a program, we can generate observations made upon a random variable Y having *any* continuous c.d.f. $F_Y(y)$ as follows:

(1) Use the random number generator to generate a value x of $X \sim U[0, 1]$.

(2) Solve the equation $F_Y(y) = x$ for y, where $F_Y(y)$ is the c.d.f. of Y. Our solution y^* is then the observed value of the simulated random variable Y. [If the solution is not unique, we use the smallest value y^* of y that satisfies $F_Y(y) = x$.]

Analytic and graphical methods of solving the equation $F_Y(y) = x$ were discussed in Chapter 5, Section 3. By repeating steps (1) and (2) over and over, we can simulate any desired number of observations of Y.

Example 4.4 (Simulation of an Exponentially Distributed Random Variable). Suppose that we desire to simulate a random variable Y having an exponential distribution with parameter $\theta = 1$. The c.d.f. $F_Y(y)$ of this distribution for possible values of y is $F_Y(y) = 1 - e^{-y}$. Using our random number generator, we generate a value x of $X \sim U[0, 1]$, and solve

$$F_Y(y) = 1 - e^{-y} = x$$

for y. Thus, $y^* = -\log(1 - x)$. For example, if we generate $x = 0.93$ for X, then $y^* = -\log(0.07) = 2.65926$.

5. THE BETA DISTRIBUTION

There are many continuous quantitative phenomena that take on values bounded above and below by known numbers a and b. We have given three examples of such phenomena in Section 4. Other examples include (i) the distance from one end of a steel bar of known length to the point at which failure occurs when the bar is placed under stress, (ii) the proportion of total farm acreage spoiled by a certain kind of fungus, (iii) the percentage of defective items in a given shipment of items, (iv) the ratio of the length of the femur to the total length of the leg of a given individual, and (v) the fraction of individuals in a given population who are able to answer a given question correctly at a given time. Of these last examples, notice that examples (ii), (iv), and (v) all involve random variables that are fractions; thus, these variables have possible values between $a = 0$ and $b = 1$. The random variable described in example (iii) is a percentage and has possible values between $a = 0$ and $b = 100$. Finally, the random variable mentioned in example (i) has possible values between $a = 0$ and $b = $ length of the steel bar.

A class of distributions that includes the uniform distribution and is rich enough to provide models for most random variables having a restricted range of possible values is the class of *beta distributions*. Because the class of beta distributions is defined only for random variables Y in $[0, 1]$, if $X \in [a, b]$, we first find a distribution to fit the transformed random variable

$$Y = \frac{X - a}{b - a},$$

which has possible values between 0 and 1. Once we have found a beta distribution that describes the variability of Y, we transform back to obtain the distribution of X.

Example 5.1. Let X be the distance (in inches), from the end of a 12-inch steel bar to the point at which the bar breaks under stress. Then X is a random variable with possible values between $a = 0$ and $b = 12$. To find a distribution for X, we first find a beta distribution that describes the variation of $Y = (X - 0)/(12 - 0) = X/12$.

Example 5.2. Let X denote the proportion of a geographical region that receives rain when a certain type of cloud formation is observed over that region. Since X is a proportion, the possible values of X are between $a = 0$ and $b = 1$. In this case, $Y = (X - 0)/(1 - 0) = X$, so that no transformation is needed.

If Y has a beta distribution, then the probability density function $f_Y(y)$ of Y has the form

$$(5.1) \qquad f_Y(y) = \begin{cases} \dfrac{y^{r-1}(1 - y)^{s-1}}{B(r, s)}, & \text{if } 0 \le y \le 1, \\ 0, & \text{if } y < 0 \text{ or } y > 1. \end{cases}$$

Here r and s are positive numbers and are the parameters of the beta distribution. The constant $B(r, s)$, called the *beta function* (see Appendix A), is defined by

$$(5.2) \qquad B(r, s) = \frac{\Gamma(r)\Gamma(s)}{\Gamma(r + s)}.$$

To illustrate the great diversity of shapes taken on by the graph of the probability density function $f_Y(y)$ of the beta distribution, Figure 5.1 gives graphs of $f_Y(y)$ for various choices of r and s. The case $r = s = 1$, not shown in Figure 5.1, yields the $U[0, 1]$ distribution. For $r < 1$ and/or $s < 1$, the density $f_Y(y)$ is infinitely large at $y = 0$ and/or $y = 1$, respectively. When $r < 1$ and $s \ge 1$, the graph of $f_Y(y)$ is decreasing in y and concave, while it is increasing and concave when $r \ge 1$ and $s < 1$. When r and s are both less than 1, the graph is "U-shaped," while when $r > 1$ and $s > 1$, the graph has a single hump. Finally, when $r = s$, the graph of $f_Y(y)$ is symmetric about $y = \frac{1}{2}$; otherwise, it is skewed.

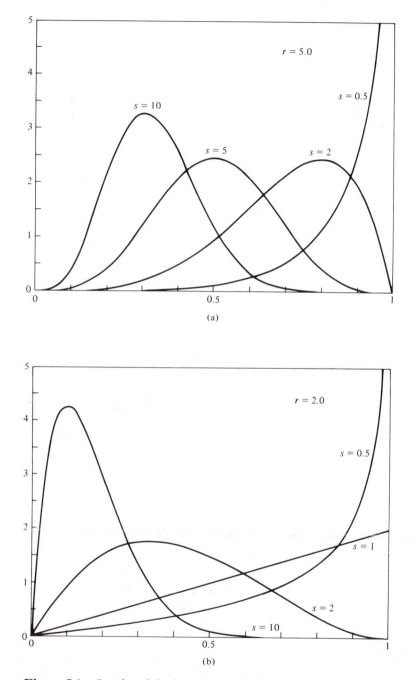

Figure 5.1: *Graphs of the beta density function for* (a) *r* = 5.0, *s* = 0.5, 2.0, 5.0, 10.0; (b) *r* = 2.0, *s* = 0.5, 1.0, 2.0, 10.0.

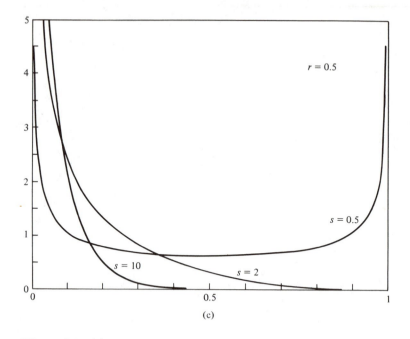

Figure 5.1: (c) $r = 0.5$, $s = 0.5$, 2.0, 10.0.

Mean and Variance of the Beta Distribution

It is shown in the Appendix that when $a \geq 0$, $b \geq 0$,

$$(5.3) \qquad B(a, b) = \int_0^1 u^{a-1}(1 - u)^{b-1} \, du.$$

Thus, when the random variable Y has a beta distribution with parameters r and s,

$$(5.4) \quad E(Y^k) = \frac{1}{B(r, s)} \int_0^1 y^{r+k-1}(1 - y)^{s-1} \, dy = \frac{B(r + k, s)}{B(r, s)} = \frac{\Gamma(r + k)\Gamma(r + s)}{\Gamma(r)\Gamma(r + s + k)},$$

for $k \geq 0$. Applying this result when $k = 0$ shows that $\int_0^1 f_Y(y) \, dy = 1$. Since

$f_Y(y) \geq 0$ for all y, $f_Y(y)$ is indeed a probability density function. Letting $k = 1$ and $k = 2$ in (5.4) yields

$$E(Y) = \frac{\Gamma(r + 1)}{\Gamma(r)} \frac{\Gamma(r + s)}{\Gamma(r + s + 1)} = \frac{r}{r + s},$$

$$E(Y^2) = \frac{\Gamma(r + 2)}{\Gamma(r)} \frac{\Gamma(r + s)}{\Gamma(r + s + 2)} = \frac{(r + 1)(r)}{(r + s + 1)(r + s)},$$

and thus

$$(5.5) \quad \mu_Y = E(Y) = \frac{r}{r + s}, \qquad \sigma_Y^2 = E(Y^2) - (\mu_Y)^2 = \frac{rs}{(r + s)^2(r + s + 1)},$$

from which

$$(5.6) \quad r = \mu_Y \left[\frac{(\mu_Y)(1 - \mu_Y)}{\sigma_Y^2} - 1 \right], \qquad s = (1 - \mu_Y) \left[\frac{(\mu_Y)(1 - \mu_Y)}{\sigma_Y^2} - 1 \right].$$

Fitting the Beta Distribution to Data

Suppose that a random variable Y has possible values between 0 and 1 and that we believe that Y has a beta distribution. If we have n statistically independent observations Y_1, Y_2, ..., Y_n upon Y and want to "fit" a beta distribution to these observations, we can estimate μ_Y by the sample average $\hat{\mu}_Y$ and σ_Y^2 by the sample variance $\hat{\sigma}_Y^2$ and then use (5.6) to obtain estimates of r and s.

Example 5.3. Let Y be the proportion of change in wholesale prices of commodities from one year to the next when the prices fall. That is, if the first year price is d_1 and the second year price is $d_2 \leq d_1$, then $Y = (d_1 - d_2)/d_1$. If prices rise (i.e., $d_1 < d_2$), no value of Y is recorded. A total of 2314 cases of falling commodity prices were observed. These data are summarized in the form of a frequency distribution in Table 5.1.

From the data in Table 5.1,

$$\hat{\mu}_Y = 0.089, \qquad \hat{\sigma}_Y^2 = 0.0064,$$

and using (5.6), estimates of r and s are

$$\hat{r} = (0.089) \left[\frac{(0.089)(1 - 0.089)}{0.0064} - 1 \right] = 1.038,$$

$$\hat{s} = (1 - 0.089) \left[\frac{(0.089)(1 - 0.089)}{0.0064} - 1 \right] = 10.63.$$

In Figure 5.2, a relative frequency histogram based on Table 5.1 is graphed

Table 5.1: *Frequency Distribution of the Proportion Y of Change in the Wholesale Prices of Commodities From One Year to the Next*

(Only falling prices are considered.)

y	Observed frequency of the event $\{y - 0.02 \leq Y \leq y + 0.02\}$
0.02	780
0.06	567
0.10	373
0.14	227
0.18	147
0.22	84
0.26	49
0.30	43
0.34	17
0.38	12
0.42	9
0.46	3
0.50	2
0.54	1
	2314

Figure 5.2: *Fitting a beta distribution with parameters $r = 1.038$ and $s = 10.63$ to the data of Table 5.1. [Note that to make the comparison, the frequencies in Table 5.1 should be converted to modified relative frequencies by dividing each frequency by (2314)(0.04).]*

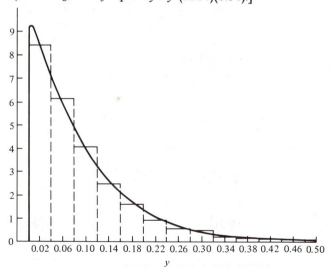

along with the probability density function of a beta distribution with parameters $r = 1.038$ and $s = 10.63$. From this figure, it appears that the beta distribution provides a reasonable fit to the observed variation of Y.

Probability Calculations

If Y has a beta distribution with parameters r and s, the cumulative distribution function $F_Y(y)$ of Y is given by

(5.7)
$$F_Y(y) = \begin{cases} 1, & \text{if } y > 1, \\ I_y(r, s), & \text{if } 0 \leq y \leq 1, \\ 0, & \text{if } y < 0, \end{cases}$$

where

(5.8)
$$I_y(r, s) = \int_0^y \frac{t^{r-1}(1 - t)^{s-1}}{B(r, s)} \, dt$$

is known as the *incomplete beta function*. Except in special cases (e.g., $s = 1$), this function cannot be expressed in terms of a simple formula involving r, s, and y, and thus must be tabled. Because we need a separate table for every pair (r, s) of parameter values, a complete book of tables is required to give $F_Y(y)$ for all pairs (r, s) that might be met in practice. Such a book is available [see Remark at the end of this section]; thus, rather than give tables for $F_Y(y)$ for a range of values of r and s, we instead discuss how to obtain $F_Y(y)$ for certain special cases.

Note first that when $r > 0$ and $s = 1$,

(5.9)
$$I_y(r, 1) = \int_0^y \frac{t^{r-1} \, dt}{B(r, 1)} = \int_0^y r t^{r-1} \, dt = t^r \Big|_0^y = y^r,$$

since $B(r, 1) = \Gamma(r + 1)/\Gamma(r)\Gamma(1) = r$. When $r > 0$ and $s \geq 1$, integration by parts in (5.8) yields the recursion relation:

$$I_y(r, s) = \int_0^y \frac{t^{r-1}(1 - t)^{s-1}}{B(r, s)} \, dt$$

(5.10)
$$= \frac{t^r(1 - t)^{s-1}}{r B(r, s)} \Big|_0^y + \int_0^y \frac{(s - 1)t^r(1 - t)^{s-2}}{r B(r, s)} \, dt$$

$$= \frac{\Gamma(r + s)y^r(1 - y)^{s-1}}{\Gamma(r + 1)\Gamma(s)} + I_y(r + 1, s - 1),$$

since

$$\frac{rB(r, s)}{s - 1} = \frac{r\Gamma(r)\Gamma(s)}{(s - 1)\Gamma(r + s)} = \frac{\Gamma(r + 1)\Gamma(s)}{(s - 1)\Gamma(r + s)} = \frac{\Gamma(r + 1)\Gamma(s - 1)}{\Gamma(r + s)}$$

$$= B(r + 1, s - 1).$$

Now when s is a positive integer, we can apply (5.10) repeatedly to obtain

$$I_y(r, s) = \frac{\Gamma(r + s)y^r(1 - y)^{s - 1}}{\Gamma(r + 1)\Gamma(s)} + I_y(r + 1, s - 1)$$

$$= \frac{\Gamma(r + s)y^r(1 - y)^{s - 1}}{\Gamma(r + 1)\Gamma(s)} + \frac{\Gamma(r + s)y^{r + 1}(1 - y)^{s - 2}}{\Gamma(r + 2)\Gamma(s - 1)}$$

$$+ I_y(r + 2, s - 2)$$

(5.11) $$= \cdots$$

$$= \sum_{i = r}^{r + s - 2} \frac{\Gamma(r + s)y^i(1 - y)^{r + s - 1 - i}}{\Gamma(i + 1)\Gamma(r + s - i)} + I_y(r + s - 1, 1)$$

$$= \sum_{i = r}^{r + s - 2} \frac{\Gamma(r + s)y^i(1 - y)^{r + s - 1 - i}}{\Gamma(i + 1)\Gamma(r + s - i)} + y^{r + s - 1}$$

$$= \sum_{i = r}^{r + s - 1} \frac{\Gamma(r + s)y^i(1 - y)^{r + s - 1 - i}}{\Gamma(i + 1)\Gamma(r + s - i)}.$$

Noting that when $r + s$ is a positive integer,

$$\frac{\Gamma(r + s)}{\Gamma(i + 1)\Gamma(r + s - i)} = \frac{(r + s - 1)!}{i!\,(r + s - 1 - i)!} = \binom{r + s - 1}{i},$$

we obtain from (5.11) the following result: If r and s are both positive integers, then for $0 \leq y \leq 1$,

(5.12) $$I_y(r, s) = \sum_{i = r}^{r + s - 1} \binom{r + s - 1}{i} y^i(1 - y)^{r + s - 1 - i},$$

which we recognize as being the probability that a random variable U with a binomial distribution with parameters $n = r + s - 1$ and $p = y$ lies between (and including) r and $r + s - 1$. We summarize this result as follows.

Relationship Between Beta and Binomial Distributions

If Y has a beta distribution with parameters r and s, where r and s are both positive integers, then for $0 \leq y \leq 1$,

(5.13) $$F_Y(y) = P\{r \leq U \leq r + s - 1\} = 1 - F_U(r - 1),$$

where U has a binomial distribution with parameters $n = r + s - 1$ and $p = y$.

Example 5.4. Let Y be the fraction of saturation of dissolved oxygen in a river at a given point. The distribution of Y in pure water is a beta distribution with parameters $r = 3$, $s = 2$. A water quality alarm system notes when the observed value of Y goes below a certain threshold level y^*. This alarm system should not falsely declare the river water to be impure too often (a "false alarm") when water is pure. If $y^* = 0.40$, then from (5.13) and Table T.1 with $n = 3 + 2 - 1 = 4$ and $p = 0.40$,

$$P\{\text{false alarm}\} = P\{Y \leq 0.40\} = F_Y(0.40) = P\{3 \leq U \leq 3 + 2 - 1\}$$
$$= P\{U = 3\} + P\{U = 4\} = 0.15360 + 0.02560 = 0.17920.$$

If this probability of false alarm is too large, the threshold can be lowered. For $y^* = 0.20$,

$$P\{Y \leq 0.20\} = F_Y(0.20) = P\{3 \leq U \leq 4\} = 0.02560 + 0.00160$$
$$= 0.02720,$$

which is satisfactorily low.

Using the cumulative distribution function $F_Y(y)$ and either (5.13) or tables of $I_y(r, s)$, probabilities of events such as $\{a \leq Y \leq b\}$ and $\{Y > a\}$ can be obtained.

Example 5.5. A certain stage of sleep, called "REM sleep," is highly associated with dreaming. A psychologist assumes that the fraction Y of the total sleeping time of normal sleepers spent in "REM sleep" has a beta distribution with parameters $r = 12$, $s = 48$. A "typical range of values" for Y is defined to be an interval $[y_1, y_2]$ such that $P\{y_1 \leq Y \leq y_2\} = 0.95$. It is suggested that $y_1 = 0.10$ and $y_2 = 0.30$ meet this requirement. Although $r = 12$ and $s = 48$ are both integers, so that (5.13) applies, $r + s - 1 = 59$ is too large for us to use Table T.1. Table 5.2 is a condensed table of values of $I_y(12, 48)$ for $y = 0.00(0.02)0.48$. From Table 5.2 and (5.7),

$$F_Y(0.30) - F_Y(0.10) = I_{0.30}(12, 48) - I_{0.10}(12, 48)$$
$$= 0.96507 - 0.01280 = 0.95227.$$

Table 5.2: Values of the Incomplete Beta Function I_y (12, 48)

	0.00	0.02	0.04	0.06	0.08
0.0	0.00000	0.00000	0.00000	0.00017	0.00219
0.1	0.01280	0.04552	0.11496	0.22646	0.37037
0.2	0.52582	0.67041	0.78857	0.87468	0.93127
0.3	0.96507	0.98353	0.99279	0.99707	0.99889
0.4	0.99961	0.99987	0.99996	0.99999	1.00000

One final, and useful, result should be noted: *If Y has a beta distribution with parameters r and s, then* $1 - Y$ *has a beta distribution with parameters s and r*. Indeed, if $V = 1 - Y$, then since $B(r, s) = B(s, r)$,

$$f_V(v) = \frac{1}{|-1|} f_Y(1 - v) = \begin{cases} \dfrac{(1 - v)^{r-1} v^{s-1}}{B(r, s)}, & \text{if } 0 \leq 1 - v \leq 1, \\ 0, & \text{otherwise,} \end{cases}$$

$$= \begin{cases} \dfrac{v^{s-1}(1 - v)^{r-1}}{B(s, r)}, & \text{if } 0 \leq v \leq 1, \\ 0, & \text{otherwise,} \end{cases}$$

which we recognize as being the probability density function of a beta distribution with parameters s and r. It follows from this result and (5.7) that for $0 < y < 1$,

(5.14) $I_y(r, s) = 1 - I_{1-y}(s, r),$

a result that is very useful for tabling the incomplete beta function.

As an example of the use of (5.14), suppose that Y has a beta distribution with parameters $r = 48$ and $s = 12$, and that we wish to calculate $P\{0.58 \leq Y \leq 0.72\}$. Since Table 5.2 gives values of the cumulative distribution function of a beta distribution with parameters $r = 12$ and $s = 48$, and $1 - Y$ has such a distribution,

$$P\{0.58 \leq Y \leq 0.72\} = P\{1 - 0.72 \leq 1 - Y \leq 1 - 0.58\}$$

$$= P\{0.28 \leq 1 - Y \leq 0.42\}$$

$$= F_{1-Y}(0.42) - F_{1-Y}(0.28)$$

$$= 0.99987 - 0.93127 = 0.06860.$$

Remark. Extensive tables of the incomplete beta function appear in *Tables of the Incomplete Beta-function* (1934), edited by Karl Pearson. In these tables a different set of symbols is used: in place of our r, they use p; in place of our s, they use q.

6. THE LOGNORMAL AND CAUCHY DISTRIBUTIONS

Even though the normal, exponential, gamma, Weibull, uniform, and beta distributions are frequently used to model the variability of continuous random phenomena, many other distributions have proved to be useful in certain contexts. A recent survey of continuous distributions [Johnson and Kotz (1970)] lists over 50 different families of such distributions. Thus, no short list of distributions can be expected to serve as models for all quantitative random phenomena.

In this section, we briefly describe two additional classes of continuous distributions—the lognormal distributions and the Cauchy distributions—not solely because of their importance as models of the variation of quantitative random phenomena, but also because these distributions illustrate facts and concepts about continuous probability distributions that are worthy of attention.

The Lognormal Distribution

The applicability of the normal distribution to many quantitative random phenomena can be theoretically justified by assuming that such quantitative random phenomena arise from the summation of many probabilistically independent and identically distributed random causes. The *lognormal distribution* arises from the *product* of many independent and identically distributed random variables. A random variable Z has approximately a lognormal distribution if we can conceive of Z as being equal to the product $\prod_1^n W_i$, where $W_1, W_2, \ldots, W_n$ are probabilistically independent realizations of the same *nonnegative* random variable W. The name "lognormal" given to such a distribution comes from noting that

$$(6.1) \qquad Y = \log Z = \log \prod_{i=1}^{n} W_i = \sum_{i=1}^{n} \log W_i.$$

Thus, $Y = \log Z$ is the sum of a large number n of independent, identically distributed random variables, and hence by the Central Limit Theorem, $Y = \log Z$ is (approximately) normally distributed. *A nonnegative random variable Z has a lognormal distribution whenever $Y = \log Z$ has a normal distribution.*

Suppose that the random variable Z has a lognormal distribution and that $Y = \log Z$ is normally distributed with expected value $\mu_Y = \xi$ and variance $\sigma_Y^2 = \delta^2$. Then, from (6.7) of Chapter 4, the probability density function $f_Z(z)$ of Z is given by

$$f_Z(z) = \left| \frac{d(\log Z)}{dz} \right| f_Y(\log Z)$$

$$(6.2) \qquad = \begin{cases} \dfrac{1}{\sqrt{2\pi z^2 \delta^2}} \exp\left[-\dfrac{1}{2\delta^2} (\log z - \xi)^2 \right], & \text{if } z \geq 0, \\ 0, & \text{if } z < 0. \end{cases}$$

The constants ξ and δ^2 are parameters of the lognormal distribution.

In contrast to the graph of the density function of the normal distribution, the graph of the probability density function $f_Z(z)$ of the lognormal distribution is not a symmetric, bell-shaped curve. Rather, the lognormal distribution is positively skewed. In Figure 6.1, graphs of the probability density function (6.2) of the lognormal distribution are given for various values of ξ and δ^2.

Mean and Variance of the Lognormal Distribution

If Z has a lognormal distribution with parameters ξ and δ^2, then changing variables from z to $y = \log z - \xi$ (so that $z = e^{\xi}e^{y}$, $dz = e^{\xi}e^{y}\,dy$, and $-\infty < y < \infty$), we have

$$E(Z^r) = \int_0^\infty z^r f_Z(z)\,dz = \int_0^\infty \frac{z^{r-1}}{\sqrt{2\pi\delta^2}} \exp\left[-\frac{1}{2\delta^2}(\log z - \xi)^2\right] dz$$

$$= \int_{-\infty}^\infty \frac{1}{\sqrt{2\pi\delta^2}} \exp\left[r\xi + ry - \frac{y^2}{2\delta^2}\right] dy$$

$$= \exp\left[r\xi + \tfrac{1}{2}\delta^2 r^2\right] \int_{-\infty}^\infty \frac{1}{\sqrt{2\pi\delta^2}} \exp\left[-\frac{1}{2\delta^2}(y^2 - 2\delta^2 ry + \delta^4 r^2)\right] dy$$

$$= \exp\left[r\xi + \tfrac{1}{2}\delta^2 r^2\right] \int_{-\infty}^\infty \frac{1}{\sqrt{2\pi\delta^2}} \exp\left[-\frac{1}{2\delta^2}(y - \delta^2 r^2)^2\right] dy$$

$$= \exp\left[r\xi + \tfrac{1}{2}\delta^2 r^2\right]$$

for $r \geq 0$. We have thus shown that

(6.3) $$E(Z^r) = \exp\left[r\xi + \tfrac{1}{2}\delta^2 r^2\right], \quad \text{for } r \geq 0.$$

In particular,

(6.4) $$\mu_Z = E(Z) = \exp\left[\xi + \tfrac{1}{2}\delta^2\right], \qquad \sigma_Z^2 = \exp\left[2\xi + \delta^2\right](\exp\left[\delta^2\right] - 1).$$

Computing Probabilities for the Lognormal Distribution

Since when Z has the lognormal distribution with parameters ξ and δ^2, $Y = \log Z \sim N(\xi, \delta^2)$, it follows that for $0 \leq a \leq b$,

(6.5) $$P\{a \leq Z \leq b\} = P\{\log a \leq Y \leq \log b\}$$

can be computed by finding the probability that a $N(\xi, \delta^2)$ random variable lies between $\log a$ and $\log b$.

Uses of the Lognormal Distribution

The lognormal distribution is used and known under a variety of alternative names. Because Galton (1879) and McAlister (1879) were perhaps the

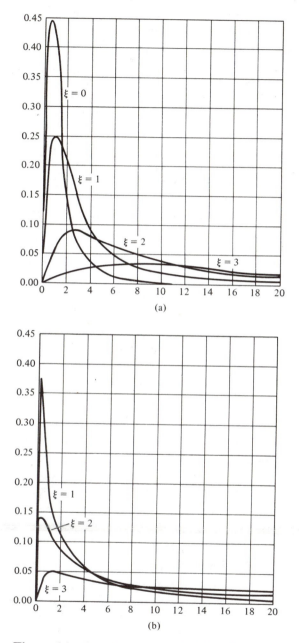

(a)

(b)

Figure 6.1: *Graphs of the probability density function of the lognormal distribution for* (a) $\delta^2 = 1, \xi = 0, 1, 2, 3$; (b) $\delta^2 = 3, \xi = 1, 2, 3$.

first scientists to use the lognormal distribution, the lognormal distribution is sometimes known as the Galton–McAlister distribution. In economics, the lognormal distribution is often applied to production data, and in such contexts it is called the Cobb–Douglas distribution [e.g., see Dhrymes (1962)]. In psychophysical studies, the lognormal distribution has been mentioned by Fechner (1897), while Gaddum (1945) and Bliss (1934) have applied this distribution to the study of critical doses of drugs in human beings and animals. Among other nonnegative quantitative random phenomena whose variation has been modeled by the lognormal distribution are

(a) particle sizes in naturally occurring aggregates [Hatch and Choute (1929); Krumbein (1936); Herdan (1960)];
(b) lengths of words [Herdan (1958)], and sentences [Williams (1940)];
(c) concentrations of the chemical elements in geological materials [Ahrens (1954a, b; 1957); Chayes (1954)];
(d) lifetimes of mechanical and electrical systems [Epstein (1947; 1948)] and other survival data [Feinlieb (1960); Goldthwaite (1961); Adams (1962)];
(e) the abundance of species of animals [Grundy (1951)];
(f) the incubation periods of infectious diseases [Sartwell (1950)].

From this list, it is apparent that the lognormal distributions are important competitors to the exponential, gamma, or Weibull distributions as models for nonnegative phenomena.

Suppose that we observe probabilistically independent observations Z_1, Z_2, ..., Z_n upon a random variable Z that has a lognormal distribution. To estimate the parameters ξ and δ^2 from these data, *transform* the data to Y_1, Y_2, ..., Y_n where $Y_j = \log Z_j$, $j = 1, 2, ..., n$. Since $Y_1, Y_2, ..., Y_n$ are then independent observations from a normal distribution $N(\xi, \delta^2)$, we estimate ξ by the sample average $\hat{\mu}_Y$ of the transformed data, and we estimate δ^2 by the sample variance $\hat{\sigma}_Y^2$. What we have suggested here illustrates the use of transformations of observations into new observations in the hope that the new transformed observations have a convenient distribution (in this case, the normal distribution).

The Cauchy Distribution

Mathematical difficulties that arise in connection with models of real-world phenomena are often regarded as theoretical curiosities of little relevance to experimental practice. While there may often be some truth in this belief, it is not necessarily always correct.

Within the context of probability theory, the *Cauchy distribution* [named after the French mathematician Augustin Cauchy (1789–1857)] exhibits certain nonintuitive behavior, behavior that seems to be inconsistent with certain consequences of the frequency interpretation of probability theory.

The probability density function $f_X(x)$ of a random variable X having a Cauchy distribution has the form

(6.6) $$f_X(x) = \frac{1}{\pi} \frac{1}{1 + (x - \theta)^2}$$

for all real numbers x. The graph of this density function for $\theta = -1, 0, 1$ appears in Figure 6.2. The graph of a Cauchy probability density function, like that of the normal probability density function, is symmetric about a central value (in this case, θ) that is both the unique median and the unique mode of the distribution. Thus, the parameter θ is a measure of location for the Cauchy distribution.

Apparently, then, the Cauchy distribution shares many properties with the normal distribution. In Figure 6.3, the density function of a $N(0, 1)$ and $N(0, 2)$ distribution, and of a Cauchy distribution with parameter $\theta = 0$, are graphed together. Notice that the Cauchy distribution is flatter and more "spread out" than the $N(0, 1)$ distribution, but not as flat as the $N(0, 2)$ distribution. Graphically, it would appear that the Cauchy distribution is mathematically well behaved.

However, if X has a Cauchy distribution, its mean μ_X is not well defined. To see this, note that if μ_X is well defined, then

$$\mu_X = \int_{-\infty}^{\infty} t f_X(t)\, dt = \int_{-\infty}^{\infty} (t - \theta) f_X(t)\, dt + \theta \int_{-\infty}^{\infty} f_X(t)\, dt$$

$$= \int_{-\infty}^{\infty} (t - \theta) f_X(t)\, dt + \theta.$$

Figure 6.2: *Graphs of the Cauchy density function for* $\theta = -1, 0, 1.$

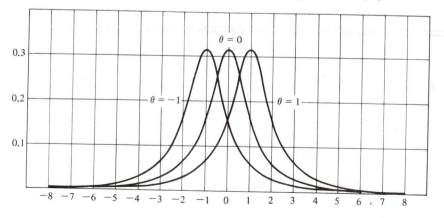

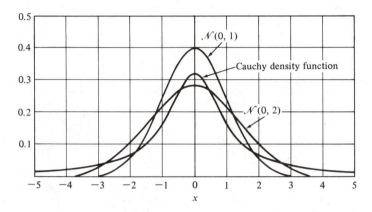

Figure 6.3: *Comparison of the $\mathcal{N}(0, 1)$ and $\mathcal{N}(0, 2)$ density functions with the Cauchy density function with $\theta = 0$.*

However, the integral of $(t - \theta)f_X(t)$ over $-\infty < t < \infty$ is not well defined, as we now demonstrate. If we make the change of variables from t to $v = t - \theta$, then

$$\int_{-\infty}^{\infty} (t - \theta)f_X(t)\, dt = \frac{1}{\pi} \int_{-\infty}^{\infty} \frac{t - \theta}{1 + (t - \theta)^2}\, dt = \frac{1}{\pi} \int_{-\infty}^{\infty} \frac{v}{1 + v^2}\, dv$$

$$= \frac{1}{\pi}\left(\int_{-\infty}^{0} \frac{v}{1 + v^2}\, dv + \int_{0}^{\infty} \frac{v}{1 + v^2}\, dv \right).$$

However, since $v/(1 + v^2) \geq 0$ for all $v \geq 0$ and $v/(1 + v^2) \geq \frac{1}{2}$ when $v > 1$,

$$\int_{0}^{\infty} \frac{v}{1 + v^2}\, dv \geq \int_{0}^{1} 0\, dv + \int_{1}^{\infty} \tfrac{1}{2}\, dv = \infty,$$

and thus the integral of $v/(1 + v^2)$ over the interval $(0, \infty)$ is infinite. Since $v/(1 + v^2)$ is an odd function of v,

$$\infty = \int_{0}^{\infty} \frac{v}{1 + v^2}\, dv = -\int_{-\infty}^{0} \frac{v}{1 + v^2}\, dv,$$

so that the integral in (6.7) is equal to the difference of two infinitely large numbers and thus cannot be well defined.

The fact that the mean of a Cauchy distribution is not well defined may be viewed as a mathematical peculiarity that is irrelevant in practice. However, suppose that we take independent observations $X_1, X_2, \ldots, X_n$ of a random variable X which has a Cauchy distribution with parameter θ and compute the sample average $\hat{\mu}_X = \sum_1^n X_i/n$. The frequency interpretation of probability leads us to believe that the variability of $\hat{\mu}_X$ about some central value, say θ, decreases as the number of observations increases, and that for very large n, $\hat{\mu}_X \simeq \theta$. In Chapters 5 and 7, we saw how this fact could be of use in experimental practice. However, for the Cauchy distribution, it is shown in Chapter 10 that *the sample average $\hat{\mu}_X$ has the same distribution as X* (i.e., the Cauchy distribution) *no matter how many observations we obtain upon X*, and thus the variability of $\hat{\mu}_X$ about θ remains the same even if we take an infinitely large number n of observations. This apparent contradiction to the frequency interpretation of probabilities is explained by the fact that the Cauchy distribution does not have a well-defined mean. A sufficient condition for the frequency interpretation of sample averages to be valid for a given probability distribution is that the mean of this distribution be well defined and finite.

It is tempting to dismiss the Cauchy distribution as a mathematical invention having "no practical importance." However, this distribution is useful in certain scientific contexts: in mechanics and electrical theory, in psychophysics [Urban (1909)], in physical anthropology [Fieller (1932)], and in measurement and calibration problems in many fields of science. Thus, the warnings which the Cauchy distribution provides about the interpretation and estimation of the mean of a distribution must be given serious attention in practice.

SUMMARY OF CONTINUOUS DISTRIBUTIONS

The various continuous distributions discussed in Sections 1 through 6 of this chapter are listed below, together with their parameters, means, and variances. The values of the probability density function are given only for possible values of the associated random variable.

Distribution	Probability density function	Parameters	Mean	Variance
Exponential	$\theta e^{-\theta x}, \quad x > 0$	$\theta > 0$	$\dfrac{1}{\theta}$	$\dfrac{1}{\theta^2}$
Gamma	$\theta^r x^{r-1} e^{-\theta x}/\Gamma(r), \quad x > 0$	$\theta > 0, r > 0$	$\dfrac{r}{\theta}$	$\dfrac{r}{\theta^2}$

(continued)

Distribution	Probability density function	Parameters	Mean	Variance
Weibull	$\dfrac{\beta}{\alpha}\left(\dfrac{x-\nu}{\alpha}\right)^{\beta-1}$ $\times \exp\left[-\left(\dfrac{x-\nu}{\alpha}\right)^{\beta}\right],$ $x > \nu$	$0 \le \nu,$ $0 < \beta,$ $0 < \alpha$	$\nu + \alpha\Gamma\left(1+\dfrac{1}{\beta}\right)$	$\alpha^2\left\{\Gamma\left(1+\dfrac{2}{\beta}\right) - \left[\Gamma\left(1+\dfrac{1}{\beta}\right)\right]^2\right\}$
Uniform	$\dfrac{1}{b-a}, \quad a \le x \le b$	$-\infty < a < b < \infty$	$\dfrac{a+b}{2}$	$\dfrac{(b-a)^2}{12}$
Beta	$\dfrac{x^{r-1}(1-x)^{s-1}}{B(r,s)},$ $0 \le x \le 1$	$r > 0, s > 0$	$\dfrac{r}{r+s}$	$\dfrac{rs}{(r+s)^2(r+s+1)}$
Lognormal	$(2\pi x^2\delta^2)^{-1/2}$ $\times \exp\left[-\dfrac{(\log x - \xi)^2}{2\delta^2}\right],$ $x > 0$	$-\infty < \xi < \infty,$ $\delta^2 > 0$	$\exp\left[\xi + \dfrac{1}{2}\delta^2\right]$	$\exp[2\xi + \delta^2](\exp[\delta^2] - 1)$
Cauchy	$\dfrac{1}{\pi[1 + (x-\theta)^2]},$ $-\infty < x < \infty$	$-\infty < \theta < \infty$	Not well defined	∞

EXERCISES

1. Rasch (1960) considers a stochastic model for reading time in which the time X in minutes required for an individual to read a given passage is a random variable following an exponential distribution with parameter θ. Experimental evidence indicates that for a certain reading exercise the value of θ is 2.0.
   (a) Find the probability of the event that it takes an individual more than 1 minute to read the given reading exercise.
   (b) Find the median and (0.90)th quantile of the distribution of the time X required to read the given reading exercise.

2. The distribution of the duration of pauses (and the duration of vocalizations) that occur in a monologue has an exponential distribution [Jaffe and Feldstein (1970)]. If the expected value of the duration of pauses is 0.70 second, what is the variance of the duration of pauses? What is the (0.90)th quantile?

3. Using the intervals for grouping shown in Figure 1.3 and the data in

Table 1.1, construct a table of observed frequencies, theoretical probabilities, and theoretical frequencies for these intervals.

4. If the random variable X has an exponential distribution with parameter θ, find $E(X - \mu_X)^3$ and $E(X - \mu_X)^4$.

5. The magnitude M of an earthquake, as measured on the Richter scale, is a random variable of considerable practical importance. As a consequence of studies of earthquakes with large magnitudes (say, magnitudes exceeding 3.25), it has been suggested that the "excess" $X = M - 3.25$ follows an exponential distribution. Over the period January 1934 to May 1943, the magnitudes of earthquakes were recorded. The resulting data are summarized by Gutenberg and Richter (1944). From these data, the frequency distribution in Table E.1 has been constructed.

Table E.1: *Observed Frequency Distribution of Earthquakes with Large Magnitudes, January 1934 to May 1943*

Midpoint m of interval of magnitudes	Midpoint x of interval of excesses	Observed frequency of earthquakes whose "excess" $X = M - 3.25$ is between $x - 0.25$ and $x + 0.25$
3.5	0.25	579
4.0	0.75	311
4.5	1.25	108
5.0	1.75	32
5.5	2.25	13
6.0	2.75	5
6.5	3.25	2
		1050

(a) Fit an exponential distribution to these data.
(b) Assuming that $X = M - 3.25$ has the exponential distribution found in part (a), what is the probability that the magnitude M of an earthquake with large magnitude will exceed 6 on the Richter scale?

6. A trucking company knows that the length of time X in hours required for one of its trucks to complete a typical haul (round trip) is exponentially distributed with parameter $\theta = \frac{1}{48}$. Given that a truck has already been gone for 48 hours, what is the conditional probability that this truck will arrive within the next hour? What is the mean additional length of time that the company must wait until the truck completes its route, given that the truck has already been gone for 48 hours?

7. Suppose that X and Y are probabilistically independent random variables, and that both X and Y have the same exponential distribution with parameter $\theta = 0.10$.
    (a) Find $P\{X > 5.0 \text{ and } Y > 5.0\}$.
    (b) Find a general formula for $P\{X > t \text{ and } Y > t\}$.
    (c) Let $W = \min\{X, Y\}$. Find the probability density function of W, and find μ_W.

8. Consider again the reading model of Exercise 1. Suppose that a student is asked to read N similar, but not necessarily identical, exercises one after the other. Also assume that the student's reading times for these exercises are probabilistically independent random variables, each having an exponential distribution with parameter θ. Let Y be the total time required for the student to read all N exercises.
    (a) What is the distribution of Y?
    (b) Find μ_Y and σ_Y^2.
    (c) Find the median and (0.90)th quantile $Q_{0.90}(Y)$ of the distribution of Y when $N = 5$ and $\theta = 2.0$.
    (d) When $N = 5$ and $\theta = 2.0$, find $P\{Y \le 2.5\}$, $P\{Y > 5\}$, and $P\{1 \le Y \le 5\}$.

9. Let Y have a gamma distribution with parameters r and θ. Show that the mode of Y is equal to 0 when $r \le 1$. When $r > 1$, find a formula for the mode of Y. Then show that for all values of r and θ, the mode of Y is less than μ_Y.

10. The gamma distribution has been used in a medical application by Masuyama and Kuroiwa (1952). They considered the sedimentation rate X at various stages during normal pregnancy. In the 30th week after commencement of pregnancy, their data were fit approximately by a gamma distribution with $r = 5.07$ and $\theta = 1/9.98$. Instead, assume that X has a gamma distribution with parameters $r = 5$ and $\theta = 0.1$. Find Med (X), the median sedimentation rate in the 30th week after commencement of pregnancy. Compute $P\{X > 50\}$ and $P\{10 \le X \le 90\}$.

11. The gamma distribution has been used as a model for the lifetimes of metals subjected to stress. In one experiment, 101 rectangular strips of aluminum of standardized dimensions were submitted to repeated alternating stresses (at a frequency of 18 cycles per second). The lifetimes of these strips of aluminum (expressed in thousands of cycles) are summarized in Table E.2.

    The sample mean and sample variance of these data are 1400.91 and 151,618.36, respectively. Estimate values of r and θ needed to fit the gamma distribution to this data. Round $\hat{r}$ to the nearest integer and $\hat{\theta}$ to the second decimal place. Using grouping intervals of length 180 starting at 370 and ending at 2530, construct a table of observed frequencies,

Table E.2: *Lifetimes Under Periodic Loading, Maximum Stress 21,000 psi, 18 Cycles Per Second, 6061-T6 Aluminum Coupon Cut Parallel to the Direction of Rolling*

(Observations listed in increasing order)

370	988	1120	1269	1420	1530	1730	1893
706	990	1134	1270	1420	1540	1750	1895
716	1000	1140	1290	1450	1560	1750	1910
746	1010	1199	1293	1452	1567	1763	1923
785	1016	1200	1300	1475	1578	1768	1940
797	1018	1200	1310	1478	1594	1781	1945
844	1020	1203	1313	1481	1602	1782	2023
855	1055	1222	1315	1485	1604	1792	2100
858	1085	1235	1330	1502	1608	1820	2130
886	1102	1238	1355	1505	1630	1868	2215
886	1102	1252	1390	1513	1642	1881	2268
930	1108	1258	1416	1522	1674	1890	2440
960	1115	1262	1419	1522			

Source: Z. W. Birnbaum and S. C. Saunders (1958). A statistical model for the life-length of materials. *Journal of the American Statistical Association*, vol. 53, pp. 151–60. Reprinted by permission.

theoretical probabilities, and theoretical frequencies for the data in Table E.2.

12. Let X have an exponential distribution with parameter $\theta = 1$ and let Y have a gamma distribution with parameters $r = 2$ and $\theta = 2$. Note that $\mu_X = \mu_Y$.
    (a) Compare σ_X^2 with σ_Y^2.
    (b) Compare $P\{X - \mu_X > 2\sigma_X\}$ with $P\{Y - \mu_Y > 2\sigma_Y\}$.

13. Use Table T.3 to find $P\{5 \leq Y \leq 10\}$ when Y has a gamma distribution with parameters $r = 3$ and $\theta = 0.5$.

14. Suppose that Y has a gamma distribution with parameters r and θ. Find $\mu_Y^{(k)} = E(Y - \mu_Y)^k$ for $k = 3, 4$. As $r \to \infty$, show that $\mu_Y^{(3)}$ tends to 0 and that $\mu_Y^{(4)}/\sigma_Y^4$ tends to 3. What does this suggest about the relationship between the gamma distributions for large r and the normal distributions?

15. It is known that the duration X, in days, of an epidemic of a certain infectious disease has a Weibull distribution with parameters $\beta = 2$, $\alpha = 10$, and $\nu = 2$. Find

    (a) μ_X, (b) σ_X^2, (c) Med(X), (d) the mode.

    If an epidemic of the disease has just been reported, what prediction would you give as to the length of time (in days) that the epidemic will last?

16. In the context of Exercise 15, find $P\{X > 14\}$ and $P\{4 \leq X \leq 7\}$.

17. The Weibull distribution has been used by Indow (1971) as a model for determining the effect of advertising. Let X be the length of time (in days) after the end of an advertising campaign that an individual is able to remember the name of the product being advertised. In a particular experiment a brand of chocolate was used, and Indow fit a Weibull distribution with $\beta = 0.98$, $\alpha = 7360.0$, and $v = 1.0$ to the variation of X.
    (a) What is the median, Med(X), of X? What is μ_X?
    (b) What proportion of all individuals could be expected to remember the brand of chocolate advertised 1 week after the advertising campaign ended?

18. Using the values of $\beta = 2.331$, $\alpha = 6.41$, and $v = 1.50$ determined in Example 3.1, find theoretical probabilities and theoretical frequencies to compare with the observed frequencies of Table 3.2.

19. Find general formulas for the median, Med(X), and pth quantile $Q_p(X)$ of a random variable X having a Weibull distribution with parameters β, α, and v.

20. Find a general formula for the mode of a Weibull distribution with parameters β, α, and v.

21. If X has a Weibull distribution with parameters $\beta = 3$, $\alpha = 5$, $v = 1$, what is the distribution of

    (a) $Y = X - 1$,   (b) $Z = \dfrac{X - 1}{5}$,   (c) $W = \dfrac{(X - 1)^2}{125}$,

    (d) $V = \dfrac{(X - 1)^2}{25}$,   (e) $Q = (X - 1)^2$?

22. Suppose that X has a Weibull distribution with parameters β, α, and $v = 0$. Find $\mu_X^{(3)} = E(X - \mu_X)^3$ and $\mu_X^{(4)} = E(X - \mu_X)^4$.

23. The monthly fire incidence (monthly total ÷ yearly total) in buildings in England and Wales in 1961 is given in Table E.3. Let X equal the exact time in days (after 12:00 A.M. on the morning of January 1, 1961) that a given fire occurs in a building in England and Wales. Assume that $X \sim U[0, 365]$ and compute theoretical incidence figures to compare with the observed incidences in Table E.3. (For example, the theoretical incidence for February would be $\frac{28}{365} = 0.0767$.) Does it appear that the $U[0, 365]$ distribution provides a good model for the variation of X?

24. A bus travels between two cities, A and B, which are 100 miles apart. If the bus has a breakdown, the distance X of the point of the breakdown from city A has a $U[0, 100]$ distribution.

Table E.3: *Monthly Fire Incidences in Buildings in England and Wales in 1961*

Month	Incidence of fires
January	0.0936
February	0.0739
March	0.0917
April	0.0704
May	0.0861
June	0.0809
July	0.0771
August	0.0699
September	0.0705
October	0.0842
November	0.0942
December	0.1075

(a) There are service garages in city A, city B, and midway between cities A and B. If a breakdown occurs, a tow truck is sent from the garage closest to the point of breakdown. What is the probability that the tow truck has to travel more than 10 miles to reach the bus?

(b) Would it be more "efficient" if the three service garages were placed 25, 50, and 75 miles from city A? Explain.

25. If buses run on the half-hour, and on any given day a commuter has a time X of arrival uniformly distributed between 8:15 A.M. and 8:45 A.M., what is the probability that the commuter will have to wait for a bus
(a) more than 15 minutes?
(b) more than 20 minutes?
(c) What is the mean length of time that the commuter must wait for a bus?

26. Let X and Y be independent $U[1, 2]$ random variables.
(a) Find $P\{X > 1.5$ *and* $Y > 1.5\}$, $P\{X \le 1.5$ *and* $Y \le 1.5\}$, and $P\{X \le 1.5$ *and* $Y > 1.5\}$.
(b) For $1 \le x \le 2$, find $P\{X > x$ *and* $Y > x\} = P\{U > x\}$ for $U = \min\{X, Y\}$. Let $V = U - 1$. Show that V has a beta distribution, and find the parameters r, s of this distribution.

27. If $X \sim U[a, b]$, find $\mu_X^{(3)} = E(X - \mu_X)^3$ and $\mu_X^{(4)} = E(X - \mu_X)^4$. Comment on the skewness and kurtosis of the uniform distribution.

28. Suppose that you wish to simulate a random variable W having a Weibull distribution with parameters $\beta = 4.0$, $\alpha = 3.0$, $v = 1.6$. You have a random number generator that simulates $X \sim U[0, 1]$. How would you simulate W? Suppose that your random number generator yields the value $x = 0.77$ for X. What value would be simulated for W?

29. Suppose that X has a beta distribution with parameters $r = 4$ and $s = 3$. Find
    (a) $P\{X \le 0.5\}$,   (b) $P\{X > 0.7\}$,   (c) $P\{0.4 \le X \le 0.6\}$,
    (d) μ_X,   (e) σ_X^2,   (f) the mode of X.

30. Find a general formula giving the mode (or modes) of a beta distribution with parameters r and s.

31. Show that when X has a beta distribution with parameters r and s, and $r = s > 1$, then the mean, median, and mode of X are all equal to $\frac{1}{2}$. What happens when $r = s < 1$?

32. Let X have a beta distribution with parameters r and s. Find $\mu_X^{(3)} = E(X - \mu_X)^3$ and $\mu_X^{(4)} = E(X - \mu_X)^4$.

33. In Example 5.3, $\hat{r} = 1.0$ and $\hat{s} = 10.6$ to one decimal place of accuracy. Assuming that Y in Example 5.3 has a beta distribution with parameters $r = 1.0$ and $s = 10.6$, compute theoretical probabilities and theoretical frequencies for the intervals shown in Table 5.1.

34. Ore samples from the Transvaal vein of level 10 of the Frisco mine were assayed for metal content. For each of 1000 ore samples, the ratios X of the weight of copper (Cu) to the sum of the weights of Cu and lead (Pb) were measured. Table E.4 gives a frequency distribution for X. For these data, $\hat{\mu}_X = 0.130760$ and $\hat{\sigma}_X^2 = 0.020384$.
    (a) Show that to two-decimal-place accuracy, $\hat{s} = 4.00$. Find $\hat{r}$.
    (b) Show that the beta distribution with the values of r and s calculated in part (a) gives a poor fit to the data by calculating theoretical frequencies for the intervals $[0.00, 0.04]$, $[0.04, 0.08]$ using (5.11), and comparing these frequencies to the observed frequencies in Table E.4.
    (c) The poor fit of the beta distribution may be due to the method of obtaining estimates of r and s. Another estimation method yielded $\hat{r} = 1.13$ and $\hat{s} = 3.00$. Using these values of r and s, compute theoretical frequencies for the intervals $[0.00, 0.04]$ and $[0.04, 0.08]$, again using (5.11) to obtain theoretical probabilities.

35. Let Y have a beta distribution with parameters r and s. Find the probability density function of $X = (b - a)Y + a$, where $-\infty < a < b < \infty$. Note that X takes on values in $[a, b]$.

Table E.4: *Frequency Distribution of the Ratio Cu/(Cu + Pb) in 1000 Samples of Ore*

Midpoint x of interval of values	Observed frequency of interval $[x - 0.02, x + 0.02]$	Midpoint x of interval of values	Observed frequency of interval $[x - 0.02, x + 0.02]$
0.02	209	0.54	5
0.06	264	0.58	4
0.10	170	0.62	8
0.14	105	0.66	1
0.18	58	0.70	5
0.22	53	0.74	4
0.26	33	0.78	5
0.30	22	0.82	2
0.34	12	0.86	1
0.38	11	0.90	1
0.42	11	0.94	0
0.46	10	0.98	2
0.50	4		$N = 1000$

Source: Based on G. S. Koch, Jr., and R. F. Link (1971). *Statistical Analysis of Geological Data*, vol. 2. New York: John Wiley & Sons, Inc. Reprinted by permission.

36. Fit a lognormal distribution to the data of Example 2.3. Check the fit by graphically comparing the theoretical survival function $1 - F_Y(y)$ to the sample survival function $1 - F_n(y)$.

9

Multivariate Distributions

1. INTRODUCTION

Many important problems in science and technology are concerned with the question of how the values of one measurable quantity can be used to explain, predict, or control another quantity. For example, physicists may be interested in predicting the motion of a body from knowledge of the forces acting on that body, aeronautical engineers may wish to determine how well information obtained from automatic sensors can be used to control a space satellite, and biologists may be concerned with the relationship between the concentration of a certain chemical in the environment and the frequency of cancerous tumors in animals. Similarly, sociologists may be interested in the relationship of group size to problem-solving productivity, psychologists may be interested in the relationship of personality test scores to anxiety level, and educators may be interested in how high school grades can be used to predict college achievement. Indeed, virtually all scientific and technological disciplines require, as part of their development, knowledge of the interrelationships among the variables that compose their empirical framework.

Mechanistic models for natural phenomena express relationships between variables in terms of exact mathematical formulas. Probability models, on the other hand, express such relationships either in terms of probabilities of joint occurrence, or in terms of mathematical formulas representing "average" or "typical" relationships. When probability models describe the joint variability of more than one random variable, the resulting model is said to be the *joint* (or *multivariate*) *distribution* of these variables. For the sake of simplicity of notation and ease of understanding, we concentrate in this chapter on the joint distribution of two random variables (a *bivariate distribution*), and limit our consideration to cases where either both variables are discrete (discrete bivariate distributions), or where both variables are continuous (continuous bivariate distributions). The most widely used continuous bivariate distribution, the bivariate normal distribution, is discussed in some detail because of its importance as a background for classical statistical methodology. However, other bivariate distributions are used in examples. Finally, we conclude this chapter by generalizing concepts discussed for bivariate distributions to cover joint distributions of three or more random variables.

2. BIVARIATE DISTRIBUTIONS

By the *bivariate distribution* of two random variables X and Y, we mean a probability model for the relative frequencies with which values of X occur jointly with values of Y. In analogy with the case of one random variable (*univariate distributions*), the bivariate distribution of X and Y can always be determined from the *joint cumulative distribution function*:

$$(2.1) \qquad F_{X,Y}(x, y) = P\{X \leq x, Y \leq y\}.$$

However, for practical applications, it is usually more convenient to work with functions analogous to the univariate mass functions and density functions discussed in Chapter 4. Thus, when X and Y are both discrete random variables, their joint distribution is determined by the *joint probability mass function*:

$$(2.2) \qquad p_{X,Y}(x, y) = P\{X = x, Y = y\}.$$

When X and Y are both continuous random variables with a joint cumulative distribution function $F_{X,Y}(x, y)$ satisfying

$$(\partial^2/\partial x\,\partial y)F_{X,Y}(x, y) = (\partial^2/\partial y\,\partial x)F_{X,Y}(x, y),$$

then the joint distribution of X and Y is determined by the *joint probability density function*:

$$(2.3) \qquad f_{X,Y}(x, y) = \frac{\partial^2}{\partial x\,\partial y} F_{X,Y}(x, y) = \frac{\partial^2}{\partial y\,\partial x} F_{X,Y}(x, y).$$

The double subscript on (2.1), (2.2), and (2.3) is cumbersome; when no confusion is possible, we simplify notation and write $F(x, y)$ for $F_{X,Y}(x, y)$, $p(x, y)$ for $p_{X,Y}(x, y)$, and $f(x, y)$ for $f_{X,Y}(x, y)$.

Contingency Tables

The joint probability mass function $p(x, y)$ of discrete variables X and Y can always be written in functional form. However, when X and Y each have a finite number of possible values, it is helpful to table the probabilities $p(x, y)$ in the form of a *contingency table* such as Table 2.1. Since the pairs (x_i, y_j) are outcomes of the random experiment in which X and Y are jointly observed, the sum of the entries $p(x_i, y_j)$ in a contingency table must always equal 1.

Table 2.1: *Contingency Table for Two Discrete Random Variables X and Y, Where X Has Three Possible Values, x_1, x_2, x_3, and Y Has Four Possible Values, y_1, y_2, y_3, y_4*

x	y			
	y_1	y_2	y_3	y_4
x_1	$p(x_1, y_1)$	$p(x_1, y_2)$	$p(x_1, y_3)$	$p(x_1, y_4)$
x_2	$p(x_2, y_1)$	$p(x_2, y_2)$	$p(x_2, y_3)$	$p(x_2, y_4)$
x_3	$p(x_3, y_1)$	$p(x_3, y_2)$	$p(x_3, y_3)$	$p(x_3, y_4)$

As an alternative to a contingency table, we can graph the joint probability mass function $p(x, y)$ as in Figure 2.1. The possible pairs (x_i, y_j) of values are represented as points in the plane, and from these points lines are drawn perpendicular to the plane with height equal to $p(x, y)$.

Example 2.1. Two professional wine tasters are asked to rate wines using a rating scale of 1, 2, 3, 4, or 5. Suppose that the wines being sampled are randomly chosen from among table wines marketed nationally. Here, wine taster I's rating is a random variable X, wine taster II's rating is a random variable Y, and we can think of both wine tasters announcing their ratings

Figure 2.1: *Graph of the joint probability mass function $p(x, y)$.*

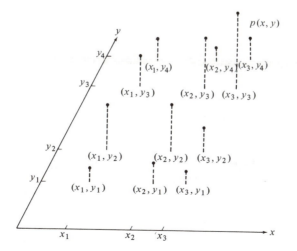

Table 2.2: *Bivariate Probability Mass Function for Two Wine Tasters Exhibited in the Form of a Contingency Table*

Rating of wine taster I, x	Rating of wine taster II, y					Totals
	1	2	3	4	5	
1	0.03	0.02	0.01	0.00	0.00	0.06
2	0.02	0.08	0.05	0.02	0.01	0.18
3	0.01	0.05	0.25	0.05	0.01	0.37
4	0.00	0.02	0.05	0.20	0.02	0.29
5	0.00	0.01	0.01	0.02	0.06	0.10
Totals	0.06	0.18	0.37	0.29	0.10	1.00

simultaneously. A possible contingency table for this experiment is given in Table 2.2. The row and column sums of contingency tables are known as *marginal totals* or *marginal probabilities*. For example, the marginal total for row 1 in Table 2.2 gives the probability $P\{X = 1\} = \sum_{y=1}^{5} p(1, y)$ that $X = 1$. A graph of the joint probability mass function $p(x, y)$ defined by Table 2.2 appears in Figure 2.2.

Figure 2.2: *Graph of the bivariate mass function $p(x, y)$ corresponding to the probability model for the joint ratings of wine tasters I and II associated with Table 2.2.*

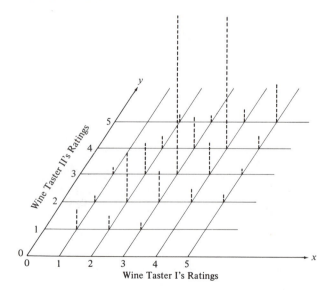

Probability Calculations

In obtaining probabilities for jointly distributed random variables X and Y, it is helpful to think of the pair (X, Y) as a random point in the plane, and to describe any event E of interest concerning X and Y in terms of a region R in the plane. For example, the event E that "X equals Y" can be written as $E = \{(X, Y) \text{ in } R\}$, where

$$R = \{(x, y): x = y\}$$

is the collection of points in the plane falling on the line $y = x$. Similarly, the event E that "X exceeds Y" can be written as $E = \{(X, Y) \text{ in } R\}$, where $R = \{(x, y): x > y\}$.

If R is any region in the plane and X and Y are discrete random variables with joint probability mass function $p(x, y)$, then

(2.4)
$$P\{(X, Y) \text{ in } R\} = \sum_{(x, y) \text{ in } R} p(x, y)$$

(see Figure 2.3).

Example 2.1 (continued). The contingency table, Table 2.2, shows that the highest probabilities are given to pairs (x, y) where $x = y$, indicating agreement among the wine tasters. To compute the probability $P\{X = Y\}$ that the

Figure 2.3: *A region R in the plane. The probability that the random point* (x, y) *falls in R is the sum of the heights* p(x, y) *of the dotted lines that correspond to points* (x, y) *in R.*

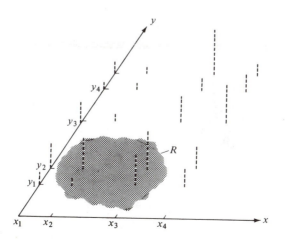

tasters give identical ratings to a sample of wine, let $R = \{(x, y): x = y\}$, so that

$$P\{X = Y\} = P(R) = \sum_{(x, y) \text{ in } R} p(x, y)$$

$$= p(1, 1) + p(2, 2) + p(3, 3) + p(4, 4) + p(5, 5)$$

$$= 0.03 + 0.08 + 0.25 + 0.20 + 0.06 = 0.62.$$

When X and Y are continuous random variables with joint probability density function $f(x, y)$, then

(2.5) $$P\{(X, Y) \text{ in } R\} = \iint_R f(x, y) \, dx \, dy,$$

provided that the double integral in (2.5) is well defined. Recall that the double integral in (2.5) is the volume over the region R between the (x, y)-plane and the surface determined by the graph of $f(x, y)$. Figures 2.4, 2.5, and 2.6 give examples of such volumes when the bivariate density function $f(x, y)$ has the form

(2.6) $$f(x, y) = \begin{cases} 1, & \text{if } 0 \le x \le 1, 0 \le y \le 1, \\ 0, & \text{otherwise.} \end{cases}$$

Example 2.2. Suppose that X and Y are continuous random variables with

Figure 2.4: *The probability $P\{0 \le X \le \frac{1}{4}$ and $\frac{1}{2} \le Y \le \frac{3}{4}\}$ equals the volume of the shaded rectangle.*

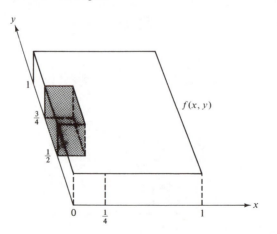

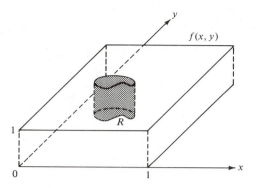

Figure 2.5: *The shaded volume equals the probability* $P\{(X, Y) \text{ in } R\}$.

the joint probability density function (2.6). Let $R = \{(x, y): 0 \le x \le \frac{1}{4} \text{ and } \frac{1}{2} \le y \le \frac{3}{4}\}$. In this case,

$$P\{(X, Y) \text{ in } R\} = P\{0 \le X \le \tfrac{1}{4} \text{ and } \tfrac{1}{2} \le Y \le \tfrac{3}{4}\}$$

$$= \int_{1/2}^{3/4} \int_{0}^{1/4} f(x, y) \, dx \, dy = \int_{1/2}^{3/4} \int_{0}^{1/4} 1 \, dx \, dy = \tfrac{1}{16},$$

as illustrated in Figure 2.4.

Of course, the region R need not be a rectangle, but can be of any shape

Figure 2.6: *The probability of the event* $\{(X - \frac{1}{2})^2 + (Y - \frac{1}{2})^2 \le \frac{1}{4}\}$ *equals the volume of the shaded cylinder whose base is the circle R.*

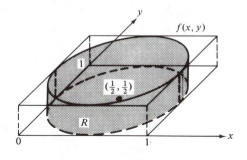

(as in Figure 2.5). For example, if R is a circle (see Figure 2.6) centered at the point $(\frac{1}{2}, \frac{1}{2})$ and with radius $\frac{1}{2}$, then

$$P\{(X - \tfrac{1}{2})^2 + (Y - \tfrac{1}{2})^2 \le \tfrac{1}{4}\} = \iint\limits_{(x-1/2)^2 + (y-1/2)^2 \le 1/4} dx\, dy$$

$$= \int_0^1 \left[\int_{1/2 - [y(1-y)]^{1/2}}^{1/2 + [y(1-y)]^{1/2}} dx \right] dy$$

$$= \int_0^1 2y^{1/2}(1 - y)^{1/2}\, dy = 2B(\tfrac{3}{2}, \tfrac{3}{2})$$

$$= \frac{2(\Gamma(\tfrac{3}{2}))^2}{\Gamma(3)} = \frac{2(\tfrac{1}{2}\sqrt{\pi})^2}{2!} = \frac{\pi}{4} = 0.78539.$$

Alternatively, this probability is the volume of a cylinder with height 1 and base $\pi(\frac{1}{2})^2$, as shown in Figure 2.6.

For two continuous variables X and Y with a joint probability density function $f(x, y)$, the interpretation of $P\{(X, Y) \text{ in } R\}$ in terms of a volume over a region R under the graph of $f(x, y)$ yields the following important consequence: *Whenever the area of R is 0, then the probability, $P\{(X, Y) \text{ in } R\}$, that (X, Y) falls in R is equal to 0.* For example, for any two numbers x_0 and y_0,

$$P\{X = x_0 \text{ and } Y = y_0\} = P(\{(x, y): (x, y) = (x_0, y_0)\}) = 0.$$

More interestingly,

$$P\{X = Y\} = P\{X - Y = 0\} = P(\{(x, y): x - y = 0\}) = 0,$$

in contrast to the nonzero probability obtained for the discrete variables X and Y in Example 2.1. In general, when X and Y are continuous random variables having a joint probability density function, then $P\{aX + bY = c\} = 0$ for any constants a, b, c; that is, the probability that (X, Y) lies along any particular straight line is zero.

Marginal Distributions

Not only does the joint distribution of the random variables X and Y contain all the relevant probabilistic information about the *joint* variation of X and Y, it also contains all information about the variation of X, and of Y, considered *individually*. For example, from the joint probability mass function $p(x, y)$ of two discrete random variables X and Y, we can obtain the individual probability mass functions $p_X(x)$ and $p_Y(y)$ of X and Y, respectively, by means of the formulas

(2.7) $$p_X(x) = \sum_y p(x, y), \qquad p_Y(y) = \sum_x p(x, y).$$

To verify (2.7), define the regions R_1 and R_2 by

$$R_1 = \{(x, y): x = x_0, y \text{ has any value}\},$$
$$R_2 = \{(x, y): y = y_0, x \text{ has any value}\}.$$

Then

$$p_X(x_0) = P\{X = x_0\} = P(R_1) = \sum_{(x, y) \text{ in } R_1} p(x, y) = \sum_y p(x_0, y),$$

$$p_Y(y_0) = P\{Y = y_0\} = P(R_2) = \sum_{(x, y) \text{ in } R_2} p(x, y) = \sum_x p(x, y_0).$$

As we have noted earlier, $p_X(x) = \sum_y p(x, y)$ and $p_Y(y) = \sum_x p(x, y)$ can be obtained from the row sums and column sums, respectively, of a contingency table (such as Table 2.2). For this reason, $p_X(x)$ and $p_Y(y)$ are referred to as *marginal probability mass functions* of the joint probability mass function $p(x, y)$.

Example 2.1 (continued). The marginal probability mass functions $p_X(x)$ and $p_Y(y)$ of the ratings X and Y of wine tasters I and II, respectively, can be obtained from the row and column totals of Table 2.2, and are summarized in Table 2.3. Note that the marginal distributions of X and Y are the same. This result might be anticipated from the fact that both wine tasters seem to be using similar standards to rate the wines.

When X and Y are continuous random variables with joint probability density function $f(x, y)$, then the *marginal probability density functions* $f_X(x)$ and $f_Y(y)$ of $f(x, y)$, which are the individual probability density functions of X and Y, respectively, are obtained by means of the formulas

$$(2.8) \qquad f_X(x) = \int_{-\infty}^{\infty} f(x, y)\, dy, \qquad f_Y(y) = \int_{-\infty}^{\infty} f(x, y)\, dx.$$

Table 2.3: *Marginal Probability Mass Functions of X and Y Based on the Joint Probability Distribution Given in Table 2.1*

Rating t	$p_X(t)$ = row sum	$p_Y(t)$ = column sum
1	0.06	0.06
2	0.18	0.18
3	0.37	0.37
4	0.29	0.29
5	0.10	0.10

To verify (2.8), we first obtain the marginal cumulative distribution functions $F_X(x_0)$ and $F_Y(y_0)$ of X and Y, respectively. To do so, we use (2.5) and the regions

$$R_1 = \{(x, y): x \le x_0, y \text{ any value}\}, \qquad R_2 = \{(x, y): y \le y_0, x \text{ any value}\}.$$

Thus

$$F_X(x_0) = P\{X \le x_0\} = P(R_1) = \int_{-\infty}^{x_0} \left[\int_{-\infty}^{\infty} f(x, y)\, dy \right] dx,$$

(2.9)

$$F_Y(y_0) = P\{Y \le y_0\} = P(R_2) = \int_{-\infty}^{y_0} \left[\int_{-\infty}^{\infty} f(x, y)\, dx \right] dy.$$

Differentiating $F_X(x_0)$ with respect to x_0 and $F_Y(y_0)$ with respect to y_0, and making use of the Fundamental Theorem of Calculus, establishes (2.8).

Example 2.2 (continued). For the bivariate probability density function $f(x, y)$ which equals 1 if $0 \le x \le 1, 0 \le y \le 1$, and equals 0 otherwise,

$$f_X(x) = \int_{-\infty}^{\infty} f(x, y)\, dy = \int_{-\infty}^{\infty} 0\, dy = 0, \quad \text{if } x < 0 \text{ or } x > 1,$$

$$f_X(x) = \int_{-\infty}^{\infty} f(x, y)\, dy = \int_{0}^{1} 1\, dy = 1, \quad \text{if } 0 \le x \le 1.$$

Thus,

$$f_X(x) = \begin{cases} 1, & \text{if } 0 \le x \le 1, \\ 0, & \text{if } x < 0 \text{ or } x > 1, \end{cases}$$

which is a $U[0, 1]$ probability density function. In a similar manner, Y has a marginal $U[0, 1]$ distribution.

Example 2.3 (Dirichlet Distributions). Consider the bivariate probability density function:

(2.10)

$$f(x, y) = \begin{cases} \dfrac{\Gamma(a + b + c)}{\Gamma(a)\Gamma(b)\Gamma(c)} x^{a-1} y^{b-1}(1 - x - y)^{c-1}, & \text{if } 0 \le x, y, x + y \le 1, \\ 0, & \text{otherwise,} \end{cases}$$

where $a > 0$, $b > 0$, $c > 0$. This distribution is called the bivariate *Dirichlet distribution* [after the German mathematician Peter Gustav Lejeune Dirichlet (1805–1859)]. Note that

$$\frac{\Gamma(a + b + c)}{\Gamma(a)\Gamma(b)\Gamma(c)} = \left(\frac{\Gamma(a + b + c)}{\Gamma(a)\Gamma(b + c)}\right)\left(\frac{\Gamma(b + c)}{\Gamma(b)\Gamma(c)}\right) = \frac{1}{B(a, b + c)B(b, c)}.$$

To obtain the marginal density $f_X(x)$, note that when $x < 0$ or $x > 1$, $f(x, y) = 0$ for all y, so $f_X(x) = 0$. When $0 \le x \le 1$, then since $f(x, y) \ne 0$ only when $0 \le y \le 1 - x$,

$$f_X(x) = \int_{-\infty}^{\infty} f(x, y)\, dy = \frac{x^{a-1}}{B(a, b + c)} \int_0^{1-x} \frac{y^{b-1}(1 - x - y)^{c-1}\, dy}{B(b, c)}$$

$$= \frac{x^{a-1}(1 - x)^{b+c-1}}{B(a, b + c)} \int_0^1 \frac{u^{b-1}(1 - u)^{c-1}\, du}{B(b, c)}$$

$$= \frac{x^{a-1}(1 - x)^{b+c-1}}{B(a, b + c)},$$

where we have changed variable of integration from y to $u = y/(1 - x)$ and used the definition of $B(b, c)$ [see (A.16) of the Appendix]. Thus,

(2.11a)
$$f_X(x) = \begin{cases} \dfrac{x^{a-1}(1 - x)^{b+c-1}}{B(a, b + c)}, & \text{if } 0 \le x \le 1, \\ 0, & \text{otherwise,} \end{cases}$$

which we recognize as being the density of a beta distribution with parameters $r = a$, $s = b + c$. Similarly,

(2.11b)
$$f_Y(y) = \begin{cases} \dfrac{y^{b-1}(1 - y)^{a+c-1}}{B(b, a + c)}, & \text{if } 0 \le y \le 1, \\ 0, & \text{otherwise,} \end{cases}$$

which is the density of a beta distribution with parameters $r = b$, $s = a + c$. Thus, the bivariate Dirichlet distribution has marginal beta distributions.

It should be emphasized that the marginal distributions obtained from the joint distribution of any pair (X, Y) of jointly distributed random variables must be the same as the distributions that we would have obtained if we had considered the variables X and Y individually.

Although the individual distributions of X and of Y can be obtained in the form of marginal distributions from knowledge of the joint distribution of X and Y [see (2.7) and (2.8)], the converse is not true. That is, *the marginal distributions of two jointly distributed random variables do not determine the*

Table 2.4: *Two Joint Discrete Probability Distributions with the Same Marginal Distributions*

(a)

x	y		Totals
	1	0	
1	0.40	0.20	0.60
0	0.30	0.10	0.40
Totals	0.70	0.30	1.00

(b)

x	y		Totals
	1	0	
1	0.42	0.18	0.60
0	0.28	0.12	0.40
Totals	0.70	0.30	1.00

joint distribution of these random variables. To verify this fact, we need only show that two *different* joint distributions can have the *same* marginal distribution for X and the *same* marginal distribution for Y. Table 2.4 shows two different contingency tables that have the same marginal distributions.

Example 2.4. Suppose that X and Y have the joint probability density function:

(2.12)
$$f(x, y) = \begin{cases} \dfrac{x^{a-1}(1-x)^{b+c-1}y^{b-1}(1-y)^{a+c-1}}{B(a, b+c)B(b, a+c)}, & \text{if } 0 \leq x \leq 1, 0 \leq y \leq 1, \\ 0, & \text{otherwise,} \end{cases}$$

for $a > 0$, $b > 0$, $c > 0$. Note that for $x < 0$ or $x > 1$, $f_X(x) = 0$. Otherwise, for $0 \leq x \leq 1$,

$$f_X(x) = \int_{-\infty}^{\infty} f(x, y)\, dy = \frac{x^{a-1}(1-x)^{b+c-1}}{B(a, b+c)} \int_0^1 \frac{y^{b-1}(1-y)^{a+c-1}}{B(b, a+c)}\, dy$$

$$= \frac{x^{a-1}(1-x)^{b+c-1}}{B(a, b+c)},$$

since $f(x, y) = 0$ if $y < 0$ or $y > 1$.

Thus, the marginal probability density function $f_X(x)$ for the joint density function (2.12) is the same as the marginal density function (2.11a) for the Dirichlet distribution (2.10). Similarly, it can be shown that the marginal probability density functions $f_Y(y)$ for the joint density functions (2.10) and (2.12) are the same. However, (2.10) and (2.12) define different distributions, as can be seen from the fact that (2.10) has value 0 for $0 \leq 1 - x < y \leq 1$, while (2.12) assigns nonzero values to (x, y) points in that region.

3. CONDITIONAL DISTRIBUTIONS

In performing a composite random experiment, the value of one of two random variables may become known. Thus, we may know that the value of the random variable X is x. On the basis of this knowledge, we want to reassess the probabilities for the various possible values of Y. Conditional probabilities now become relevant.

If X and Y have joint probability mass function $p(x, y)$, then the *conditional probability mass function* of Y given $X = x$ is defined to be

$$(3.1a) \qquad p_{Y|X=x}(y) = \frac{p(x, y)}{p_X(x)},$$

assuming that $p_X(x) > 0$. [The conditional probability mass function $p_{Y|X=x}(y)$ is defined only where $p_X(x) > 0$, since only such values of x are observable.] Clearly, $p_{Y|X=x}(y) \geq 0$ for all values y, and

$$\sum_y p_{Y|X=x}(y) = \sum_y \frac{p(x, y)}{p_X(x)} = \frac{\sum_y p(x, y)}{p_X(x)} = \frac{p_X(x)}{p_X(x)} = 1.$$

Thus, $p_{Y|X=x}(y)$ is a probability mass function. Correspondingly, the conditional probability mass function of X given $Y = y$ is defined by

$$(3.1b) \qquad p_{X|Y=y}(x) = \frac{p(x, y)}{p_Y(y)},$$

assuming that $p_Y(y) > 0$. The probability mass function $p_{Y|X=x}(y)$ contains all the relevant probabilistic information concerning Y when $X = x$, whereas the probability mass function $p_{X|Y=y}(x)$ contains all the relevant probabilistic information concerning X when $Y = y$.

Example 3.1. In a study of voting habits, let $X = 1$ if a selected individual voted in a presidential election, and $X = 0$ if the individual did not vote in that election. Similarly, let $Y = 1$ if the individual voted in the next congressional election and $Y = 0$ if the individual did not vote in that election. The following is a hypothesized joint probability mass function for X and Y:

	y		
x	0	1	$p_X(x)$
0	0.225	0.125	0.350
1	0.300	0.350	0.650
$p_Y(y)$	0.525	0.475	1.000

The conditional probability mass functions $p_{Y|X=x}(y)$ are given by

$$p_{Y|X=0}(0) = \frac{p(0,0)}{p_X(0)} = \frac{0.225}{0.350} = 0.643, \qquad p_{Y|X=1}(0) = \frac{p(1,0)}{p_X(1)} = \frac{0.300}{0.650} = 0.462,$$

$$p_{Y|X=0}(1) = \frac{p(0,1)}{p_X(0)} = \frac{0.125}{0.350} = 0.357, \qquad p_{Y|X=1}(1) = \frac{p(1,1)}{p_X(1)} = \frac{0.350}{0.650} = 0.538,$$

which are presented in tabular form as follows:

y	0	1
$p_{Y|X=0}(y)$	0.643	0.357

y	0	1
$p_{Y|X=1}(y)$	0.462	0.538

A comparison of the marginal and conditional probability distributions for Y shows that knowledge of whether or not an individual has voted in a presidential election improves our prediction of whether or not that voter will vote in a subsequent congressional election.

Example 2.1 (continued). Suppose that we try to predict the rating Y given to a sample of wine by wine taster II from knowledge of the rating $X = x$ given to that wine by wine taster I. From Table 2.1 the conditional probability mass functions $p_{Y|X=x}$ for Y, given $X = 1, 2, 3, 4,$ and 5, are as shown in Table 3.1. If we do not know taster I's rating, then our best guess of taster II's rating is the mode $y = 3$ of the marginal distribution of Y, and our probability of correct prediction is $P\{Y = 3\} = 0.37$. On the other hand, from Table 3.1 if we know taster I's rating x, and predict that taster II's rating is also x, then the conditional probability $p_{Y|X=x}(x)$ of a correct prediction is never less than 0.444 and can be as high as 0.690.

In analogy to the discrete bivariate case [(3.1a) and (3.1b)], we define the conditional probability density function $f_{Y|X=x}(y)$ of Y given $X = x$, and the conditional probability density function $f_{X|Y=y}(x)$ of X given $Y = y$, in

Table 3.1: *Conditional Probability Mass Functions of Y Given X = x for the Wine-Tasting Experiment*

y	1	2	3	4	5
$p_{Y|X=1}(y)$	0.500	0.333	0.167	0.000	0.000
$p_{Y|X=2}(y)$	0.111	0.444	0.278	0.111	0.056
$p_{Y|X=3}(y)$	0.027	0.135	0.676	0.135	0.027
$p_{Y|X=4}(y)$	0.000	0.069	0.172	0.690	0.069
$p_{Y|X=5}(y)$	0.000	0.100	0.100	0.200	0.600

the case where the continuous random variables X and Y have joint probability density function $f(x, y)$. Thus,

(3.2a)
$$f_{Y|X=x}(y) = \frac{f(x, y)}{f_X(x)}, \quad \text{when } f_X(x) > 0,$$

and

(3.2b)
$$f_{X|Y=y}(x) = \frac{f(x, y)}{f_Y(y)}, \quad \text{when } f_Y(y) > 0.$$

To show that $f_{Y|X=x}(y)$ and $f_{X|Y=y}(x)$ are, indeed, probability density functions, note that each of these functions is nonnegative and that

$$\int_{-\infty}^{\infty} f_{Y|X=x}(y) \, dy = \int_{-\infty}^{\infty} \frac{f(x, y)}{f_X(x)} \, dy = \frac{\int_{-\infty}^{\infty} f(x, y) \, dy}{f_X(x)} = \frac{f_X(x)}{f_X(x)} = 1,$$

with a similar result holding for $f_{X|Y=y}(x)$.

An idea of the *shape* of the graph of $f_{Y|X=x^*}(y)$ can be obtained by drawing a plane perpendicular to the (x, y) plane along the line $x = x^*$ and observing the curves $g(y) = f_{X, Y}(x^*, y)$ cut by the plane on the surface formed by the graph of $f_{X, Y}(x, y)$. Similarly, the shape of the graph of $f_{X|Y=y^*}(x)$ can be obtained by drawing a plane perpendicular to the (x, y) plane along the line $y = y^*$ and observing the curve $h(x) = f_{X, Y}(x, y^*)$ cut by the plane on the surface formed by the graph of $f_{X, Y}(x, y)$ (see Figure 3.1).

Figure 3.1: *Shapes of the graphs* $g(y) = f(x^*, y)$ *and* $h(x) = f(x, y^*)$.

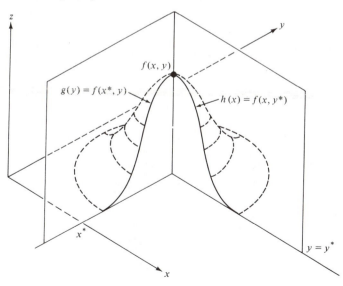

Example 2.3 (continued). If X and Y have the bivariate Dirichlet distribution (2.10), then, dividing (2.10) by (2.11a), the conditional probability density function of Y given $X = x$ is

$$(3.3) \qquad f_{Y|X=x}(y) = \begin{cases} \dfrac{y^{b-1}(1-x-y)^{c-1}}{(1-x)^{b+c-1}B(b,c)}, & \text{if } 0 \le y \le 1-x, \\ 0, & \text{otherwise,} \end{cases}$$

since $f(x, y) = 0$ when $x + y > 1$. The conditional density is undefined for $x < 0$ or $x > 1$.

Conditional Expected Values

To eliminate needless repetition, we henceforth consider only the case of conditional distributions of Y given $X = x$. Results for the conditional distribution of X given $Y = y$ are analogous, and can be easily obtained by interchanging the roles of X and Y in the relevant formulas.

Let $h(y)$ be any function of Y. Since $p_{Y|X=x}(y)$ and $f_{Y|X=x}(y)$ are legitimate probability mass functions and probability density functions, respectively, we can define the *conditional expected value of $h(Y)$ given $X = x$*, denoted $E[h(Y)|X = x]$, to be the expected value of $h(Y)$ with respect to these conditional probability distributions. That is, in the discrete case,

$$(3.4) \qquad E[h(Y)|X = x] = \sum_y h(y)p_{Y|X=x}(y),$$

whereas in the continuous case,

$$(3.5) \qquad E[h(Y)|X = x] = \int_{-\infty}^{\infty} h(y)f_{Y|X=x}(y)\, dy,$$

provided that the sum in (3.4) or the integral in (3.5) is well defined. Because $p_{Y|X=x}(y)$ and $f_{Y|X=x}(y)$ define legitimate univariate distributions, the conditional expected values $E[h(Y)|X = x]$ satisfy Properties 4.1 and 4.2 (Chapter 5, Section 4) of expected values. That is, for any functions $h_1(y)$ and $h_2(y)$ for which (3.4) or (3.5) are well defined, and for any constants a and b,

$$(3.6) \quad E[ah_1(Y) + bh_2(Y)|X = x] = aE[h_1(Y)|X = x] + bE[h_2(Y)|X = x],$$

and if $h(y) \ge 0$ for all y, then $E[h(Y)|X = x] \ge 0$.

Important special cases of the expected value computation are

$$(3.7) \qquad \mu_{Y|X=x} = E[Y|X = x],$$

which is the *conditional mean of Y given that $X = x$*, and

$$(3.8) \qquad \sigma^2_{Y|X=x} = E[(Y - \mu_{Y|X=x})^2|X = x],$$

which is the *conditional variance of Y given that* $X = x$. The two indices $\mu_{Y|X=x}$ and $\sigma^2_{Y|X=x}$ play the same roles in describing properties of the conditional distribution of Y given $X = x$ that μ_Y and σ^2_Y play in describing properties of the unconditional (or marginal) distribution $p_Y(y)$ or $f_Y(y)$ of Y. It follows from (3.4) that when X and Y are discrete random variables,

$$(3.9) \quad \mu_{Y|X=x} = \sum_y y p_{Y|X=x}(y), \qquad \sigma^2_{Y|X=x} = \sum_y (y - \mu_{Y|X=x})^2 p_{Y|X=x}(y),$$

whereas in the continuous case,

$$(3.10) \quad \mu_{Y|X=x} = \int_{-\infty}^{\infty} y f_{Y|X=x}(y)\, dy, \qquad \sigma^2_{Y|X=x} = \int_{-\infty}^{\infty} (y - \mu_{Y|X=x})^2 f_{Y|X=x}(y)\, dy.$$

It is often helpful to use the fact that

$$(3.11) \qquad \sigma^2_{Y|X=x} = E[Y^2 | X = x] - (\mu_{Y|X=x})^2,$$

analogous to the result proved for the unconditional variance σ^2_Y of Y in Chapter 5.

Prediction and the Regression Function

When predicting a future value of a discrete random variable Y, we can use the mode of the distribution of Y if we want the highest probability of being *exactly* right. Alternatively, we can choose the predictor c of Y to minimize $E[(Y - c)^2]$, the expected squared error. We noted that under this last criterion, the mean μ_Y of Y is the best predictor of Y, in that $\sigma^2_Y = E[(Y - \mu_Y)^2]$ is the smallest expected squared error of prediction.

When we know that the random variables X and Y are jointly distributed, and have observed $X = x$, the probability model upon which we base our predictions has changed, and $c = \mu_{Y|X=x}$ provides the best predictor of Y. That is,

$$(3.12) \quad \begin{aligned} \sigma^2_{Y|X=x} &= E[(Y - \mu_{Y|X=x})^2 | X = x] \\ &\le E[(Y - c)^2 | X = x], \quad \text{for all values of } c. \end{aligned}$$

In general, the values of the conditional mean $\mu_{Y|X=x}$ of Y given $X = x$ differ for different values of x. The function that assigns to every possible value of x of X the corresponding value of $\mu_{Y|X=x}$ is called the *regression function* of Y on X. In practical applications, the regression function gives a rule for predicting Y when we know the value x of X. Further, the regression function expresses, at least "on the average," the mathematical relationship existing between *the dependent variable Y and the independent variable X.*

Example 2.1 (continued). For our wine-tasting example, the regression function $\mu_{Y|X=x}$ is given by the expression

(3.13)
$$\mu_{Y|X=x} = \begin{cases} 1.667, & \text{if } x = 1, \\ 2.557, & \text{if } x = 2, \\ 3.000, & \text{if } x = 3, \\ 3.759, & \text{if } x = 4, \\ 4.300, & \text{if } x = 5, \end{cases}$$

obtained by applying (3.9) to Table 3.1. For example, the value $\mu_{Y|X=3}$ of the regression function $\mu_{Y|X=x}$ for $x = 3$ is

$$\mu_{Y|X=3} = \sum_{y=1}^{5} y p_{Y|X=3}(y)$$

$$= (1)(0.027) + (2)(0.135) + (3)(0.676) + (4)(0.135) + (5)(0.027)$$

$$= 3.000.$$

If we predict Y by $\mu_{Y|X=3}$ every time that $X = 3$ is observed, the expected squared error of prediction is

$$\sigma^2_{Y|X=3} = (1 - 3)^2(0.027) + (2 - 3)^2(0.135) + (3 - 3)^2(0.676)$$
$$+ (4 - 3)^2(0.135) + (5 - 3)^2(0.027) = 0.486.$$

The regression function (3.13) is graphed in Figure 3.2.

Figure 3.2: *The regression function* $\mu_{Y|X=x}$ *for the joint probability distribution given in Table 2.2 of Example 2.1.*

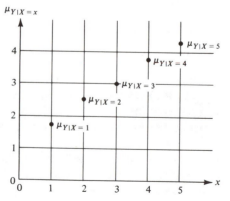

Example 2.3 (continued). If X and Y have the bivariate Dirichlet distribution (2.10), then (3.3) gives the conditional probability density function $f_{Y|X=x}(y)$ of Y given $X = x$. Thus, for $0 \le x \le 1$, the regression function $\mu_{Y|X=x}$ is given by

$$\mu_{Y|X=x} = \int_{-\infty}^{\infty} y f_{Y|X=x}(y)\, dy$$

$$= \frac{1}{(1-x)^{b+c-1} B(b, c)} \int_{0}^{1-x} y[y^{b-1}(1-x-y)^{c-1}]\, dy$$

(3.14)
$$= (1-x) \int_{0}^{1} \frac{u^b(1-u)^{c-1}\, du}{B(b, c)}$$

$$= (1-x) \frac{B(b+1, c)}{B(b, c)} = (1-x)\left(\frac{b}{b+c}\right),$$

where we have made the change of variables from y to $u = y/(1-x)$. With the same change of variables,

$$\sigma^2_{Y|X=x} = \frac{1}{(1-x)^{b+c-1} B(b, c)} \int_{0}^{1} \left[y - (1-x)\frac{b}{b+c} \right]^2 (y^{b-1}(1-x-y)^{c-1})\, dy$$

$$= (1-x)^2 \int_{0}^{1} \frac{\left(u - \dfrac{b}{b+c}\right)^2 u^{b-1}(1-u)^{c-1}}{B(b, c)}\, du$$

$$= (1-x)^2 \sigma^2_U,$$

where U has a beta distribution with $r = b$, $s = c$. Hence,

(3.15)
$$\sigma^2_{Y|X=x} = (1-x)^2 \frac{bc}{(b+c)^2(b+c+1)}.$$

Probabilistic Independence

The notion of probabilistic independence as applied to events, and to trials of a random experiment, was discussed in Chapter 3. Two jointly distributed random variables X and Y are said to be *probabilistically independent* if the conditional distribution of Y given $X = x$ is the same as the marginal (unconditional) distribution of Y for all possible values of x. Thus, if X and Y have a joint probability mass function, then X and Y are independent if

(3.16)
$$p_{Y|X=x}(y) = p_Y(y), \quad \text{all } x, \text{ all } y.$$

If X and Y have a joint probability density function, then X and Y are independent if

(3.17) $$f_{Y|X=x}(y) = f_Y(y), \quad \text{all } x, \text{ all } y.$$

Switching the roles of X and Y, we say that X and Y are independent if

(3.18) $$p_{X|Y=y}(x) = p_X(x), \quad \text{all } x, \text{ all } y,$$

in the discrete case;

(3.19) $$f_{X|Y=y}(x) = f_X(x), \quad \text{all } x, \text{ all } y,$$

in the continuous case. Indeed, (3.16) and (3.18) are equivalent definitions of independence in the discrete case, and (3.17) and (3.19) are equivalent definitions of independence in the continuous case. The essence of the concept of independence lies in the fact that X and Y are independent if and only if knowledge of the values of one of these two random variables provides no new probabilistic information whatsoever about the values of the other variable.

It follows directly from (3.16) and (3.17) that if $h(y)$ is any function of y, and X and Y are independent, then

(3.20) $$E[h(Y)|X = x] = E[h(Y)], \quad \text{for all } x.$$

In particular,

(3.21) $$\mu_{Y|X=x} = \mu_Y, \qquad \sigma^2_{Y|X=x} = \sigma^2_Y, \quad \text{all } x,$$

so that the regression function is constant at the unconditional mean μ_Y of Y. Similar results hold for $E[h(X)|Y = y]$, $\mu_{X|Y=y}$, and $\sigma^2_{X|Y=y}$ when X and Y are independent.

The following result is typically used as a computationally more convenient way of verifying whether two random variables X and Y are independent. If X and Y have joint probability mass function $p(x, y)$ or joint probability density function $f(x, y)$, then X and Y are independent if and only if

(3.22) $$p(x, y) = p_X(x)p_Y(y), \quad \text{all } x, \text{ all } y,$$

in the discrete case, or

(3.23) $$f(x, y) = f_X(x)f_Y(y), \quad \text{all } x, \text{ all } y,$$

in the continuous case.

To see that definitions (3.16), (3.18), and (3.22) are equivalent, note that if (3.16) holds, then

$$p_Y(y) = p_{Y|X=x}(y) = \frac{p(x, y)}{p_X(x)}, \quad \text{for all } x, \text{ all } y,$$

from which, multiplying this equality by $p_X(x)$, it follows that (3.22) holds. On the other hand, if (3.22) holds, then

$$p_{Y|X=x}(y) = \frac{p(x, y)}{p_X(x)} = \frac{p_X(x)p_Y(y)}{p_X(x)} = p_Y(y), \quad \text{all } x, \text{ all } y,$$

so that (3.16) holds. Interchanging the roles of X and Y, we can use the same steps to show that (3.18) and (3.22) are equivalent. A similar argument verifies the equivalence of (3.17), (3.19), and (3.23).

Two random variables that are not independent are said to be *dependent*.

Table 2.4 provides us with examples of both independent and dependent discrete random variables. In Table 2.4(b), X and Y are independent since $p_X(x)p_Y(y) = p(x, y)$ for $x, y = 0, 1$. On the other hand, in Table 2.4(a), X and Y are dependent since, for example, $p_X(1)p_Y(1) = 0.60(0.70) \neq 0.40 = p(1, 1)$.

Example 3.2. Suppose that X and Y have the joint probability density function

$$f(x, y) = \begin{cases} e^{-(x+y)}, & \text{if } x > 0 \text{ and } y > 0, \\ 0, & \text{otherwise.} \end{cases}$$

Then for $x > 0$,

$$f_X(x) = \int_{-\infty}^{\infty} f(x, y) \, dy = \int_{0}^{\infty} e^{-(x+y)} \, dy = e^{-x} \int_{0}^{\infty} e^{-y} \, dy = e^{-x},$$

while for $x < 0, f_X(x) = 0$. Thus,

$$f_X(x) = \begin{cases} e^{-x}, & \text{if } x > 0, \\ 0, & \text{otherwise,} \end{cases}$$

and in similar fashion

$$f_Y(y) = \begin{cases} e^{-y}, & \text{if } y > 0, \\ 0, & \text{otherwise,} \end{cases}$$

so that the marginal distributions of X and Y are exponential distributions with parameter $\theta = 1$. Now when $x < 0$ or $y < 0$, $f_X(x)f_Y(y) = 0 = f(x, y)$. If $x > 0, y > 0$, then

$$f_X(x)f_Y(y) = e^{-x}e^{-y} = e^{-(x+y)} = f(x, y).$$

Since $f(x, y) = f_X(x)f_Y(y)$ for all x, y, X and Y are independent. (Example 2.4 provides another example of independence for continuous random variables.)

Let A be any collection of possible values for X and let B be any collection of possible values for Y. If X and Y are independent, then

(3.24) $P\{X \text{ in } A \text{ and } Y \text{ in } B\} = P\{X \text{ in } A\}P\{Y \text{ in } B\}.$

To prove (3.24) in the continuous case, let $R = \{(x, y): x \text{ in } A, y \text{ in } B\}$. Then from (3.23),

$$P\{X \text{ in } A \text{ and } Y \text{ in } B\} = P(R) = \iint\limits_{(x, y) \text{ in } R} f(x, y) \, dx \, dy$$

$$= \int\limits_{x \text{ in } A} \int\limits_{y \text{ in } B} f_X(x)f_Y(y) \, dy \, dx$$

$$= \int\limits_{x \text{ in } A} f_X(x) \, dx \int\limits_{y \text{ in } B} f_Y(y) \, dy$$

$$= P\{X \text{ in } A\}P\{Y \text{ in } B\}.$$

Proof of (3.24) in the discrete case follows similarly from (3.22). It follows directly from (3.24), letting $A = \{x: x \le x_0\}$, $B = \{y: y \le y_0\}$, that if X and Y are independent random variables, then

(3.25) $F_{X, Y}(x_0, y_0) = F_X(x_0)F_Y(y_0), \quad \text{all } x_0, \text{ all } y_0.$

The converse of this assertion is also true (Exercise 10), so that (3.25) provides still another equivalent definition of independence.

Note that when two random variables X and Y are independent, their joint distribution can be obtained from their marginal distributions [see (3.22), (3.23), and (3.25)]. This fact is in contrast to the general case, where we have seen that the marginal distributions of two random variables X and Y cannot, by themselves, determine the joint distribution of X and Y.

An example of independent random variables occurs when a random experiment is performed in physically independent stages and the random variable X depends only on the first stage, while the random variable Y depends only on the second stage. For example, assume that two individuals are selected, *with replacement* between selections, at random from a population. Let X denote the age of the first individual and Y denote the age of the second individual. Then it can be reasonably assumed that the two random variables are independent.

However, X and Y may be independent even when the experiment is not performed in physically independent stages. For example, suppose that one individual is selected at random from a given population. Let X denote yearly income and Y the length of time since the last birthday of the individual selected. It would seem reasonable to expect that these two random variables would be independent.

4. *DESCRIPTIVE INDICES OF BIVARIATE DISTRIBUTIONS*

For univariate distributions, the mean and variance are frequently used as descriptive indices of location and of dispersion, respectively. The measure of location usually used for bivariate distributions of random variables X and Y is the pair (μ_X, μ_Y), where μ_X is the mean of the marginal distribution of X and μ_Y is the mean of the marginal distribution of Y. The pair (μ_X, μ_Y) is called the *joint mean* or *mean vector* of the joint distribution of X and Y.

   Although μ_X and μ_Y can be computed from the respective marginal distributions of X and of Y, these quantities can also be computed directly from the joint distribution of X and Y. To do so, it is helpful to define the bivariate generalization of the notion of the expected value of a function.

Expected Value of a Function of Two Jointly Distributed Random Variables

   Let $h(x, y)$ be any function of x and y. Then if the random variables X and Y have joint probability mass function $p(x, y)$, the *expected value* of $h(X, Y)$ is defined to be

(4.1)
$$E[h(X, Y)] = \sum_{(x, y)} h(x, y)p(x, y)$$
$$= \sum_x \sum_y h(x, y)p(x, y) = \sum_y \sum_x h(x, y)p(x, y),$$

provided that one of the sums in (4.1) is well defined and finite. If X and Y have joint probability density function $f(x, y)$, then

$$(4.2) \quad E[h(X, Y)] = \int_{-\infty}^{\infty} \int_{-\infty}^{\infty} h(x, y)f(x, y)\, dx\, dy = \int_{-\infty}^{\infty} \int_{-\infty}^{\infty} h(x, y)f(x, y)\, dy\, dx,$$

provided that one of the integrals in (4.2) is well defined and finite.

   It follows directly from properties of integrals and sums (as in the proofs of Properties 4.1 and 4.2 in Chapter 5) that for any two functions $h_1(x, y)$ and $h_2(x, y)$ having expected values, and any constants a and b,

$$(4.3) \qquad E[ah_1(X, Y) + bh_2(X, Y)] = aE[h_1(X, Y)] + bE[h_2(X, Y)],$$

and that if $h(x, y) > 0$ for all x and y, then

$$(4.4) \qquad\qquad\qquad E[h(X, Y)] \geq 0.$$

   When $h(x, y)$ is a function solely of one variable (say x), then the bivariate generalization of the definition of an expected value given by (4.1) or (4.2) yields the same result as the definition of the expected value relative to the marginal distribution (of X). That is, if $h(x, y) = g(x)$, then

$$E[g(X)] = \int\limits_{-\infty}^{\infty} \int\limits_{-\infty}^{\infty} g(x)f(x, y)\, dy\, dx = \int\limits_{-\infty}^{\infty} g(x)\left[\int\limits_{-\infty}^{\infty} f(x, y)\, dy\right] dx$$

$$= \int\limits_{-\infty}^{\infty} g(x)f_X(x)\, dx$$

in the continuous case, and

$$E[g(X)] = \sum_x \sum_y g(x)p(x, y) = \sum_x g(x)\left[\sum_y p(x, y)\right] = \sum_x g(x)p_X(x)$$

in the discrete case. Hence, μ_X, σ_X^2, μ_Y, and σ_Y^2 can either be computed as expected values relative to the appropriate marginal distribution, or as expected values

$$\mu_X = E[X], \qquad \mu_Y = E[Y], \qquad \sigma_X^2 = E[(X - \mu_X)^2], \qquad \sigma_Y^2 = E[(Y - \mu_Y)^2],$$

relative to the joint distribution of X and Y. Using the latter characterization of μ_X and μ_Y, and (4.3) with $h_1(x, y) = x$, $h_2(x, y) = y$, we obtain

(4.5) $$\mu_{aX+bY} = E[aX + bY] = a\mu_X + b\mu_Y.$$

The Covariance

In the search for measures of dispersion for bivariate distributions, the variances σ_X^2, σ_Y^2 of X and Y come naturally to our attention. However, these indices are determined only by the marginal distributions of X and Y, respectively. Since the marginal distributions of X and Y do not determine the joint distribution, it is clear that σ_X^2 and σ_Y^2 by themselves cannot tell us how X and Y vary jointly. That is, we lack a measure of the *interrelationship* or *covariation* of X and Y.

Note that σ_X^2 measures dispersion along the x-direction in the plane (Figure 4.1) and that σ_Y^2 measures dispersion along the y-direction in the plane (Figure 4.2). Correspondingly, σ_{aX+bY}^2 should measure dispersion along the $(ax + by)$-direction in the plane. A knowledge of σ_{aX+bY}^2 for all values of a and b would allow us to infer considerable information about the joint variability of X and Y. However, using (4.3) repeatedly,

$$\sigma_{aX+bY}^2 = E[(aX + bY - \mu_{aX+bY})^2] = E[(aX + bY - a\mu_X - b\mu_Y)^2]$$

(4.6) $$= E[a^2(X - \mu_X)^2 + 2ab(X - \mu_X)(Y - \mu_Y) + b^2(Y - \mu_Y)^2]$$

$$= a^2\sigma_X^2 + 2abE[(X - \mu_X)(Y - \mu_Y)] + b^2\sigma_Y^2.$$

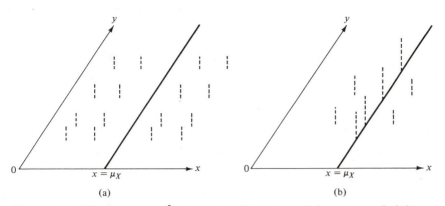

Figure 4.1: *The variance σ_X^2 shows the dispersion of the joint probability mass function $p(x, y)$ around the line $x = \mu_X$. The dispersion in* (a) *is larger than the dispersion in* (b).

Since σ_X^2 and σ_Y^2 measure how X and Y vary individually, the index

(4.7) $$\sigma_{X, Y} = E[(X - \mu_X)(Y - \mu_Y)],$$

called the *covariance* of X and Y, measures how X and Y covary jointly.
Note that

(4.8)
$$\begin{aligned}
\sigma_{aX+b, \, cY+d} &= E[(aX + b - \mu_{aX+b})(cY + d - \mu_{cY+d})] \\
&= E[(aX + b - a\mu_X - b)(cY + d - c\mu_Y - d)] \\
&= E[ac(X - \mu_X)(Y - \mu_Y)] = ac\sigma_{X, Y}.
\end{aligned}$$

Figure 4.2: *The variance σ_Y^2 shows the dispersion of the joint probability mass function $p(x, y)$ around the line $y = \mu_Y$. The dispersion in* (a) *is larger than the dispersion in* (b).

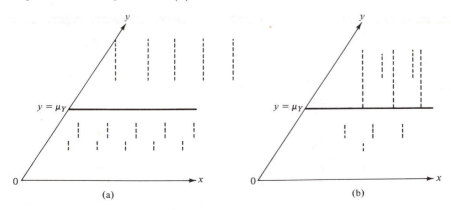

It is also frequently convenient to use the alternative computational formula

(4.9) $$\sigma_{X,Y} = E[XY] - \mu_X \mu_Y$$

for computing the covariance $\sigma_{X,Y}$. To verify (4.9), note that

$$\sigma_{X,Y} = E[(X - \mu_X)(Y - \mu_Y)] = E[XY] - \mu_X E[Y] - \mu_Y E[X] + \mu_X \mu_Y$$
$$= E[XY] - \mu_X \mu_Y.$$

If $X = Y$, then by substitution,

$$\sigma_{X,X} = E[(X - \mu_X)(X - \mu_X)] = E[(X - \mu_X)^2] = \sigma_X^2,$$

and similarly $\sigma_{Y,Y} = \sigma_Y^2$.

The covariance can be positive or negative. It is positive when large values of X go with large values of Y and small values of X with small values of Y. It is negative when X and Y move in opposite directions.

To illustrate this point, if X and Y denote a student's high school and college grade-point averages, respectively, then we expect the covariance of X and Y to be positive. If X denotes the amount of minerals in a water supply and Y is a measure of pipe corrosion, we also expect a positive covariance. On the other hand, if X denotes the amount of alcohol in the blood and Y is a measure of motor coordination, the covariance between X and Y should be negative.

The following examples illustrate the computation of $\sigma_{X,Y}$.

Example 4.1. Suppose that X and Y have the joint probability mass function $p(x, y)$ summarized by a contingency table:

			y		
x		1	2	3	$p_X(x)$
1		0.20	0.10	0.01	0.31
2		0.15	0.30	0.06	0.51
3		0.03	0.05	0.10	0.18
$p_Y(y)$		0.38	0.45	0.17	

From the marginal probability mass functions $p_X(x)$ and $p_Y(y)$,

$$\mu_X = 1.87, \qquad \mu_Y = 1.79.$$

Also,

$$E[XY] = \sum_{x=1}^{3} \sum_{y=1}^{3} xy \, p(x, y)$$
$$= (1)(1)(0.20) + (1)(2)(0.10) + \cdots + (3)(2)(0.05) + (3)(3)(0.10)$$
$$= 3.58,$$

so that using (4.9),

$$\sigma_{X,Y} = E[XY] - \mu_X \mu_Y = 3.58 - 1.87(1.79) = 0.2327.$$

Example 2.3 (continued). When X and Y have a joint Dirichlet distribution with joint probability density function

$$f(x, y) = \frac{\Gamma(a + b + c)}{\Gamma(a)\Gamma(b)\Gamma(c)} x^{a-1} y^{b-1} (1 - x - y)^{c-1}, \qquad 0 \leq x, y, x + y \leq 1,$$

and 0 otherwise, then we have seen that the marginal distributions of X and Y are each beta distributions with parameters $r = a$, $s = b + c$, and $r = b$, $s = a + c$, respectively. Since the mean of a beta distribution with parameters r and s is $r/(r + s)$,

$$\mu_X = \frac{a}{a + b + c}, \qquad \mu_Y = \frac{b}{a + b + c}.$$

Let

$$K = \frac{\Gamma(a + b + c)}{\Gamma(a)\Gamma(b)\Gamma(c)}.$$

Then

$$E[XY] = K \int_{-\infty}^{\infty} \int_{-\infty}^{\infty} xy f(x, y) \, dy \, dx$$

$$= K \int_{0}^{1} \left[\int_{0}^{1-x} x^a y^b (1 - x - y)^{c-1} \, dy \right] dx$$

$$= K \left[\int_{0}^{1} x^a (1 - x)^{b+c} \, dx \right] \left[\int_{0}^{1} u^b (1 - u)^{c-1} \, du \right]$$

$$= KB(a + 1, b + c + 1) B(b + 1, c)$$

$$= \frac{\Gamma(a + b + c)}{\Gamma(a)\Gamma(b)\Gamma(c)} \frac{\Gamma(a + 1)\Gamma(b + c + 1)}{\Gamma(a + b + c + 2)} \frac{\Gamma(b + 1)\Gamma(c)}{\Gamma(b + c + 1)}$$

$$= \frac{ab}{(a + b + c + 1)(a + b + c)},$$

so that, using (4.9),

$$\sigma_{X,Y} = E[XY] - \mu_X \mu_Y$$

$$= \frac{ab}{(a + b + c)(a + b + c + 1)} - \frac{ab}{(a + b + c)^2}$$

$$= \frac{-ab}{(a + b + c)^2(a + b + c + 1)},$$

which is always negative. It is not surprising that $\sigma_{X,Y}$ is negative, since $f(x, y)$ puts nonzero density only on points (x, y) for which $0 \le x + y \le 1$. Thus, if X is large (near 1), Y must be small (near 0) in order for $X + Y$ to remain ≤ 1. Similarly, if Y is large, X must be small. Since X and Y move in opposite directions, we expect that $\sigma_{X,Y} < 0$.

The Correlation Coefficient

As a measure of how closely X and Y covary, the covariance possesses one main defect: it is sensitive to the scales of measurement used for X and Y. For example, if X represents the length in feet of a randomly selected rod and Y denotes an individual's guess of that length, measured in feet, we might find that $\sigma_{X,Y} = 0.01$. However, if we measure in inches, so that we observe $X^* = 12X$, $Y^* = 12Y$, then, using (4.8),

$$\sigma_{X^*, Y^*} = \sigma_{12X, 12Y} = (12)(12)\sigma_{X,Y} = 144(0.01) = 1.44.$$

Although the experiment has not changed, the index of covariance has changed from 0.01 to 1.44. To remove this dependence on scale of measurement, consider the index

(4.10) $$\rho_{X,Y} = \frac{\sigma_{X,Y}}{\sigma_X \sigma_Y}.$$

Then for $a > 0$, $c > 0$,

(4.11) $$\rho_{aX+b, cY+b} = \frac{\sigma_{aX+b, cY+d}}{\sigma_{aX+b}\sigma_{cY+d}} = \frac{ac\sigma_{XY}}{a\sigma_X c\sigma_Y} = \frac{\sigma_{XY}}{\sigma_X \sigma_Y} = \rho_{X,Y}.$$

The index $\rho_{X,Y}$ is called the *correlation coefficient*. From (4.11) we see that $\rho_{X,Y}$ has a value independent of the units of measurement used for X and Y. Since the signs of $\rho_{X,Y}$ and $\sigma_{X,Y}$ are the same [see (4.10)], interpretation of the nature of the relation between X and Y using the sign of $\rho_{X,Y}$ is the same as that previously outlined for $\sigma_{X,Y}$.

The correlation coefficient $\rho_{X,Y}$ assumes values between -1 and 1. To demonstrate this fact, consider the random variable

$$Z = \left(\frac{1}{\sigma_Y}\right)Y - \left(\frac{\rho_{X,Y}}{\sigma_X}\right)X.$$

Using (4.6) with $a = (1/\sigma_Y)$, $b = (-\rho_{X,Y}/\sigma_X)$, the variance of Z is

$$\sigma_Z^2 = \left(\frac{1}{\sigma_Y}\right)^2 \sigma_Y^2 + 2\left(\frac{1}{\sigma_Y}\right)\left(\frac{-\rho_{X,Y}}{\sigma_X}\right)\sigma_{X,Y} + \left(\frac{-\rho_{X,Y}}{\sigma_X}\right)^2 \sigma_X^2$$

$$= 1 - \rho_{X,Y}^2.$$

Since σ_Z^2 must be nonnegative, it follows that $1 - \rho_{X,Y}^2 \ge 0$, or

(4.12) $$-1 \le \rho_{X,Y} \le 1.$$

Note that if $\rho_{X,Y} = \pm 1$, then $\sigma_Z^2 = 0$ and Z does not vary from its mean μ_Z. That is, $P\{Z = \mu_Z\} = 1$ or, equivalently,

$$Y = \left(\frac{\sigma_Y \rho_{X,Y}}{\sigma_X}\right) X + \sigma_Y \mu_Z = \left(\pm \frac{\sigma_Y}{\sigma_X}\right) X + \sigma_Y \mu_Z$$

with probability equal to one. Thus, if $\rho_{X,Y} = \pm 1$, Y is (with probability equal to one) equal to a nonconstant linear function of X. On the other hand, if $Y = aX + b$, $a \neq 0$, then

$$\rho_{X,Y} = \rho_{X, aX+b} = \frac{\sigma_{X, aX+b}}{\sigma_X \sigma_{aX+b}} = \frac{a\sigma_X^2}{|a|\sigma_X^2} = \frac{a}{|a|},$$

so that $\rho_{X,Y} = 1$ if $a > 0$, and $\rho_{X,Y} = -1$ if $a < 0$. We conclude that $\rho_{X,Y} = \pm 1$ *if and only if* Y is, with probability equal to one, a nonconstant linear function of X.

Of course, as we already remarked in Section 2, two jointly continuous random variables X and Y have zero probability of exactly satisfying *any* linear function. For such variables, and in general for any pair (X, Y) of jointly distributed random variables, the foregoing interpretation of the extreme values of $\rho_{X,Y}$ serves only to help interpret the intermediate values of $\rho_{X,Y}$. That is, this analysis of the meaning of the extreme values ± 1 of $\rho_{X,Y}$ indicates that *the closer the probability mass or density concentrates around a line $L = \{(x, y): y = ax + b\}$ with $a > 0$, the closer the value of $\rho_{X,Y}$ is to $+1$, while the closer the probability mass or density concentrates around a line L with $a < 0$, the closer the value of $\rho_{X,Y}$ is to -1.*

Example 4.2. Two individuals rate a common list of items. If one individual's ratings are closely predictable from the ratings of the other individual, then there is good reason to believe that both raters use the same criteria and perceptions in making judgments. If the ratings of the two individuals are independent of one another, then the judgments of the individuals are not based on a common behavioral dimension.

The following contingency tables (Table 4.1) give some possible bivariate probability distributions for the ranks X and Y on a three-point scale given by two raters. The marginal probability mass functions $p_X(x)$ and $p_Y(y)$ are kept constant from table to table; the differences between the contingency tables (and the associated values of the correlation coefficient $\rho_{X,Y}$) thus reflect differences in the interrelationships between the variables X and Y. Comparing these tables, we see that $\rho_{X,Y}$ becomes more positive as there is more probability mass close to the rising line $L = \{(x, y): y = x\}$, and $\rho_{X,Y}$ becomes more negative as more probability mass is assigned to points (x, y) near the falling line $L = \{(x, y): y = 4 - x\}$.

Similar analyses of the correspondence between the values of $\rho_{X,Y}$ and the strength of the probabilistic relationship between X and Y are given in Section 5 for the bivariate normal distribution.

Table 4.1: *Joint Probability Model for the Ratings of Two Individuals*

(a) $\rho_{X,Y} = 0$

x	y 1	2	3	Totals
3	$\frac{1}{9}$	$\frac{1}{9}$	$\frac{1}{9}$	$\frac{1}{3}$
2	$\frac{1}{9}$	$\frac{1}{9}$	$\frac{1}{9}$	$\frac{1}{3}$
1	$\frac{1}{9}$	$\frac{1}{9}$	$\frac{1}{9}$	$\frac{1}{3}$
Totals	$\frac{1}{3}$	$\frac{1}{3}$	$\frac{1}{3}$	1

(b) $\rho_{X,Y} = \frac{1}{2}$

x	y 1	2	3	Totals
3	$\frac{1}{18}$	$\frac{1}{18}$	$\frac{4}{18}$	$\frac{1}{3}$
2	$\frac{1}{18}$	$\frac{4}{18}$	$\frac{1}{18}$	$\frac{1}{3}$
1	$\frac{4}{18}$	$\frac{1}{18}$	$\frac{1}{18}$	$\frac{1}{3}$
Totals	$\frac{1}{3}$	$\frac{1}{3}$	$\frac{1}{3}$	1

(c) $\rho_{X,Y} = -\frac{1}{2}$

x	y 1	2	3	Totals
3	$\frac{4}{18}$	$\frac{1}{18}$	$\frac{1}{18}$	$\frac{1}{3}$
2	$\frac{1}{18}$	$\frac{4}{18}$	$\frac{1}{18}$	$\frac{1}{3}$
1	$\frac{1}{18}$	$\frac{1}{18}$	$\frac{4}{18}$	$\frac{1}{3}$
Totals	$\frac{1}{3}$	$\frac{1}{3}$	$\frac{1}{3}$	1

(d) $\rho_{X,Y} = \frac{5}{9}$

x	y 1	2	3	Totals
3	$\frac{1}{27}$	$\frac{2}{27}$	$\frac{6}{27}$	$\frac{1}{3}$
2	$\frac{2}{27}$	$\frac{5}{27}$	$\frac{2}{27}$	$\frac{1}{3}$
1	$\frac{6}{27}$	$\frac{2}{27}$	$\frac{1}{27}$	$\frac{1}{3}$
Totals	$\frac{1}{3}$	$\frac{1}{3}$	$\frac{1}{3}$	1

(e) $\rho_{X,Y} = -\frac{5}{9}$

x	y 1	2	3	Totals
3	$\frac{6}{27}$	$\frac{2}{27}$	$\frac{2}{27}$	$\frac{1}{3}$
2	$\frac{2}{27}$	$\frac{5}{27}$	$\frac{2}{27}$	$\frac{1}{3}$
1	$\frac{1}{27}$	$\frac{2}{27}$	$\frac{6}{27}$	$\frac{1}{3}$
Totals	$\frac{1}{3}$	$\frac{1}{3}$	$\frac{1}{3}$	1

(f) $\rho_{X,Y} = \frac{2}{3}$

x	y 1	2	3	Totals
3	$\frac{1}{36}$	$\frac{2}{36}$	$\frac{9}{36}$	$\frac{1}{3}$
2	$\frac{2}{36}$	$\frac{8}{36}$	$\frac{2}{36}$	$\frac{1}{3}$
1	$\frac{9}{36}$	$\frac{2}{36}$	$\frac{1}{36}$	$\frac{1}{3}$
Totals	$\frac{1}{3}$	$\frac{1}{3}$	$\frac{1}{3}$	1

(g) $\rho_{X,Y} = -\frac{2}{3}$

x	y 1	2	3	Totals
3	$\frac{9}{36}$	$\frac{2}{36}$	$\frac{1}{36}$	$\frac{1}{3}$
2	$\frac{2}{36}$	$\frac{8}{36}$	$\frac{2}{36}$	$\frac{1}{3}$
1	$\frac{1}{36}$	$\frac{2}{36}$	$\frac{9}{36}$	$\frac{1}{3}$
Totals	$\frac{1}{3}$	$\frac{1}{3}$	$\frac{1}{3}$	1

Independence and Correlation

The correlation coefficient can often be used to establish that two random variables X and Y are dependent. Indeed, *if X and Y are independent, then $\rho_{X, Y} = 0$.* Thus, if we find that $\rho_{X, Y} \neq 0$, it must be the case that X and Y are dependent.

Assume that X and Y are discrete random variables with joint probability mass function $p(x, y)$. Let $g_1(x)$ be any function of x and $g_2(y)$ be any function of y for which $E[g_1(X)]$ and $E[g_2(Y)]$ are well defined. If X and Y are independent, then

$$E[g_1(X)g_2(Y)] = \sum_x \sum_y g_1(x)g_2(y)p(x, y)$$

$$= \sum_x \sum_y g_1(x)g_2(y)p_X(x)p_Y(y)$$

$$= \left[\sum_x g_1(x)p_X(x)\right]\left[\sum_y g_2(y)p_Y(y)\right]$$

$$= E[g_1(X)]E[g_2(Y)].$$

Similarly, if X and Y are continuous random variables with a joint probability density function $f(x, y)$, and X and Y are independent, then

$$E[g_1(X)g_2(Y)] = \int_{-\infty}^{\infty} \int_{-\infty}^{\infty} g_1(x)g_2(y)f(x, y)\, dx\, dy$$

$$= \int_{-\infty}^{\infty} \int_{-\infty}^{\infty} g_1(x)g_2(y)f_X(x)f_Y(y)\, dx\, dy$$

$$= \left[\int_{-\infty}^{\infty} g_1(x)f_X(x)\, dx\right]\left[\int_{-\infty}^{\infty} g_2(x)f_Y(y)\, dy\right]$$

$$= E[g_1(X)]E[g_2(Y)].$$

Applying these results for $g_1(x) = x$, $g_2(y) = y$, we see that if X and Y are independent, then

$$E[XY] = E[X]E[Y] = \mu_X\mu_Y$$

so that $\sigma_{X, Y} = 0$ and

$$\rho_{X, Y} = \frac{\sigma_{X, Y}}{\sigma_X \sigma_Y} = 0.$$

This proves that independence of X and Y implies that $\rho_{X, Y} = 0$.

On the other hand, if we know that $\rho_{X,Y} = 0$, we cannot assert in general that X and Y are independent. For example, consider the case where X and Y have joint probability mass function

$$p(x, y) = \begin{cases} \frac{1}{3}, & \text{if } (x, y) = (-1, 1), (1, 1), \\ \frac{1}{6}, & \text{if } (x, y) = (-2, 4), (2, 4), \\ 0, & \text{otherwise.} \end{cases}$$

Note that for this probability distribution,

$$P\{Y = X^2\} = 1,$$

so that X and Y are not only probabilistically dependent, they are also functionally dependent. [To show that X and Y are probabilistically dependent, note that $p_X(-1) = \frac{1}{3}$, $p_Y(1) = \frac{1}{3} + \frac{1}{3} = \frac{2}{3}$, but $p(-1, 1) = \frac{1}{3} \neq \left(\frac{1}{3}\right) \times \left(\frac{2}{3}\right) = p_X(-1)p_Y(1)$.] However, it is easily shown that $\mu_X = 0$ and $E[XY] = 0$. Hence,

$$\sigma_{X,Y} = E[XY] - \mu_X \mu_Y = 0,$$

and thus $\rho_{X,Y} = 0$. Thus, we have shown that *uncorrelated random variables are not necessarily independent.*

5. *THE BIVARIATE NORMAL DISTRIBUTION*

Two continuous random variables X and Y have a bivariate normal distribution if the graph of their joint probability density function $f(x, y)$ has a shape similar to that shown in Figure 5.1.

Figure 5.1: *The bivariate normal probability density function.*

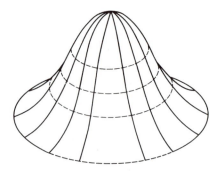

To be precise, the joint probability density function $f(x, y)$ of a bivariate normal distribution has the form:

(5.1)

$$f(x, y) = \frac{\exp\left\{-\frac{1}{2}\left[\frac{1}{1 - \rho^2}\right]\left[\frac{(x - \mu_1)^2}{\sigma_1^2} - 2\rho\frac{(x - \mu_1)(y - \mu_2)}{\sigma_1\sigma_2} + \frac{(y - \mu_2)^2}{\sigma_2^2}\right]\right\}}{2\pi\sigma_1\sigma_2\sqrt{1 - \rho^2}}$$

for $-\infty < x, y < \infty$. The quantities $\mu_1, \mu_2, \sigma_1, \sigma_2, \rho$ are the parameters of the bivariate normal distribution. The pair (μ_1, μ_2) tells us where the center of the "mountain" shown in Figure 5.1 is located in the (x, y)-plane, while σ_1^2 and σ_2^2 measure the spread of this "mountain" in the x-direction and y-direction, respectively. Finally, the parameter ρ, $-1 < \rho < 1$, determines the shape and orientation on the (x, y)-plane of the "mountain" shown in Figure 5.1. As may be inferred from the notation, μ_1 and σ_1^2 are the mean and variance of X, μ_2 and σ_2^2 are the mean and variance of Y, and ρ is the correlation coefficient between X and Y. If X and Y have the joint bivariate probability density function (5.1), we write $(X, Y) \sim \text{BVN}((\mu_1, \mu_2), (\sigma_1^2, \sigma_2^2, \rho))$, in a notation analogous to that used for the univariate normal distribution in Chapter 7.

Marginal Distributions

Consider the quadratic function

(5.2) $\quad Q(x, y) = \dfrac{1}{1 - \rho^2}\left[\dfrac{(x - \mu_1)^2}{\sigma_1^2} - \dfrac{2\rho(x - \mu_1)(y - \mu_2)}{\sigma_1\sigma_2} + \dfrac{(y - \mu_2)^2}{\sigma_2^2}\right],$

which appears in the exponent in (5.1). Note that

$$Q(x, y) - \frac{(x - \mu_1)^2}{\sigma_1^2}$$

(5.3)

$$= \frac{1}{\sigma_2^2(1 - \rho^2)}\left[(y - \mu_2)^2 - 2(y - \mu_2)\frac{\rho\sigma_2}{\sigma_1}(x - \mu_1) + \frac{\rho^2\sigma_2^2}{\sigma_1^2}(x - \mu_1)^2\right]$$

$$= \frac{[y - \mu_2 - (\rho\sigma_2/\sigma_1)(x - \mu_1)]^2}{\sigma_2^2(1 - \rho^2)},$$

and similarly

(5.4) $\quad Q(x, y) - \dfrac{(y - \mu_2)^2}{\sigma_2^2} = \dfrac{[x - \mu_1 - (\rho\sigma_1/\sigma_2)(y - \mu_2)]^2}{\sigma_1^2(1 - \rho^2)}.$

From (5.1), (5.2), and (5.3), it follows that

$$
(5.5) \quad f(x, y) = \frac{\exp\left[-\frac{1}{2}(x - \mu_1)^2/\sigma_1^2\right]}{\sqrt{2\pi\sigma_1^2}}
$$

$$
\times \frac{\exp\left\{-\frac{1}{2}[y - \mu_2 - (\rho\sigma_2/\sigma_1)(x - \mu_1)]^2/\sigma_2^2(1 - \rho^2)\right\}}{\sqrt{(2\pi)\sigma_2^2(1 - \rho^2)}},
$$

so that

$$
f_X(x) = \int_{-\infty}^{\infty} f(x, y)\, dy = \frac{\exp\left[-\frac{1}{2}(x - \mu_1)^2/\sigma_1^2\right]}{\sqrt{(2\pi)\sigma_1^2}}
$$

$$
(5.6) \quad \times \int_{-\infty}^{\infty} \frac{\exp\left\{-\frac{1}{2}[y - \mu_2 - (\rho\sigma_2/\sigma_1)(x - \mu_1)]^2/\sigma_2^2(1 - \rho^2)\right\}\, dy}{\sqrt{(2\pi)\sigma_2^2(1 - \rho^2)}}
$$

$$
= \frac{\exp\left[-\frac{1}{2}(x - \mu_1)^2/\sigma_1^2\right]}{\sqrt{(2\pi)\sigma_1^2}},
$$

since we recognize the integral in (5.6) as being the density function of a certain normal distribution. Thus, the marginal probability density function $f_X(x)$ has the form of a $N(\mu_1, \sigma_1^2)$ probability density function, and we conclude that the marginal distribution of X is $N(\mu_1, \sigma_1^2)$. It follows from this result that X has mean μ_1 and variance σ_1^2, as asserted. In similar fashion, using (5.1), (5.2), and (5.4), we can show that the marginal distribution of Y is $N(\mu_2, \sigma_2^2)$, so that $\mu_Y = \mu_2$, $\sigma_Y^2 = \sigma_2^2$.

We have shown that if $(X, Y) \sim \mathrm{BVN}((\mu_1, \mu_2), (\sigma_1^2, \sigma_2^2, \rho))$, then X and Y each have marginal normal distributions. However, not only are X and Y each normally distributed, but also *any linear combination $aX + bY + c$ is normally distributed*:

$$
(5.7) \quad aX + bY + c \sim N(a\mu_1 + b\mu_2 + c, \ a^2\sigma_1^2 + 2ab\rho\sigma_1\sigma_2 + b^2\sigma_2^2).
$$

The bivariate normal distribution is the *only* joint distribution for two random variables, X and Y, that has the property (5.7) for all constants a, b, c. This property indeed *characterizes* the bivariate normal distribution. On the other hand, many bivariate nonnormal distributions possess normal marginal distributions (Exercise 14). Thus, *the property of having normal marginal distributions does not characterize the bivariate normal distribution.*

The Correlation Coefficient

We have shown that the parameters $\mu_1, \mu_2, \sigma_1^2, \sigma_2^2$ of a bivariate normal distribution are the mean of X, the mean of Y, the variance of X, and the variance of Y, respectively. We now show that $\rho_{X,Y} = \rho$. To do so, we first find $\sigma_{X,Y}$. Thus,

$$\sigma_{X,Y} = \int_{-\infty}^{\infty} \int_{-\infty}^{\infty} (x - \mu_1)(y - \mu_2) f(x, y) \, dx \, dy$$

(5.8)

$$= \int_{-\infty}^{\infty} \int_{-\infty}^{\infty} \frac{(x - \mu_1)(y - \mu_2) \exp\left[-\tfrac{1}{2} Q(x, y)\right]}{2\pi \sigma_1 \sigma_2 \sqrt{1 - \rho^2}} \, dx \, dy,$$

where $Q(x, y)$ is defined by (5.2). Make the change of variables from x to $u = (x - \mu_1)/\sigma_1$, and from y to $v = (y - \mu_2)/\sigma_2$ in (5.8). We obtain

$$\sigma_{X,Y} = \sigma_1 \sigma_2 \int_{-\infty}^{\infty} \int_{-\infty}^{\infty} \frac{uv \exp\left[-\tfrac{1}{2}(u^2 - 2\rho uv + v^2)/(1 - \rho^2)\right]}{2\pi\sqrt{(1 - \rho^2)}} \, du \, dv$$

(5.9)

$$= \sigma_1 \sigma_2 \int_{-\infty}^{\infty} \frac{v \exp\left[-\tfrac{1}{2}v^2\right]}{\sqrt{2\pi}} \left[\int_{-\infty}^{\infty} \frac{u \exp\left[-\tfrac{1}{2}(u - \rho v)^2/(1 - \rho^2)\right]}{\sqrt{2\pi(1 - \rho^2)}} \, du \right] dv.$$

We recognize the inner integral (over u) as being that defining the mean of a $N(\rho v, 1 - \rho^2)$ distribution, so that (5.9) becomes

$$\sigma_{X,Y} = \sigma_1 \sigma_2 \int_{-\infty}^{\infty} \frac{v \exp\left(-\tfrac{1}{2}v^2\right)}{\sqrt{2\pi}} (\rho v) \, dv$$

$$= \sigma_1 \sigma_2 \rho \int_{-\infty}^{\infty} \frac{v^2 \exp\left(-\tfrac{1}{2}v^2\right)}{\sqrt{2\pi}} \, dv = \sigma_1 \sigma_2 \rho E[V^2],$$

where $V \sim N(0, 1)$. Since $E[V^2] = \sigma_V^2 + \mu_V^2 = 1$,

$$\sigma_{X,Y} = \sigma_1 \sigma_2 \rho$$

and hence

(5.10) $$\rho_{X,Y} = \frac{\sigma_{X,Y}}{\sigma_X \sigma_Y} = \frac{\sigma_1 \sigma_2 \rho}{\sigma_1 \sigma_2} = \rho.$$

Thus, if $(X, Y) \sim \text{BVN}((\mu_1, \mu_2), (\sigma_1^2, \sigma_2^2, \rho))$, the value of the correlation coefficient, $\rho_{X,Y}$, between X and Y is ρ.

Conditional Distributions

When $(X, Y) \sim \text{BVN}((\mu_1, \mu_2), (\sigma_1^2, \sigma_2^2, \rho))$, it follows from (5.5) that the conditional probability density function of Y given $X = x$ is

$$(5.11) \quad f_{Y|X=x}(y) = \frac{f(x, y)}{f_X(x)} = \frac{\exp\left\{-\frac{1}{2}[y - \mu_2 - (\rho\sigma_2/\sigma_1)(x - \mu_1)]^2\right\}}{\sqrt{(2\pi)\sigma_2^2(1 - \rho^2)}},$$

which we recognize as being the probability density function of a $N(\mu_2 + (\rho\sigma_2/\sigma_1)(x - \mu_1), \sigma_2^2(1 - \rho^2))$ distribution. Hence, the regression function of Y on X is

$$\mu_{Y|X=x} = \mu_2 + \frac{\rho\sigma_2}{\sigma_1}(x - \mu_1)$$

(5.12)

$$= \frac{\rho\sigma_2}{\sigma_1}x + \left(\mu_2 - \frac{\rho\sigma_2}{\sigma_1}\mu_1\right),$$

which is a linear function $ax + b$ of x with slope and intercept

$$a = \frac{\rho\sigma_2}{\sigma_1}, \qquad b = \mu_2 - \frac{\rho\sigma_2}{\sigma_1}\mu_1,$$

respectively. It also follows from (5.11) that the conditional variance of Y given $X = x$ is

$$(5.13) \qquad \sigma_{Y|X=x}^2 = \sigma_2^2(1 - \rho^2) = \sigma_Y^2(1 - \rho_{X,Y}^2),$$

which is the same quantity regardless of the value of x. Similarly, interchanging the roles of X and Y, it can be shown that the conditional distribution of X given $Y = y$ is normal with

$$(5.14) \qquad \mu_{X|Y=y} = \left(\mu_1 - \frac{\rho\sigma_1}{\sigma_2}\mu_2\right) + \frac{\rho\sigma_1}{\sigma_2}y$$

and

$$(5.15) \qquad \sigma_{X|Y=y}^2 = \sigma_1^2(1 - \rho^2) = \sigma_X^2(1 - \rho_{X,Y}^2).$$

When $\rho_{X,Y} = \rho = 0$, the regression lines (5.12) and (5.14) have zero slopes, and are thus constant.

Indeed, if (X, Y) has a bivariate normal distribution with parameters μ_1, μ_2, σ_1^2, σ_2^2, and ρ, and if $\rho_{X,Y} = \rho = 0$, then the conditional distribution of Y given $X = .x$ is a $N(\mu_2, \sigma_2^2)$ distribution for all x. Since the marginal distribution of Y is also a $N(\mu_2, \sigma_2^2)$ distribution, it follows that $f_{Y|X=x}(y) = f_Y(y)$ for all values of x and y, and hence X and Y are independent. On the other hand, we have seen that independent random variables X and Y have a zero correlation coefficient ($\rho_{X,Y} = 0$). Hence, *when random variables X and Y have a bivariate normal distribution, the properties of being independent and uncor-*

related are equivalent. If $(X, Y) \sim \text{BVN}((\mu_1, \mu_2), (\sigma_1^2, \sigma_2^2, \rho))$, then X and Y are independent if and only if $\rho = 0$.

Note that when X and Y have a $\text{BVN}((\mu_1, \sigma_2), (\sigma_1^2, \sigma_2^2, \rho))$ distribution with $0 < |\rho| < 1$, then

$$\sigma_{Y|X=x}^2 = \sigma_2^2(1 - \rho^2) < \sigma_2^2 = \sigma_Y^2.$$

That is, the expected squared error of prediction $\sigma_{Y|X=x}^2$ for Y given that $X = x$ is always strictly less than the expected squared error of prediction σ_Y^2 assuming ignorance of the value of X, and hence our predictions of Y are always more accurate when we know the value x of X than they are when we do not know x. That a similar assertion is generally false for nonnormal distributions can be seen by comparing $\sigma_{Y|X=x}^2 = (1 - x)^2/12$ with $\sigma_Y^2 = 1/18$ for the bivariate Dirichlet distribution (Example 2.3) when $a = b = c = 1$. If $0 \leq x < 1 - \sqrt{2/3}$, then $\sigma_{Y|X=x}^2 > \sigma_Y^2$.

The Probability Contours of a Bivariate Normal Distribution

If we slice the surface formed by the graph of the bivariate normal probability density function (see Figure 5.2) by a plane parallel to the (x, y) plane, then the cross section (*probability contour*) created by the intersection of the plane and the surface is ellipsoidal in shape, no matter what horizontal plane we use. Knowledge of this ellipsoidal shape is equivalent to knowledge of how probability is distributed in the plane by the bivariate normal distribution.

Figures 5.3, 5.4, and 5.5 represent some different elliptical probability contours that define the bivariate normal distribution. They illustrate

Figure 5.2: *Probability contours of the bivariate normal density function given in full view and in cross section.*

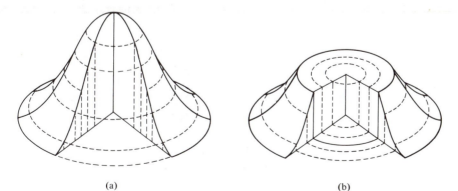

(a)   (b)

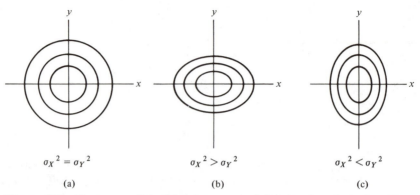

$\sigma_X{}^2 = \sigma_Y{}^2$

(a)

$\sigma_X{}^2 > \sigma_Y{}^2$

(b)

$\sigma_X{}^2 < \sigma_Y{}^2$

(c)

Figure 5.3: *Contours of the bivariate normal density function when X and Y are uncorrelated.*

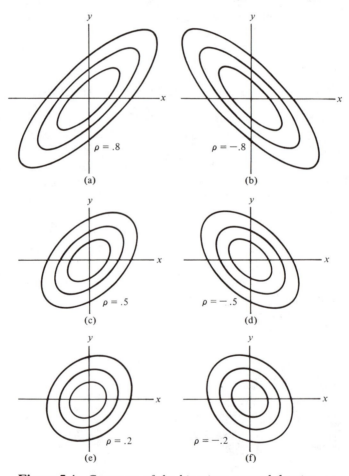

$\rho = .8$

(a)

$\rho = -.8$

(b)

$\rho = .5$

(c)

$\rho = -.5$

(d)

$\rho = .2$

(e)

$\rho = -.2$

(f)

Figure 5.4: *Contours of the bivariate normal density function for different values of the correlation coefficient ρ when $\sigma_X^2 = \sigma_Y^2 = 1$.*

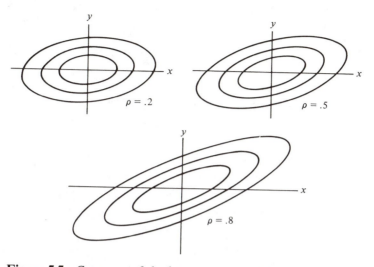

Figure 5.5: *Contours of the bivariate normal density function for different values of ρ when $\sigma_X^2 = 4$ and $\sigma_Y^2 = 1$.*

how the correlation coefficient ρ determines the orientation and shape of the surface formed by a graph of the bivariate normal probability density function.

The Standard Bivariate Normal Distribution

For the univariate normal distribution, if $X \sim N(\mu, \sigma^2)$, then $F_X(a) = F_Z((a - \mu)/\sigma)$, where $Z \sim N(0, 1)$ has the standard normal distribution. It was this result that permitted us to compute the probabilities for any normal distribution from only one table, that of the cumulative distribution function of the standard normal distribution.

In the bivariate case, we have a similar key result, namely if $(X, Y) \sim \text{BVN}((\mu_1, \mu_2), (\sigma_1^2, \sigma_2^2, \rho))$, then

$$(5.16) \qquad F_{X, Y}(a, b) = F_{Z_1, Z_2}\left(\frac{a - \mu_1}{\sigma_1}, \frac{b - \mu_2}{\sigma_2}\right),$$

where $(Z_1, Z_2) \sim \text{BVN}((0, 0), (1, 1, \rho))$. Put another way, every pair of bivariate normal random variables X and Y can be transformed to standardized bivariate normal random variables

$$Z_1 = \frac{X - \mu_1}{\sigma_1}, \qquad Z_2 = \frac{Y - \mu_2}{\sigma_2},$$

so that $\mu_{Z_1} = \mu_{Z_2} = 0$, $\sigma_{Z_1} = \sigma_{Z_2} = 1$, and $\rho_{Z_1, Z_2} = \rho_{X, Y} = \rho$. Although there is only one standard univariate normal distribution, there are many standard bivariate normal distributions, one for each value of the correlation coefficient ρ. Since tables of $F_{Z_1, Z_2}(h, k)$ have three variable arguments, h, k, and ρ, they require considerable space.

Let $A = \{Z_1 \leq h\}$, $B = \{Z_2 \leq k\}$, and note that

$$F_{Z_1, Z_2}(h, k) = P(A \cap B) = 1 - P((A \cap B)^c) = 1 - P(A^c \cup B^c)$$
$$= 1 - P(A^c) - P(B^c) + P(A^c \cap B^c)$$
$$(5.17) \qquad = 1 - (1 - F_{Z_1}(h)) - (1 - F_{Z_2}(k)) + P\{Z_1 > h \text{ and } Z_2 > k\}$$
$$= F_{Z_1}(h) + F_{Z_2}(k) + P\{Z_1 > h \text{ and } Z_2 > k\} - 1.$$

Thus, if we have tables of $P\{Z_1 > h \text{ and } Z_2 > k\}$ for the standard bivariate normal distribution, these tables together with Table T.5 allow us to calculate values of $F_{Z_1, Z_2}(h, k)$. Tables of probabilities of events of the form $P\{Z_1 > h \text{ and } Z_2 > k\}$ for $(Z_1, Z_2) \sim \text{BVN}((0, 0), (1, 1, \rho))$, $h, k = 0.0(0.1)4.0$, $\pm\rho = 0.00(0.05)0.95(0.01)1.00$ are provided in "Tables of the Bivariate Normal Distribution Function and Related Functions" PB 176 520, June 15, 1959, of the National Bureau of Standards. (Note: In these tables, ρ is denoted by r.) Note that only nonnegative values of h and k appear in these tables, while we may have need to calculate $F_{Z_1, Z_2}(h, k)$ when h or k are negative. For such calculations, it is helpful to note that:

(5.18) (i) If $(Z_1, Z_2) \sim \text{BVN}((0, 0), (1, 1, \rho))$, then $(Z_1, -Z_2)$ and $(-Z_1, Z_2) \sim \text{BVN}((0, 0), (1, 1, -\rho))$.

(ii) If $(Z_1, Z_2) \sim \text{BVN}((0, 0), (1, 1, \rho))$, then $(-Z_1, -Z_2) \sim \text{BVN}((0, 0), (1, 1, \rho))$.

Let us adopt the notation

$$(5.19) \quad F(h, k; \rho) = F_{Z_1, Z_2}(h, k), \qquad \bar{F}(h, k; \rho) = P\{Z_1 > h \text{ and } Z_2 > k\},$$

where $(Z_1, Z_2) \sim \text{BVN}((0, 0), (1, 1, \rho))$. The function $\bar{F}(h, k; \rho)$ is the function tabled by the National Bureau of Standards for $h, k \geq 0$. Table 5.1 indicates how values of $F(h, k; \rho)$ and $\bar{F}(h, k; \rho)$ can be obtained from the National Bureau of Standards tables.

Table 5.1: *Formulas for Obtaining $F(h, k; \rho)$ and $\bar{F}(h, k; \rho)$ from Values of $\bar{F}(h, k; \rho)$ for $h, k \geq 0$ [$F_Z(a)$ is the cumulative distribution function of the $N(0, 1)$ distribution]*

Cases	$F(h, k; \rho)$	$\bar{F}(h, k; \rho)$
$h \geq 0, k \geq 0$	$F_Z(h) + F_Z(k) + \bar{F}(h, k; \rho) - 1$	$\bar{F}(h, k; \rho)$
$h \geq 0, k < 0$	$1 - F_Z(-k) - \bar{F}(h, -k; -\rho)$	$1 - \bar{F}(h, -k; -\rho) - F_Z(h)$
$h < 0, k \geq 0$	$1 - F_Z(-h) - \bar{F}(-h, k; -\rho)$	$1 - \bar{F}(-h, k; -\rho) - F_Z(k)$
$h < 0, k < 0$	$\bar{F}(-h, -k; +\rho)$	$F_Z(-h) + F_Z(-k) + \bar{F}(-h, -k; +\rho) - 1$

To show how the entries in Table 5.1 are obtained from (5.17) and (5.18), note first that the entries for $h \geq 0$, $k \geq 0$ follow directly from (5.19) and (5.17). For $h \geq 0$, $k < 0$, it follows from (5.18i) that

$$F(h, k; \rho) = P\{Z_1 \leq h \text{ and } Z_2 \leq k\} = P\{Z_1 \leq h \text{ and } -Z_2 \geq -k\}$$

$$= P\{-Z_2 > -k\} - P\{Z_1 > h \text{ and } -Z_2 > -k\}$$

$$= 1 - F_Z(-k) - \bar{F}(h, -k; -\rho),$$

and then (5.17) can be used to obtain

$$\bar{F}(h, k; \rho) = 1 + F(h, k; \rho) - F_Z(h) - F_Z(k)$$

$$= 1 + [1 - F_Z(-k) - \bar{F}(h, -k; -\rho)] - F_Z(h) - F_Z(k)$$

$$= 1 - \bar{F}(h, -k; -\rho) - F_Z(h).$$

Similar arguments establish the formulas for the case $h < 0$, $k \geq 0$, whereas when $h < 0$, $k < 0$, (5.18ii) is used to show that

$$F(h, k; \rho) = P\{Z_1 \leq h \text{ and } Z_2 \leq k\} = P\{-Z_1 \geq -h \text{ and } -Z_2 \geq -k\}$$

$$= P\{-Z_1 > -h \text{ and } -Z_2 > -k\} = \bar{F}(-h, -k; -\rho)$$

and then (5.17) is used to obtain the formula for $\bar{F}(h, k; \rho)$ given in Table 5.1.

Let $(X, Y) \sim \text{BVN}((\mu_1, \mu_2), (\sigma_1^2, \sigma_2^2, \rho))$, and suppose that we want to calculate the probability, $P\{(X, Y) \text{ in } R\}$, of the rectangle

$$R = \{(X, Y): a \leq x \leq b \text{ and } c \leq y \leq d\}.$$

Since X and Y have a joint density function, the probability of any of the line segments making up the boundary of R is zero, and the probability of R equals the probability of any rectangle R' formed from R by deleting one or more of its boundaries. In particular,

$$P\{(X, Y) \text{ in } R\} = P\{a < X \leq b \text{ and } c < Y \leq d\}$$

$$= F_{X,Y}(b, d) + F_{X,Y}(a, c) - F_{X,Y}(a, d) - F_{X,Y}(b, c)$$

(5.20)
$$= F\left(\frac{b - \mu_1}{\sigma_1}, \frac{d - \mu_2}{\sigma_2}; \rho\right) + F\left(\frac{a - \mu_1}{\sigma_1}, \frac{c - \mu_2}{\sigma_2}; \rho\right)$$

$$- F\left(\frac{a - \mu_1}{\sigma_1}, \frac{d - \mu_2}{\sigma_2}; \rho\right) - F\left(\frac{b - \mu_1}{\sigma_1}, \frac{c - \mu_2}{\sigma_2}; \rho\right),$$

where the second equality in (5.20) is the result of Exercise 11, and the last equality follows from (5.16) and (5.19). Also, substituting (5.17) into (5.20),

(5.21)
$$P\{(X, Y) \text{ in } R\} = \bar{F}\left(\frac{b - \mu_1}{\sigma_1}, \frac{d - \mu_2}{\sigma_2}; \rho\right) + \bar{F}\left(\frac{a - \mu_1}{\sigma_1}, \frac{c - \mu_2}{\sigma_2}; \rho\right)$$

$$- \bar{F}\left(\frac{a - \mu_1}{\sigma_1}, \frac{d - \mu_2}{\sigma_2}; \rho\right) - \bar{F}\left(\frac{b - \mu_1}{\sigma_1}, \frac{c - \mu_2}{\sigma_2}; \rho\right),$$

enabling us to use the National Bureau of Standards tables to calculate the probability of R.

Example 5.1. Table 5.2, taken from Pearson and Lee (1903), provides grouped data on the heights (in inches) of mothers (Y) and of their sons (X). To fit a bivariate normal distribution to these data, we need to obtain estimates of the parameters $\mu_X = \mu_1$, $\mu_Y = \mu_2$, $\sigma_X^2 = \sigma_1^2$, $\sigma_Y^2 = \sigma_2^2$, and $\rho_{X,Y} = \rho$, and then calculate theoretical probabilities and theoretical frequencies of rectangles such as $\{(x, y): 68 \le x \le 69 \ and \ 62 \le y \le 63\}$. The estimates $\hat{\mu}_X$, $\hat{\sigma}_X^2$, $\hat{\mu}_Y$, $\hat{\sigma}_Y^2$ have been obtained from the marginal frequency distributions (*raw data*) of the data (not shown), and the covariance $\sigma_{X,Y}$ is estimated from the (raw data) relative frequency distribution of the pairs (X, Y) through the formula

$$\hat{\sigma}_{X,Y} = \sum_x \sum_y (x - \hat{\mu}_X)(y - \hat{\mu}_Y) \ \text{r.f.}\{X = x \ and \ Y = y\},$$

where the double sum is taken over the distinct pairs (x, y) of values observed for (X, Y). (Similar estimates can be obtained from the grouped data in Table 5.2; these will differ somewhat from the estimates obtained from the raw data.) The estimate of $\rho_{X,Y}$ is then obtained from $\hat{\rho}_{X,Y} = \hat{\sigma}_{X,Y}/\hat{\sigma}_X \hat{\sigma}_Y$. The estimates obtained from the data are

$$\hat{\mu}_X = 68.65, \qquad \hat{\sigma}_X = 2.71, \qquad \hat{\mu}_Y = 62.48, \qquad \hat{\sigma}_Y = 2.39, \qquad \hat{\rho}_{X,Y} = 0.494.$$

In order to avoid excessive interpolations in tables, we take $\hat{\rho} = 0.50$.

Suppose that we want to calculate the theoretical probability and theoretical frequency of the rectangle

$$R = \{(x, y): 68 \le x \le 69 \ and \ 62 \le y \le 63\}.$$

Then

$P\{(X, Y) \ \text{in} \ R\}$

$$= \bar{F}\left(\frac{69 - 68.65}{2.71}, \frac{63 - 62.48}{2.39}; 0.5\right) + \bar{F}\left(\frac{68 - 68.65}{2.71}, \frac{62 - 62.48}{2.39}; 0.5\right)$$

$$\text{(5.22)} \qquad - \bar{F}\left(\frac{69 - 68.65}{2.71}, \frac{62 - 62.48}{2.39}; 0.5\right) - \bar{F}\left(\frac{68 - 68.65}{2.71}, \frac{63 - 62.48}{2.39}; 0.5\right)$$

$$= \bar{F}(0.13, 0.22; 0.5) + \bar{F}(-0.24, -0.20; 0.5) - \bar{F}(0.13, -0.20; 0.5)$$

$$- \bar{F}(-0.24, 0.22; 0.5).$$

Table 5.3 gives the values of $\bar{F}(h, k; 0.5)$ and $\bar{F}(h, k; -0.5)$ needed to calculate the quantities in (5.22).

Table 5.2: *Mother's Stature and Son's Stature in Inches*

Son's stature	\	\	\	\	\	\	\	Mother's stature	\	\	\	\	\	\	\	\	\	\	\
	52–53	53–54	54–55	55–56	56–57	57–58	58–59	59–60	60–61	61–62	62–63	63–64	64–65	65–66	66–67	67–68	68–69	69–70	70–71	Totals
59–60	—	—	—	—	—	—	—	0.25	0.25	0.25	0.25	—	—	—	—	—	—	—	—	1
60–61	—	—	—	—	—	—	—	0.75	0.25	0.25	0.25	—	—	—	—	—	—	—	—	1.5
61–62	—	—	—	—	—	—	1	0.5	—	—	—	—	—	—	—	—	—	—	—	1.5
62–63	—	—	—	—	—	1.25	—	3.5	1.5	2	0.5	1.25	—	0.5	—	—	—	—	—	8
63–64	0.5	0.5	—	—	0.25	3.25	2.5	6.75	9	5	3.75	6.75	1.75	0.5	—	—	—	—	—	30
64–65	—	—	0.25	0.25	1	5.5	5.75	13	7.75	7	6.75	10	1.75	—	1	—	—	—	—	49
65–66	—	—	0.5	1	3.75	2.75	4.5	9	14.25	18.25	7.25	27	7.75	2	1.25	0.25	0.5	—	—	74
66–67	—	—	0.25	0.75	1.75	4.25	8.5	15	19.5	20	22.25	26.75	12.75	4	2.75	0.75	0.5	—	—	114.5
67–68	—	—	—	—	0.25	—	5	11.25	27.25	31	36	20.5	23.25	12.75	2.25	1.5	0.75	—	—	163
68–69	—	—	—	—	—	—	6.75	12.25	27.5	27.25	31.5	23.5	20	13.25	5	1.75	0.75	—	—	175.5
69–70	—	—	—	—	—	—	—	6	15.25	17	23.25	16	15.75	14.25	4.5	1	0.5	0.5	—	124
70–71	—	—	—	—	—	2	—	2.75	9.5	22	21.5	11	13.25	12.75	7.5	2.75	—	0.5	—	122
71–72	—	—	—	—	—	1	0.5	1	2.75	10.25	13.5	6.25	7.5	4	6	1	1	—	—	78
72–73	—	—	—	—	—	—	0.5	1	5	4.25	6	2.5	3.75	7	4.75	2.5	1	—	—	47.5
73–74	—	—	—	—	—	—	0.5	—	1.25	4	6.75	0.5	1	3	1.5	1.75	1.25	—	—	36
74–75	—	—	—	—	—	—	—	—	—	2.5	2.5	1	1	1	2	0.75	1.25	1.5	1	17
75–76	—	—	—	—	—	—	—	—	—	0.5	—	—	—	1	1.5	1	1.5	1.5	—	6.5
76–77	—	—	—	—	—	—	—	—	—	—	—	—	—	—	0.75	0.25	—	—	—	3.5
77–78	—	—	—	—	—	—	—	—	—	0.5	—	—	—	0.5	0.25	0.75	—	—	0.5	1.5
78–79	—	—	—	—	—	—	—	—	—	0.5	—	—	—	—	0.5	0.5	—	—	0.5	2
79–80	—	—	—	—	—	—	—	—	—	—	—	—	—	—	—	1	—	—	—	1
Totals	0.5	0.5	1	2	7	20	35.5	83	141	172.5	182	153	108.5	76.5	41.5	17.5	9	4	2	1057

Table 5.3: Values of $\bar{F}(h, k; 0.5)$ and $\bar{F}(h, k; -0.5)$ for $h = 0.1(0.1)0.3$, $k = 0.2$, 0.3

| | | $\rho = 0.5$ | | | | $\rho = -0.5$ | |
| | | h | | | | h | |
k	0.1	0.2	0.3	k	0.1	0.2	0.3
0.2	0.275161	0.257709	0.239718	0.2	0.1130216	0.0982164	0.0845419
0.3	0.255392	0.239718	0.223488	0.3	0.0976550	0.0845419	0.0724876

We must interpolate in Table 5.3 to obtain the quantities in (5.22). For example, to obtain $\bar{F}(0.13, 0.22; 0.5)$ we first linearly interpolate by rows, and then by columns:

| | | h | |
k	0.1	0.13	0.2
0.2	0.275161	—	0.257709
0.22	0.271207	0.266078	0.254111
0.3	0.255392	—	0.239718

We thus obtain $\bar{F}(0.13, 0.22; 0.5) = 0.266078$. Proceeding similarly, we obtain

$$\bar{F}(0.24, 0.20; 0.5) = 0.2505126, \qquad \bar{F}(0.13, 0.20; -0.5) = 0.1085801,$$

$$\bar{F}(0.24, 0.22; -0.5) = 0.0901414,$$

and from Table T.5:

$$F_Z(0.13) = 0.55172, \qquad F_Z(0.20) = 0.57926,$$

$$F_Z(0.22) = 0.58706, \qquad F_Z(0.24) = 0.59483.$$

Thus, from Table 5.1,

$$\bar{F}(-0.24, -0.20; 0.5) = F_Z(0.24) + F_Z(0.20) + \bar{F}(0.24, 0.20; 0.5) - 1$$

$$= 0.59483 + 0.57926 + 0.2505126 - 1 = 0.4246026,$$

$$\bar{F}(0.13, -0.20; 0.5) = 1 - \bar{F}(0.13, 0.20; -0.5) - F_Z(0.13)$$

$$= 1 - 0.1085801 - 0.55172 = 0.3396999,$$

and

$$\bar{F}(-0.24, 0.22; 0.5) = 1 - \bar{F}(0.24, 0.22; -0.5) - F_Z(0.22)$$

$$= 1 - 0.0901414 - 0.58706 = 0.3227986.$$

Substituting these results into (5.22) yields

$$P\{68 \leq X \leq 69 \text{ and } 62 \leq Y \leq 63\} = 0.266078 + 0.4246026$$
$$- 0.3396999 - 0.3227986$$
$$= 0.0281821.$$

The theoretical frequency of the event $\{68 \leq X \leq 69$ *and* $62 \leq Y \leq 63\}$ is equal to $(1057)(0.0281821) = 29.79$. The observed frequency of this event is, from Table 5.2, equal to 31.5.

The procedure illustrated above is tedious if we wish to fit a complete distribution, but is not too troublesome if we need only a few cells. If a complete distribution is needed, then we may have to resort to a computer program.

6. MULTIVARIATE GENERALIZATIONS

In the previous sections, the notion of a bivariate distribution was introduced, the essential idea being that if two random variables are of interest in an experiment, the probability model should reflect how they jointly vary. When several random variables $X_1, X_2, \ldots, X_k$ are considered, we again require knowledge of their joint variation. This knowledge is presented in terms of a *multivariate distribution*, usually given either by a *multivariate probability mass function*,

(6.1) $\quad p(x_1, x_2, \ldots, x_k) = P\{X_1 = x_1 \text{ and } X_2 = x_2 \text{ and } \cdots \text{ and } X_k = x_k\},$

when the random variables all are discrete, or in terms of a *multivariate probability density function*, $f(x_1, x_2, \ldots, x_k)$, when the random variables $X_1, X_2, \ldots, X_k$ are all continuous.

The individual marginal distributions are obtained from the multivariate distributions by summation or integration of the multivariate probability mass function or multivariate probability density function, respectively. For example, if X_1, X_2, X_3 have a *trivariate probability mass function* $p(x_1, x_2, x_3)$, then

$$p_{X_1}(x_1) = \sum_{x_2} \sum_{x_3} p(x_1, x_2, x_3),$$

(6.2) $\qquad p_{X_2}(x_2) = \sum_{x_1} \sum_{x_3} p(x_1, x_2, x_3),$

$$p_{X_3}(x_3) = \sum_{x_1} \sum_{x_2} p(x_1, x_2, x_3).$$

Similarly, if X_1, X_2, X_3 have a *trivariate probability density function* $f(x_1, x_2, x_3)$, then

$$(6.3) \qquad f_{X_1}(x_1) = \int_{-\infty}^{\infty} \int_{-\infty}^{\infty} f(x_1, x_2, x_3)\, dx_2\, dx_3,$$

with analogous formulas defining $f_{X_2}(x_2)$ and $f_{X_3}(x_3)$.

We can also obtain *marginal joint distributions* of any subcollection X_{i_1}, X_{i_2}, ..., X_{i_r}, $r < k$, of the random variables $X_1, X_2, ..., X_k$ by summation or integration of their multivariate probability mass functions or multivariate probability density functions, respectively. Thus, if X_1, X_2, X_3 have the trivariate probability mass function $p(x_1, x_2, x_3)$, the marginal bivariate probability mass function $p_{X_1, X_2}(x_1, x_2)$ of X_1 and X_2 is obtained from

$$(6.4) \qquad p_{X_1, X_2}(x_1, x_2) = \sum_{x_3} p(x_1, x_2, x_3),$$

while if X_1, X_2, X_3 have the trivariate probability density function $f(x_1, x_2, x_3)$, the marginal probability density function $f_{X_1, X_3}(x_1, x_3)$ of X_1 and X_3 is obtained from

$$(6.5) \qquad f_{X_1, X_3}(x_1, x_3) = \int_{-\infty}^{\infty} f(x_1, x_2, x_3)\, dx_2.$$

Formulas for the other marginal bivariate mass functions or bivariate density functions are obtained in a fashion analogous to (6.4) or (6.5), respectively.

For any function $h(x_1, x_2, ..., x_k)$ of $x_1, x_2, ..., x_k$, we may define the *expected value* $E[h(X_1, X_2, ..., X_k)]$ by

$$(6.6) \quad E[h(X_1, X_2, ..., X_k)] = \sum_{x_1} \sum_{x_2} \cdots \sum_{x_k} h(x_1, x_2, ..., x_k) p(x_1, x_2, ..., x_k)$$

in the discrete case, and

$$(6.7) \quad \begin{aligned} &E[h(X_1, X_2, ..., X_k)] \\ &= \int_{-\infty}^{\infty} \int_{-\infty}^{\infty} \cdots \int_{-\infty}^{\infty} h(x_1, x_2, ..., x_k) f(x_1, x_2, ..., x_k)\, dx_1\, dx_2 \cdots dx_k \end{aligned}$$

in the continuous case. The expected value of $h(X_1, X_2, ..., X_k)$ obeys the two important properties: (i) For any two functions $h_1(x_1, x_2, ..., x_k)$ and $h_2(x_1, x_2, ..., x_k)$, and any constants a and b,

$$(6.8) \quad \begin{aligned} &E[a h_1(X_1, X_2, ..., X_k) + b h_2(X_1, X_2, ..., X_k)] \\ &= a E[h_1(X_1, X_2, ..., X_k)] + b E[h_2(X_1, X_2, ..., X_k)]; \end{aligned}$$

(ii) If $h(x_1, x_2, \ldots, x_k) \geq 0$, then $E[h(X_1, X_2, \ldots, X_k)] \geq 0$. Important special cases of the expected value computation are the means $\mu_{X_1}, \mu_{X_2}, \ldots, \mu_{X_k}$, the variances $\sigma_{X_1}^2, \sigma_{X_2}^2, \ldots, \sigma_{X_k}^2$, and the covariances σ_{X_i, X_j}, $i, j = 1, 2, \ldots, k$. The means μ_{X_i}, $i = 1, 2, \ldots, k$ can be computed as

$$\mu_{X_i} = E[X_i] = \sum_{x_1} \sum_{x_2} \cdots \sum_{x_k} x_i p(x_1, x_2, \ldots, x_k)$$

(6.9)

$$= \int_{-\infty}^{\infty} \int_{-\infty}^{\infty} \cdots \int_{-\infty}^{\infty} x_i f(x_1, x_2, \ldots, x_k) \, dx_1 \, dx_2 \cdots dx_k$$

depending on whether we are in the discrete or the continuous case. Similarly,

$$\sigma_{X_i}^2 = E[(X_i - \mu_{X_i})^2]$$
$$= \sum_{x_1} \sum_{x_2} \cdots \sum_{x_k} (x_i - \mu_{X_i})^2 p(x_1, x_2, \ldots, x_k)$$

(6.10)

$$= \int_{-\infty}^{\infty} \int_{-\infty}^{\infty} \cdots \int_{-\infty}^{\infty} (x_i - \mu_{X_i})^2 f(x_1, x_2, \ldots, x_k) \, dx_1 \, dx_2 \cdots dx_k$$

and

$$\sigma_{X_i, X_j} = E[(X_i - \mu_{X_i})(X_j - \mu_{X_j})]$$
$$= \sum_{x_1} \sum_{x_2} \cdots \sum_{x_k} (x_i - \mu_{X_i})(x_j - \mu_{X_j}) p(x_1, x_2, \ldots, x_k)$$

(6.11)

$$= \int_{-\infty}^{\infty} \int_{-\infty}^{\infty} \cdots \int_{-\infty}^{\infty} (x_i - \mu_{X_i})(x_j - \mu_{X_j}) f(x_1, x_2, \ldots, x_k) \, dx_1 \, dx_2 \cdots dx_k.$$

Although (6.9), (6.10), and (6.11) are helpful for theoretical purposes, the indices μ_{X_i}, $\sigma_{X_i}^2$, and σ_{X_i, X_j} are usually computed from the appropriate marginal distributions.

The *mean vector* $(\mu_{X_1}, \mu_{X_2}, \ldots, \mu_{X_k})$ locates the probability distribution in the k-dimensional space of points $(x_1, x_2, \ldots, x_k)$. The individual variances $\sigma_{X_i}^2$ serve as measures of the dispersion of probability in the x_i-direction in k-dimensional space, $i = 1, 2, \ldots, k$. Through use of the formulas

(6.12)
$$\mu_{\Sigma_i a_i X_i} = \sum_{i=1}^{k} a_i \mu_{X_i},$$

and

(6.13)
$$\sigma_{\Sigma_i a_i X_i}^2 = \sum_{i=1}^{k} \sum_{j=1}^{k} a_i a_j \sigma_{X_i, X_j},$$

where $\sigma_{X_i, X_i} = \sigma_{X_i}^2$, we can use the means, variances, and covariances to identify the dispersion of probability in the $\sum_{i=1}^{k} a_i x_i$-direction in k-dimensional space. Because dispersion in any direction can be obtained from the means, variances, and covariances, such possible measures of higher-order covariation as $E[(X_1 - \mu_{X_1})(X_2 - \mu_{X_2}) \cdots (X_k - \mu_{X_k})]$ are rarely used or needed.

Conceptually, probabilities of any event concerning $X_1, X_2, \ldots, X_k$ of the form $\{(X_1, X_2, \ldots, X_k)$ is in $R\}$, where R is a region in k-dimensional space, can be computed by an appropriate summation or integration of probabilities over that region. However, such probabilities are generally quite difficult to compute. When such probabilities are needed, they are frequently worked out using methods of numerical approximation. One special case that is sometimes tabled is the *multivariate cumulative distribution function*:

$$F_{X_1, X_2, \ldots, X_k}(x_1, x_2, \ldots, x_k) = P\{X_1 \le x_1 \text{ and } X_2 \le x_2 \text{ and } \cdots \text{ and } X_k \le x_k\}$$

$$= \sum_{t_1 \le x_1} \sum_{t_2 \le x_2} \cdots \sum_{t_k \le x_k} p(t_1, t_2, \ldots, t_k)$$

(6.14)

$$= \int_{-\infty}^{x_1} \int_{-\infty}^{x_2} \cdots \int_{-\infty}^{x_k} f(t_1, t_2, \ldots, t_k) \, dt_1 \, dt_2 \cdots dt_k.$$

Using the multivariate cumulative distribution function, probabilities of *k-dimensional rectangles*:

$$R = \{(x_1, x_2, \ldots, x_k) : a_i \le x_i \le b_i, \quad i = 1, 2, \ldots, k\}$$

can be computed through formulas analogous to (5.20).

Finally, any random variable X_i, or group of random variables $X_{i_1}, X_{i_2}, \ldots, X_{i_r}$, can be predicted from the values of any of the remaining random variables through knowledge of the *conditional probability distributions* and associated *regression functions*. For example, suppose that X_1, X_2, X_3 have a trivariate probability mass function $p(x_1, x_2, x_3)$. If we want to predict X_1 from the values x_2, x_3 of X_2 and X_3, we can construct the conditional probability mass function

(6.15)
$$p_{X_1 | X_2 = x_2, X_3 = x_3}(x_1) = \frac{p(x_1, x_2, x_3)}{p_{X_2, X_3}(x_2, x_3)},$$

for X_1 given $X_2 = x_2$ *and* $X_3 = x_3$, and from that function calculate the regression function

(6.16)
$$\mu_{X_1 | X_2 = x_2, X_3 = x_3} = \sum_{x_1} x_1 \, p_{X_1 | X_2 = x_2, X_3 = x_3}(x_1)$$

which for every pair (x_2, x_3) of possible values of X_2 and X_3 gives the *conditional mean* of X_1 given that $X_2 = x_2$ and $X_3 = x_3$. The regression function (6.16) permits prediction of X_1 when we know the values x_2, x_3 of X_2 and X_3 (namely, X_1 is predicted by $\mu_{X_1 | X_2 = x_2, X_3 = x_3}$), and also gives a formula that "on the average" relates the values of X_1 to those of X_2 and X_3.

We can also predict X_1 from knowledge of only the values of X_2 through the conditional probability mass function:

$$(6.17) \qquad p_{X_1|X_2=x_2}(x_1) = \frac{p_{X_1, X_2}(x_1, x_2)}{p_{X_2}(x_2)}$$

and the associated regression function $\mu_{X_1|X_2=x_2}$, or we can predict X_1 from X_3 using $p_{X_1|X_3=x_3}(x_1)$ and $\mu_{X_1|X_3=x_3}$. Similarly, we can define conditional distributions of X_2 given $X_1 = x_1$ and $X_3 = x_3$, of X_2 given $X_1 = x_1$, of X_2 given $X_3 = x_3$, of X_3 given $X_1 = x_1$ and $X_2 = x_2$, of X_3 given $X_1 = x_1$, and of X_3 given $X_2 = x_2$ using formulas analogous to (6.15) and (6.17). Replacing probability mass functions by probability density functions, analogous conditional probability density functions can be defined in the case where X_1, X_2, X_3 have a trivariate probability density function $f(x_1, x_2, x_3)$.

When X_1, X_2, X_3 have a trivariate probability mass function $p(x_1, x_2, x_3)$, we may also wish to predict X_1 and X_2 from the values x_3 of X_3. We can do so through the *conditional bivariate probability mass function* of X_1 and X_2 given $X_3 = x_3$ defined by

$$(6.18) \qquad p_{X_1, X_2|X_3=x_3}(x_1, x_2) = \frac{p(x_1, x_2, x_3)}{p_{X_3}(x_3)},$$

when $p_{X_3}(x_3) > 0$. The mass function (6.18) is a legitimate bivariate probability mass function. Thus, we can define the conditional mean vector $(\mu_{X_1|X_3=x_3}, \mu_{X_2|X_3=x_3})$ of X_1 and X_2 given $X_3 = x_3$ as a measure of the location of the conditional probability distribution of X_1 and X_2 given $X_3 = x_3$. The elements $\mu_{X_1|X_3=x_3}$ and $\mu_{X_2|X_3=x_3}$ of this conditional mean vector are defined by

$$(6.19) \qquad \mu_{X_i|X_3=x_3} = \sum_{x_1}\sum_{x_2} x_i p_{X_1, X_2|X_3=x_3}(x_1, x_2), \quad i = 1, 2.$$

Dispersion of this conditional probability distribution is measured by the conditional variances

$$(6.20) \qquad \sigma^2_{X_i|X_3=x_3} = \sum_{x_1}\sum_{x_2} (x_i - \mu_{X_i|X_3=x_3})^2 p_{X_1, X_2|X_3=x_3}(x_1, x_2)$$

and conditional covariance

$$(6.21) \qquad \begin{aligned} &\sigma_{X_1, X_2|X_3=x_3} \\ &= \sum_{x_1}\sum_{x_2} (x_1 - \mu_{X_1|X_3=x_3})(x_2 - \mu_{X_2|X_3=x_3}) p_{X_1, X_2|X_3=x_3}(x_1, x_2). \end{aligned}$$

The *conditional* or *partial correlation coefficient*

$$(6.22) \qquad \rho_{X_1, X_2|X_3=x_3} = \frac{\sigma_{X_1, X_2|X_3=x_3}}{(\sigma_{X_1|X_3=x_3})(\sigma_{X_2|X_3=x_3})}$$

then measures the extent of the *linear* relationship between X_1 and X_2 after accounting for the predictive influence of X_3 on X_1 and X_2. The values of the *partial correlation coefficient* are interpreted precisely as we would interpret the values of the correlation coefficient ρ_{X_1, X_2}. However, if both X_1 and X_2 have a linear relationship with X_3, it is quite possible for ρ_{X_1, X_2} to be near 1 or -1, and yet $\rho_{X_1, X_2 | X_3 = x_3}$ to be near 0, suggesting that X_1 and X_2 are probabilistically related only because both are probabilistically related to X_3. It is also possible for ρ_{X_1, X_2} to be near 0 and yet $\rho_{X_1, X_2 | X_3 = x_3}$ to be near $+1$ or -1, suggesting that the relationships of X_1 and of X_2 with X_3 tend to " mask " the relationship between X_1 and X_2.

When X_1, X_2, X_3 have trivariate probability density function $f(x_1, x_2, x_3)$, we can predict X_1 and X_2 from the values of X_3 through the *conditional bivariate probability density function* of X_1 and X_2 given $X_3 = x_3$ defined by

$$(6.23) \qquad f_{X_1, X_2 | X_3 = x_3}(x_1, x_2) = \frac{f(x_1, x_2, x_3)}{f_{X_3}(x_3)}$$

when $f_{X_3}(x_3) > 0$. Since $f_{X_1, X_2 | X_3 = x_3}(x_1, x_2)$ is a legitimate bivariate probability density function, the various conditional means, variances, covariances, and correlations can be obtained from $f_{X_1, X_2 | X_3 = x_3}(x_1, x_2)$ through formulas analogous to (6.18) through (6.22), using integration in place of summation. It is worth remarking that in both the discrete and continuous cases, an alternative way of obtaining $p_{X_1 | X_3 = x_3}(x_1)$ or $f_{X_1 | X_3 = x_3}(x_1)$, which can be used in place of a formula such as (6.17) or its density function analogy, is through the expressions

$$(6.24) \qquad p_{X_1 | X_3 = x_3}(x_1) = \sum_{x_2} p_{X_1, X_2 | X_3 = x_3}(x_1, x_2)$$

and

$$(6.25) \qquad f_{X_1 | X_3 = x_3}(x_1) = \int_{-\infty}^{\infty} f_{X_1, X_2 | X_3 = x_3}(x_1, x_2) \, dx_2,$$

respectively. To show that (6.24) and the formula for $p_{X_1 | X_3 = x_3}(x_1)$ analogous to (6.17) yield the same result, note that

$$p_{X_1 | X_3 = x_3}(x_1) = \frac{p_{X_1, X_3}(x_1, x_2)}{p_{X_3}(x_3)} = \frac{\sum_{x_2} p(x_1, x_2, x_3)}{p_{X_3}(x_3)}$$

$$= \sum_{x_2} \frac{p(x_1, x_2, x_3)}{p_{X_3}(x_3)} = \sum_{x_2} p_{X_1, X_2 | X_3 = x_3}(x_1, x_2).$$

From the brief outline given above, it should be apparent that multivariate distributions can be quite complex, and offer many avenues for analysis and interpretation. In dealing with multivariate distributions, a clear idea of the structure of the model is needed—both in terms of a specification of the functional form of the multivariate probability mass function or multivariate probability density function that determines the distribution, and also in terms of the types of relationships and interrelationships that will be looked for among the variables.

Mutual (Probabilistic) Independence

One powerful structural assumption that is often made in constructing multivariate probability models is that of *mutual independence* among the random variables. If X_1, X_2, ..., X_k have a multivariate probability mass function $p(x_1, x_2, ..., x_k)$, then we say that X_1, X_2, ..., X_k are *mutually independent if*

(6.26) $$p(x_1, x_2, ..., x_k) = p_{X_1}(x_1)p_{X_2}(x_2) \cdots p_{X_k}(x_k)$$

for all possible values $(x_1, x_2, ..., x_k)$ of $(X_1, X_2, ..., X_k)$. Similarly, if X_1, X_2, ..., X_k have a multivariate probability density function $f(x_1, x_2, ..., x_k)$, then X_1, X_2, ..., X_k are mutually probabilistically independent if

(6.27) $$f(x_1, x_2, ..., x_k) = f_{X_1}(x_1)f_{X_2}(x_2) \cdots f_{X_k}(x_k).$$

As can be seen from (6.26) and (6.27), the great advantage of the assumption of mutual independence is that we can obtain the multivariate distribution from knowledge of the marginal distributions, which provides a great simplification.

In analogy to the case of independence of two random variables, mutual independence reflects the notion that knowledge of the values of any variable or combination of variables cannot give additional probabilistic information for predicting the values of any other variable, or other group of variables. For example, if we are interested in the three random variables X_1, X_2, X_3, then it can be shown that X_1, X_2, X_3 are mutually independent if and only if the conditional distribution of X_1 given $X_2 = x_2$, $X_3 = x_3$ is the same as the marginal distribution of X_1 *and also* the conditional distribution of X_2 given $X_3 = x_3$ is the same as the marginal distribution of X_2. Mutually independent random variables often arise from physically independent components of an experiment (such as the draws when we sample with replacement), but the validity of the model of mutual independence is not confined to such situations.

Some consequences of the assumption of mutual independence are the following:

(1) For any event of the form $\{X_1$ in A_1, X_2 in $A_2, ...,$ *and* X_k in $A_k\}$, we have

$$P\{X_1 \text{ in } A_1, X_2 \text{ in } A_2, ..., X_k \text{ in } A_k\}$$
$$= P\{X_1 \text{ in } A_1\}P\{X_2 \text{ in } A_2\} \cdots P\{X_k \text{ in } A_k\}$$

and in particular

$$F_{X_1, X_2, \ldots, X_k}(x_1, x_2, \ldots, x_k) = F_{X_1}(x_1)F_{X_2}(x_2) \cdots F_{X_k}(x_k).$$

(2) $\sigma_{X_i, X_j} = 0$ and $\rho_{X_i, X_j} = 0$, for all $i \neq j$.

(3) Every conditional distribution is identical to the corresponding unconditional distribution.

(4) All regression functions are constant in value.

Violation of any one of these consequences of mutual independence can be used to show that $X_1, X_2, \ldots, X_k$ are not mutually independent, in which case we say that these variables are *mutually dependent*. We remark that the concepts of independence for two random variables (called *pairwise independence*) and of mutual independence are not equivalent. Using (5.22) and (5.23), we can show that *if $X_1, X_2, \ldots, X_k$ are mutually independent, then X_i and X_j are pairwise independent for all $i \neq j$.* However, the converse of this assertion is false in general: Even if every pair X_i and X_j, $i \neq j$, of random variables is pairwise independent, it need not be the case that $X_1, X_2, \ldots, X_k$ are mutually independent (Exercise 23).

   We conclude this section by illustrating, using one discrete example and one continuous example, the concepts and calculations that we have discussed.

The Multinomial Distribution

   One of the most important discrete multivariate probability mass functions is that belonging to the *multinomial distribution*. This distribution is a generalization of the binomial distribution. The binomial distribution results from n independent repetitions of a random experiment having two possible outcomes: success or failure. The multinomial distribution arises from a random experiment in which there are k possible distinct outcomes. If this experiment is independently repeated n times, we can let X_1 denote the number of occurrences of outcome 1, X_2 denote the number of occurrences of outcome 2, $\ldots$, X_k denote the number of occurrences of outcome k. Since some outcome occurs on each repetition, $X_1 + X_2 + \cdots + X_k = n$. Further suppose that p_i is the probability of observing outcome i on any repetition of the experiment, $i = 1, 2, \ldots, k$. Thus, $\sum_{i=1}^{k} p_i = 1$ and $0 \leq p_i \leq 1$, $i = 1, 2, \ldots, k$. It can be shown that $(X_1, X_2, \ldots, X_k)$ have the multivariate probability mass function

(6.28) $$p(x_1, x_2, \ldots, x_k) = \frac{n!}{x_1! x_2! \cdots x_k!} p_1^{x_1} p_2^{x_2} \cdots p_k^{x_k}$$

for $x_1 + x_2 + \cdots + x_k = n$, $x_1, x_2, \ldots, x_k$ nonnegative integers, and otherwise $p(x_1, x_2, \ldots, x_k) = 0$.

To illustrate, suppose that 10 fish are randomly chosen with replacement from a large lake, and classified as to whether they are bass, trout, or some other kind of fish. Let X_1 be the number of bass sampled, X_2 the number of trout, and X_3 the number of other kinds of fish, and assume that the proportion of bass in the lake is $p_1 = 0.25$, the proportion of trout in the lake is $p_2 = 0.30$, and that the proportion of other kinds of fish in the lake is $p_3 = 1 - 0.25 - 0.30 = 0.45$. Then

$$P\{X_1 = 3, X_2 = 4, \text{ and } X_3 = 3\} = p(3, 4, 3)$$

$$= \frac{10!}{3!\,4!\,3!}(0.25)^3(0.30)^4(0.45)^3 = 0.0484.$$

Let us restrict ourselves to the case $k = 3$ (the *trinomial* distribution) and find the marginal probability mass function $p_{X_1}(x_1)$ of X_1. From (6.2),

$$p_{X_1}(x_1) = \sum_{x_2} \sum_{x_3} p(x_1, x_2, x_3) = \sum_{x_2=0}^{n-x_1} p(x_1, x_2, n - x_1 - x_2),$$

since $p(x_1, x_2, x_3) = 0$ unless $x_3 = n - x_1 - x_2$ and $0 \le x_3 = n - x_1 - x_2 \le n$ (i.e., $0 \le x_1 + x_2 \le n$). Thus,

$$p_{X_1}(x_1) = \sum_{x_2=0}^{n-x_1} \frac{n!}{x_1!\,x_2!\,(n - x_1 - x_2)!} p_1^{x_1} p_2^{x_2}(1 - p_1 - p_2)^{n - x_1 - x_2}$$

$$= \frac{n!}{x_1!\,(n - x_1)!} p_1^{x_1} \left[\sum_{x_2=0}^{n-x_1} \binom{n - x_1}{x_2} p_2^{x_2}(1 - p_1 - p_2)^{(n - x_1) - x_2} \right]$$

(6.29)

$$= \frac{n!}{x_1!\,(n - x_1)!} p_1^{x_1} [(p_2 + 1 - p_1 - p_2)^{n - x_1}]$$

$$= \frac{n!}{x_1!\,(n - x_1)!} p_1^{x_1}(1 - p_1)^{n - x_1},$$

which is the probability mass function of the binomial distribution with parameters n and $p = p_1$. This result is intuitively reasonable since X_1 is the number of times outcome 1 appears, and $X_2 + X_3 = n - X_1$ is the number of times that outcome 1 does not appear in the n repetitions of the experiment. Similarly, X_2 has a (marginal) binomial distribution with parameters n and p_2, and X_3 has a (marginal) binomial distribution with parameters n and p_3. It follows from these results that

(6.30) $$\mu_{X_i} = np_i, \qquad \sigma_{X_i}^2 = np_i(1 - p_i), \quad i = 1, 2, 3.$$

We have already noted that $p(x_1, x_2, x_3) = 0$ unless $x_3 = n - x_1 - x_2$ and $0 \leq x_1 + x_2 \leq n$. It follows that

$$p_{X_1, X_2}(x_1, x_2) = \sum_{x_3} p(x_1, x_2, x_3) = p(x_1, x_2, n - x_1 - x_2)$$

(6.31)

$$= \frac{n!}{x_1! x_2! (n - x_1 - x_2)!} p_1^{x_1} p_2^{x_2}(1 - p_1 - p_2)^{n - x_1 - x_2}$$

for $0 \leq x_1 + x_2 \leq n$, x_1 and x_2 nonnegative integers, and otherwise $p_{X_1, X_2}(x_1, x_2) = 0$. Using (4.9) and $p_{X_1, X_2}(x_1, x_2)$, we can find the covariance σ_{X_1, X_2} by first computing

$$E[X_1 X_2] = \sum_{x_1} \sum_{x_2} p_{X_1, X_2}(x_1, x_2).$$

From (6.31), dropping off terms in which $x_1 = 0$ or $x_2 = 0$,

$$E[X_1 X_2]$$

$$= \sum_{x_1 = 0}^{n} \sum_{x_2 = 0}^{n - x_1} x_1 x_2 \frac{n!}{x_1! x_2! (n - x_1 - x_2)!} p_1^{x_1} p_2^{x_2}(1 - p_1 - p_2)^{n - x_1 - x_2}$$

$$= \sum_{x_1 = 1}^{n - 1} \sum_{x_2 = 1}^{n - x_1} \frac{n!}{(x_1 - 1)! (x_2 - 1)! (n - x_1 - x_2)!} p_1^{x_1} p_2^{x_2}(1 - p_1 - p_2)^{n - x_1 - x_2}$$

$$= n(n - 1)p_1 p_2 \left[\sum_{i = 0}^{n - 2} \sum_{j = 0}^{n - 2 - i} \frac{(n - 2)!}{i! j! (n - 2 - i - j)!} p_1^i p_2^j (1 - p_1 - p_2)^{n - 2 - i - j} \right]$$

$$= n(n - 1)p_1 p_2,$$

since the sum in brackets is the sum over x_1 and x_2 of a joint probability mass function of the form (6.31) based on $n - 2$ observations, and thus must equal 1. Hence,

(6.32) $\sigma_{X_1, X_2} = E[X_1 X_2] - \mu_{X_1} \mu_{X_2} = n(n - 1)p_1 p_2 - n^2 p_1 p_2 = -np_1 p_2.$

The result that $\sigma_{X_1, X_2} = -np_1 p_2 \leq 0$ is reasonable, considering that $0 \leq X_1 + X_2 \leq n$. If X_1 is large, then X_2 must be small, and vice versa, or else $X_1 + X_2$ can exceed n. In general,

$$\sigma_{X_i, X_j} = -np_i p_j, \quad i \neq j,$$

and

(6.33) $\rho_{X_i, X_j} = \dfrac{-np_i p_j}{\sqrt{np_i(1 - p_i)np_j(1 - p_j)}} = -\sqrt{\dfrac{p_i p_j}{(1 - p_i)(1 - p_j)}}.$

Finally, since $X_1 + X_2 + X_3 = n$, it is easy to see that

$$p_{X_1 | X_2 = x_2, X_3 = x_3}(x_1) = \begin{cases} 1, & \text{if } x_1 = n - x_2 - x_3, \\ 0, & \text{otherwise,} \end{cases}$$

and less trivially, noting that $x_2 + x_3 = n - x_1$, $1 - p_1 = p_2 + p_3$, that

$$p_{X_2, X_3 | X_1 = x_1}(x_2, x_3) = \frac{p(x_1, x_2, x_3)}{p_{X_1}(x_1)}$$

(6.34)
$$= \frac{\dfrac{n!}{x_1! x_2! x_3!} p_1^{x_1} p_2^{x_2} p_3^{x_3}}{\dfrac{n!}{x_1! (n - x_1)!} p_1^{x_1} (p_2 + p_3)^{x_2 + x_3}}$$

$$= \frac{(n - x_1)!}{x_2! x_3!} \left(\frac{p_2}{p_2 + p_3} \right)^{x_2} \left(\frac{p_3}{p_2 + p_3} \right)^{x_3}$$

for $x_2 + x_3 = n - x_1$, x_2 and x_3 nonnegative integers, and otherwise $p_{X_2, X_3 | X_1 = x_1}(x_2, x_3) = 0$. Thus, the conditional distribution of X_2 and X_3 given $X_1 = x_1$ is again a multinomial distribution. Also, from (6.24) and (6.34), noting that $p_{X_2, X_3 | X_1 = x_1}(x_2, x_3) = 0$ unless $x_2 = n - x_1 - x_2$, we have

$$p_{X_2 | X_1 = x_1}(x_2) = \sum_{x_3} p_{X_2, X_3 | X_1 = x_1}(x_2, x_3)$$

(6.35)
$$= p_{X_2, X_3 | X_1 = x_1}(x_2, n - x_1 - x_2)$$

$$= \frac{(n - x_1)!}{x_2! (n - x_1 - x_2)!} \left(\frac{p_2}{p_2 + p_3} \right)^{x_2} \left(1 - \frac{p_2}{p_2 + p_3} \right)^{n - x_1 - x_2},$$

which is a binomial distribution with parameters $n - x_1$ and $p = (p_2/(p_2 + p_3))$. Thus, if in x_1 of our repetitions we observe outcome 1, the remainder of the $n - x_1$ repetitions act as $n - x_1$ independent Bernoulli trials where outcome 2 is a success and outcome 3 is a failure.

Remark. Since in the conditional distribution (6.34), we must have $X_2 + X_3 = n - X_1$ with conditional probability equal to 1, it follows from the meaning of a correlation of -1 that the partial correlation coefficient $\rho_{X_2, X_3 | X_1 = x_1}$ must equal -1. We also remark that it can be seen in many ways [e.g., from (6.33), or by comparing (6.35) to $p_{X_2}(x_2)$], that if X_1, X_2, and X_3 have a trinomial distribution, then X_1, X_2, and X_3 are not mutually independent. Of course, since it must be the case that $X_1 + X_2 + X_3 = n$, this fact is obvious a priori.

The Multivariate Dirichlet Distribution

The multivariate Dirichlet distribution, which is a generalization of the bivariate Dirichlet distribution discussed in Example 2.3, has wide applications in statistical inference. The continuous random variables $X_1, X_2, \ldots, X_k$

have a multivariate Dirichlet distribution if they possess a multivariate probability density function of the form

$$f(x_1, x_2, \ldots, x_k)$$

(6.36)
$$= \frac{\Gamma\left(\sum\limits_{i=1}^{k+1} a_i\right)}{\Gamma(a_1)\Gamma(a_2)\cdots\Gamma(a_k)\Gamma(a_{k+1})} \prod_{i=1}^{k} x_i^{a_i-1} \left(1 - \sum_{i=1}^{k} x_i\right)^{a_{k+1}-1}$$

for $0 \leq x_i$, $i = 1, \ldots, k$, $\sum_{i=1}^{k} x_i \leq 1$. Here, the $k+1$ positive constants $a_1, a_2, \ldots, a_k, a_{k+1}$ are the parameters of the Dirichlet distribution. For convenience in writing formulas, we sometimes let $A = \sum_{i=1}^{k+1} a_i$.

If X_1, X_2, X_3 have a trivariate Dirichlet distribution, then

$$f_{X_1, X_2, X_3}(x_1, x_2, x_3)$$

(6.37)
$$= \frac{\Gamma(A)}{\Gamma(a_1)\Gamma(a_2)\Gamma(a_3)\Gamma(a_4)} x_1^{a_1-1} x_2^{a_2-1} x_3^{a_3-1} (1 - x_1 - x_2 - x_3)^{a_4-1}$$

for $0 \leq x_1, x_2, x_3, x_1 + x_2 + x_3 \leq 1$, and $f_{X_1, X_2, X_3}(x_1, x_2, x_3) = 0$ otherwise. To obtain the marginal joint distribution of X_1 and X_2, note that unless $0 \leq x_1, x_2, x_1 + x_2 \leq 1$, $f_{X_1, X_2, X_3}(x_1, x_2, x_3) = 0$ and thus $f_{X_1, X_2}(x_1, x_2) = \int_{-\infty}^{\infty} f_{X_1, X_2, X_3}(x_1, x_2, x_3)\, dx_3 = 0$. If $0 \leq x_1, x_2, x_1 + x_2 \leq 1$, then it follows from (6.33) that

(6.38)

$$f_{X_1, X_2}(x_1, x_2)$$

$$= \frac{\Gamma(A)}{\Gamma(a_1)\Gamma(a_2)\Gamma(a_3)\Gamma(a_4)} \int_0^{1-x_1-x_2} x_1^{a_1-1} x_2^{a_2-1} x_3^{a_3-1} (1 - x_1 - x_2 - x_3)^{a_4-1}\, dx_3$$

$$= \frac{\Gamma(A) x_1^{a_1-1} x_2^{a_2-1} (1 - x_1 - x_2)^{a_3+a_4-1}}{\Gamma(a_1)\Gamma(a_2)\Gamma(a_3)\Gamma(a_4)} \int_0^1 u^{a_3-1} (1 - u)^{a_4-1}\, du$$

$$= \frac{\Gamma(a_1 + a_2 + a_3 + a_4)}{\Gamma(a_1)\Gamma(a_2)\Gamma(a_3 + a_4)} x_1^{a_1-1} x_2^{a_2-1} (1 - x_1 - x_2)^{a_3+a_4-1},$$

where we have changed variables from x_3 to $u = x_3/(1 - x_1 - x_2)$, and used the fact that

$$\int_0^1 u^{a_3-1} (1 - u)^{a_4-1}\, du = B(a_3, a_4) = \frac{\Gamma(a_3)\Gamma(a_4)}{\Gamma(a_3 + a_4)}.$$

We recognize (6.38) as being the joint probability density function of a bivariate Dirichlet distribution with parameters $a = a_1$, $b = a_2$, $c = a_3 + a_4$. It then follows from the results obtained in Example 2.3 that X_1 and X_2 both have marginal distributions which are beta distributions—for example, X_1 has a beta distribution with parameters $r = a_1$ and $s = a_2 + a_3 + a_4$. Also,

$$\mu_{X_1} = \frac{a_1}{A}, \qquad \sigma^2_{X_1} = \frac{a_1(A - a_1)}{A^2(A + 1)}, \qquad \sigma_{X_1, X_2} = \frac{-a_1 a_2}{A^2(A + 1)},$$

where here $A = \sum_{i=1}^{4} a_i$.

These results can be extended to the general multivariate Dirichlet distribution to show that

(1) each X_i has a marginal beta distribution with parameters $r = a_i$, $s = A - a_i$, mean $\mu_{X_i} = a_i/A$, and variance

$$\sigma^2_{X_i} = a_i(A - a_i)/A^2(A + 1);$$

(2) X_i and X_j, $i \neq j$, have a marginal bivariate Dirichlet distribution with parameters $a = a_i$, $b = a_j$, $c = A - a_i - a_j$, and

$$\sigma_{X_i, X_j} = \frac{-a_i a_j}{A^2(A + 1)}, \qquad \rho_{X_i, X_j} = \frac{-a_i a_j}{\sqrt{a_i(A - a_i)a_j(A - a_j)}}.$$

Since $\sigma_{X_i, X_j} < 0$, X_i and X_j cannot be pairwise independent, and hence X_1, $X_2, \ldots, X_k$ are mutually dependent.

When X_1, X_2, and X_3 have a trivariate Dirichlet distribution, the conditional probability density function of X_3 given $X_1 = x_1$, $X_2 = x_2$ can be obtained by dividing (6.37) by (6.38). Thus,

$$(6.39) \quad f_{X_3|X_1=x_1, X_2=x_2}(x_3) = \frac{\Gamma(a_3 + a_4)}{\Gamma(a_3)\Gamma(a_4)} \frac{x_3^{a_3-1}(1 - x_1 - x_2 - x_3)^{a_4-1}}{(1 - x_1 - x_2)^{a_3+a_4-1}}$$

for $0 \leq x_3 \leq 1 - x_1 - x_2$, and otherwise $f_{X_3|X_1=x_1, X_2=x_2}(x_3) = 0$. If we make the same change of variables used in obtaining (6.38), we find that the regression function of X_3 on X_1 and X_2 is

$$\mu_{X_3|X_1=x_1, X_2=x_2} = \int_0^{1-x_1-x_2} x_3 \, f_{X_3|X_1=x_1, X_2=x_2}(x_3) \, dx_3$$

(6.40)

$$= (1 - x_1 - x_2)\frac{a_3}{a_3 + a_4},$$

which is linear in x_1 and x_2.

We can also find the conditional joint probability density function of, say, X_1 and X_2 given that $X_3 = x_3$. Since X_3 has a marginal beta distribution with parameters $r = a_3$, $s = A - a_3$, the marginal probability density function of X_3 is

$$f_{X_3}(x_3) = \begin{cases} \dfrac{\Gamma(A)}{\Gamma(a_3)\Gamma(A - a_3)} x_3^{a_3 - 1}(1 - x_3)^{A - a_3 - 1}, & \text{if } 0 \le x_3 \le 1, \\[2mm] 0, & \text{otherwise.} \end{cases}$$

Thus, for $0 \le x_3 \le 1$, dividing (6.37) by $f_{X_3}(x_3)$, we obtain

(6.41)
$$\begin{aligned} &f_{X_1, X_2 | X_3 = x_3}(x_1, x_2) \\[2mm] &= \frac{\Gamma(A - a_3)}{\Gamma(a_1)\Gamma(a_2)\Gamma(a_4)} \frac{x_1^{a_1 - 1} x_2^{a_2 - 1}(1 - x_3 - x_1 - x_2)^{a_4 - 1}}{(1 - x_3)^{A - a_3 - 1}} \end{aligned}$$

for $0 \le x_1, x_2, x_1 + x_2 \le 1$, and $f_{X_1, X_2 | X_3 = x_3}(x_1, x_2) = 0$ otherwise. Note that for $r_1, r_2 \ge 0$,

$$\begin{aligned} &E[X_1^{r_1} X_2^{r_2} | X_3 = x_3] \\[2mm] &= \frac{\Gamma(A - a_3)}{\Gamma(a_1)\Gamma(a_2)\Gamma(a_4)} \\[2mm] &\quad \times \int_0^{1 - x_2} \int_0^{1 - x_3 - x_2} \frac{x_1^{r_1} x_2^{r_2} x_1^{a_1 - 1} x_2^{a_2 - 1}(1 - x_3 - x_1 - x_2)^{a_4 - 1}}{(1 - x_3)^{A - a_3 - 1}} \, dx_1 \, dx_2 \\[2mm] &= (1 - x_3)^{r_1 + r_2} \frac{\Gamma(A - a_3)\Gamma(a_1 + r_1)\Gamma(a_2 + r_2)}{\Gamma(A - a_3 + r_1 + r_2)\Gamma(a_1)\Gamma(a_2)} \\[2mm] &\quad \times \int_0^1 \int_0^{1 - v} \frac{\Gamma(A - a_3 + r_1 + r_2) u^{a_1 + r_1 - 1} v^{a_2 + r_2 - 1}(1 - u - v)^{a_4 - 1} \, du \, dv}{\Gamma(a_1 + r_1)\Gamma(a_2 + r_2)\Gamma(a_4)} \\[2mm] &= (1 - x_3)^{r_1 + r_2} \frac{\Gamma(A - a_3)\Gamma(a_1 + r_1)\Gamma(a_2 + r_2)}{\Gamma(A - a_3 + r_1 + r_2)\Gamma(a_1)\Gamma(a_2)}, \end{aligned}$$

where we have made the changes of variables from x_1 to $u = x_1/(1 - x_3)$, and from x_2 to $v = x_2/(1 - x_3)$, and then used the fact that the integral over u and v is the total probability of a bivariate Dirichlet distribution with parameters $a = a_1 + r_1$, $b = a_2 + r_2$, and $c = A - a_3 - a_1 - a_2 = a_4$, and thus must be 1. Hence,

$$\mu_{X_1 | X_3 = x_3} = E[X_1 X_2^0 | X_3 = x_3] = (1 - x_3)\frac{a_1}{A - a_3}$$

and similarly $\mu_{X_2|X_3=x_3} = (1 - x_3)a_2/(A - a_3)$. Also,

$$\sigma^2_{X_1|X_3=x_3} = E[X_1^2 X_2^0 \mid X_3 = x_3] - (\mu_{X_1|X_3=x_3})^2$$

$$= (1 - x_3)^2 \frac{a_1(A - a_3 - a_1)}{(A - a_3)^2(A - a_3 + 1)}$$

and

$$\sigma^2_{X_2|X_3=x_3} = (1 - x_3)^2 \frac{a_2(A - a_3 - a_2)}{(A - a_3)^2(A - a_3 + 1)}.$$

Finally,

$$\sigma_{X_1, X_2|X_3=x_3} = E[X_1 X_2 \mid X_3 = x_3] - \mu_{X_1|X_3=x_3}\mu_{X_2|X_3=x_3}$$

$$= (1 - x_3)^2 \frac{-a_1 a_2}{(A - a_3)^2(A - a_3 + 1)},$$

from which it follows that the partial correlation coefficient of X_1 and X_2 given $X_3 = x_3$ is

$$\rho_{X_1, X_2|X_3=x_3} = \frac{\sigma_{X_1, X_2|X_3=x_3}}{(\sigma_{X_1|X_3=x_3})(\sigma_{X_2|X_3=x_3})} = \frac{-a_1 a_2}{\sqrt{a_1(A - a_3 - a_1)a_2(A - a_3 - a_2)}}.$$

Compared to the correlation coefficient

$$\rho_{X_1, X_2} = \frac{-a_1 a_2}{\sqrt{a_1(A - a_1)a_2(A - a_2)}},$$

the partial correlation coefficient $\rho_{X_1, X_2|X_3=x_3}$ is more negative (closer to -1), indicating that the covariation of X_1 and of X_2 with X_3 tends to "mask" the relationship of X_1 with X_2.

EXERCISES

1. A sample of 455 married couples were asked to (i) rate the happiness of their marriage, and (ii) indicate the extent to which they agree on ways of dealing with their in-laws [Burgess and Cottrell (1955)]. Let X be the rating $(-2, -1, 0, 1, 2)$ of the happiness of the marriage of a given couple. Let Y be a rating $(0, 1, 2, 3, 4, 5)$ of the extent of agreement between the marriage partners on ways of dealing with in-laws. The joint distribution of X and Y is summarized in Table E.1.
   Suppose that the relative frequencies shown in Table E.1 are the actual probabilities for the joint distribution of X and Y.
   (a) Find the marginal distributions of X and of Y.
   (b) Find $P\{X = Y\}$.

Table E.1: *Comparison of Ratings of Happiness in Marriage with Extent of Agreement on Ways of Dealing with In-laws*

(*Observed relative frequencies for 455 couples*)

Extent of agreement, y	Rating of happiness, x				
	-2	-1	0	1	2
5	0.015	0.011	0.046	0.077	0.251
4	0.011	0.015	0.042	0.073	0.103
3	0.011	0.028	0.033	0.029	0.042
2	0.018	0.024	0.022	0.015	0.015
1	0.009	0.022	0.009	0.002	0.002
0	0.028	0.029	0.009	0.007	0.002

    (c) Find the conditional distributions of Y given that $X = x$, for $x = -2, -1, 0, 1,$ and 2.

    (d) Find $E[Y \mid X = x_1]$ for $x = -2, -1, 0, 1,$ and 2.

    (e) Find $\rho_{X,Y}$. Are X and Y independent?

    (f) What value would you predict for Y in ignorance of X? What value would you predict for Y if you knew that $X = -2$?

2. Let X be the number of successes in 4 independent Bernoulli trials with probability of success $p = 0.6$. Let Y be the length of the longest "run" in the 4 trials. A "run" is a consecutive sequence of 1 or more identical outcomes. For example, the outcome FSSF has 3 runs, and the longest run has length 2.

    (a) Find the joint probability mass function $p_{X,Y}(x, y)$ of X and Y.

    (b) Find the marginal probability mass function $p_Y(y)$ of Y. Also find μ_Y and σ_Y^2.

    (c) Find the conditional probability mass function $p_{X \mid Y=2}(x)$ of X given that $Y = 2$.

    (d) Find the conditional probability mass function $p_{Y \mid X=2}(y)$ of Y given that $X = 2$.

    (e) Find $\rho_{X,Y}$.

3. Let X and Y have joint probability mass function:

$$p(x, y) = \begin{cases} p^2(1-p)^y, & \text{if } x, y = 0, 1, 2, \ldots, \text{ and } x \le y, \\ 0, & \text{otherwise,} \end{cases}$$

where $0 < p < 1$.

    (a) Show that the marginal distribution of X is a negative binomial distribution with parameters $r = 1$ and p, while the marginal distribution of Y is a negative binomial distribution with parameters $r = 2$ and p.

(b) Find $\sigma_{X,Y}$ and $\rho_{X,Y}$. Are X and Y probabilistically independent?
(c) Find $p_{X|Y=y}(x)$, $\mu_{X|Y=y}$, and $\sigma^2_{X|Y=y}$.
(d) Find $p_{Y|X=x}(y)$, $\mu_{Y|X=x}$, and $\sigma^2_{Y|X=x}$.
(e) The joint distribution of X and Y given above can arise when we observe independent Bernoulli trials with probability p of success, X the number of failures before the first success is observed, and Y the total number of failures before the second success is observed. If we observe $X = 5$, what prediction should we make for the total number of failures observed before the second success is observed?

4. Let X and Y have joint probability density function

$$f(x, y) = \begin{cases} \frac{1}{3} y^3 e^{-xy-y}, & \text{for } x > 0 \text{ and } y > 0, \\ 0, & \text{otherwise.} \end{cases}$$

(a) Find $f_X(x)$, $f_Y(y)$, μ_X, μ_Y, σ^2_X, and σ^2_Y.
(b) Find $f_{Y|X=x}(y)$, $\mu_{Y|X=x}$, and $\sigma^2_{Y|X=x}$.
(c) Find $f_{X|Y=y}(x)$, $\mu_{X|Y=y}$, and $\sigma^2_{X|Y=y}$.
(d) Find $\sigma_{X,Y}$ and $\rho_{X,Y}$. Are X and Y independent?
(e) If you know that $X = 3$, what value would you predict for Y?

5. Let X and Y have joint probability density function

$$f(x, y) = \begin{cases} 2e^{-x-y}, & \text{if } 0 \le x \le y < \infty, \\ 0, & \text{otherwise.} \end{cases}$$

(a) Find $P\{X = Y\}$.
(b) Find $P\{XY = 1\}$.
(c) Find $P\{X > 2X\}$.
(d) Are X and Y independent?

6. Let X and Y have joint probability density function

$$f(x, y) = \begin{cases} c, & \text{if } x^2 + y^2 \le 1, \\ 0, & \text{otherwise.} \end{cases}$$

(a) Find the value of c.
(b) Show that $\mu_X = \mu_Y = \rho_{X,Y} = 0$.
(c) Show that X and Y are dependent.

7. If X and Y have a bivariate Dirichlet probability density function $f(x, y)$, show that $\int_{-\infty}^{\infty} \int_{-\infty}^{\infty} f(x, y) \, dx \, dy = 1$.

8. Two random variables X and Y have a joint probability distribution with $\mu_X = 0.5$, $\mu_Y = 2.0$, $\sigma^2_X = 4.0$, $\sigma^2_Y = 16.0$, $\sigma_{X,Y} = 4.0$. Let $Z = (0.5)X - 0.25$ and $W = (0.25)Y - 0.5$. Find
(a) μ_Z, μ_W, σ^2_Z, σ^2_W,   (b) $\sigma_{Z,Y}$, $\sigma_{X,W}$, $\sigma_{Z,W}$,
(c) $\rho_{X,Y}$, $\rho_{Z,Y}$, $\rho_{X,W}$, $\rho_{Z,W}$.

9. If X and Y are discrete random variables with possible values

$x_1 < x_2 < \cdots < x_r$ and $y_1 < y_2 < \cdots < y_s$, respectively, and joint cumulative distribution function $F_{X,Y}(x, y)$, show that

$$p(x_i, y_j) = F_{X,Y}(x_i, y_j) - F_{X,Y}(x_{i-1}, y_j)$$
$$- F_{X,Y}(x_i, y_{j-1}) + F_{X,Y}(x_{i-1}, y_{j-1}),$$

where $p(x, y)$ is the joint probability mass function of X and Y.

10. Let X and Y have joint cumulative distribution function $F_{X,Y}(x, y)$ and suppose that

$$F_{X,Y}(x, y) = F_X(x)F_Y(y), \quad \text{all } x, \text{ all } y.$$

Show separately (a) in the discrete case and (b) in the continuous case that X and Y are independent.

11. Let X, Y be jointly distributed random variables with joint cumulative distribution $F_{X,Y}(x, y)$. Show that

$$P\{a < X \le b, c < Y \le d\} = F_{X,Y}(b, d) - F_{X,Y}(b, c)$$
$$- F_{X,Y}(a, d) + F_{X,Y}(a, c).$$

[Hint: Draw a picture.]

12. Let X, Y be jointly distributed random variables, and let $W = \max(X, Y)$, $V = \min(X, Y)$. Suppose that (X, Y) has joint cumulative distribution function $F_{X,Y}(x, y)$. Show that
(a) $F_W(w) = F_{X,Y}(w, w)$,
(b) $F_V(v) = F_X(v) + F_Y(v) - F_{X,Y}(v, v)$,

(c) $F_{V,W}(v, w) = \begin{cases} F_{XY}(w, w) & \text{if } v > w, \\ F_{X,Y}(v, w) + F_{X,Y}(w, v) - F_{X,Y}(v, v), & \text{if } v \le w. \end{cases}$

[Hint: $W \le w$ if and only if $X \le w$ and $Y \le w$, while $V > v$ if and only if $X \le v$ and $Y \le v$. Further, $v < V \le W \le w$ if and only if $v < X, Y \le w$.]

13. Suppose that X and Y are independent random variables, and that both X and Y have a gamma distribution with parameters r and θ. Let $V = \min(X, Y)$, $W = \max(X, Y)$. Find
(a) $f_{V,W}(v, w)$,   (b) $f_V(v)$,   (c) $f_W(w)$,   (d) $f_{V|W=w}(v)$.
(e) Are V and W independent?

14. Suppose that X and Y have joint probability density function

$$f(x, y) = \begin{cases} [\pi\sqrt{1 - \rho^2}]^{-1} \exp\left[-\frac{1}{2}\left(\frac{x^2 - 2\rho xy + y^2}{1 - \rho^2}\right)\right], & \text{if } xy \ge 0, \\ 0, & \text{if } xy < 0. \end{cases}$$

(a) Show that X and Y each have marginal $N(0, 1)$ distributions, but that $f(x, y)$ is not the joint probability density function of a bivariate normal distribution.
(b) Find $\rho_{X,Y}$.
(c) Find $f_{X|Y=y}(x)$ and $\mu_{X|Y=y}$.
(d) Find $f_{Y|X=x}(y)$ and $\mu_{Y|X=x}$.

15. Suppose that the heights of mothers (Y) and sons (X) followed the distribution fitted in Example 5.1; that is,

$$(X, Y) \sim \text{BVN}((68.65, 62.48), (7.344, 5.712, 0.5)).$$

(a) Find $\mu_{X|Y=y}$ and $\sigma^2_{X|Y=y}$. If you were told that a mother's height is $Y = 65$ inches, what value would you predict for her son's height? If you were not told the mother's height, what height would you predict for her son? Compare $\sigma^2_{X|Y=65}$ with σ^2_X.

(b) Find the probability $P\{Y > X\}$ that a mother will be taller than her son.

16. The relationship, in German words, between the number X of syllables and the number Y of phonemes (letters) was studied for 20,453 headwords in Viëtor-Meyer's *Deutches Aussprachewörterbuch* [see Herdan (1966, p. 301)]. It was found that

$$\hat{\mu}_X = 2.780, \qquad \hat{\mu}_Y = 7.296, \qquad \hat{\sigma}^2_X = 1.2122,$$
$$\hat{\sigma}^2_Y = 6.6793, \qquad \hat{\sigma}_{X,Y} = 2.4040.$$

If we regard X and Y as if they are continuous random variables that have been rounded off to the nearest integer, then a bivariate normal distribution provides a reasonably good fit to the data.

(a) Find $\hat{\rho}_{X,Y}$.

(b) Find $\mu_{Y|X=x}$ and $\sigma^2_{Y|X=x}$. If you are told that a German headword has $X = 5$ syllables, how many letters would you predict that it had? If you were not told the number of syllables that the headword has, how many letters Y would you predict the word has? Compare $\sigma_{Y|X=5}$ with σ_Y.

17. Suppose that $(X, Y) \sim \text{BVN}((5, 3), (49, 16, 0.5))$. Use Table 5.3 to find

(a) $P\{X > 6 \text{ and } Y > 4\}$,   (b) $P\{X \le 6 \text{ and } Y \le 4\}$,

(c) $P\{X > 4 \text{ and } Y > 2\}$,   (d) $P\{X \le 4 \text{ and } Y \le 2\}$,

(e) $P\{X \ge 4 \text{ and } Y \ge 4\}$,   (f) $P\{X \le 6 \text{ and } Y \le 2\}$,

(g) $P\{4 \le X \le 6 \text{ and } 2 \le Y \le 4\}$.

18. An industrial process simultaneously produces two interlocking steel pipes. The lengths X and Y of the pipes have a bivariate normal distribution with parameters $\mu_X = \mu_Y = 3.25$ inches, $\sigma_X = \sigma_Y = 0.05$ inch, and ρ. After the pipes are produced, they are fitted together. Assuming that the pipes are joined without their combined length being lessened, find the probability that the combined length of the two joined pipes is less than 6.60 inches when

(a) $\rho = 0.0$,   (b) $\rho = 0.5$,   (c) $\rho = -0.5$,   (d) $\rho = 1.0$,

(e) $\rho = -1.0$.

19. For three numbers x, y, z, it is a fact that $x \ge y$ and $y \ge z$ implies that $x \ge z$. Suppose that the discrete random variables X, Y, Z have the following joint probability mass function:

(x, y, z)	$(1, 1, 1)$	$(1, 1, 0)$	$(1, 0, 1)$	$(0, 1, 1)$
$p(x, y, z)$	0.25	0.25	0.25	0.25

Find $P\{X \geq Y\}$, $P\{Y \geq Z\}$, and $P\{Z \geq X\}$.

20. A "random walk" is the name given to a certain sequence of move-
     ments (steps) of a particle, where at each step the particle either moves
     one unit forward with probability p, or one unit backward with probabi-
     lity $1 - p$. The results of the various steps are mutually probabilistically
     independent. Let X_k represent the result of the kth step of the particle:
     $X_k = 1$ if the particle moves one step forward, and $X_k = -1$ if the
     particle moves one step backward.
     (a) Find the joint probability mass function of X_1, X_2, and X_3.
     (b) Let U_k be the position relative to the starting point of the particle
         after k steps, so that $U_k = \sum_{i=1}^{k} X_i$. Find the joint probability mass
         function of U_1, U_2, and U_3.
     (c) Find the marginal joint probability mass function of U_2 and U_3.
     (d) Find the marginal probability mass function of U_3.

21. Let X, Y, and Z have joint probability density function
     $$f(x, y, z) = \begin{cases} z^2 e^{-z(1+x+y)}, & \text{if } x, y, z > 0, \\ 0, & \text{otherwise.} \end{cases}$$
     (a) Show that $\int_{-\infty}^{\infty} \int_{-\infty}^{\infty} \int_{-\infty}^{\infty} f(x, y, z) \, dx \, dy \, dz = 1$.
     (b) Are X and Y pairwise independent?
     (c) Find $f_Z(z)$.
     (d) Find $f_{X, Z}(x, z)$, $f_{X|Z=z}(x)$.
     (e) Find $f_{X, Y|Z=z}(x, y)$ and show that $f_{X, Y|Z=z}(x, y) = f_{X|Z=z}(x) f_{Y|Z=z}(y)$.
         What does this result imply about the relationship between X and
         Y? What is the value of $\rho_{X, Y|Z=z}$?

22. If X, Y, and Z are mutually independent random variables, show in (a)
     the discrete case and (b) the continuous case that X and Y, X and Z,
     and Y and Z are pairwise independent.

23. Let X, Y, and Z have the joint probability mass function $p(x, y, z)$
     defined by

(x, y, z)	$(1, 0, 0)$	$(0, 1, 0)$	$(0, 0, 1)$	$(1, 1, 1)$
$p(x, y, z)$	0.25	0.25	0.25	0.25

     (a) Find $p_{X, Y}(x, y)$, $p_{X, Z}(x, z)$, $p_{Y, Z}(y, z)$ and show that X and Y, X and
         Z, and Y and Z are pairwise independent.
     (b) However, show that X, Y, and Z are *not* mutually independent.
     (c) Find $p_{X|Y=1}(x)$.
     (d) Find $p_{X, Y|Z=1}(x, y)$.

10

Joint Transformations of

Several Random Variables

In Chapter 4, we discussed transformations $Y = g(X)$ of a single random variable X. In a multivariable experiment, we may wish to transform our original measured variables $X_1, X_2, ..., X_p$ into new variables $Y_i = g_i(X_1, X_2, ..., X_p)$, $i = 1, 2, ..., p$. For example, we may observe the weights $X_1, X_2, ..., X_p$ of an organism at times $t_1, t_2, ..., t_p$, but desire to model the covariability of the initial weight $Y_1 = X_1$, and the growths $Y_j = X_j - X_{j-1}$ in weight, $j = 2, ..., p$, during each succeding time period.

In the present chapter, we study methods of obtaining the joint distribution of $Y_i = g_i(X_1, X_2, ..., X_p)$, $i = 1, 2, ..., p$, from a knowledge of the joint distribution of $X_1, X_2, ..., X_p$. The case $p = 2$ is studied in Section 1, and the case $p > 2$ in Section 2. Section 3 treats the case where we are interested in reducing the number of variables under consideration by transforming $X_1, X_2, ..., X_p$ to a single random variable $Y = g(X_1, X_2, ..., X_p)$ (a *many-to-one transformation*). Finally, Section 4 deals with the very important special case

$$Y = \sum_{i=1}^{p} a_i X_i.$$

1. TRANSFORMATIONS OF TWO RANDOM VARIABLES

Suppose that the random variables X_1 and X_2 have a bivariate distribution. We are interested in finding the joint distribution of Y_1 and Y_2 where these new random variables are the result of a *joint one-to-one transformation*

(1.1) $$Y_1 = g_1(X_1, X_2), \qquad Y_2 = g_2(X_1, X_2),$$

of X_1 and X_2. For (1.1) to define a joint one-to-one transformation, there must exist two functions $h_1(y_1, y_2)$ and $h_2(y_1, y_2)$ such that

(1.2) $y_i = g_i(x_1, x_2)$, $i = 1, 2$ if and only if $x_i = h_i(y_1, y_2)$, $i = 1, 2$.

If X_1 and X_2 are discrete random variables with bivariate probability mass function $p_{X_1, X_2}(x_1, x_2)$, then it follows from (1.2) that the bivariate probability mass function $p_{Y_1, Y_2}(y_1, y_2)$ of Y_1 and Y_2 is given by

(1.3) $p_{Y_1, Y_2}(y_1, y_2) = p_{X_1, X_2}(h_1(y_1, y_2), h_2(y_1, y_2))$.

If X_1, X_2 have a joint probability density function, and if the first partial derivatives of $h_1(y_1, y_2)$ and $h_2(y_1, y_2)$ with respect to y_1 and to y_2 exist, then

(1.4) $f_{Y_1, Y_2}(y_1, y_2) = |J(y_1, y_2)| f_{X_1, X_2}(h_1(y_1, y_2), h_2(y_1, y_2))$,

where

(1.5) $J(y_1, y_2) = \left(\dfrac{\partial h_1(y_1, y_2)}{\partial y_1}\right)\left(\dfrac{\partial h_2(y_1, y_2)}{\partial y_2}\right) - \left(\dfrac{\partial h_1(y_1, y_2)}{\partial y_2}\right)\left(\dfrac{\partial h_2(y_1, y_2)}{\partial y_1}\right)$

is the Jacobian of the transformation $x_1 = h_1(y_1, y_2)$, $x_2 = h_2(y_1, y_2)$.

To verify (1.4), recall from Chapter 9, Section 2, that the *bivariate cumulative distribution function* $F_{Y_1, Y_2}(y_1, y_2)$ of two jointly distributed continuous random variables Y_1, Y_2 is determined from the *bivariate probability density function* $f_{Y_1, Y_2}(y_1, y_2)$ of Y_1, Y_2 by

$$F_{Y_1, Y_2}(y_1, y_2) = P\{Y_1 \le y_1, Y_2 \le y_2\}$$

(1.6)

$$= \int_{-\infty}^{y_2} \int_{-\infty}^{y_1} f_{Y_1, Y_2}(t_1, t_2) \, dt_1 \, dt_2,$$

and that $f_{Y_1, Y_2}(y_1, y_2)$ can be obtained from $F_{Y_1, Y_2}(y_1, y_2)$ through the expression

(1.7) $\dfrac{\partial^2}{\partial y_1 \partial y_2} F_{Y_1, Y_2}(y_1, y_2) = f_{Y_1, Y_2}(y_1, y_2)$.

Now

$$F_{Y_1, Y_2}(y_1, y_2) = P\{Y_1 \le y_1, Y_2 \le y_2\}$$

$$= P\{g_1(X_1, X_2) \le y_1, g_2(X_1, X_2) \le y_2\}$$

(1.8)

$$= \iint_{\substack{g_1(x_1, x_2) \le y_1 \\ g_2(x_1, x_2) \le y_2}} f_{X_1, X_2}(x_1, x_2) \, dx_1 \, dx_2$$

$$= \int_{-\infty}^{y_2} \int_{-\infty}^{y_1} |J(t_1, t_2)| f_{X_1, X_2}(h_1(t_1, t_2), h_2(t_1, t_2)) \, dt_1 \, dt_2,$$

where we have changed variable of integration from (x_1, x_2) to $t_1 = g_1(x_1, x_2)$, $t_2 = g_2(x_1, x_2)$, and made use of the theory of change of variables in multiple integration. Comparing (1.8) and (1.6), or making use of (1.7), demonstrates the validity of (1.4).

Example 1.1. Suppose that $Y_1 = X_1 + X_2$, $Y_2 = X_2$. Then

$$h_1(y_1, y_2) = y_1 - y_2$$

and $h_2(y_1, y_2) = y_2$, so that

$$p_{Y_1, Y_2}(y_1, y_2) = p_{X_1, X_2}(y_1 - y_2, y_2).$$

Example 1.2. In Chapter 8, we remarked that waiting times often have gamma distributions. Suppose that X_1 and X_2 are independent measurements of the times taken to complete two similar tasks. Assuming that X_1 and X_2 each have a gamma distribution with parameters θ and r, the joint density function of X_1 and X_2 is

$$f_{X_1, X_2}(x_1, x_2) = f_{X_1}(x_1)f_{X_2}(x_2)$$

$$= \begin{cases} \dfrac{\theta^{2r}x_1^{r-1}x_2^{r-1}e^{-\theta x_1}e^{-\theta x_2}}{[\Gamma(r)]^2}, & \text{if } 0 < x_1 < \infty, 0 < x_2 < \infty, \\ 0, & \text{otherwise.} \end{cases}$$

Consider new variables $Y_1 = X_1 + X_2$, $Y_2 = X_1/X_2$, which give the total time required and the ratio of the times required to do the tasks, respectively. The transformation $y_1 = x_1 + x_2$, $y_2 = x_1/x_2$ is one-to-one, with

$$x_1 = h_1(y_1, y_2) = \frac{y_1 y_2}{1 + y_2}, \qquad x_2 = h_2(y_1, y_2) = \frac{y_1}{1 + y_2}.$$

Applying (1.5), the Jacobian of this transformation is

$$J(y_1, y_2) = \left(\frac{y_2}{1 + y_2}\right)\left(\frac{-y_1}{(1 + y_2)^2}\right) - \left(\frac{y_1}{(1 + y_2)^2}\right)\left(\frac{1}{1 + y_2}\right) = -\frac{y_1}{(1 + y_2)^2}$$

Hence, from Equation (1.4), the joint density function of (Y_1, Y_2) is

$$f_{Y_1, Y_2}(y_1, y_2) = |J(y_1, y_2)| f_{X_1, X_2}(h_1(y_1, y_2), h_2(y_1, y_2))$$

$$= \begin{cases} \dfrac{y_1}{(1 + y_2)^2} \dfrac{\theta^{2r}y_1^{2r-2}y_2^{r-1}e^{-\theta y_1}}{(1 + y_2)^{2r-2}[\Gamma(r)]^2}, & \text{if } 0 < \dfrac{y_1 y_2}{1 + y_2} < \infty, \\ & 0 < \dfrac{y_1}{1 + y_2} < \infty, \\ 0, & \text{otherwise.} \end{cases}$$

$$= \begin{cases} \dfrac{\theta^{2r}y_1^{2r-1}e^{-\theta y_1}}{\Gamma(2r)} \dfrac{y_2^{r-1}}{B(r, r)(1 + y_2)^{2r}}, & \text{if } 0 < y_1 < \infty, \\ & 0 < y_2 < \infty, \\ 0, & \text{otherwise.} \end{cases}$$

It is interesting to note that $Y_1 = X_1 + X_2$ and $Y_2 = X_1/X_2$ have a joint density function which is the product of their marginal density functions, thus showing that Y_1 and Y_2 are independent.

Bivariate Linear Transformations

Let X_1, X_2 have the joint probability density function $f_{X_1, X_2}(x_1, x_2)$. A *bivariate linear transformation* from X_1, X_2 to Y_1, Y_2 has the form

(1.9)
$$Y_1 = a_{11}X_1 + a_{12}X_2,$$
$$Y_2 = a_{21}X_1 + a_{22}X_2.$$

By solving Equation (1.9) for the X's in terms of the Y's, we see that $y_i = a_{i1} x_1 + a_{i2} x_2$, $i = 1, 2$, if and only if $x_i = b_{i1} y_1 + b_{i2} y_2$, $i = 1, 2$, where

(1.10)
$$b_{11} = \frac{a_{22}}{a_{11}a_{22} - a_{12}a_{21}}, \qquad b_{12} = \frac{-a_{12}}{a_{11}a_{22} - a_{12}a_{21}},$$

$$b_{21} = \frac{-a_{21}}{a_{11}a_{22} - a_{12}a_{21}}, \qquad b_{22} = \frac{a_{11}}{a_{11}a_{22} - a_{12}a_{21}},$$

provided that $a_{11} a_{22} - a_{12} a_{21} \neq 0$.
Note that

$$h_1(y_1, y_2) = b_{11}y_1 + b_{12}y_2, \qquad h_2(y_1, y_2) = b_{21}y_1 + b_{22}y_2$$

so that, from (1.5),

$$J(y_1, y_2) = (b_{11})(b_{22}) - (b_{12})(b_{21})$$
$$= \frac{(a_{22})(a_{11}) - (-a_{12})(-a_{21})}{(a_{11}a_{22} - a_{12}a_{21})^2} = \frac{1}{a_{11}a_{22} - a_{12}a_{21}}.$$

Hence, by (1.4),

(1.11)
$$f_{Y_1, Y_2}(y_1, y_2) = \frac{1}{|a_{11}a_{22} - a_{12}a_{21}|} f_{X_1, X_2}(b_{11}y_1 + b_{12}y_2, b_{21}y_1 + b_{22}y_2).$$

For example, if we are interested in the average $Y_1 = \frac{1}{2}(X_1 + X_2)$ and difference $Y_2 = X_1 - X_2$ of the X's, then $a_{11} = a_{12} = \frac{1}{2}$, $a_{21} = -a_{22} = 1$, $a_{11} a_{22} - a_{12} a_{21} = -1$, and

$$b_{11} = 1, \qquad b_{12} = \frac{1}{2}, \qquad b_{21} = 1, \qquad b_{22} = -\frac{1}{2}.$$

Thus, from (1.11),

$$f_{Y_1, Y_2}(y_1, y_2) = f_{X_1, X_2}(y_1 + \tfrac{1}{2}y_2, y_1 - \tfrac{1}{2}y_2).$$

Example 1.3. Suppose that X_1 and X_2 are probabilistically independent and that $X_1 \sim N(0, 1)$, $X_2 \sim N(0, 1)$. Thus,

$$f_{X_1, X_2}(x_1, x_2) = f_{X_1}(x_1)f_{X_2}(x_2) = \frac{\exp\left[-\frac{1}{2}(x_1^2 + x_2^2)\right]}{2\pi},$$

so that

$$f_{Y_1, Y_2}(y_1, y_2) = f_{X_1, X_2}(y_1 + \tfrac{1}{2}y_2, y_1 - \tfrac{1}{2}y_2)$$

$$= \frac{\exp\left\{-\frac{1}{2}[(y_1 + \frac{1}{2}y_2)^2 + (y_1 - \frac{1}{2}y_2)^2]\right\}}{2\pi}$$

$$= \frac{\exp\left[-\frac{1}{2}(2y_1^2 + \frac{1}{2}y_2^2)\right]}{2\pi}$$

$$= \left| \frac{1}{\sqrt{\pi}} \exp\left[-\tfrac{1}{2}(2y_1^2)\right] \right| \left| \frac{1}{\sqrt{4\pi}} \exp\left[-\tfrac{1}{2}(\tfrac{1}{2}y_2^2)\right] \right|.$$

A straightforward calculation yields

$$f_{Y_1}(y_1) = \int_{-\infty}^{\infty} f_{Y_1, Y_2}(y_1, y_2)\, dy_2 = \frac{1}{\sqrt{\pi}} \exp\left[-\tfrac{1}{2}(2y_1^2)\right],$$

$$f_{Y_2}(y_2) = \int_{-\infty}^{\infty} f_{Y_1, Y_2}(y_1, y_2)\, dy_1 = \frac{1}{\sqrt{4\pi}} \exp\left[-\tfrac{1}{2}(\tfrac{1}{2}y_2^2)\right],$$

so that

$$f_{Y_1, Y_2}(y_1, y_2) = f_{Y_1}(y_1)f_{Y_2}(y_2).$$

Thus, Y_1 and Y_2 are independent, and $Y_1 \sim N(0, \tfrac{1}{2})$, $Y_2 \sim N(0, 2)$.

2. ONE-TO-ONE TRANSFORMATIONS OF p RANDOM VARIABLES

The transformation from $\mathbf{x} = (x_1, x_2, \ldots, x_p)$ to $\mathbf{y} = (y_1, y_2, \ldots, y_p)$ defined by

$$y_i = g_i(x_1, x_2, \ldots, x_p), \quad i = 1, 2, \ldots, p,$$

is said to be one-to-one if there exist functions $h_i(\mathbf{y}) = h_i(y_1, y_2, \ldots, y_p)$, $i = 1, 2, \ldots, p$, such that

(2.1) $y_i = g_i(\mathbf{x})$, $i = 1, 2, \ldots, p$ if and only if $x_i = h_i(\mathbf{y})$, $i = 1, 2, \ldots, p$.

If $X_1, X_2, \ldots, X_p$ are discrete random variables with probability mass function $p_{X_1, X_2, \ldots, X_p}(x_1, x_2, \ldots, x_p)$, then, using (2.1), we obtain

(2.2) $p_{Y_1, Y_2, \ldots, Y_p}(\mathbf{y}) = p_{X_1, X_2, \ldots, X_p}(h_1(\mathbf{y}), h_2(\mathbf{y}), \ldots, h_p(\mathbf{y})).$

Example 2.1. If $Y_i = \sum_{j=1}^{i} X_j$, $i = 1, 2, \ldots, p$, then

$$x_i = h_i(y_1, y_2, \ldots, y_p) = y_i - y_{i-1}, \quad i = 2, \ldots, p,$$

and $x_1 = h_1(y_1, y_2, \ldots, y_p) = y_1$. Thus,

$$p_{Y_1, Y_2, \ldots, Y_p}(y_1, y_2, \ldots, y_p) = p_{X_1, X_2, \ldots, X_p}(y_1, y_2 - y_1, \ldots, y_p - y_{p-1}).$$

If $X_1, X_2, \ldots, X_p$ are continuous random variables with joint probability density function $f_{X_1, X_2, \ldots, X_p}(\mathbf{x})$, then the joint probability density function $f_{Y_1, Y_2, \ldots, Y_p}(\mathbf{y})$ of $Y_1, Y_2, \ldots, Y_p$ is given by

$$(2.3) \qquad f_{Y_1, Y_2, \ldots, Y_p}(\mathbf{y}) = |J(\mathbf{y})| f_{X_1, X_2, \ldots, X_p}(h_1(\mathbf{y}), h_2(\mathbf{y}), \ldots, h_p(\mathbf{y}))$$

where $J(\mathbf{y})$ is the *Jacobian* of the transformation

$$x_1 = h_1(\mathbf{y}), \, x_2 = h_2(\mathbf{y}), \ldots, x_p = h_p(\mathbf{y}),$$

and is equal to the determinant of a matrix whose (i, j)th element is $\partial h_i(\mathbf{y})/\partial y_j$.

The determination of the joint probability density function of Y_1, $Y_2, \ldots, Y_p$ using (2.3) can be quite complicated, both because of the necessity of finding the functions $h_i(\mathbf{y})$, $i = 1, 2, \ldots, p$, and because of the often complicated form of the matrix whose determinant must be found in order to calculate the Jacobian $J(y_1, y_2, \ldots, y_p)$. Fortunately, a stepwise approach, as illustrated in Example 2.2 below, often allows us to break up a complicated computation into a series of simpler steps.

Example 2.2. Suppose that (X_1, X_2, X_3) have a joint probability density function

$$f_{X_1, X_2, X_3}(x_1, x_2, x_3)$$

$$= \begin{vmatrix} \theta^3 \exp\left[-\theta(x_1 + x_2 + x_3)\right], & \text{if } 0 < x_1, x_2, x_3 < \infty, \\ 0, & \text{otherwise,} \end{vmatrix}$$

and we wish to obtain the joint distribution of

$$Y_1 = \frac{X_2}{X_1 + X_2}, \qquad Y_2 = \frac{X_3}{X_1 + X_2 + X_3}, \qquad Y_3 = X_1 + X_2 + X_3.$$

We may proceed stepwise as follows. First, obtain the joint distribution of

$$W_1 = \frac{X_2}{X_1 + X_2}, \qquad W_2 = X_1 + X_2, \qquad W_3 = X_3.$$

Note that $w_1 = x_2(x_1 + x_2)^{-1}$, $w_2 = x_1 + x_2$, $w_3 = x_3$, if and only if

$$x_1 = h_1(w_1, w_2, w_3) = w_2(1 - w_1), \qquad x_2 = h_2(w_1, w_2, w_3) = w_1 w_2,$$

and $x_3 = h_3(w_1, w_2, w_3) = w_3$. Since $h_3(w_1, w_2, w_3)$ involves *only* w_3 (i.e.,

$w_3 = x_3$ is an *identity* transformation), then for the purpose of calculating the Jacobian, the transformation from (x_1, x_2, x_3) to (w_1, w_2, w_3) can be treated as if w_3 is constant, and as if the transformation is from (x_1, x_2) to (w_1, w_2). That is,

$$J(w_1, w_2, w_3) = \left(\frac{\partial h_1(w_1, w_2, w_3)}{\partial w_1}\right)\left(\frac{\partial h_2(w_1, w_2, w_3)}{\partial w_2}\right)$$

$$- \left(\frac{\partial h_1(w_1, w_2, w_3)}{\partial w_2}\right)\left(\frac{\partial h_2(w_1, w_2, w_3)}{\partial w_1}\right)$$

$$= (-w_2)(w_1) - (1 - w_1)(w_2) = -w_2 .$$

Applying (2.3), we obtain

$$f_{W_1, W_2, W_3}(w_1, w_2, w_3)$$

$$= |-w_2| f_{X_1, X_2, X_3}(w_2(1 - w_1), w_1 w_2, w_3)$$

$$= \begin{cases} \theta^3 w_2 \exp\left[-\theta(w_2 + w_3)\right], & \text{if } 0 < w_2(1 - w_1), w_1 w_2, w_3 < \infty, \\ 0, & \text{otherwise.} \end{cases}$$

Next, we obtain the joint distribution of

$$Y_1 = W_1 = \frac{X_2}{X_1 + X_2}, \qquad Y_2 = \frac{W_3}{W_2 + W_3} = \frac{X_3}{X_1 + X_2 + X_3},$$

$$Y_3 = W_2 + W_3 = X_1 + X_2 + X_3.$$

Since $y_1 = w_1$, $y_2 = w_3(w_2 + w_3)^{-1}$, $y_3 = w_2 + w_3$ if and only if

$$w_1 = h_1(y_1, y_2, y_3) = y_1, \qquad w_2 = y_3(1 - y_2) = h_2(y_1, y_2, y_3),$$

$$w_3 = y_2 y_3 = h_3(y_1, y_2, y_3),$$

and since $h_1(y_1, y_2, y_3)$ depends *only* on y_1,

$$J(y_1, y_2, y_3) = \left(\frac{\partial h_2(y_1, y_2, y_3)}{\partial y_2}\right)\left(\frac{\partial h_3(y_1, y_2, y_3)}{\partial y_3}\right)$$

$$- \left(\frac{\partial h_2(y_1, y_2, y_3)}{\partial y_3}\right)\left(\frac{\partial h_3(y_1, y_2, y_3)}{\partial y_2}\right)$$

$$= (-y_3)(y_2) - (1 - y_2)(y_3) = (-y_3).$$

Again applying (2.3), we obtain

$$f_{Y_1, Y_2, Y_3}(y_1, y_2, y_3) = |-y_3| f_{W_1, W_2, W_3}(y_1, y_3(1 - y_2), y_2 y_3)$$

$$= \begin{cases} \theta^3 y_3^2(1 - y_2)e^{-\theta y_3}, & \text{if } 0 < y_1, y_2 < 1, 0 < y_3 < \infty, \\ 0, & \text{otherwise.} \end{cases}$$

Multivariate Linear Transformations

A one-to-one transformation of p random variables that is used frequently in applications is the *multivariate linear transformation*:

$$Y_1 = a_{11}X_1 + a_{12}X_2 + \cdots + a_{1p}X_p$$

(2.4)
$$Y_2 = a_{21}X_1 + a_{22}X_2 + \cdots + a_{2p}X_p$$

$$\vdots \qquad \vdots \qquad \vdots \qquad \qquad \vdots$$

$$Y_p = a_{p1}X_1 + a_{p2}X_2 + \cdots + a_{pp}X_p.$$

We say that this transformation is "invertible," "nonsingular," or "one-to-one onto," if there exist constants b_{ij}, $i = 1, 2, \ldots, p$; $j = 1, 2, \ldots, p$, such that $y_i = \sum_{k=1}^{p} a_{ik} x_k$, $i = 1, 2, \ldots, p$, if and only if $x_i = \sum_{k=1}^{p} b_{ik} y_k$. In this case,

$$h_i(\mathbf{y}) = h_i(y_1, y_2, \ldots, y_p) = \sum_{k=1}^{p} b_{ik} y_k, \quad i = 1, 2, \ldots, p,$$

and $J(\mathbf{y})$ is the determinant of a matrix B whose (i, j)th element is b_{ij}. It follows from results of matrix theory that the determinant of B is the inverse of the determinant of the *coefficient matrix* A of the linear transformation; this coefficient matrix A has (i, j)th element equal to a_{ij} in (2.4). Let det A denote the determinant of A. It follows from (2.3) that the joint density function $f_{Y_1, Y_2, \ldots, Y_p}(\mathbf{y})$ of the transformed values $Y_1, Y_2, \ldots, Y_p$ defined by (2.4) can be obtained from the joint density $f_{X_1, X_2, \ldots, X_p}(\mathbf{x})$ of the original variables $X_1, X_2, \ldots, X_p$ through the equation

$$(2.5) \quad f_{Y_1, Y_2, \ldots, Y_p}(\mathbf{y}) = \frac{1}{|\det A|} f_{X_1, X_2, \ldots, X_p}\left(\sum_{k=1}^{p} b_{1k} y_k, \sum_{k=1}^{p} b_{2k} y_k, \ldots, \sum_{k=1}^{p} b_{pk} y_k \right).$$

Example 2.3. Suppose that X_1, X_2, and X_3 are mutually probabilistically independent random variables, each having the standard normal $N(0, 1)$ distribution. Let a trivariate linear transformation be defined by

$$Y_1 = X_1 + X_2 + X_3, \qquad Y_2 = X_1 - X_2, \qquad Y_3 = X_2 - X_3.$$

Then

$$A = \begin{pmatrix} a_{11} & a_{12} & a_{13} \\ a_{21} & a_{22} & a_{23} \\ a_{31} & a_{32} & a_{33} \end{pmatrix} = \begin{pmatrix} 1 & 1 & 1 \\ 1 & -1 & 0 \\ 0 & 1 & -1 \end{pmatrix},$$

and it can easily be seen that

$$x_1 = \frac{y_1 + 2y_2 + y_3}{3}, \qquad x_2 = \frac{y_1 - y_2 + y_3}{3}, \qquad x_3 = \frac{y_1 - y_2 - 2y_3}{3}.$$

Further,

$$|\det A| = 3.$$

Finally, since X_1, X_2, and X_3 are mutually independent $N(0, 1)$ variables,

(2.6)
$$f_{X_1, X_2, X_3}(x_1, x_2, x_3) = \prod_{i=1}^{3} \left(\frac{1}{\sqrt{2\pi}} e^{-\frac{1}{2}x_i^2} \right)$$
$$= \left(\frac{1}{\sqrt{2\pi}} \right)^3 e^{-\frac{1}{2}(x_1^2 + x_2^2 + x_3^2)},$$

for $-\infty < x_1, x_2, x_3 < \infty$. Applying (2.5), we find that for all real y_1, y_2, y_3,

(2.7)
$$f_{Y_1, Y_2, Y_3}(y_1, y_2, y_3) = \frac{1}{3} \left(\frac{1}{\sqrt{2\pi}} \right)^3 e^{-\frac{1}{2}Q(y_1, y_2, y_3)}$$

where

$$Q(y_1, y_2, y_3) = \left(\frac{y_1 + 2y_2 + y_3}{3} \right)^2 + \left(\frac{y_1 - y_2 + y_3}{3} \right)^2 + \left(\frac{y_1 - y_2 - 2y_3}{3} \right)^2$$
$$= \frac{y_1^2}{3} + \frac{2y_2^2}{3} + \frac{2y_3^2}{3} + \frac{2y_2 y_3}{3}.$$

From (2.7), it can be shown (see Chapter 9) that Y_1 has a marginal $N(0, 3)$ distribution, that (Y_2, Y_3) have a marginal $BVN((0, 0), (2, 2, -0.5))$ distribution, and that Y_1 is independent of either Y_2 or Y_3.

We may also obtain (2.7) by stepwise methods. First, make the transformation

$$W_1 = X_1 + X_2, \qquad W_2 = X_1 - X_2, \qquad W_3 = X_3.$$

Noting that $X_3 = h_3(w_1, w_2, w_3) = w_3$ involves only w_3 and that

$$x_1 = h_1(w_1, w_2, w_3) = \frac{w_1 + w_2}{2}, \qquad x_2 = h_2(w_1, w_2, w_3) = \frac{w_1 - w_2}{2},$$

we have

$$J(w_1, w_2, w_3) = \frac{\partial h_1}{\partial w_1} \frac{\partial h_2}{\partial w_2} - \frac{\partial h_1}{\partial w_2} \frac{\partial h_2}{\partial w_1} = -\frac{1}{2}.$$

It then follows from (2.3) and (2.6) that

$$f_{W_1, W_2, W_3}(w_1, w_2, w_3)$$

(2.8)
$$= \left| -\frac{1}{2} \right| \left(\frac{1}{\sqrt{2\pi}} \right)^3 \exp\left\{ -\frac{1}{2} \left[\left(\frac{w_1 + w_2}{2} \right)^2 + \left(\frac{w_1 - w_2}{2} \right)^2 + w_3^2 \right] \right\}$$
$$= \frac{1}{2(\sqrt{2\pi})^3} \exp\left\{ -\frac{1}{2} \left[\frac{1}{2} (w_1^2 + w_2^2) + w_3^2 \right] \right\}$$

for $-\infty < w_1, w_2, w_3 < \infty$. Next, we make the transformation

$$Y_1 = W_1 + W_3 = (X_1 + X_2) + X_3, \qquad Y_2 = W_2 = X_1 - X_2,$$

$$Y_3 = \frac{W_1 - W_2 - 2W_3}{2} = X_2 - X_3,$$

for which $w_2 = h_2(y_1, y_2, y_3) = y_2$ is a function only of y_2,

$$w_1 = h_1(y_1, y_2, y_3) = \frac{2y_1 + y_2 + 2y_3}{3},$$

$$w_3 = h_3(y_1, y_2, y_3) = \frac{y_1 - y_2 - 2y_3}{3},$$

and

$$J(y_1, y_2, y_3) = \frac{\partial h_1}{\partial y_1} \frac{\partial h_3}{\partial y_3} - \frac{\partial h_1}{\partial y_3} \frac{\partial h_3}{\partial y_1} = -\frac{2}{3}.$$

Thus, using (2.3) again and (2.8), we find that

$$f_{Y_1, Y_2, Y_3}(y_1, y_2, y_3) = \left(\frac{2}{3}\right) \frac{1}{2(\sqrt{2\pi})^3}$$

$$\times \exp\left\{-\frac{1}{2}\left[\frac{1}{2}\left(\frac{2y_1 + y_2 + 2y_3}{3}\right)^2 + \frac{1}{2}y_2^2 + \left(\frac{y_1 - y_2 - 2y_3}{3}\right)^2\right]\right\},$$

which, after simplification, yields (2.7).

3. MANY-TO-ONE TRANSFORMATIONS

Suppose that $X_1, X_2, \ldots, X_p$ have a joint distribution and we want to find the distribution of

$$(3.1) \qquad\qquad Y = g(X_1, X_2, \ldots, X_p).$$

The theory of one-to-one transformations does not hold here, since many values of $(x_1, x_2, \ldots, x_p)$ may yield the same value of $y = g(x_1, x_2, \ldots, x_p)$. One method of obtaining the distribution of Y is to find $p - 1$ further transformations

$$y_i = g_i(x_1, x_2, \ldots, x_p), \quad i = 1, \ldots, p - 1,$$

so that these, together with $y_p = g(X_1, X_2, \ldots, X_p)$, yield a one-to-one transformation from $(x_1, x_2, \ldots, x_p)$ to $(y_1, y_2, \ldots, y_p)$. Using (2.2) or (2.3), depending on whether the random variables $X_1, X_2, \ldots, X_p$ are discrete or continuous, we can find the joint distribution $p_{Y_1, Y_2, \ldots, Y_p}(y_1, y_2, \ldots, y_p)$ or

$f_{Y_1, Y_2, ..., Y_p}(y_1, y_2, ..., y_p)$ of $(Y_1, Y_2, ..., Y_p)$. From this joint distribution, we can then find the marginal distribution, $p_{Y_p}(y_p) = p_Y(y)$ or $f_{Y_p}(y_p) = f_Y(y)$, of Y.

Example 3.1. In Example 1.2, X_1 and X_2 were continuous random variables arising from independent measurements of the lengths of time needed to complete two similar tasks. To obtain the distribution of the ratio $Y = X_1/X_2$, recall that

$$Y_1 = X_1 + X_2, \qquad Y_2 = \frac{X_1}{X_2},$$

is a one-to-one transformation, for which the joint density function is

$$f_{Y_1, Y_2}(y_1, y_2) = \begin{cases} \dfrac{\theta^{2r} y_1^{2r-1} e^{-\theta y_1}}{\Gamma(2r)} \dfrac{y_2^{r-1}}{B(r, r)(1 + y_2)^{2r}}, & \text{if } 0 < y_1 < \infty, \\ & 0 < y_2 < \infty, \\ 0, & \text{otherwise.} \end{cases}$$

The marginal density function $f_{Y_2}(y_2) = f_Y(y)$ of Y can be obtained directly from the joint distribution. Thus, if $0 < y < \infty$,

$$f_Y(y) = \int_0^\infty f_{Y_1, Y_2}(y_1, y)\, dy_1 = \frac{y^{r-1}}{B(r, r)(1 + y)^{2r}} \int_0^\infty \frac{\theta^{2r} y_1^{2r-1} e^{-\theta y_1}}{\Gamma(2r)}\, dy_1$$

$$= \frac{y^{r-1}}{B(r, r)(1 + y)^{2r}},$$

and $f_Y(y) = \int_0^\infty 0\, dy_1 = 0$ for $y \le 0$.

In this example, we might instead have been interested in the total time $Y_1 = X_1 + X_2$ needed to complete the two tasks. We can obtain the distribution of Y_1 by finding its marginal density function

$$f_{Y_1}(y_1) = \int_0^\infty f_{Y_1, Y_2}(y_1, y_2)\, dy_2 = \begin{cases} \dfrac{\theta^{2r} y_1^{2r-1} e^{-\theta y_1}}{\Gamma(2r)}, & 0 < y_1 < \infty, \\ 0, & \text{otherwise.} \end{cases}$$

We observe that Y_1 has a gamma distribution with parameters θ and $2r$.

It is not always easy to find $Y_i = g_i(X_1, X_2, ..., X_p)$, $i = 1, 2, ..., p - 1$, the remaining functions that, together with $Y = g(X_1, X_2, ..., X_p)$, will make up a workable one-to-one transformation. If in Example 3.1 we had not already obtained the joint distribution of $Y_1 = X_1 + X_2$ and $Y_2 = X_1/X_2$, we might have had some trouble finding a transformation $Y_1 = g_1(X_1, X_2)$ to pair with $Y_2 = X_1/X_2$ in order to obtain a one-to-one transformation suitable for obtaining the marginal distribution of Y_2.

One way to bypass this indirect approach is through the cumulative distribution function. If $Y = g(X_1, X_2, \ldots, X_p)$, then

$$F_Y(y) = P\{Y \leq y\} = P\{g(X_1, X_2, \ldots, X_p) \leq y\}$$

(3.2)
$$= \int \cdots \int_{g(x_1, x_2, \ldots, x_p) \leq y} f_{X_1, \ldots, X_p}(x_1, \ldots, x_p)\, dx_1 \cdots dx_p.$$

Suppose that there exists a function $h(y; x_2, \ldots, x_p)$ such that

(3.3) $g(x_1, x_2, \ldots, x_p) \leq y$ if and only $x_1 \leq h(y; x_2, \ldots, x_p)$.

Then

$$F_Y(y) = \int \cdots \int_{g(x_1, x_2, \ldots, x_p) \leq y} f_{X_1, \ldots, X_p}(x_1, \ldots, x_p)\, dx_1 \cdots dx_p$$

(3.4)
$$= \int_{-\infty}^{\infty} \cdots \int_{-\infty}^{\infty} \left[\int_{-\infty}^{h(y; x_2, \ldots, x_p)} f_{X_1, \ldots, X_p}(x_1, \ldots, x_p)\, dx_1 \right] dx_2 \cdots dx_p$$

$$= \int_{-\infty}^{\infty} \cdots \int_{-\infty}^{\infty} q(y; x_2, \ldots, x_p)\, dx_2 \cdots dx_p,$$

say, where

(3.5) $\displaystyle q(y; x_2, \ldots, x_p) = \int_{-\infty}^{h(y; x_2, \ldots, x_p)} f_{X_1, X_2, \ldots, X_p}(x_1, x_2, \ldots, x_p)\, dx_1.$

Note that

(3.6)
$$\frac{\partial q(y; x_2, \ldots, x_p)}{\partial y}$$
$$= \frac{\partial h(y; x_2, \ldots, x_p)}{\partial y} [f_{X_1, X_2, \ldots, X_p}(h(y; x_2, \ldots, x_p), x_2, \ldots, x_p)]$$

Thus, if we can show that

(3.7)
$$\frac{\partial}{\partial y} \int_{-\infty}^{\infty} \cdots \int_{-\infty}^{\infty} q(y; x_2, \ldots, x_p)\, dx_2 \cdots dx_p$$

$$= \int_{-\infty}^{\infty} \cdots \int_{-\infty}^{\infty} \frac{\partial}{\partial y} q(y; x_2, \ldots, x_p)\, dx_2 \cdots dx_p,$$

it will then follow from (3.4) and (3.6) that

$$(3.8) \qquad f_Y(y) = \frac{\partial}{\partial y} F_Y(y) = \int_{-\infty}^{\infty} \cdots \int_{-\infty}^{\infty} \frac{\partial h(y; x_2, \ldots, x_p)}{\partial y}$$

$$\times [f_{X_1, X_2, \ldots, X_p}(h(y, x_2, \ldots, x_p), x_2, \ldots, x_p)] \, dx_2 \cdots dx_p.$$

Of course, equality (3.7) need not always hold. However, if $q(y; x_2, \ldots, x_p)$ and $\partial q(y; x_2, \ldots, x_p)/\partial y$ are well-defined and jointly continuous functions of $y, x_2, \ldots, x_p$ for $-\infty < y, x_2, x_3, \ldots, x_p < \infty$, then equality (3.7) holds, and thus (3.8) will be valid.

Example 3.2. A distributional problem of great importance to the theory of statistical inference is the following. Suppose that the random variables X_1 and X_2 are independent, with $X_1 \sim \mathcal{N}(0, 1)$ and X_2 having a gamma distribution with parameters $\theta = \frac{1}{2}$ and $r = n/2$, where $n \geq 1$ is an integer. We wish to find the distribution of

$$Y = \frac{X_1}{\sqrt{X_2/n}}.$$

Note that $g(x_1, x_2) = x_1/\sqrt{x_2/n}$, and that

$$\frac{x_1}{\sqrt{x_2/n}} \leq y \quad \text{if and only if} \quad x_1 \leq y\sqrt{x_2/n},$$

so that $h(y; x_2) = y\sqrt{x_2/n}$. From the facts given above,

$$f_{X_1, X_2}(x_1, x_2) = f_{X_1}(x_1) f_{X_2}(x_2)$$

$$= \left(\frac{1}{\sqrt{2\pi}} e^{-\frac{1}{2}x_1^2} \right) \frac{x_2^{n/2 - 1} e^{-\frac{1}{2}x_2}}{2^{n/2} \Gamma(n/2)},$$

when $-\infty < x_1 < \infty, 0 < x_2 < \infty$, and $f_{X_1, X_2}(x_1, x_2) = 0$, otherwise. Thus, from (3.5),

$$q(y; x_2) = \int_{-\infty}^{y\sqrt{x_2/n}} \frac{1}{\sqrt{2\pi}} e^{-\frac{1}{2}x_1^2} \frac{x_2^{n/2 - 1} e^{-\frac{1}{2}x_2}}{2^{n/2} \Gamma(n/2)} \, dx_1,$$

and from (3.6),

$$\frac{\partial}{\partial y} q(y; x_2) = \left(\sqrt{\frac{x_2}{n}} \right) \frac{1}{\sqrt{2\pi}} e^{-\frac{1}{2}(y\sqrt{x_2/n})^2} \frac{x_2^{n/2 - 1} e^{-\frac{1}{2}x_2}}{2^{n/2} \Gamma(n/2)}.$$

Both of these functions are well-defined and continuous functions of y and x_2 for all $-\infty < y < \infty$, $0 \leq x_2 < \infty$. From (3.8), it follows that

$$f_Y(y) = \frac{1}{\sqrt{2\pi n}\ \Gamma(n/2)2^{n/2}} \int_0^\infty x_2^{\frac{1}{2}(n-1)} e^{-\frac{1}{2}x_2(1+y^2/n)}\, dx_2$$

(3.9)

$$= \frac{\Gamma\left(\dfrac{n+1}{2}\right)}{\Gamma(n/2)\sqrt{\pi n}(1+y^2/n)^{\frac{1}{2}(n+1)}},$$

for $-\infty < y < \infty$.

Remark. The range of x_2 can be taken to be $0 \leq x_2 < \infty$ rather than $-\infty < x_2 < \infty$ in the preceding computations since $f_{X_1, X_2}(x_1, x_2) = 0$ for $x_2 < 0$.

The distribution corresponding to the density function $f_Y(y)$ in (3.9) is called *Student's t-distribution with n degrees of freedom*. This distribution was originally studied in connection with a method of statistical inference concerning the mean of a normal distribution (called the "*t*-test") proposed by the English statistician William Sealy Gosset (1876–1937) writing under the pseudonym "Student."

It is worth mentioning that the distribution of the random variable X_2 in this example (namely, the gamma distribution with parameter $\theta = 1/2$ and $r = n/2$, where n is a positive integer) is called a *chi-squared distribution with n degrees of freedom* in the statistical literature, and the notation $X \sim \chi_n^2$ is used to indicate that the random variable X has a chi-squared distribution with n degrees of freedom. This distribution originally arose in connection with work by the English statistician Karl Pearson (1857–1936) on the statistical analysis of sample contingency tables, and also arises in connection with statistical inferences about the variance σ^2 of a normal distribution. The χ_n^2 distribution was originally discovered by the German physicist F. R. Helmert (1843–1917).

Another distribution that is of statistical importance is the *F-distribution with n_1 and n_2 degrees of freedom*, which is used in connection with the statistical methodology now called the analysis of variance. Suppose that X_1 and X_2 are independent random variables, and that $X_1 \sim \chi_{n_1}^2$, $X_2 \sim \chi_{n_2}^2$. Let

$$Y = \frac{X_1/n_1}{X_2/n_2}.$$

Then Y is said to have the *F*-distribution with n_1 and n_2 degrees of freedom. The distribution of Y can be obtained indirectly, using the one-to-one transformation $Y_1 = Y = (X_1/n_1)/(X_2/n_2)$, $Y_2 = X_1 + X_2$, by finding the joint probability density function $f_{Y_1, Y_2}(y_1, y_2)$ of Y_1 and Y_2, and then integrating this joint density to find the marginal probability density function of Y_1. This is the approach used in Examples 1.2 and 3.1. Alternatively, we can use the direct method of Example 3.2, making use of Equation (3.8), as illustrated below.

Note that

$$g(x_1, x_2) = \frac{x_1/n_1}{x_2/n_2} \le y \quad \text{if and only if} \quad x_1 \le y\left(\frac{n_1}{n_2}\right)x_2,$$

so that $h(y; x_2) = y(n_1/n_2)x_2$. From the facts that X_1 and X_2 are independent and that $X_i \sim \chi^2_{n_i}$, $i = 1, 2$, we have

$$f_{X_1, X_2}(x_1, x_2) = f_{X_1}(x_1)f_{X_2}(x_2)$$

$$= \frac{x_1^{\frac{1}{2}n_1 - 1}e^{-\frac{1}{2}x_1}}{2^{\frac{1}{2}n_1}\Gamma(n_1/2)} \frac{x_2^{\frac{1}{2}n_2 - 1}e^{-\frac{1}{2}x_2/2}}{2^{\frac{1}{2}n_2}\Gamma(n_2/2)},$$

where $0 \le x_1, x_2 < \infty$ and $f_{X_1, X_2}(x_1, x_2) = 0$ otherwise. Thus,

$$q(y; x_2) = \int_0^{y(n_1/n_2)x_2} \frac{x_1^{\frac{1}{2}n_1 - 1}x_2^{\frac{1}{2}n_2 - 1}e^{-\frac{1}{2}(x_1 + x_2)}}{2^{\frac{1}{2}(n_1 + n_2)}\Gamma(n_1/2)\Gamma(n_2/2)}\,dx_1$$

and

$$\frac{\partial}{\partial y}q(y; x_2) = \frac{y^{\frac{1}{2}n_1 - 1}(n_1/n_2)^{\frac{1}{2}n_1}x_2^{\frac{1}{2}(n_1 + n_2) - 1}e^{-\frac{1}{2}x_2(1 + n_1y/n_2)}}{2^{\frac{1}{2}(n_1 + n_2)}\Gamma(n_1/2)\Gamma(n_2/2)}$$

are both well-defined and continuous functions of y and x_2 for all $0 \le y$, $x_2 < \infty$.

Remark. $f_{X_1, X_2}(x_1, x_2) \ne 0$ only when $0 \le x_1, x_2 < \infty$.

From (3.8), it follows that for $y \ge 0$,

$$f_Y(y) = \frac{(n_1/n_2)^{\frac{1}{2}n_1}y^{\frac{1}{2}n_1 - 1}}{2^{\frac{1}{2}(n_1 + n_2)}\Gamma(n_1/2)\Gamma(n_2/2)}\int_0^\infty x_2^{\frac{1}{2}(n_1 + n_2) - 1}e^{-\frac{1}{2}x_2(1 + n_1y/n_2)}\,dx_2$$

(3.10)

$$= \frac{(n_1/n_2)^{\frac{1}{2}n_1}\Gamma((n_1 + n_2)/2)y^{\frac{1}{2}n_1 - 1}}{\Gamma(n_1/2)\Gamma(n_2/2)(1 + n_1y/n_2)^{\frac{1}{2}(n_1 + n_2)}},$$

while clearly $f_Y(y) = 0$, for $y < 0$. The probability density function (3.10) thus defines the F-distribution with n_1 and n_2 degrees of freedom.

4. SUMS AND WEIGHTED SUMS OF RANDOM VARIABLES

Perhaps the most frequently used many-to-one transformation of p random variables $X_1, X_2, \ldots, X_p$ is the weighted sum $Y = \sum_{i=1}^p a_i X_i$ of these variables.

Example 4.1. Suppose that $X_1, X_2, \ldots, X_p$ are p independent observations of a given random variable X. Then the sample mean $\hat{\mu}_X = \sum_{i=1}^{p} (1/p)X_i$ has been previously used as an estimate of the population mean μ_X of X.

Example 4.2. In a "before–after" type of statistical study, measurements X_1 and X_2 are made before and after the application of a treatment. The difference $Y = X_2 - X_1$ is used to measure the change brought about by the treatment.

Example 4.3. A final grade for a course in which there were two midterm exams X_1 and X_2, homework X_3, and a final examination X_4. might be determined by the total score $Y = \sum_{i=1}^{4} X_i$, or by a weighted total score, say $Y = 2X_1 + 2X_2 + 2X_3 + 3X_4$.

To find the distribution of a weighted sum $Y = \sum_{i=1}^{p} a_i X_i$ of the random variables $X_1, X_2, \ldots, X_p$, we first identify a coefficient a_i not equal to zero. For the sake of notational convenience, suppose that $a_1 \neq 0$. Next, we consider the one-to-one transformation

$$Y_1 = Y = \sum_{i=1}^{p} a_i X_i, \qquad Y_2 = X_2, \ldots, Y_p = X_p,$$

for which $x_i = h_i(y_1, y_2, \ldots, y_p) = y_i$, $i = 2, \ldots, p$, and

$$x_1 = h_1(y_1, y_2, \ldots, y_p) = \frac{y_1 - \sum_{i=2}^{p} a_i y_i}{a_1}.$$

We can now find the joint distribution of $Y_1, Y_2, \ldots, Y_p$, and from this joint distribution find the marginal distribution of $Y_1 = Y$. Hence, when X_1, $X_2, \ldots, X_1$ are discrete random variables with joint probability mass function

$$p_{X_1, X_2, \ldots, X_p}(x_1, x_2, \ldots, x_p) = p(x_1, x_2, \ldots, x_p),$$

the steps above yield

$$(4.1) \qquad p_Y(y) = \sum_{y_2} \cdots \sum_{y_p} p\left(\frac{y - \sum_{i=2}^{p} a_i y_i}{a_1}, y_2, \ldots, y_p\right).$$

Similarly, when $X_1, X_2, \ldots, X_p$ are continuous random variables with joint probability density function

$$f_{X_1, X_2, \ldots, X_p}(x_1, x_2, \ldots, x_p) = f(x_1, x_2, \ldots, x_p),$$

then

$$(4.2) \quad f_Y(y) = \int_{-\infty}^{\infty} \cdots \int_{-\infty}^{\infty} \left|\frac{1}{a_1}\right| f\left(\frac{y - \sum_{i=2}^{p} a_i y_i}{a_1}, y_2, \ldots, y_p\right) dy_2 \cdots dy_p.$$

Of course, if $a_j \neq 0$ for some $j \neq 1$, identical arguments yield the alternative expressions

$$(4.1') \quad p_Y(y) = \sum_{y_1} \cdots \sum_{y_{j-1}} \sum_{y_{j+1}} \cdots \sum_{y_p} p\left(y_1, \ldots, y_{j-1}, \frac{y - \sum_{i \neq j} a_i y_i}{a_j}, y_{j+1}, \ldots, y_p\right)$$

and

$$(4.2') \quad \begin{aligned} f_Y(y) = \int_{-\infty}^{\infty} \cdots \int_{-\infty}^{\infty} \left|\frac{1}{a_j}\right| f\left(y_1, \ldots, y_{j-1}, \frac{y - \sum_{i \neq j} a_i y_i}{a_j}, y_{j+1}, \ldots, y_p\right) \\ \times \, dy_1 \cdots dy_{j-1} \, dy_{j+1} \cdots dy_p \end{aligned}$$

for the probability mass function or probability density function of $Y = \sum_{i=1}^{p} a_i X_i$.

Example 4.4. Suppose that (X_1, X_2) have the joint probability mass function

$$p_{X_1, X_2}(x_1, x_2) = \frac{n!}{x_1! \, x_2! \, (n - x_1 - x_2)!} \, p_1^{x_1} p_2^{x_2} (1 - p_1 - p_2)^{n - x_1 - x_2}$$

if $x_1, x_2 = 0, 1, \ldots, n,$ $0 \leq x_1 + x_2 \leq n,$ and $p_{X_1, X_2}(x_1, x_2) = 0$ otherwise. Then, Equation (4.1) tells us that the probability mass function of $Y = X_1 + X_2$ is

$$p_Y(y) = \sum_{y_2 = 0}^{n} p_{X_1, X_2}(y - y_2, y_2) = \sum_{y_2 = 0}^{y} p_{X_1, X_2}(y - y_2, y_2)$$

if $y = 0, 1, \ldots, n,$ and $p_y(y) = 0$ otherwise, since $p_{X_1, X_2}(x_1, x_2) = 0$ for x_1 or x_2 not equal to $0, 1, 2, \ldots, n$. Now,

$$\begin{aligned} \sum_{y_2 = 0}^{y} p_{X_1, X_2}(y - y_2, y_2) &= \frac{n! \, (1 - p_1 - p_2)^{n-y}}{(n - y)! \, y!} \left[\sum_{y_2 = 0}^{y} \frac{y!}{y_2! \, (y - y_2)!} p_1^{y - y_2} p_2^{y_2}\right] \\ &= \binom{n}{y} (1 - p_1 - p_2)^{n - y} (p_1 + p_2)^y. \end{aligned}$$

Thus,

$$p_Y(y) = \begin{cases} \binom{n}{y} (p_1 + p_2)^y (1 - p_1 - p_2)^{n - y}, & \text{if } y = 0, 1, \ldots, n, \\ 0, & \text{otherwise,} \end{cases}$$

which is the probability mass function of a binomial distribution with parameters n and $p = p_1 + p_2$.

Example 4.5. Let X_1 and X_2 be statistically independent random variables, where X_1 and X_2 both have exponential distributions with parameter $\theta = 1$. Thus,

$$f_{X_1, X_2}(x_1, x_2) = \begin{cases} e^{-(x_1 + x_2)}, & \text{if } x_1 > 0, \ x_2 > 0, \\ 0, & \text{otherwise.} \end{cases}$$

Let us find the distribution of $Y = 2X_1 + 3X_2$. From Equation (4.2), the probability density function, $f_Y(y)$, of Y is

$$f_Y(y) = \int_{-\infty}^{\infty} \left| \frac{1}{2} \right| f_{X_1, X_2}\left(\frac{y - 3y_2}{2}, y_2 \right) dy_2.$$

Since $f(x_1, x_2) = 0$ if $x_1 < 0$, it follows that $f_Y(y) = 0$ if $y < 0$. If $y > 0$, then

$$f_Y(y) = \tfrac{1}{2} \int_0^{y/3} \exp\left(-\frac{y - 3y_2}{2} - y_2 \right) dy_2 + \tfrac{1}{2} \int_{y/3}^{\infty} 0 \ dy_2$$

$$= \tfrac{1}{2} \exp\left(-\tfrac{1}{2}y \right) \int_0^{y/3} \exp\left(-\tfrac{1}{2}y_2 \right) dy_2$$

$$= \tfrac{1}{2} \exp\left(-\frac{y}{2} \right) \left[2 \exp\left(\frac{y}{6} \right) - 2 \right]$$

$$= \exp\left(-\frac{y}{3} \right) - \exp\left(-\frac{y}{2} \right).$$

Although Equation (4.1) is sometimes useful in finding the distribution of a weighted sum of discrete random variables, in many cases a direct approach is easier.

Example 4.6. Assume that the joint distribution of the random variables X_1, X_2, and X_3 is given by the following joint probability mass function.

(x_1, x_2, x_3)	$(0, 0, 0)$	$(1, 0, 0)$	$(1, 1, 0)$	$(1, 0, 1)$	$(0, 1, 0)$	$(0, 1, 1)$	$(0, 0, 1)$	$(1, 1, 1)$
$p(x_1, x_2, x_3)$	$\frac{1}{8}$	$\frac{1}{4}$	$\frac{1}{16}$	$\frac{1}{8}$	$\frac{1}{16}$	$\frac{1}{16}$	$\frac{1}{16}$	$\frac{1}{4}$

Let $Y = \frac{1}{3}(X_1 + X_2 + X_3)$. Since $P\{Y = 0\} = p(0, 0, 0) = \frac{1}{8}$,

$$P\{Y = 1\} = p(1, 1, 1) = \frac{2}{8},$$

$$P\{Y = \tfrac{1}{3}\} = p(1, 0, 0) + p(0, 1, 0) + p(0, 0, 1) = \tfrac{1}{4} + \tfrac{1}{16} + \tfrac{1}{16} = \tfrac{3}{8},$$

$$P\{Y = \tfrac{2}{3}\} = p(1, 1, 0) + p(1, 0, 1) + p(0, 1, 1) = \tfrac{1}{16} + \tfrac{1}{8} + \tfrac{1}{16} = \tfrac{2}{8},$$

the distribution of Y is

y	0	$\frac{1}{3}$	$\frac{2}{3}$	1
$P\{Y = y\}$	$\frac{1}{8}$	$\frac{3}{8}$	$\frac{2}{8}$	$\frac{2}{8}$

Sums of Random Variables

We have already seen (in Example 4.4) an example of the derivation of the distribution of the (unweighted) sum $Y = \sum_{i=1}^{p} X_i$ of p random variables. In this special case, Equations (4.1) and (4.2) become

$$(4.3) \qquad p_Y(y) = \sum_{y_2} \cdots \sum_{y_p} p\left(y - \sum_{i=2}^{p} y_i, y_2, \ldots, y_p\right)$$

and

$$(4.4) \qquad f_Y(y) = \int_{-\infty}^{\infty} \cdots \int_{-\infty}^{\infty} f\left(y - \sum_{i=2}^{p} y_i, y_2, \ldots, y_p\right) dy_2 \cdots dy_p$$

for the discrete and continuous cases, respectively. By symmetry, there are formulas similar to (4.3) and (4.4) where the sum or integral is over different sets of $(p - 1)$ of the y's. For example, in the special case of $p = 2$ random variables X_1, X_2,

$$(4.5) \qquad p_Y(y) = \sum_{y_1} p_{X_1, X_2}(y_1, y - y_1) = \sum_{y_2} p_{X_1, X_2}(y - y_2, y_2)$$

and

$$(4.6) \qquad f_Y(y) = \int_{-\infty}^{\infty} f_{X_1, X_2}(y_1, y - y_1)\, dy_1 = \int_{-\infty}^{\infty} f_{X_1, X_2}(y - y_2, y_2)\, dy_2,$$

where $Y = X_1 + X_2$.

Example 4.7. Let $(X_1, X_2) \sim BVN((0, 0), (1, 1, \rho))$. Then for $-\infty < x_1, x_2 < \infty$, the joint density function of (X_1, X_2) is

$$f_{X_1, X_2}(x_1, x_2) = \frac{1}{2\pi\sqrt{1 - \rho^2}} \exp\left[-\frac{x_1^2 - 2\rho x_1 x_2 + x_2^2}{2(1 - \rho^2)}\right],$$

and thus, applying (4.6), the probability density function of $Y = X_1 + X_2$ is

$$f_Y(y) = \frac{1}{2\pi\sqrt{1 - \rho^2}} \int_{-\infty}^{\infty} \exp\left[-\frac{y_1^2 - 2\rho y_1(y - y_1) + (y - y_1)^2}{2(1 - \rho^2)}\right] dy_1$$

$$= \frac{\exp\left[-\dfrac{y^2}{4(1 + \rho)}\right]}{\sqrt{2\pi(2)(1 + \rho)}} \int_{-\infty}^{\infty} \frac{\exp\left[-\dfrac{1}{2}\dfrac{(y_1 - \frac{1}{2}y)^2}{(1 - \rho)/2}\right]}{\sqrt{2\pi(1 - \rho)/2}} dy_1$$

$$= \frac{\exp\left[-\dfrac{1}{2}\dfrac{y^2}{2(1 + \rho)}\right]}{\sqrt{2\pi(2)(1 + \rho)}}, \quad -\infty < y < \infty,$$

which is the probability density function of a $N(0, 2(1 + \rho))$ random variable.

When X_1 and X_2 are independent discrete random variables, then $p_{X_1, X_2}(x_1, x_2) = p_{X_1}(x_1)p_{X_2}(x_2)$ and (4.5) becomes

(4.7) $$p_Y(y) = \sum_{y_1} p_{X_1}(y_1)p_{X_2}(y - y_1) = \sum_{y_2} p_{X_1}(y - y_2)p_{X_2}(y_2).$$

Similarly, when X_1 and X_2 are independent random variables, then (4.6) becomes

(4.8) $$f_Y(y) = \int_{-\infty}^{\infty} f_{X_1}(y_1)f_{X_2}(y - y_1)\, dy_1 = \int_{-\infty}^{\infty} f_{X_1}(y - y_2)f_{X_2}(y_2)\, dy_2.$$

Formulas (4.7) and (4.8) are said to be *convolutions* of the probability mass functions $p_{X_1}(x_1)$, $p_{X_2}(x_2)$ or probability density functions $f_{X_1}(x_1)$, $f_{X_2}(x_2)$, respectively, and are well-known formulas for calculating the distribution of the sum of two independent random variables.

Example 4.8. Suppose that X_1 and X_2 are independent discrete random variables, and that $X_1 + 1$, $X_2 + 1$, each have a geometric distribution with parameter p. That is,

$$p_{X_i}(x_i) = \begin{cases} p(1 - p)^{x_i}, & \text{if } x_i = 0, 1, 2, \ldots, \\ 0, & \text{otherwise.} \end{cases}$$

To obtain the distribution of $Y = X_1 + X_2$, we can use the convolution formula (4.7). Thus, for $y = 0, 1, 2, \ldots,$

$$p_Y(y) = \sum_{y_1 = 0}^{\infty} p_{X_1}(y_1)p_{X_2}(y - y_1) = \sum_{y_1 = 0}^{y} p_{X_1}(y_1)p_{X_2}(y - y_1)$$

$$= \sum_{y_1 = 0}^{y} p^2(1 - p)^y = (y + 1)p^2(1 - p)^y,$$

and otherwise $p_Y(y) = 0$. Since $y + 1 = \binom{y+2-1}{2-1}$, $p_Y(y)$ is the probability mass function of the negative binomial distribution with parameters $r = 2$ and p.

Example 4.9. Suppose that X_1 and X_2 are independent random variables and that $X_i \sim \chi_{n_i}^2$, $i = 1, 2$. That is,

$$f_{X_i}(x_i) = \begin{cases} \dfrac{x_i^{\frac{1}{2}n_i-1} e^{-\frac{1}{2}x_i}}{2^{\frac{1}{2}n_i}\Gamma(n_i/2)}, & \text{if } 0 < x_i < \infty, \\ 0, & \text{otherwise} \end{cases}$$

for $i = 1, 2$. To obtain the distribution of $Y = X_1 + X_2$, use the convolution formula (4.8). When $y \leq 0$, $f_Y(y) = 0$; when $y > 0$,

$$f_Y(y) = \frac{e^{-\frac{1}{2}y}}{2^{\frac{1}{2}(n_1+n_2)}\Gamma(n_1/2)\Gamma(n_2/2)} \int_0^y (y - y_2)^{\frac{1}{2}n_1-1} y_2^{\frac{1}{2}n_2-1}\, dy_2$$

$$= \frac{y^{\frac{1}{2}(n_1+n_2)-1} e^{-\frac{1}{2}y}}{2^{\frac{1}{2}(n_1+n_2)}\Gamma[(n_1+n_2)/2]} \int_0^1 \frac{z^{\frac{1}{2}n_2-1}(1-z)^{\frac{1}{2}n_1-1}}{B(n_1/2, n_2/2)}\, dz$$

$$= \frac{y^{\frac{1}{2}(n_1+n_2)-1} e^{-\frac{1}{2}y}}{2^{\frac{1}{2}(n_1+n_2)}\Gamma[(n_1+n_2)/2]},$$

where we have made the change of variable from y_2 to $z = y_2/y$. The density function of Y is that of a gamma distribution with parameters $\theta = \frac{1}{2}$, $r = (n_1 + n_2)/2$, and thus $Y \sim \chi_{n_1+n_2}^2$. We have shown that the convolution of two chi-squared distributions is a chi-squared distribution with degrees of freedom equal to the sum of the degrees of freedom of the original chi-squared distributions.

When $X_1, X_2, \ldots, X_p$ are independent, it is often possible to obtain the distribution of $Y = X_1 + X_2 + \cdots + X_p$ iteratively. That is, first we use a convolution formula [(4.7) or (4.8)] to obtain the distribution of $(X_1 + X_2) + X_3$. Continuing in this fashion, we use the convolution formulas to obtain the distributions of $(X_1 + X_2 + X_3) + X_4$, $(X_1 + X_2 + X_3 + X_4) + X_5$, and so on, until at last, one final convolution gives us the distribution of $(X_1 + X_2 + \cdots + X_{p-1}) + X_p = Y$.

Remark. In applying the foregoing iterative method for finding the distribution of $Y = X_1 + X_2 + \cdots + X_p$, we implicitly make use of the fact that if the random variables $X_1, X_2, \ldots, X_p$ are independent, then $\sum_{i=1}^j X_i$ and $X_{j+1}, \ldots, X_p$ are independent. Although intuitively reasonable, this fact requires rigorous proof. The proof can be given by making the one-to-one transformation from $X_1, X_2, \ldots, X_p$ to

$$Z_1 = X_1, \quad Z_2 = X_2, \ldots, Z_{j-1} = X_{j-1},$$

$$Z_j = \sum_{i=1}^{j} X_i, \quad Z_{j+1} = X_{j+1}, \ldots, Z_p = X_p,$$

obtaining the joint distribution of $Z_1, Z_2, \ldots, Z_p$, and then finding the marginal joint distribution of $Z_j, Z_{j+1}, \ldots, Z_p$. It will then be seen that the marginal joint distribution of $Z_j, Z_{j+1}, \ldots, Z_p$ is the product of the marginal distributions of $Z_j, Z_{j+1}, \ldots, Z_p$.

Example 4.10. Suppose that $X_1, X_2, \ldots, X_p$ are independent, and that $X_i \sim \chi^2_{n_i}$, $i = 1, 2, \ldots, p$. We wish to find the distribution of $Y = \sum_{i=1}^{p} X_i$. From Example 4.9, we know that $X_1 + X_2 \sim \chi^2_{n_1 + n_2}$. Next, $\sum_{i=1}^{3} X_i = (X_1 + X_2) + X_3 \sim \chi^2_m$, where $m = (n_1 + n_2) + n_3 = \sum_{i=1}^{3} n_i$, $\sum_{i=1}^{4} X_i = \sum_{i=1}^{3} X_i + X_4 \sim \chi^2_m$ with $m = \sum_{i=1}^{3} n_i + n_4 = \sum_{i=1}^{4} n_i$, and so on. Finally, $Y = \sum_{i=1}^{p} X_i \sim \chi^2_m$ with $m = \sum_{i=1}^{p} n_i$; that is, the convolution of independent random variables, each of which has a chi-squared distribution, is a chi-squared distribution with degrees of freedom equal to the sum of the individual degrees of freedom.

p-Fold Convolutions

If we use the iterative approach to find the distribution of the sum $Y = X_1 + X_2 + X_3$ of the independent random variables X_1, X_2, X_3, we find, using the convolution formulas (4.7) and (4.8),

(4.9)
$$\begin{aligned} p_Y(y) &= \sum_{y_2} \left[\sum_{y_1} p_{X_1}(y_1) p_{X_2}(y_2 - y_1) \right] p_{X_3}(y - y_2) \\ &= \sum_{y_2} \sum_{y_1} p_{X_1}(y_1) p_{X_2}(y_2 - y_1) p_{X_3}(y - y_2) \end{aligned}$$

and

(4.10)
$$f_Y(y) = \int_{-\infty}^{\infty} \int_{-\infty}^{\infty} f_{X_1}(y_1) f_{X_2}(y_2 - y_1) f_{X_3}(y - y_2) \, dy_1 \, dy_2.$$

These formulas give the *convolutions* $p_Y(y)$ and $f_Y(y)$ of the probability mass functions $p_{X_1}(x_1)$, $p_{X_2}(x_2)$, $p_{X_3}(x_3)$ and the probability density functions $f_{X_1}(x_1)$, $f_{X_2}(x_2)$, $f_{X_3}(x_3)$, respectively. In the literature of probability theory, a special notation is used for the operations (4.9) and (4.10); the convolution

$p_Y(y)$ is written $p_Y = p_{X_1} * p_{X_2} * p_{X_3}$, and the convolution $f_Y(y)$ is written $f_Y = f_{X_1} * f_{X_2} * f_{X_3}$. In general,

$$(p_{X_1} * p_{X_2} * \cdots * p_{X_p})(y)$$

(4.11)
$$= \sum_{y_{p-1}} \sum_{y_{p-2}} \cdots \sum_{y_1} p_{X_1}(y_1) p_{X_2}(y_2 - y_1) \cdots p_{X_p}(y - y_{p-1}),$$

and

$$(f_{X_1} * f_{X_2} * \cdots * f_{X_p})(y)$$

(4.12)
$$= \int_{-\infty}^{\infty} \int_{-\infty}^{\infty} \cdots \int_{-\infty}^{\infty} f_{X_1}(y_1) f_{X_2}(y_2 - y_1) \cdots f_{X_p}(y - y_{p-1}) \, dy_1 \cdots dy_{p-1},$$

denote the *p-fold convolution* of the probability mass functions $p_{X_i}(x_i)$, $i = 1$, $2, \ldots, p$, or probability density functions $f_{X_i}(x_i)$, $i = 1, 2, \ldots, p$, respectively. If we can find

(4.13)
$$p_Y(y) = (p_{X_1} * p_{X_2} * \cdots * p_{X_p})(y)$$

in the discrete case, or if we can find

(4.14)
$$f_Y(y) = (f_{X_1} * f_{X_2} * \cdots * f_{X_p})(y)$$

in the case when $X_1, X_2, \ldots, X_p$ are independent continuous random variables, then we have determined the distribution of the sum $Y = X_1 + X_2 + \cdots + X_p$. Since *p-fold convolutions* are extensively discussed in the literature, it is often possible to avoid actually performing the computation (4.11) or (4.12) by looking up the needed result.

It is well worth noting that the operation of convolution has many of the properties possessed by the familiar mathematical operations of summation and multiplication—namely, commutivity and associativity:

(4.15) $\quad p_{X_1} * p_{X_2} = p_{X_2} * p_{X_1}, \qquad (p_{X_1} * p_{X_2}) * p_{X_3} = p_{X_1} * (p_{X_2} * p_{X_3}),$

in the discrete case, with corresponding definitions for the continuous case. Convolutions of probability mass functions also have a unity, namely, $p_0(x) = 1$ if $x = 0$ and $p_0(x) = 0$, otherwise. Thus, $p_X * p_0 = p_X$ for any probability mass function $p_X(x)$.

5. MEANS AND VARIANCES OF WEIGHTED SUMS, AND THE LAW OF LARGE NUMBERS

In Chapter 9, we showed that if X_1 and X_2 are jointly distributed random variables with means μ_{X_1} and μ_{X_2}, variances $\sigma_{X_1}^2$ and $\sigma_{X_2}^2$, and covariance σ_{X_1, X_2}, and if $Y = a_1 X_1 + a_2 X_2$, then

(5.1)
$$\mu_Y = a_1 \mu_{X_1} + a_2 \mu_{X_2},$$

(5.2)
$$\sigma_Y^2 = a_1^2 \sigma_{X_1}^2 + 2a_1 a_2 \sigma_{X_1, X_2} + a_2^2 \sigma_{X_2}^2,$$

These results can be extended to the case of p variables $X_1, \ldots, X_p$. That is, if $Y = \sum_{i=1}^{p} a_i X_i$, then

$$(5.3) \qquad\qquad \mu_Y = \sum_{i=1}^{p} a_i \mu_{X_i},$$

$$(5.4) \qquad\qquad \sigma_Y^2 = \sum_{i=1}^{p} a_i^2 \sigma_{X_i}^2 + 2 \sum_{1 \le i < j \le p} a_i a_j \sigma_{X_i, X_j},$$

where $\mu_{X_1}, \mu_{X_2}, \ldots, \mu_{X_p}, \sigma_{X_1}^2, \sigma_{X_2}^2, \ldots, \sigma_{X_p}^2$, and $\sigma_{X_i, X_j}, i \ne j$, are the means, variances, and covariances of $X_1, X_2, \ldots, X_p$. Notice that to obtain the mean and variance of the weighted sum Y, the precise joint distribution of the X_i's need not be known.

When $X_1, X_2, \ldots, X_p$ are pairwise (but not necessarily mutually) independent, then $\sigma_{X_i, X_j} = 0$ for all $i \ne j$, and (5.4) becomes

$$(5.5) \qquad\qquad \sigma_Y^2 = \sum_{i=1}^{p} a_i^2 \sigma_{X_i}^2.$$

If $X_1, X_2, \ldots, X_p$ also have common mean μ and common variance σ^2, then

$$(5.6) \qquad\qquad \mu_Y = \mu \sum_{i=1}^{p} a_i, \qquad \sigma_Y^2 = \sigma^2 \sum_{i=1}^{p} a_i^2.$$

Example 5.1. Suppose that X_1 and X_2 are jointly distributed random variables with

$$\mu_{X_1} = 7, \qquad \mu_{X_2} = 5, \qquad \sigma_{X_1}^2 = 9, \qquad \sigma_{X_2}^2 = 16.$$

Then from (5.1) or (5.3),

$$\mu_{X_1 + X_2} = \mu_{X_1} + \mu_{X_2} = 7 + 5 = 12,$$

$$\mu_{X_1 - X_2} = \mu_{X_1} - \mu_{X_2} = 7 - 5 = 2,$$

while from (5.2) or (5.4),

$$\sigma_{X_1 + X_2}^2 = \sigma_{X_1}^2 + 2\sigma_{X_1, X_2} + \sigma_{X_2}^2 = 25 + 2\sigma_{X_1, X_2},$$

$$\sigma_{X_1 - X_2}^2 = \sigma_{X_1}^2 - 2\sigma_{X_1, X_2} + \sigma_{X_2}^2 = 25 - 2\sigma_{X_1, X_2}.$$

It is interesting to note that when (and *only* when) $\sigma_{X_1, X_2} = 0$, then $\sigma_{X_1 + X_2}^2 = \sigma_{X_1 - X_2}^2$.

Example 5.2. Suppose that $X_1, X_2, \ldots, X_p$ are mutually independent random variables, each having an exponential distribution with parameter θ. Then

$$\mu_{X_i} = \frac{1}{\theta}, \qquad \sigma_{X_i}^2 = \frac{1}{\theta^2}, \qquad i = 1, 2, \ldots, p.$$

Since mutual independence implies pairwise independence, $\sigma_{X_i, X_j} = 0$, $i \ne j$.

Thus, if $Y = \sum_{i=1}^{p} a_i X_i$, we have

$$\mu_Y = \frac{1}{\theta} \sum_{i=1}^{p} a_i, \qquad \sigma_Y^2 = \frac{1}{\theta^2} \sum_{i=1}^{p} a_i^2.$$

In the special case where $a_1 = a_2 = \cdots = a_p = 1$, so that $Y = \sum_{i=1}^{p} X_i$,

$$\mu_Y = \frac{1}{\theta} \sum_{i=1}^{p} (1) = \frac{p}{\theta}, \qquad \sigma_Y^2 = \frac{1}{\theta^2} \sum_{i=1}^{p} (1)^2 = \frac{p}{\theta^2}.$$

These are precisely the mean and variance of a gamma distribution with parameters θ and $r = p$, suggesting that $Y = \sum_{i=1}^{p} X_i$ might have that distribution. (This is, in fact, the case.)

The Law of Large Numbers

Perhaps the most important weighted sum in terms of practical use is the sample average

(5.7)
$$\bar{X} = \hat{\mu}_X = \sum_{i=1}^{n} \frac{X_i}{n}$$

of n independent observations $X_1, X_2, \ldots, X_n$ of a random variable X. If X has a mean $\mu_X = \mu$ and variance $\sigma_X^2 = \sigma^2$, then

(5.8)
$$\mu_{\bar{X}} = \mu \sum_{i=1}^{n} \left(\frac{1}{n}\right) = \mu, \qquad \sigma_{\bar{X}}^2 = \sigma^2 \sum_{i=1}^{n} \left(\frac{1}{n}\right)^2 = \frac{\sigma^2}{n}.$$

Consequently, the *variance of the sample average $\bar{X}$ decreases as the sample size n increases*. As the sample size n increases, it is a consequence of the Bienaymé–Chebyshev Inequality (Chapter 5) that for any positive number c (no matter how small), the probability that $\bar{X}$ lies in the interval from $\mu - c$ to $\mu + c$ goes to 1 as N goes to infinity. To see this, note that by the Bienaymé–Chebyshev Inequality,

$$P\{\mu - c \le \bar{X} \le \mu + c\} = P\{|\bar{X} - \mu| \le c\}$$

(5.9)
$$= 1 - P\left\{|\bar{X} - \mu| > \frac{c}{\sqrt{\sigma^2/n}} \sigma_{\bar{X}}\right\}$$

$$\ge 1 - \frac{1}{(c/\sqrt{\sigma^2/n})^2}$$

$$= 1 - \frac{\sigma^2}{nc^2}.$$

Since any probability is bounded above by 1, and since $P\{\mu - c \le \bar{X} \le \mu + c\}$ is bounded below by a number $1 - (\sigma^2/nc^2)$ which tends to 1 as $n \to \infty$, we conclude that for any $c > 0$,

(5.10)
$$\lim_{n \to \infty} P\{\mu - c \le \bar{X} \le \mu + c\} = 1.$$

Table 5.1: *Values of the Sample Size* n *Required to Assure at Least a Probability of* P *That* $\bar{X}$ *Falls Within a Specified Number* c *of Units from Its Mean* μ *When* $\sigma^2 = 1$

Number of units, c	P = 0.95: n	P = 0.99: n
0.01	200,000	1,000,000
0.10	2,000	10,000
0.50	80	400
1.00	20	100
1.50	9	45
2.00	5	25
2.50	4	16
3.00	3	12

To summarize, *as the sample size n becomes large, the sample average $\bar{X}$ concentrates more and closely around its mean μ.* Using Inequality (5.9), Table 5.1 gives values of the sample size n required to guarantee with probability at least P (for $P = 0.95$, 0.99) that $\bar{X}$ is within various distances c from its mean μ when $\sigma^2 = 1$.

The conclusion we can draw from this analysis is that by making n large enough, we ensure a high probability of having $\bar{X}$ fall as close as we please to its mean μ. This assertion, which is a somewhat informal statement of the Law of Large Numbers, can be derived from the axioms of probability theory; it corresponds to the frequency interpretation of the mean we gave in Chapter 5, Section 2.

A popular misconception concerning the law of large numbers is that it is a statement about the sample total $X_1 + X_2 + \cdots + X_n$, rather than about the sample average $(1/n)(X_1 + X_2 + \cdots + X_n)$. For example, in tossing a fair coin, the so-called "man in the street" feels that the number of heads observed should equal the number of tails. Thus, if he observed 50 heads in a row, he tends to believe that a tail will appear on the 51st toss, claiming that the "law of averages" says that the total number of heads must equal the total number of tails. But if the coin tosses are independent, as assumed, observing 50 heads on the first 50 tosses tells us nothing new about the probability of the 51st toss resulting in a tail: the probability of tail is still $\frac{1}{2}$. The law of large numbers does not say that the total number of heads must eventually equal the total number of tails; rather, it says that the proportion of heads eventually equals the proportion of tails. The difference between these two statements can be illustrated by assuming that the first 10 tosses are heads and that after the first 10 tosses, the number of heads does equal the number of tails. Thus, in 1000 tosses we have 505 heads and 495 tails, in

10,000 tosses we have 5005 heads and 4995 tails, and so on. Always, the number of heads is 10 more than the number of tails, but the proportions are 0.505 as against 0.495 (after 10,000 tosses), 0.5005 as against 0.4995 (after 10,000) tosses), 0.50005 as against 0.4995 (after 100,000 tosses), and so on. That is, when we compute proportions rather than totals, the excess of heads is overwhelmed by the total number of tosses; the difference between the total number of heads and tails observed becomes a smaller and smaller fraction of the total number of tosses as the number of tosses gets large.

Remark. The sample average being discussed here is the proportion of heads; this proportion can be given as the average of the observations X_1, $X_2, \ldots, X_n$, where X_i is 1 if the ith toss is heads and 0 otherwise.

6. DISTRIBUTIONS OF SUMS (AND WEIGHTED SUMS) IN SPECIAL CASES

The following are special cases where the distributions of certain weighted sums are known.

(1) *The Normal Distribution.* If $X_1, X_2, \ldots, X_m$ are mutually independent, and $X_i \sim \mathcal{N}(\mu_i, \sigma_i^2)$, $i = 1, 2, \ldots, m$, then $Y = \sum_{i=1}^{m} a_i X_i \sim \mathcal{N}(\sum_{i=1}^{m} a_i \mu_i, \sum_{i=1}^{m} a_i^2 \sigma_i^2)$. In particular, if $\mu_i = \mu$, $\sigma_i^2 = \sigma^2$, $i = 1, 2, \ldots, m$, then

$$\bar{X} = \sum_{i=1}^{m} \frac{X_i}{m} \sim \mathcal{N}\left(\mu, \frac{\sigma^2}{m}\right).$$

(2) *The Binomial Distribution.* If $X_1, X_2, \ldots, X_m$ are mutually independent, and if each X_i has a binomial distribution with parameters n_i and p, where p is the probability of success, then

$$Y = \sum_{i=1}^{m} X_i$$

also has a binomial distribution with parameters $n = \sum_{i=1}^{m} n_i$ and p.

(3) *The Poisson Distribution.* If $X_1, X_2, \ldots, X_m$ are mutually independent, and if each X_i has a Poisson distribution with parameter λ_i, then

$$Y = \sum_{i=1}^{m} X_i$$

also has a Poisson distribution with parameter $\lambda = \sum_{i=1}^{m} \lambda_i$.

(4) *The Exponential Distribution.* If X_1, X_2, ..., X_m are mutually independent, and if each X_i has an exponential distribution with parameter θ, then

$$Y = \sum_{i=1}^{m} X_i$$

has a gamma distribution with parameters θ and m.

(5) *The Gamma Distribution.* If X_1, X_2, ..., X_m are mutually independent, and if each X_i has a gamma distribution with parameters θ and d_i, then

$$Y = \sum_{i=1}^{m} X_i$$

has a gamma distribution with parameters θ and $d = \sum_{i=1}^{m} d_i$.

(6) *The Cauchy Distribution.* If X_1, X_2, ..., X_m are mutually independent, and if each X_i has a Cauchy distribution with parameter θ, then

$$\bar{X} = \sum_{i=1}^{m} \frac{X_i}{m}$$

has a Cauchy distribution with the same parameter θ.

Cases (1), (2), (3), and (5) can be obtained by first using the convolution formulas (4.5) or (4.6) to prove the result in the case $m = 2$, and then using inductive methods (see Example 4.6) to extend the results to the situation where $m > 2$. Since the exponential distribution is also a gamma distribution with parameters θ and $r = 1$, case (4) is a direct application of case (5). Cases (1) through (5) can also be obtained through use of generating functions (Chapter 11).

Case (6) is more difficult. To verify case (6), we first establish a more general result for weighted sums of two independent and identically distributed Cauchy random variables. If X_1 and X_2 are independent random variables, each having a Cauchy distribution with parameter θ, then for any positive constants a_1 and a_2 with $a_1 + a_2 = 1$, $Y = a_1 X_1 + a_2 X_2$ has a Cauchy distribution with parameter θ. It follows from this general result that $\frac{1}{2}(X_1 + X_2)$ has a Cauchy distribution with parameter θ; that $\frac{1}{3}(X_1 + X_2 + X_3) = \frac{2}{3}[\frac{1}{2}(X_1 + X_2)] + \frac{1}{3}X_3$ has a Cauchy distribution with parameter θ; and in general that

$$\bar{X} = \frac{1}{m} \sum_{i=1}^{m} X_i = \frac{m-1}{m}\left(\frac{1}{m-1} \sum_{i=1}^{m-1} X_i\right) + \frac{1}{m} X_m$$

has a Cauchy distribution with parameter θ. The proof of the assertion that $a_1 X_1 + a_2 X_2$, where $a_1 > 0$, $a_2 > 0$, $a_1 + a_2 = 1$, has a Cauchy distribution with parameter θ involves straightforward, but complicated, algebra. The basic steps of the computation, however, can be illustrated in the special case where $\theta = 0$ and $a_1 = a_2 = \frac{1}{2}$.

Example 6.1. Suppose that X_1 and X_2 are independent random variables, where X_1 and X_2 each have a Cauchy distribution with parameter $\theta = 0$. Let us find the distribution of $Y = \frac{1}{2}X_1 + \frac{1}{2}X_2$. The joint probability density function of X_1, X_2 is

$$f_{X_1, X_2}(x_1, x_2) = f_{X_1}(x_1)f_{X_2}(x_2) = \frac{1}{\pi^2(1 + x_1^2)(1 + x_2^2)}$$

for $-\infty < x_1, x_2 < \infty$. Using Equation (4.2), the probability density function $f_Y(y)$ of Y is

$$f_Y(y) = \int_{-\infty}^{\infty} \frac{2\,dy_1}{\pi^2[1 + (2y - y_1)^2](1 + y_1^2)}.$$

Now

$$\frac{1}{[1 + (2y - y_1)^2](1 + y_1^2)} = \frac{Ay_1 + B}{(1 + y_1)^2} + \frac{Cy_1 + D}{1 + (2y - y_1)^2},$$

where

$$A = \frac{1}{4y(y^2 + 1)}, \qquad C = -A, \quad B = Ay, \quad D = 3Ay.$$

Thus,

$$f_Y(y) = \frac{2}{\pi^2} \int_{-\infty}^{\infty} \frac{Ay_1 + B}{1 + y_1^2}\,dy_1 + \int_{-\infty}^{\infty} \frac{Cy_1 + D}{1 + (2y - y_1)^2}\,dy_1$$

$$= \frac{2}{\pi^2} \int_{-\infty}^{\infty} \frac{Ay_1 + B}{1 + y_1^2}\,dy_1 + \int_{-\infty}^{\infty} \frac{Cz + 2Cy + D}{1 + z^2}\,dz ,$$

where in the second integral we have made a change of variable from y_1 to $z = y_1 - 2y$. Combining integrals and noting that $A + C = 0$, we find that

$$f_Y(y) = \frac{2}{\pi} \int_{-\infty}^{\infty} \frac{B - 2Ay + D}{\pi(1 + z^2)}\,dz = \frac{4Ay}{\pi} \int_{-\infty}^{\infty} \frac{1}{\pi(1 + z^2)}\,dz = \frac{1}{\pi(1 + y^2)},$$

which is the probability density function of a Cauchy distribution with parameter $\theta = 0$.

All of the special cases (1) through (6) listed above deal with weighted sums of independent random variables. In Chapter 9, Section 5, we gave one example where the weighted sum of dependent random variables is known.

(7) *Bivariate Normal Distribution.* If X and Y have a bivariate normal distribution with parameters μ_X, μ_Y, σ_X^2, σ_Y^2, and ρ, then the weighted sum $aX + bY + c$ has a normal distribution with mean

$$\mu_{aX+bY+c} = a\mu_X + b\mu_Y + c$$

and variance

$$\sigma_{aX+bY+c}^2 = a^2\sigma_X^2 + b^2\sigma_Y^2 + 2ab\sigma_X\sigma_Y\rho.$$

A particular example of case (7) was verified in Example 4.3, where we showed that if $(X, Y) \sim BVN((0, 0), (1, 1, \rho))$, then $X + Y \sim \mathcal{N}(0, 2(1 + \rho))$. We can use this example to prove the general case. Thus, if $(X, Y) \sim BVN((\mu_X, \mu_Y), (\sigma_X^2, \sigma_Y^2, \rho))$, make the transformation

$$Z_1 = \frac{X - \mu_X}{\sigma_X}, \qquad Z_2 = \frac{Y - \mu_Y}{\sigma_Y}.$$

Using Equation (1.11), it is readily shown that $(Z_1, Z_2) \sim \mathcal{N}((0, 0), (1, 1, \rho))$. But

$$aX + bY + c = a\sigma_X\left(\frac{X - \mu_X}{\sigma_X}\right) + b\sigma_Y\left(\frac{Y - \mu_Y}{\sigma_Y}\right) + a\mu_X + b\mu_Y + c$$

$$= (a\sigma_X)Z_1 + (b\sigma_Y)Z_2 + (a\mu_X + b\mu_Y + c).$$

Using techniques similar to those used in Example 4.3, we find that

$$a\sigma_X Z_1 + b\sigma_Y Z_2 \sim \mathcal{N}(0, a^2\sigma_X^2 + b^2\sigma_Y^2 + 2ab\rho\sigma_X\sigma_Y),$$

and thus

$$aX + bY + c = (a\sigma_X Z_1 + b\sigma_Y Z_2) + (a\mu_X + b\mu_Y + c)$$

$$\sim \mathcal{N}(a\mu_X + b\mu_Y + c, a^2\sigma_X^2 + b^2\sigma_Y^2 + 2ab\rho\sigma_X\sigma_X).$$

EXERCISES

1. Let X and Y be independent $N(0, 1)$ random variables.
    (a) Find the joint probability density function of $V = X/Y$ and Y.
    (b) Find the (marginal) probability density function of V.

2. Let $X \sim \mathcal{N}(0, 1)$ and $Y \sim \chi_r^2$ be independent.
    (a) Find the joint probability density function of

$$V = \frac{X}{\sqrt{Y + X^2}}, \qquad W = Y + X^2.$$

(b) Find the marginal probability density functions of V and W. Show that V and W are independent.

(c) Show that V^2 has a beta distribution.

3. Let X_1 and X_2 be independent random variables, where X_i has a chi-squared distribution with n_i degrees of freedom, $i = 1, 2$. Find the distribution of $Y = (X_1/n_1)/(X_2/n_2)$ as follows:

(a) First find the joint probability density function of $Y_1 = X_1/X_2$ and $Y_2 = X_1 + X_2$.

(b) Then find the marginal probability density function of Y_1.

(c) Finally, let $Y = (n_2/n_1)Y_1$ and find the probability density function of Y from that of Y_1.

(d) What is the marginal distribution of Y_2?

4. Let X_1 have a negative binomial distribution with parameters u and p, where u is an integer. Let X_2 have a geometric distribution with parameter p. Let X_1 and X_2 be independent. Find the joint probability mass function $p_{Y_1, Y_2}(y_1, y_2)$ of $Y_1 = X_1 + X_2 - 1$ and $Y_2 = X_2$. Then, find the marginal distribution of Y_1.

5. Let $(X_1, X_2) \sim \mathcal{N}((\mu_1, \mu_2), (\sigma_1^2, \sigma_2^2, \rho))$. Find the joint probability density function of

$$ Y_1 = \frac{1}{\sqrt{2}}\left(\frac{X_1}{\sigma_1} + \frac{X_2}{\sigma_2}\right), \qquad Y_2 = \frac{1}{\sqrt{2}}\left(\frac{X_1}{\sigma_1} - \frac{X_2}{\sigma_2}\right). $$

Are Y_1 and Y_2 independent? Find the marginal distributions of Y_1 and Y_2.

6. Let X and Y be independent random variables with X having a beta distribution with parameters r and s and Y having a gamma distribution with parameters $r + s$ and $\theta = 1$.

(a) Find the joint probability density function of $V = XY$ and $W = (1 - X)Y$.

(b) Find the marginal distributions of V and W, and show that V and W are independent.

7. Let X and Y have a bivariate Dirichlet distribution with parameters a, b, $c > 0$.

(a) Find the joint probability density function of $V = X + Y$ and $W = X/Y$.

(b) Find the marginal distributions of V and of $W^* = (b/a)W$. Are V and W^* independent?

8. Let X_1 and X_2 be independent $U[0, 1]$ random variables. Let $W = \max(X_1, X_2)$ and $V = \min(X_1, X_2)$.

(a) Show that

$$F_{V,W}(v, w) = \begin{cases} 1, & \text{if } 1 \le v \le w < \infty, \\ 2v - v^2, & \text{if } 0 \le v \le 1 < w < \infty, \\ 2vw - v^2, & \text{if } 0 \le v \le w \le 1, \\ 0, & \text{otherwise.} \end{cases}$$

(b) Find the joint probability density function of V and $R = W - V$. Verify that V and R have a bivariate Dirichlet distribution.

(c) Find the marginal probability density functions of V and R. Are V and R independent?

9. Let X_1, X_2 be independent random variables, each with an exponential distribution with parameter $\theta = 1$. Let $W = \max(X_1, X_2)$, $V = \min(X_1, X_2)$.

   (a) Let $Y_1 = 1 - \exp(-X_1)$, $Y_2 = 1 - \exp(-X_2)$. Show that Y_1 and Y_2 are independent $U[0, 1]$ random variables.

   (b) Note that $\max(Y_1, X_2) = 1 - \exp(-W)$ and $\min(Y_1, Y_2) = 1 - \exp(-V)$. Find the joint probability density function of V and W, and the marginal probability density functions of V and W.

10. Let X_1, X_2, X_3 be mutually independent random variables, where X_i has a gamma distribution with parameters r_i and $\theta_i = 1$, $i = 1, 2, 3$. Let

$$Y_i = \frac{X_i}{X_1 + X_2 + X_3}, \quad i = 1, 2, \qquad Y_3 = X_1 + X_2 + X_3$$

   (a) Find the joint probability density function of Y_1, Y_2, and Y_3.

   (b) Show that the marginal joint distribution of Y_1 and Y_2 is a bivariate Dirichlet distribution.

11. Let X_1, X_2 be independent $U[0, 1]$ random variables. Find the probability density function of $Y = X_1 X_2$ in each of the following ways:

    (a) By finding the joint probability density function of $Y_1 = X_1 X_2$ and $Y_2 = X_2$.

    (b) By use of (3.7), verifying needed conditions.

    (c) By showing that $V_i = -\log(X_i)$, $i = 1, 2$, are independent exponentially distributed random variables with parameter $\theta = 1$, verifying that $V_1 + V_2$ has a gamma distribution with parameters $r = 2$ and $\theta = 1$, and then finding the probability density function of $Y = \exp[-(V_1 + V_2)] = X_1 X_2$.

12. Let X_1 and X_2 be independent random variables, each with a Cauchy distribution with parameter θ. Show that $Y = aX_1 + bX_2$ for general a, $b > 0$, $a + b = 1$, has a Cauchy distribution with parameter θ.

13. Show that when deriving the probability density function of the sum $Y = X_1 + X_2$ of two jointly distributed random variables X_1, X_2 with

joint probability density function $f_{X_1, X_2}(x_1, x_2)$, use of Equation (3.7) yields the second integral in Equation (4.6).

14. Let X_1 and X_2 be independent random variables, each having a Weibull distribution with parameters $\alpha = 1$, $\beta = 2$, $v = 0$. Let $Y = X_1^2 + X_2^2$, $Z = X_1 + X_2$.
    (a) Find the probability density function of Y, using (3.7).
    (b) Find the probability distribution of Y by first finding the joint probability density function of $V_1 = X_1^2$ and $V_2 = X_2^2$, and then finding the probability density function of $Y = V_1 + V_2$.
    (c) Find the probability density function of Z.

15. Show that if $X_1, X_2, \ldots, X_r$ are mutually independent, and if X_i has a binomial distribution with parameters n_i and p, $i = 1, 2, \ldots, r$, then $Y = \sum_{i=1}^{r} X_i$ has a binomial distribution with parameters $n = \sum_{i=1}^{r} n_i$ and p.

16. Show that if $X_1, X_2, \ldots, X_r$ are mutually independent, and if X_i has a Poisson distribution with parameter λ_i, $i = 1, 2, \ldots, r$, then $Y = \sum_{i=1}^{r} X_i$ has a Poisson distribution with parameter $\lambda = \sum_{i=1}^{r} \lambda_i$.

17. Show that if $X_1, X_2, \ldots, X_r$ are mutually independent random variables, and if X_i has a negative binomial distribution with parameters u_i and p, where u_i is an integer, $i = 1, 2, \ldots, r$, then $Y = \sum_{i=1}^{r} X_i$ has a negative binomial distribution with parameters $u = \sum_{i=1}^{r} u_i$ and p.

18. Show that if $X_1, X_2, \ldots, X_r$ are mutually independent random variables, and if X_i has a geometric distribution with parameter $p_i = p$, $i = 1, 2, \ldots, r$, then $Y = \sum_{i=1}^{r} X_i - r$ has a negative binomial distribution with parameters r and p.

19. Let $X_1, X_2, \ldots, X_s$ be mutually probabilistically independent random variables and let X_i have a gamma distribution with parameters r_i and θ, $i = 1, 2, \ldots, r$.
    (a) Show that $Y = \sum_{i=1}^{s} X_i$ has a gamma distribution with parameters $r = \sum_{i=1}^{s} r_i$ and θ.
    (b) Hence, show that if $X_1, X_2, \ldots, X_s$ are mutually independent random variables, where X_i has an exponential distribution with parameter θ, then $Y = \sum_{i=1}^{s} X_i$ has a gamma distribution with parameters s and θ.

20. Let $X_1, X_2, \ldots, X_r$ be mutually independent random variables, and let $X_i \sim U[0, 1]$, $i = 1, 2, \ldots, r$.
    (a) For $r = 2$, find the probability density function of $Y = X_1 + X_2$.
    (b) For $r = 3$, find the probability density function of $Y = X_1 + X_2 + X_3$.
    (c) Find the probability density function of $Y = \sum_{i=1}^{r} X_i$ for general r.

(d) Find the probability density function $f_{Z_r}(z)$ of

$$Z_r = \frac{\sum_{i=1}^{r} (X_i - \frac{1}{2})}{\sqrt{r/12}}$$

for $r = 1, 2, \ldots$.

(e) Show that the limit of $f_{Z_r}(z)$ as $r \to \infty$ is

$$\lim_{r \to \infty} f_{Z_r}(z) = \frac{1}{\sqrt{2\pi}} e^{-\frac{1}{2}z^2}.$$

21. The number of automobile accidents occurring on any given nonholiday weekend at a particular intersection has a Poisson distribution with parameter $\lambda = 2$. The number of accidents occurring on any one weekend is independent of the number of accidents occurring on all other weekends.

(a) What is the distribution of the total number of accidents that occur on weekends in a month (4 weekends) that has no holiday weekends?

(b) In a year there are 52 weekends, of which 6 are holiday weekends and 46 are nonholiday weekends. On any holiday weekend, the number of accidents that occur has a Poisson distribution with parameter $\lambda = 3.5$. What is the distribution of the total number of weekend accidents Y that occur in a year?

(c) Suppose that $X_1, X_2, \ldots, X_{52}$ are mutually independent random variables each of which has a Poisson distribution with parameter $\lambda = \frac{113}{52}$. Show that $Z = X_1 + X_2 + \cdots + X_{52}$ has the same distribution as the random variable Y described in part (b). Thus, show that $P\{Y \geq 120\}$ and $P\{Z \geq 120\}$ are equal.

(d) Use the result in part (c) and the Central Limit Theorem to find the probability that at least 120 automobile accidents occur on weekends at the given intersection in a year.

22. Let X_1 and X_2 be independent random variables, with X_i having an exponential distribution with parameter θ_i, $i = 1, 2$. Let $Y = X_1 + X_2$.

(a) Show, by finding the probability density function of Y, that Y does not have a gamma distribution unless $\theta_1 = \theta_2$.

(b) Find μ_Y and σ_Y^2.

(c) A common method for finding a distribution with known or easily calculatable probabilities to approximate a given distribution, such as that of Y in this problem, is to match moments. In the present case, we might look for a gamma distribution with parameters r and θ so that this distribution has the same mean and variance as Y. Show that such a distribution has parameters

$$r = \frac{(\theta_1 + \theta_2)^2}{\theta_1^2 + \theta_2^2}, \qquad \theta = \frac{(\theta_1 + \theta_2)\theta_1\theta_2}{\theta_1^2 + \theta_2^2}.$$

Show that if $\theta_1 = \theta_2$, then $r = 2$ and $\theta = \theta_1 = \theta_2$.

23. Show that the observations X_1, X_2, ..., X_n on X need only be (pair-wise) uncorrelated (i.e., $\rho_{X_i, X_j} = 0$, all $i \neq j$) in order for (5.8) and inequality (5.9) to hold.

24. Two randomly selected groups of individuals are given a reaction time test, the first group after taking a particular drug, and the second group after taking a placebo. The reaction times X_1, X_2, ..., X_n of the individuals who took the drug are assumed to be mutually independent, and each X_i has mean μ and variance σ^2. The reaction times Y_1, Y_2, ..., Y_m of the individuals who took the placebo are assumed to be mutually independent, and each Y_j has mean v and variance σ^2.

   (a) To evaluate the effect on reaction time of taking the drug, we desire to estimate $\Delta = \mu - v$. We know that $\hat{\mu}_X = (X_1 + X_2 + \cdots + X_n)/n$ is a reasonable approximation to the mean μ of any X_i when n is large. We know that $\hat{\mu}_Y = (Y_1 + Y_2 + \cdots + Y_m)/m$ is a reasonable approximation to the mean v of any Y_j when m is large. This suggests estimating Δ by $\hat{\mu}_X - \hat{\mu}_Y$. Find the mean and variance of $\hat{\mu}_X - \hat{\mu}_Y$.
   (b) Use the Bienaymé–Chebyshev Inequality (see Chapter 5) and the answer to part (a) to show that the probability distribution of $\hat{\mu}_X - \hat{\mu}_Y$ concentrates more and more probability around the value Δ as the sample sizes n and m become large.
   (c) If the X_i's and the Y_j's are all normally distributed, find the distribution of $\hat{\mu}_X - \hat{\mu}_Y$.
   (d) Under the assumptions in part (c), find the probability that $\hat{\mu}_X - \hat{\mu}_Y$ is within one unit of Δ when $n = m = 100$, $\sigma = 4$.
   (e) Assume that $n = m$ and that $\sigma = 4$. How many individuals must take the drug (and how many individuals must take the placebo) in order that

   $$P\{-1 \leq \hat{\mu}_X - \hat{\mu}_Y - \Delta \leq 1\} \geq 0.95?$$

   Answer using the assumption that the X_i's and Y_i's are all normally distributed [see part (c)]. Then answer using the Bienaymé–Chebyshev Inequality. Compare your results.

11

Generating Functions

The method of generating functions is a powerful tool in probability theory. Through the use of generating functions, we often can obtain distributions (or moments of distributions) indirectly in a far less complicated fashion than by a direct computation.

There are many different generating functions, each of which is useful for certain kinds of problems and certain types of random variables. The *factorial moment generating function* (Section 1) is generally the most useful tool for dealing with nonnegative integer-valued random variables. The *moment generating function* (Section 2) can be used profitably with any random variable, discrete or continuous, nonnegative or not, for which this function can be defined, while the *characteristic function* (Section 2) is defined for all random variables. All of these generating functions are particularly appropriate for finding the distribution and moments of sums and weighted averages of independent random variables. In contrast, the *Mellin transform* (Section 3) can be used to find the distribution and moments of products or ratios of independent, nonnegative random variables.

In this chapter, we pay greatest attention to the factorial moment generating function and the moment generating function. Although these generating functions are restricted in their application, their applications are extensive enough to resolve a wide variety of problems. In particular, the moment generating function can be used to provide a rigorous proof of the Central Limit Theorem in certain cases (Section 2).

1. THE FACTORIAL MOMENT GENERATING FUNCTION

Let X be a discrete random variable whose possible values are the nonnegative integers. For example, X can have the binomial, hypergeometric, Poisson, geometric, or negative binomial distributions. For any real number θ,

$0 \leq \theta \leq 1$, define

(1.1)
$$G_X(\theta) = E(\theta^X) = \sum_{k=0}^{\infty} \theta^k p_X(k).$$

Since $0 \leq \theta^k \leq 1$ for $0 \leq \theta \leq 1$ and $0 \leq k < \infty$, it follows that $0 \leq G_X(\theta) \leq 1$, and thus $G_X(\theta)$ is well defined. The function $G_X(\theta)$ is called the *factorial moment generating function* (*factorial m.g.f.*) of the distribution of the random variable X.

Example 1.1. Suppose that X has a binomial distribution with parameters n and p. Then the factorial m.g.f. of X is

$$
\begin{aligned}
G_X(\theta) = E[\theta^X] &= \sum_{k=0}^{n} \theta^k \binom{n}{k} p^k (1-p)^{n-k} \\
&= \sum_{k=0}^{n} \binom{n}{k} (\theta p)^k (1-p)^{n-k} \\
&= [\theta p + (1-p)]^n.
\end{aligned}
$$

Example 1.2. Suppose that X has a Poisson distribution with parameter λ. Then the factorial m.g.f. of X is

$$
\begin{aligned}
G_X(\theta) = E[\theta^X] &= \sum_{k=0}^{\infty} \theta^k \frac{\lambda^k}{k!} e^{-\lambda} \\
&= e^{-\lambda + \lambda\theta} \left[\sum_{k=0}^{\infty} \frac{(\theta\lambda)^k}{k!} e^{-\theta\lambda} \right] \\
&= e^{-\lambda + \lambda\theta}.
\end{aligned}
$$

Factorial moment generating functions have several very useful properties that aid us in investigating properties of their associated distributions.

Property 1.1 (Uniqueness). The factorial m.g.f. $G_X(\theta)$ uniquely determines the probability mass function of any random variable X defined on the nonnegative integers.

Indeed,

$$G_X(\theta) = p_X(0) + \sum_{k=1}^{\infty} \theta^k p_X(k),$$

so that $G_X(0) = p_X(0)$, and

(1.2)
$$\frac{1}{k!} G_X^{[k]}(0) = p_X(k), \quad k = 1, 2, \ldots,$$

where $G_X^{[k]}(\theta) = (d/d\theta)^k G_X(\theta)$. To prove (1.2) for $k = 1$, note that for $0 < \theta \le 1$,

$$\sum_{k=1}^{\infty} \theta^{k-1} p_X(k) = \frac{G_X(\theta) - p_X(0)}{\theta} = \frac{G_X(\theta) - G_X(0)}{\theta}.$$

Letting $\theta \to 0$ in this equality yields

$$p_X(1) = \lim_{\theta \to 0} \sum_{k=1}^{\infty} \theta^{k-1} p_X(k) = \lim_{\theta \to 0} \frac{G_X(\theta) - G_X(0)}{\theta} = G_X^{[1]}(0).$$

Using this result, we can prove (1.2) for $k = 2$. For $0 < \theta \le 1$,

$$\sum_{k=2}^{\infty} \theta^{k-2} p_X(k) = \frac{\frac{G_X(\theta) - p_X(0)}{\theta} - p_X(1)}{\theta} = \frac{\frac{G_X(\theta) - G_X(0)}{\theta} - G_X^{[1]}(0)}{\theta}.$$

Taking $\theta \to 0$ in this equality, and applying L'Hospital's Rule, we obtain

$$p_X(2) = \tfrac{1}{2} G_X^{[2]}(0).$$

The proof of (1.2) for $k > 2$ now proceeds recursively in the manner shown above.

Example 1.3. Suppose that the factorial m.g.f. of X is

$$G_X(\theta) = \frac{p\theta}{1 - (1 - p)\theta},$$

for $0 < p < 1$, $0 \le \theta \le 1$. Then $p_X(0) = G_X(0) = 0$,

$$p_X(1) = G_X^{[1]}(0) = \frac{p}{[1 - (1 - p)0]^2} = p,$$

$$p_X(2) = \tfrac{1}{2} G_X^{[2]}(0) = \frac{1}{2} \frac{2p(1 - p)}{[1 - (1 - p)0]^3} = p(1 - p).$$

In general, we can show that

$$p_X(k) = p(1 - p)^{k-1}, \quad k = 1, 2, \ldots,$$

which we recognize as the probability mass function of the geometric distribution with parameter p.

Property 1.2 (Factorial Moments). Let X be a nonnegative integer-valued random variable with probability mass function $p_X(k)$ and factorial m.g.f. $G_X(\theta)$. If the rth factorial moment of X,

$$E[X(X - 1) \cdots (X - r + 1)] = \sum_{k=0}^{\infty} k(k - 1) \cdots (k - r + 1) p_X(k)$$

$$= \sum_{k=r}^{\infty} k(k - 1) \cdots (k - r + 1) p_X(k)$$

is finite, then

(1.3) $G_X^{[r]}(1) = E[X(X - 1) \cdots (X - r + 1)].$

To verify (1.3), note that

$$\left(\frac{d}{d\theta}\right)^r \theta^k = \begin{cases} k(k - 1) \cdots (k - r + 1)\theta^{k-r}, & \text{if } k \geq r, \\ 0, & \text{if } k < r. \end{cases}$$

Thus,

$$G_X^{(r)}(\theta) = \left(\frac{d}{d\theta}\right)^r \sum_{k=0}^{\infty} \theta^k p_X(k) = \sum_{k=0}^{\infty} \left(\frac{d}{d\theta}\right)^r \theta^k p_X(k)$$

(1.4)

$$= \sum_{k=r}^{\infty} k(k - 1) \cdots (k - r + 1)\theta^{k-r} p_X(k),$$

provided that we can justify interchanging the derivative $(d/d\theta)^r$ with the infinite sum $\sum_{k=0}^{\infty}$ in (1.4). This interchange can be shown to be valid whenever the last sum in (1.4) is finite. Since $0 \leq \theta \leq 1$ implies that $0 \leq \theta^{k-r} \leq 1$,

$$\sum_{k=r}^{\infty} k(k - 1) \cdots (k - r + 1)\theta^{k-r} p_X(k) \leq \sum_{k=r}^{\infty} k(k - 1) \cdots (k - r + 1) p_X(k)$$

$$= E[X(X - 1) \cdots (X - r + 1)]$$

and $E[X(X - 1) \cdots (X - r + 1)] < \infty$ by our assumption. Letting $\theta = 1$ in (1.4) verifies (1.3).

Example 1.2 (continued). When X has a Poisson distribution with parameter λ, then

$$G_X(\theta) = e^{-\lambda + \lambda\theta}$$

and

$$G_X^{(r)}(\theta) = \left(\frac{d}{d\theta}\right)^r (e^{-\lambda + \lambda\theta}) = \lambda^r e^{-\lambda + \lambda\theta}.$$

Thus,

$$E[X(X - 1) \cdots (X - r + 1)] = G_X^{[r]}(1) = \lambda^r.$$

In particular, $E[X] = \lambda$, $E[X(X - 1)] = \lambda^2$, so that $\mu_X = \lambda$ and

$$\sigma_X^2 = E[X(X - 1)] + E[X] - (E[X])^2 = \lambda^2 + \lambda - \lambda^2 = \lambda.$$

Property 1.3. If a and b are nonnegative integers, then

(1.5) $G_{aX+b}(\theta) = E[\theta^{aX+b}] = \theta^b E[(\theta^a)^X] = \theta^b G_X(\theta^a).$

Example 1.3 (continued). If X has a geometric distribution with parameter p, then $G_X(\theta) = p\theta/[1 - (1 - p)\theta]$. Recall that $Y = X - 1$ has a negative binomial distribution with parameters $u = 1$ and p. Since

$$G_X(\theta) = G_{Y+1}(\theta) = \theta G_Y(\theta),$$

it follows that the factorial m.g.f. of a negative binomial distribution with parameters $u = 1$ and p is

$$G_Y(\theta) = \frac{1}{\theta} G_X(\theta) = \frac{1}{\theta} \frac{p\theta}{1 - (1 - p)\theta} = \frac{p}{1 - (1 - p)\theta}.$$

Property 1.4 (Convolutions). Let X and Y be independent nonnegative integer-valued random variables with factorial m.g.f.'s $G_X(\theta)$ and $G_Y(\theta)$, respectively. Then

(1.6)
$$G_{X+Y}(\theta) = G_X(\theta)G_Y(\theta).$$

To verify (1.6), recall that the probability mass function $p_{X+Y}(k)$ of $X + Y$ is given by the convolution

(1.7)
$$p_{X+Y}(k) = \sum_{x=0}^{\infty} p_X(x)p_Y(k - x) = \sum_{x=0}^{k} p_X(x)p_Y(k - x),$$

since $p_Y(k - x) = 0$ when $k - x < 0$. Now

$$G_{X+Y}(\theta) = \sum_{k=0}^{\infty} \theta^k p_{X+Y}(k) = \sum_{k=0}^{\infty} \theta^k \sum_{x=0}^{k} p_X(x)p_Y(k - x).$$

On the other hand,

$$G_X(\theta)G_Y(\theta) = \left[\sum_{x=0}^{\infty} \theta^x p_X(x) \right]\left[\sum_{y=0}^{\infty} \theta^y p_Y(y) \right]$$

$$= \sum_{y=0}^{\infty} \sum_{x=0}^{\infty} \theta^{x+y} p_X(x)p_Y(y)$$

$$= \sum_{x+y=0}^{\infty} \theta^{x+y} \sum_{x=0}^{\infty} p_X(x)p_Y((x + y) - x)$$

$$= \sum_{k=0}^{\infty} \theta^k \sum_{x=0}^{\infty} p_X(x)p_Y(k - x) = G_{X+Y}(\theta),$$

which establishes (1.6).

We can use Equation (1.6) and the uniqueness property (Property 1.1) of factorial moment generating functions to derive the distributions of sums of independent, nonnegative integer-valued random variables.

Example 1.4. Suppose that the independent random variables X and Y have binomial distributions with parameters n_1 and p, and n_2 and p, respectively. Then, from Example 1.1 and (1.6),

$$G_{X+Y}(\theta) = G_X(\theta)G_Y(\theta) = [p\theta + (1-p)]^{n_1}[p\theta + (1-p)]^{n_2}$$
$$= [p\theta + (1-p)]^{n_1+n_2},$$

which is the factorial m.g.f. of the binomial distribution with parameters $n_1 + n_2$ and p. Hence, by Property 1.1, $X + Y$ has a binomial distribution with parameters $n_1 + n_2$ and p.

Let X_1, X_2, ..., X_n be mutually independent, nonnegative integer-valued random variables with factorial m.g.f.'s $G_{X_1}(\theta)$, $G_{X_2}(\theta)$, ..., $G_{X_n}(\theta)$, respectively. Then repeatedly applying (1.6) to the sums $S_m = X_1 + \cdots + X_m$, $m = 1, 2, \ldots, n$, we find that

(1.8)
$$G_{S_n}(\theta) = G_{S_{n-1}}(\theta)G_{X_n}(\theta) = G_{S_{n-2}}(\theta)G_{X_{n-1}}(\theta)G_{X_n}(\theta)$$
$$= \cdots = G_{X_1}(\theta)G_{X_2}(\theta) \cdots G_{X_n}(\theta).$$

Example 1.2 (continued). If X_1, X_2, ..., X_n are mutually independent random variables, with X_i having a Poisson distribution with parameter λ_i, $i = 1, 2, \ldots, n$, and if $Y = \sum_{i=1}^{n} X_i$, then

$$G_Y(\theta) = G_{X_1}(\theta)G_{X_2}(\theta) \cdots G_{X_n}(\theta)$$
$$= (e^{-\lambda_1 + \lambda_1\theta})(e^{-\lambda_2 + \lambda_2\theta}) \cdots (e^{-\lambda_n + \lambda_n\theta})$$
$$= \exp\left[-\left(\sum_{i=1}^{n}\lambda_i\right) + \left(\sum_{i=1}^{n}\lambda_i\right)\theta\right]$$

so that Y has a Poisson distribution with parameter $\lambda = \sum_{i=1}^{n}\lambda_i$.

Property 1.5 (Preservation of Limits). For each $n = 1, 2, \ldots$, let $p_X(k; n)$ be a probability mass function on the nonnegative integers k, and let $G_X(\theta; n)$ be the corresponding factorial m.g.f. If

(1.9) $$\lim_{n\to\infty} G_X(\theta; n) = G_X(\theta) \quad \text{for all } \theta, \quad 0 \le \theta \le 1,$$

where $G_X(\theta)$ is a factorial m.g.f. corresponding to the probability mass function $p_X(k)$, then

(1.10) $$\lim_{n\to\infty} p_X(k; n) = p_X(k), \quad k = 0, 1, 2, \ldots.$$

We postpone the proof of Property 1.5 until after we illustrate its use in the following example.

Example 1.5. In Chapter 6, we showed that the probabilities $p(k; n, p_n)$ for a binomial distribution with parameters n and $p_n = \lambda/n$ can, when n is large, be approximated by the probabilities $p(k) = (\lambda^k e^{-\lambda})/k!$ for the Poisson distribution with parameter λ. Let us prove this same result using factorial moment generating functions. The factorial m.g.f. $G(\theta; n)$ of the binomial distribution with parameters n and $p_n = \lambda/n$ is given by

$$G(\theta; n) = [p_n \theta + (1 - p_n)]^n = [1 + (\theta - 1)p_n]^n.$$

When $|(\theta - 1)p_n| \leq 1$,

$$\log G(\theta; n) = n \log [1 + (\theta - 1)p_n]$$

(1.11)
$$= n(\theta - 1)p_n - \frac{n(\theta - 1)^2 p_n^2}{2} + \frac{n(\theta - 1)^2 p_n^3}{3} - \cdots.$$

Since $\lim_{n \to \infty} np_n = \lambda$, $\lim_{n \to \infty} np_n^k = 0$ for $k > 1$, it follows from (1.11) that

$$\lim_{n \to \infty} \log G(\theta; n) = (\theta - 1)\lambda.$$

Consequently,

$$\lim_{n \to \infty} G(\theta; n) = \exp \left[\lim_{n \to \infty} \log G(\theta; n) \right] = \exp [(\theta - 1)\lambda]$$

(1.12)
$$= \exp (-\lambda + \theta\lambda) = G(\theta),$$

where $G(\theta)$ is the factorial m.g.f. of the Poisson distribution with parameter λ. It now is a consequence of Property 1.5 that

$$\lim_{n \to \infty} p(k; n, p_n) = p(k) = \frac{\lambda^k e^{-\lambda}}{k!},$$

which shows that the binomial probabilities $p(k; n, p_n)$ for large n are approximated by the Poisson probabilities $p(k)$. Note that our proof did not use the fact that $p_n = \lambda/n$ (as assumed in Chapter 6), but only the assumption that $\lim_{n \to \infty} np_n = \lambda$.

To verify Property 1.5, recall that $G_X(0; n) = p_X(0; n)$ for $n = 1, 2, \ldots$, and that $G_X(0) = p_X(0)$. Hence, from (1.9) it follows that

$$\lim_{n \to \infty} p_X(0; n) = \lim_{n \to \infty} G_X(0; n) = G_X(0) = p_X(0),$$

which verifies (1.10) for $k = 0$. We now use mathematical induction to show that (1.10) holds for general values of k. Thus, assume that (1.10) holds for $0 \leq k < K$, and let us try to show that $\lim_{n \to \infty} p_X(K; n) = p_X(K)$. First note that since $\lim_{n \to \infty} p_X(k; n) = p_X(k)$ for $k = 0, 1, \ldots, K - 1$, and since $\lim_{n \to \infty} G_X(\theta; n) = G_X(\theta)$, it follows that for $0 < \theta < 1$,

$$\lim_{n \to \infty} \sum_{k=K}^{\infty} p_X(k; n)\theta^{k-K} = \lim_{n \to \infty} \frac{G_X(\theta; n) - \sum_{k=0}^{K-1} p_X(k; n)\theta^k}{\theta^k}$$

$$= \frac{G_X(\theta) - \sum_{k=0}^{K-1} p_X(k)\theta^k}{\theta^K} = \sum_{k=K}^{\infty} p_X(k)\theta^{k-K}.$$

Thus, since $\theta^{K-K} = 1$,

(1.13) $\displaystyle \lim_{n \to \infty} p_X(K; n) + \lim_{n \to \infty} \sum_{k=K+1}^{\infty} p_X(k; n)\theta^{k-K} = p_X(K) + \sum_{k=K+1}^{\infty} p_X(k)\theta^{k-K}.$

Now since $0 \leq p_X(k; n) \leq 1$ and $0 \leq p_X(k) \leq 1$, all k,

$$0 \leq \sum_{k=K+1}^{\infty} p_X(k; n)\theta^{k-K} \leq \sum_{k=K+1}^{\infty} \theta^{k-K} = \sum_{i=1}^{\infty} \theta^i = \frac{\theta}{1-\theta}$$

and

$$0 \leq \sum_{k=K+1}^{\infty} p_X(k)\theta^{k-K} \leq \sum_{k=K+1}^{\infty} \theta^{k-K} = \frac{\theta}{1-\theta}.$$

Thus as $\theta \to 0$, the second terms in each of the left-hand and right-hand sides of (1.13) go to 0, and we conclude that $\lim_{n \to \infty} p_X(K; n) = p_X(K)$. Thus Equation (1.10) holds for $k = K$, and consequently, by induction, must hold for all values $k = 0, 1, 2, \ldots$.

Remark. It is worth mentioning that the converse of Property 1.5 also holds; that is, $\lim_{n \to \infty} p_X(k; n) = p_X(k)$, for all $k = 0, 1, 2, \ldots$, implies that $\lim_{n \to \infty} G_X(\theta; n) = G_X(\theta)$ for all θ, $0 \leq \theta \leq 1$. For $0 \leq \theta < 1$ this result is verified by noting that for any K,

(1.14) $\displaystyle \sum_{k=0}^{K} p_X(k; n)\theta^k \leq G_X(\theta; n) \leq \sum_{k=0}^{K} p_X(k; n)\theta^k + \sum_{k=K+1}^{\infty} \theta^k.$

Since $\sum_{k=K+1}^{\infty} \theta^k = \theta^{K+1}/(1 - \theta)$, and $\lim_{n \to \infty}$ can be interchanged with the finite sum $\sum_{k=0}^{K}$, we can take the limit as $n \to \infty$ in (1.14) to obtain

$$\sum_{k=0}^{K} p_X(k)\theta^k \leq \lim_{n \to \infty} G_X(\theta; n) \leq \sum_{k=0}^{K} p_X(k)\theta^k + \frac{\theta^{K+1}}{1-\theta}.$$

Taking $K \to \infty$ in this last result shows that $\lim_{n \to \infty} G_X(\theta; n) = G_X(\theta)$ for $0 \leq \theta < 1$. For any factorial m.g.f. $G(\theta)$, it is easy to see that $G(1) = 1$, and thus trivially

$$\lim_{n \to \infty} G_X(1; n) = 1 = G_X(1).$$

2. *MOMENT GENERATING FUNCTIONS*

The *moment generating function (m.g.f.)* of any random variable X, whether discrete or continuous, is formally defined to be

$$(2.1) \qquad M_X(\theta) = E(e^{\theta X}),$$

for all values θ for which the expected value in (2.1) exists. Thus, if X is a discrete random variable,

$$(2.2) \qquad M_X(\theta) = E(e^{\theta X}) = \sum_x e^{\theta x} p_X(x),$$

while if X is a continuous random variable,

$$(2.3) \qquad M_X(\theta) = \int_{-\infty}^{\infty} e^{\theta x} f_X(x)\, dx.$$

Since $e^0 = 1$, the m.g.f. $M_X(\theta)$ is always defined at $\theta = 0$, and $M_X(0) = E[e^{0X}] = E[1] = 1$. When $M_X(\theta)$ is defined for an open interval (θ_L, θ_U) of values that contains $\theta = 0$, the m.g.f. has properties for describing the distribution of X similar to the properties possessed by the factorial m.g.f. $G_X(\theta)$. Indeed, if X is a nonnegative integer-valued random variable, then $M_X(\theta)$ exists for $-\infty < \theta \le 0$, and

$$(2.4) \qquad M_X(\theta) = E[e^{\theta X}] = E[(e^{\theta})^X] = G_X(e^{\theta}).$$

An advantage of the m.g.f. over the factorial m.g.f. is that the m.g.f. has meaning over a wider class of random variables. A disadvantage of the m.g.f. is that it does not always exist in an open interval (θ_L, θ_U) of θ-values containing $\theta = 0$ (Example 2.4).

Example 2.1. Since the factorial m.g.f. of the binomial distribution with parameters n and p is $G(\theta) = [\theta p + (1 - p)]^n$, it follows from (2.4) that the m.g.f. of the binomial distribution is

$$M(\theta) = G(e^{\theta}) = [e^{\theta} p + (1 - p)]^n,$$

which is defined for $-\infty < \theta < \infty$. Similarly, since the factorial m.g.f. of the Poisson distribution with parameter λ is $G(\theta) = \exp(-\lambda + \theta\lambda)$, the m.g.f. of the Poisson distribution is

$$M(\theta) = \exp(-\lambda + e^{\theta}\lambda).$$

The m.g.f. of the geometric distribution with parameter p is

$$M(\theta) = pe^{\theta}\left[\frac{1}{1 - (1 - p)e^{\theta}}\right],$$

which can be inferred from the fact that $G(\theta) = p\theta[1 - (1 - p)\theta]^{-1}$, or obtained directly as follows:

$$M(\theta) = Ee^{\theta X} = \sum_{k=1}^{\infty} e^{\theta k} p(1 - p)^{k-1} = pe^{\theta} \sum_{k=1}^{\infty} [(1 - p)e^{\theta}]^{k-1}$$

$$= pe^{\theta} \left[\frac{1}{1 - (1 - p)e^{\theta}}\right].$$

Again, this m.g.f. is defined for $-\infty < \theta < \infty$.

Example 2.2. If X has a standard gamma distribution with parameter r, then for $-\infty < \theta < 1$,

$$M(\theta) = E[e^{\theta X}] = \int_0^{\infty} \frac{e^{\theta x} x^{r-1} e^{-x} dx}{\Gamma(r)} = \frac{1}{(1 - \theta)^r} \int_0^{\infty} \frac{v^{r-1} e^{-v} dv}{\Gamma(r)}$$

$$= \frac{1}{(1 - \theta)^r},$$

where we changed the variable of integration from x to $v = x(1 - \theta)$.

Example 2.3. If $Z \sim N(0, 1)$, then for $-\infty < \theta < \infty$,

$$M(\theta) = \int_{-\infty}^{\infty} \frac{e^{\theta z} e^{-\frac{1}{2}z^2} dz}{\sqrt{2\pi}} = e^{\frac{1}{2}\theta^2} \int_{-\infty}^{\infty} \frac{e^{-\frac{1}{2}(z-\theta)^2} dz}{\sqrt{2\pi}} = e^{\frac{1}{2}\theta^2}.$$

Examples 2.1 through 2.3 yielded well-defined m.g.f.'s, but this need not be the case in general.

Example 2.4. Let the integer-valued random variable X have probability mass function

$$p_X(k) = \begin{cases} \frac{1}{2\zeta(2)k^2}, & \text{if } k = \pm 1, \pm 2, \pm 3, \ldots, \\ 0, & \text{otherwise,} \end{cases}$$

where $\zeta(2)$ is the Riemann zeta function evaluated at $s = 2$ [Chapter 6, Equation (7.3)]. Since for $\theta \neq 0$,

$$M_X(\theta) = E[e^{\theta X}] = \sum_{k=-\infty}^{-1} \frac{e^{\theta k}}{2\zeta(2)k^2} + \sum_{k=1}^{\infty} \frac{e^{\theta k}}{2\zeta(2)k^2}$$

$$= \sum_{k=1}^{\infty} \frac{e^{-|\theta|k} + e^{|\theta|k}}{2\zeta(2)k^2}$$

and

$$e^{-|\theta|k} + e^{|\theta|k} \geq e^{|\theta|k} = 1 + |\theta|k + \frac{|\theta|^2 k^2}{2} + \cdots \geq \frac{|\theta|^2 k^2}{2},$$

we have

$$M_X(\theta) = \sum_{k=1}^{\infty} \frac{e^{-|\theta|k} + e^{|\theta|k}}{2\zeta(2)k^2} \geq \frac{\theta^2}{4\zeta(2)} \sum_{k=1}^{\infty} \frac{k^2}{k^2} = \infty,$$

so that $M_X(\theta)$ does not exist for $\theta \neq 0$. In other words, $M_X(\theta)$ is only defined for $\theta = 0$.

Suppose that the m.g.f. $M_X(\theta)$ is defined for an open interval (θ_L, θ_U) of values containing $\theta = 0$. Then the following properties, analogous to corresponding properties for the factorial m.g.f., are possessed by $M_X(\theta)$.

Property 2.1 (Uniqueness). There is one and only one distribution that has the m.g.f. $M_X(\theta)$. To be specific, since the cumulative distribution function $F_X(x)$ determines the distribution of any random variable X, to $M_X(\theta)$ there corresponds a unique c.d.f. $F_X(x)$.

Property 2.2 (Moments). If $E[X^r]$ is well defined, then

(2.5) $$E[X^r] = M_X^{[r]}(0), \quad r = 1, 2, \ldots,$$

where $M_X^{[r]}(\theta) = (d/d\theta)^r M_X(\theta)$.

Property 2.3 (Linear Transformation). If $Y = aX + b$ is any linear transformation of X, then

(2.6) $$M_Y(\theta) = M_{aX+b}(\theta) = e^{b\theta} M_X(a\theta).$$

Property 2.4 (Convolutions). If X and Y are independent random variables with respective m.g.f.'s $M_X(\theta)$, $M_Y(\theta)$, then

(2.7) $$M_{X+Y}(\theta) = M_X(\theta) M_Y(\theta).$$

Property 2.5 (Preservation of Limits). For each $n = 1, 2, \ldots$ let $F(x; n)$ be a c.d.f. with corresponding m.g.f. $M(\theta; n)$ defined on an open interval (θ_L, θ_U) containing $\theta = 0$. If there exists a m.g.f. $M(\theta)$ defined on (θ_L, θ_U) for which

(2.8) $$\lim_{n \to \infty} M(\theta; n) = M(\theta), \quad \text{all } \theta \text{ in } (\theta_L, \theta_U),$$

then

(2.9) $$\lim_{n \to \infty} F(x; n) = F(x), \quad \text{all } x \text{ in } C,$$

where $F(x)$ is the c.d.f. corresponding to $M(\theta)$, and $C = \{x: F(x) \text{ is continuous at } x\}$.

Remark. It can also be shown that if (2.9) is true, then this implies that (2.8) is true.

Demonstrations of Properties 2.1, 2.2, and 2.5 would require introducing a great deal of mathematical theory, and thus we omit such proofs. We remark that verification of Property 2.1 is accomplished by exhibiting *inversion formulas* which show theoretically how to obtain the c.d.f. $F(x)$ corresponding to a m.g.f. $M(\theta)$. Some insight into the justification for Property 2.2 can be gained by noting that since $(d/d\theta)^r e^{\theta x} = x^r e^{\theta x}$,

(2.10)
$$M_X^{[r]}(\theta) = \left(\frac{d}{d\theta}\right)^r M_X(\theta) = \left(\frac{d}{d\theta}\right)^r E(e^{\theta X})$$
$$= E\left[\left(\frac{d}{d\theta}\right)^r e^{\theta X}\right] = E[X^r e^{\theta X}],$$

provided that we can interchange the order in which we take the derivative $(d/d\theta)^r$ and the expected value $E(e^{\theta X})$. Once (2.10) is established, Equation (2.5) follows by taking $\theta = 0$.

The proof of Property 2.3 is straightforward:

$$M_Y(\theta) = M_{aX+b}(0) = E[e^{(aX+b)\theta}] = e^{b\theta}E[e^{(a\theta)X}]$$
$$= e^{b\theta}M_X(a\theta).$$

Example 2.5. If $X \sim \mathcal{N}(\mu, \sigma^2)$, then $Z = (X - \mu)/\sigma \sim \mathcal{N}(0, 1)$. In Example 2.3, we showed that $M_Z(\theta) = \exp\left(\frac{1}{2}\theta^2\right)$. Thus, since $X = \sigma Z + \mu$, Property 2.3 implies that

$$M_X(\theta) = e^{\mu\theta}M_Z(\sigma\theta) = e^{\mu\theta + \frac{1}{2}\sigma^2\theta^2}.$$

If we wish to evaluate $E[X^3]$ and $E[X^4]$, we note that

$$M_X^{[3]}(\theta) = [3\sigma^2(\mu + \sigma^2\theta) + (\mu + \sigma^2\theta)^3] \exp\left(\mu\theta + \frac{1}{2}\sigma^2\theta^2\right),$$
$$M_X^{[4]}(\theta) = [3\sigma^4 + 6\sigma^2(\mu + \sigma^2\theta)^2 + (\mu + \sigma^2\theta)^4] \exp\left(\mu\theta + \frac{1}{2}\sigma^2\theta^2\right),$$

and thus from Property 2.2,

$$E[X^3] = 3\sigma^2\mu + \mu^3, \qquad E[X^4] = M_X^{[4]}(0) = 3\sigma^4 + 6\sigma^2\mu^2 + \mu^4.$$

The proof of Property 2.4 is also easily given. For example, if X and Y are independent continuous random variables, then

$$M_{X+Y}(\theta) = E(e^{\theta(X+Y)}) = \int_{-\infty}^{\infty} \int_{-\infty}^{\infty} e^{\theta x}e^{\theta y}f_X(x)f_Y(y)\,dx\,dy$$

$$= \left[\int_{-\infty}^{\infty} e^{\theta x}f_X(x)\,dx\right]\left[\int_{-\infty}^{\infty} e^{\theta y}f_Y(y)\,dy\right] = M_X(\theta)M_Y(\theta).$$

The proof in the case of discrete random variables X and Y is similar.

Example 2.6. Suppose that X and Y are independent random variables and that $X \sim \mathcal{N}(\mu_X, \sigma_X^2)$, $Y \sim N(\mu_Y, \sigma_Y^2)$. From Example 2.5 and Property 2.4,

$$M_{X+Y}(\theta) = M_X(\theta)M_Y(\theta) = [\exp\,(\mu_X\theta + \tfrac{1}{2}\sigma_X^2\theta^2)][\exp\,(\mu_Y\theta + \tfrac{1}{2}\sigma_Y^2\theta^2)]$$

$$= \exp\,[(\mu_X + \mu_Y)\theta + \tfrac{1}{2}(\sigma_X^2 + \sigma_Y^2)\theta^2]$$

which is the m.g.f. of the $\mathcal{N}(\mu_X + \mu_Y, \sigma_X^2 + \sigma_Y^2)$ distribution. Thus, Property 2.1 implies that $X + Y \sim \mathcal{N}(\mu_X + \mu_Y, \sigma_X^2 + \sigma_Y^2)$.

Example 2.7. Suppose that X and Y are independent random variables having a standard gamma distribution with parameters r_1 and r_2, respectively. Then from Example 2.2 and Property 2.4,

$$M_{X+Y}(\theta) = M_X(\theta)M_Y(\theta) = \frac{1}{(1-\theta)^{r_1}}\frac{1}{(1-\theta)^{r_2}} = \frac{1}{(1-\theta)^{r_1+r_2}},$$

which is the m.g.f. of a standard gamma distribution with parameter $r = r_1 + r_2$. Thus, by Property 2.1, $X + Y$ has a standard gamma distribution with parameter $r = r_1 + r_2$.

Property 2.4 can be generalized as follows. Let $X_1, X_2, \ldots, X_n$ be mutually independent random variables, and let $a_1, a_2, \ldots, a_n$ be constants. If $Y_k = \sum_{i=1}^{k} a_i X_i$, $k = 1, \ldots, n$, then

(2.11)
$$\begin{aligned}
M_Y(\theta) = M_{Y_n}(\theta) &= M_{Y_{n-1}}(\theta)M_{a_n X_n}(\theta) \\
&= M_{Y_{n-2}}(\theta)M_{a_{n-1}X_{n-1}}(\theta)M_{a_n X_n}(\theta) = \cdots \\
&= M_{a_1 X_1}(\theta)M_{a_2 X_2}(\theta) \cdots M_{a_n X_n}(\theta) \\
&= M_{X_1}(a_1\theta)M_{X_2}(a_2\theta) \cdots M_{X_n}(a_n\theta).
\end{aligned}$$

In particular, if $X_1, X_2, \ldots, X_n$ all have the same distribution with m.g.f. $M_X(\theta)$, then the m.g.f. of the sample average $\bar{X} = (1/n)\sum_{i=1}^{n} X_i$ is

(2.12)
$$M_{\bar{X}}(\theta) = \left[M_X\!\left(\frac{\theta}{n}\right)\right]^n$$

and the m.g.f. of the sample total $S_n = \sum_{i=1}^{n} X_i$ is

(2.13)
$$M_{S_n}(\theta) = [M_X(\theta)]^n.$$

Proof of the Central Limit Theorem

Recall that in Chapter 7, we asserted that

(2.14)
$$\lim_{n\to\infty} F_{Z_n}(z) = F_Z(z), \quad -\infty < z < \infty,$$

where

(2.15) $$Z_n = \frac{\bar{X} - \mu_X}{\sigma_X/\sqrt{n}} = \frac{\sum_{i=1}^n X_i - n\mu_X}{\sqrt{n}\,\sigma_X}$$

is obtained from mutually independent observations X_1, X_2, ..., X_n on a random variable X with mean μ_X and finite variance σ_X^2, $F_{Z_n}(z)$ is the c.d.f. of Z_n, and $F_Z(z)$ is the c.d.f. of $Z \sim \mathcal{N}(0, 1)$. Equation (2.14) is, indeed, the assertion of the Central Limit Theorem.

We now prove (2.14) using the method of moment generating functions—in particular, Property 2.5 will be used. To do so, we will need to assume that the random variable X on which the observations X_1, X_2, ..., X_n are taken has a moment generating function $M_X(\theta)$ defined on an open interval (θ_L, θ_U) containing $\theta = 0$. Since not all random variables have moment generating functions, this is an extra assumption. To keep the mathematical arguments simple, we will also assume that $E(X - \mu_X)^3$ is well defined (and thus finite).

Note from (2.15) that

(2.16) $$Z_n = \frac{\sum_{i=1}^n X_i - n\mu_X}{\sqrt{n}\,\sigma_X} = \frac{1}{\sqrt{n}} \sum_{i=1}^n \left(\frac{X_i - \mu_X}{\sigma_X} \right) = \frac{1}{\sqrt{n}} \sum_{i=1}^n Y_i,$$

where Y_1, Y_2, ..., Y_n are mutually independent observations on the standardization

$$Y = \frac{X - \mu_X}{\sigma_X}$$

of X. Since X has a m.g.f., so does Y; also since $E(X - \mu_X)^3$ is well defined and finite, so is $E[Y^3] = (\sigma_X)^{-3} E(X - \mu_X)^3$. Let

$$M_Y(\theta) = \exp\left(-\theta\mu_X/\sigma_X \right) M_X(\theta/\sigma_X)$$

denote the m.g.f. of Y. Property 2.3, (2.13), and (2.16) imply that

(2.17) $$M_{Z_n}(\theta) = M_{(1/\sqrt{n})\sum_{i=1}^n Y_i}(\theta) = M_{\sum_{i=1}^n Y_i}\left(\frac{\theta}{\sqrt{n}} \right) = \left[M_Y\left(\frac{\theta}{\sqrt{n}} \right) \right]^n.$$

Since Y is a standardized random variable, $\mu_Y = 0$ and $\sigma_Y^2 = 1$. Thus, from Property 2.2,

(2.18) $$M_Y^{[1]}(0) = 0, \qquad M_Y^{[2]}(0) = 1, \qquad M_Y^{[3]}(0) = E(Y^3) < \infty.$$

For large values of n, $\theta/\sqrt{n}$ is close to 0. This fact suggests expanding $M_Y(\theta/\sqrt{n})$ in a Taylor's series about $\theta/\sqrt{n} = 0$. Using (2.18),

$$M_Y\left(\frac{\theta}{\sqrt{n}} \right) = M_Y(0) + M_Y^{[1]}(0)\frac{\theta}{\sqrt{n}} + M_Y^{[2]}(0)\frac{(\theta/\sqrt{n})^2}{2!} + R_n$$

(2.19)

$$= 1 + (0)\frac{\theta}{\sqrt{n}} + (1)\frac{\theta^2}{2n} + R_n = 1 + \frac{\theta^2 + 2nR_n}{2n},$$

where

(2.20) $$R_n = \frac{(\tilde{\theta})^3}{3!(\sqrt{n})^3} M_Y^{[3]}\!\left(\frac{\tilde{\theta}}{\sqrt{n}}\right) = \frac{(\tilde{\theta})^3}{6n^{3/2}} M_Y^{[3]}\!\left(\frac{\tilde{\theta}}{\sqrt{n}}\right)$$

and $\tilde{\theta}$ is a number between 0 and θ. Note that

$$\lim_{n\to\infty} M_Y^{[3]}\!\left(\frac{\tilde{\theta}}{\sqrt{n}}\right) = M_Y^{[3]}(0) = EY^3 < \infty,$$

and thus

(2.21) $$\lim_{n\to\infty} nR_n = \lim_{n\to\infty} \frac{n\tilde{\theta}^3}{6(\sqrt{n})^3} M^{[3]}\!\left(\frac{\tilde{\theta}}{\sqrt{n}}\right) = 0.$$

It follows from (2.17), (2.19), and (2.21) that

$$\lim_{n\to\infty} M_{Z_n}(\theta) = \lim_{n\to\infty} \left[M_Y\!\left(\frac{\theta}{\sqrt{n}}\right)\right]^n$$

(2.22)

$$= \lim_{n\to\infty} \left(1 + \frac{\theta^2 + 2nR_n}{2n}\right)^n = e^{\theta^2/2} = M_Z(\theta),$$

where $Z \sim \mathcal{N}(0, 1)$. Property 2.5 and (2.22) now yield the desired result, (2.16).

The Characteristic Function

As already noted, a major defect of both the factorial m.g.f. and the m.g.f. is that these generating functions are not defined for all random variables X. (This defect in the m.g.f. limits the generality of our proof of the Central Limit Theorem.) In contrast, the *characteristic function* (or Fourier transform) has properties parallel to Properties 2.1 to 2.5 of m.g.f.'s, but is defined for *all* random variables. The characteristic function $c_X(\theta)$ of a random variable X is defined to be

(2.23) $$c_X(\theta) = \sum_x e^{i\theta x} p_X(x), \quad -\infty < \theta < \infty,$$

when X is a discrete random variable, and

(2.24) $$c_X(\theta) = \int_{-\infty}^{\infty} e^{i\theta x} f_X(x)\, dx, \quad -\infty < \theta < \infty,$$

when X is a continuous random variable. Here $i = \sqrt{-1}$ is the imaginary number. The values of characteristic functions can be complex numbers, so that understanding and use of this kind of generating function requires knowledge of the theory of complex variables. For this reason, study of characteristic functions is left to more advanced books on probability theory. The method of characteristic functions can be used to prove the Central Limit Theorem without the necessity of requiring that the m.g.f. of X exists or that $E(X - \mu_X)^3$ is finite.

3. MELLIN TRANSFORMS

Generating functions are specifically designed to allow us to find distributions of sums of independent random variables. There is a corresponding approach that allows us to obtain the distribution of *products* of independent random variables. When the distributions considered are restricted to those of non-negative random variables, we obtain functions called *Mellin transforms*. The Mellin transform $T(\theta)$ of a random variable X is given by

$$(3.1) \qquad\qquad T_X(\theta) = E(X^\theta)$$

for all values of θ for which the expected value $E(X^\theta)$ exists.

The properties of Mellin transforms are easily obtained by noting that

$$(3.2) \qquad T_X(\theta) = E(X^\theta) = E(\exp(\theta \log X)) = M_{\log X}(\theta).$$

For example, if a Mellin transform $T_X(\theta)$ is defined on an open interval containing $\theta = 0$, it has the uniqueness property (Property 2.1), since in such a case $M_{\log X}(\theta)$ is defined for an open interval of θ-values containing $\theta = 0$. From $T_X(\theta)$, we obtain $M_{\log X}(\theta)$, from which in turn we can obtain the distribution of $\log X$. From the distribution of $\log X$, we can obtain the distribution of $X = e^{\log X}$, thus relating $T_X(\theta)$ to a given distribution. On the other hand, if we have the distribution of X, we can obtain the distribution of $\log X$ and from that distribution obtain $T_X(\theta) = M_{\log X}(\theta)$. Similar arguments show that Mellin transforms preserve limits (Property 2.5). Corresponding to Property 2.4 for moment generating functions, we have the following property for Mellin transforms: *If X and Y are independent, nonnegative random variables, then*

$$(3.3) \qquad\qquad T_{XY}(\theta) = T_X(\theta)T_Y(\theta).$$

To see this, note that from (3.2),

$$T_{XY}(\theta) = M_{\log XY}(\theta) = M_{\log X + \log Y}(\theta) = M_{\log X}(\theta)M_{\log Y}(\theta) = T_X(\theta)T_Y(\theta).$$

Of course, the moments EX^k of a nonnegative random variable can easily be obtained from its Mellin transform $T_X(\theta)$, since

$$(3.4) \qquad\qquad T_X(k) = E(X^k).$$

Finally, corresponding to Property 2.3 of moment generating functions, we have the relationship

$$T_{bX^a}(\theta) = M_{\log (bX^a)}(\theta) = M_{a \log X + \log b}(\theta)$$

(3.5)
$$= e^{\theta \log b} M_{\log X}(a\theta)$$

$$= b^\theta T_X(a\theta),$$

which holds for $b > 0$ and $a\theta$ in the range of definition of $T_X(\theta)$.

As we have indicated, the most important property of the Mellin transform is its relationship to products of independent random variables, as expressed in Equation (3.3).

Example 3.1. Suppose that W has a beta distribution with parameters r and s. Then for $-r < \theta < \infty$,

(3.6)
$$T_W(\theta) = E(W^\theta) = \frac{\int_0^1 w^\theta w^{r-1}(1 - w)^{s-1} \, dw}{B(r, s)}$$

$$= \frac{B(r + \theta, s)}{B(r, s)} = \frac{\Gamma(r + \theta)\Gamma(r + s)}{\Gamma(r)\Gamma(r + s + \theta)}.$$

Now suppose that X has a beta distribution with parameters $r = a$, $s = b$, that Y has a beta distribution with parameters $r = a + b$, $s = c$, and that X and Y are independent. To find the distribution of XY, we use Equations (3.3) and (3.6) and find that

$$T_{XY}(\theta) = T_X(\theta)T_Y(\theta) = \frac{\Gamma(a + \theta)\Gamma(a + b)}{\Gamma(a)\Gamma(a + b + \theta)} \frac{\Gamma(a + b + \theta)\Gamma(a + b + c)}{\Gamma(a + b)\Gamma(a + b + c + \theta)}$$

$$= \frac{\Gamma(a + \theta)\Gamma(a + b + c)}{\Gamma(a)\Gamma(a + b + c + \theta)},$$

which from (3.6) we recognize as being the Mellin transform of a beta distribution with parameters $r = a$ and $s = b + c$. Thus, by the uniqueness property of Mellin transforms, XY has a beta distribution with parameters a and $b + c$.

The Mellin transform is a relatively unused tool of probability theory, mainly because sums, rather than products of random variables, tend to arise in practice. Tables of Mellin transforms and their associated distributions are not as extensive as might be desired. However, because of (3.2) the more extensive tables of moment generating functions can be used, so that from a Mellin transform it is often possible to obtain the corresponding distribution.

EXERCISES

1. Prove (1.2) when $k = 3, 4$.

2. Find the factorial moment generating function $G_X(\theta)$ of the negative binomial distribution with parameters u an integer and p. Then
    (a) Find $p_X(k)$ for $k = 0, 1, 2$.
    (b) Find $E[X(X - 1) \cdots (X - r + 1)]$ for $r = 1, 2$. Then find σ_X^2.
    (c) Show that if $X_1, X_2, \ldots, X_n$ are mutually independent random variables, with X_i having a negative binomial distribution with parameters u_i an integer and p, $i = 1, 2, \ldots, n$, then $Y = \sum_{i=1}^{n} X_i$ has a negative binomial distribution with parameters $y = \sum_{i=1}^{n} u_i$ and p.

3. Find the factorial m.g.f. of the truncated Poisson distribution with parameter λ. Then
    (a) Find $E[X(X - 1) \cdots (X - r + 1)]$ for $r = 1, 2$. Find σ_X^2.
    (b) Find the factorial m.g.f. of the sum $Y = \sum_{i=1}^{n} X_i$ of mutually independent random variables, where X_i has a truncated Poisson distribution with parameter λ_i, $i = 1, 2, \ldots, n$.
    (c) For the case $n = 2$ in part (b), find $p_Y(k)$ for $k = 0, 1, 2$.

4. The truncated binomial distribution with parameters n and p has probability mass function

    $$p_X(k; n, p) = \begin{cases} \dfrac{\binom{n}{k} p^k (1 - p)^{n-k}}{1 - (1 - p)^n}, & \text{if } k = 1, 2, \ldots, n, \\ 0, & \text{if } k = 0. \end{cases}$$

    (a) Find the factorial m.g.f. $G_X(\theta; n, p)$ of the truncated binomial distribution with parameters n and p.
    (b) If $p_n = \lambda/n$, show that the probabilities $p_X(k; n, p_n)$ of the truncated binomial distribution can be approximated by the probabilities $p_X(k)$ of the truncated Poisson distribution with parameter λ.

5. Find the limit of the factorial m.g.f. of the hypergeometric distribution with parameters N, M, and n as $N \to \infty$, $M \to \infty$ with $M/N = p$.

6. The random variable X is bounded if there exist numbers a, b with $-\infty < a \le b < \infty$ for which $P\{a \le X \le b\} = 1$. Show that if X is a bounded random variable, then the m.g.f. $M_X(\theta)$ is defined for all θ, $-\infty < \theta < \infty$.

7. Show that the m.g.f. of the Cauchy distribution is not defined except for $\theta = 0$.

8. Show that the m.g.f. of the zeta distribution with parameter $\alpha = 1$ is not defined for $\theta > 0$.

9. If X has a gamma distribution with parameters r and θ, then $Y = \theta X$ has a standard gamma distribution. Find the m.g.f. of X. Find $E(X^k)$ for $k = 1, 2, 3, 4$.

10. Find the m.g.f. of the $U[0, 1]$ distribution and the m.g.f. of the $U[a, b]$ distribution.

11. Find the m.g.f. of $X + Y$, where X and Y are independent $U[a, b]$ random variables.

12. The *cumulant generating function* $K_X(\theta)$ of a random variable X is defined by

$$K_X(\theta) = \log M_X(\theta).$$

Let $K_X^{[r]}(\theta) = (d/d\theta)^r K_X(\theta)$. The quantities

$$\kappa_r = K_X^{[r]}(0), \quad r = 1, 2, \ldots,$$

are known as the rth *cumulants* of the random variable X.
(a) Show that $\kappa_1 = \mu_X$, $\kappa_2 = \sigma_X^2$. Find κ_3 and κ_4.
(b) If X and Y are independent random variables, show that

$$K_{X+Y}(\theta) = K_X(\theta) + K_Y(\theta).$$

(c) If X and Y are independent random variables, show that the rth cumulant of $X + Y$ is the sum of the rth cumulants for X and Y.

13. Let X have a negative binomial distribution with parameters u an integer and p.
(a) Find the m.g.f. $M_X(\theta)$ of X.
(b) Find μ_X, σ_X^2, and $E(X^3)$.
(c) Let $Y_1, Y_2, \ldots, Y_n$ be mutually probabilistically independent random variables, where Y_i has a geometric distribution with parameter p. Find the m.g.f. of $\sum_{i=1}^n Y_i$. Then argue that $\sum_{i=1}^n Y_i - n$ has a negative binomial distribution with parameters $u = n$ and p.

14. The probability density function

$$f(x) = \tfrac{1}{2}e^{-|x-v|}, \quad -\infty < x < \infty,$$

defines the *double exponential distribution with location parameter v*.
(a) Show that if X has a double exponential distribution with location parameter v, then $Y = X - v$ has a double exponential distribution with location parameter $v = 0$.
(b) Find the m.g.f. of the double exponential distribution with parameter $v = 0$. [Hint: Break up the integral from $-\infty$ to ∞ into the part from $-\infty$ to 0, and the part from 0 to ∞.]
(c) Find the m.g.f. of the double exponential distribution with location parameter v.
(d) Find the first four moments of the double exponential distribution with location parameter v.

15. Let X and Y be independent random variables, each having an exponential distribution with parameter $\theta = 1$. Show that $X - Y$ has a double exponential distribution with location parameter $v = 0$.

16. Let $X_1, X_2, \ldots, X_n$ be mutually independent random variables, where X_i has a Poisson distribution with parameter λ_i, $i = 1, 2, \ldots, n$.
    (a) Show, using m.g.f.'s, that $S_n = \sum_{i=1}^{n} X_i$ has a Poisson distribution with parameter $\lambda = \sum_{i=1}^{n} \lambda_i$.
    (b) Let

    $$Z_n = \frac{S_n - \sum_{i=1}^{n} \lambda_i}{\sqrt{\sum_{i=1}^{n} \lambda_i}}.$$

    Show that if

    $$\lim_{n \to \infty} \frac{1}{n} \sum_{i=1}^{n} \lambda_i = c$$

    exists, and $c \neq 0$, then

    $$\lim_{n \to \infty} F_{Z_n}(z) = F_Z(z), \quad -\infty < z < \infty,$$

    where $Z \sim \mathcal{N}(0, 1)$. [Show that $\lim_{n \to \infty} \log M_{Z_n}(\theta) = \tfrac{1}{2}\theta^2$.]

17. Find the Mellin transform $T(\theta)$ of the lognormal distribution with parameters ξ and δ^2. Use $T(\theta)$ to find a formula for the rth moment of the lognormal distribution, $r = 1, 2, \ldots$.

18. Show that if X and Y are independent random variables, where X and Y have lognormal distributions with parameters ξ_1, σ_1^2, and ξ_2, σ_2^2, respectively, then XY has a lognormal distribution with parameters $\xi_1 + \xi_2$ and $\delta_1^2 + \delta_2^2$.

19. Find the Mellin transform of the Weibull distribution with parameters β, α, and $v = 0$.

12

Markov Chains

1. INTRODUCTION

In previous chapters, we have discussed ways of modeling the variation of a single random phenomenon and the joint variation of two or more random phenomena. In the present chapter we consider a special type of dependence when observations are taken over time.

To illustrate, consider the size of a population of, say, seals. At an initial point in time, time 0, there may be X_0 such seals. Between this point and the next observed point in time, time 1, fatal accidents may occur to some of the seals, but other seals in the population may mate and produce new offspring. The size X_1 of the population of seals at time 1 is a random variable, which depends on the number X_0 of seals that were alive at time 0. The size X_2 of the population of seals at the next observed time, time 2, is a random variable depending probabilistically on the sizes X_0 and X_1 of the population of seals at times 0 and 1, and so on. A probabilistic model for this process should specify how the dynamics of growth (or decrease) of the population of seals between any two time points, time i and time $j > i$, depends on the sizes X_0, X_1, X_2, ..., X_i at or previous to time point i. Then, using such a model and a knowledge of the initial size of the population of seals, we can determine the probability model that describes the variability of the size X_i of the population of seals at any specified time i, the probability model that describes the joint variability of the sizes X_i and X_j of the population of seals at any pair of time points i and j, and so on.

Scientists are increasingly interested in modeling and understanding dynamic random processes. Biologists are interested in the growth and decline of animal and human populations. Traffic engineers study vehicular traffic. Physicists study the movement and collisions of small particles, while on a larger scale, astronomers observe the evolution of the universe. Chemists investigate the dynamics of chemical reactions, and psychologists have embarked upon the study of the process of learning and understanding. Although science has dealt with dynamic mechanistic processes since

Newton's time, only recently have scientists been actively concerned with dynamic *random* or *stochastic* processes.

Stochastic processes can be classified into two categories. In *continuous-time* stochastic processes, the process is considered to be instantaneously observable at any time t after some initial time. Because we must consider the outcomes of a continuous-time stochastic process at all possible times t, the modeling and analysis of such processes is quite complicated. *Discrete-time* stochastic processes, on the other hand, are observable only at specified times. For example, the voting behavior of Americans is observable only at election times, the growth of the population of a given city is observed only at census times, the transmission of genetic material is observable in mice only when new mice are born, and the extent of learning in a child is observed only when the child is tested. Thus, the task of modeling and analyzing discrete-time processes is not as complex as the corresponding task for continuous-time processes. Nevertheless, discrete-time stochastic processes are of interest and importance in a wide diversity of disciplines. In the remainder of this chapter, we discuss only discrete-time stochastic processes.

2. FINITE MARKOV CHAINS

It is often possible to describe the status of a stochastic process at any time as being a member of a collection of possible states for the process. For example, the process under observation may be the political affiliation of a randomly chosen individual. In this case the "states" of the process are the various political parties, plus the "state" of having no party affiliation at all. An observation of this process is described by listing the political affiliation of the individual at all times at which the individual is observed.

As another example, consider the population of seals mentioned in Section 1. At any given time the "state" of the seal population is the number X_j of seals in the population. The possible "states" of this process are then any of the nonnegative integers. An outcome of the seal population process is a list of the numbers $X_0, X_1, X_2, \ldots$ of seals at each of the various times of observation.

We call a collection of possible states of a process *denumerable* if this collection can be indexed in a one-to-one fashion by the positive integers. In both the political affiliation and seal population processes mentioned above, the collection of states is denumerable.

To construct a probability model for a discrete-time stochastic process with a denumerable collection of states, let $X_n = i$ if at time n the process is in state S_i. The outcomes of the stochastic process can now be described by listing the indices $X_0, X_1, X_2, \ldots$, of the states observed for the process at each time n, $n = 0, 1, 2, \ldots$. Thus, the outcomes of the process are sequences $(X_0, X_1, X_2, \ldots)$ of integers. The outcome $(4, 3, 6, \ldots)$, for example, means that the process is in state S_4 at time 0, in state S_3 at time 1, in state S_6 at time 2, and so on. Alternatively, the outcomes in the process can be described simply in terms of the sequence $(S_4, S_3, S_6, \ldots)$ of states that are observed over time.

Example 2.1 A randomly selected individual is given a yearly medical examination, starting at age 30, in which his or her blood cholesterol level is measured and classified into one of the following five categories (states):

State S_1: very low cholesterol level,

State S_2: low cholesterol level,

State S_3: normal cholesterol level,

State S_4: high cholesterol level,

State S_5: very high cholesterol level.

The resulting yearly observations are an outcome of the process in which blood cholesterol levels fluctuate over time: time 0 = age 30, time 1 = age 31, and so on. If the cholesterol level in the individual's blood is low at age 30, low at age 31, normal at age 32, very low at age 33, high at age 34, ..., then $X_0 = 2$, $X_1 = 2$, $X_2 = 3$, $X_4 = 1$; and so on.

Example 2.2. Starting with January 1965, an economist notes the number of New York Stock Exchange seats sold each month. In this process, the states are the number of seats sold, and the times of observation are: time 0 = January 1965, time 1 = February 1965, and so on. If 3 seats were sold in January, 1 seat was sold in February, 0 seats were sold in March, and so on, then $X_0 = 3$, $X_1 = 1$, $X_2 = 0$, and so on.

To construct probability models for discrete-time stochastic processes with a denumerable number of states, we need to assign probabilities to the events of the process. In particular, for all times n, we need to determine probabilities for events of the form

(2.1) $\{X_0 = i_0, X_1 = i_1, X_2 = i_2, X_3 = i_3, \ldots, X_n = i_n\}$,

where $i_0, i_1, i_2, i_3, \ldots$ are integers indexing those states of the process which are observed at times 0, 1, 2, 3, ..., n. Probabilities of events of the form (2.1) can always be calculated from the unconditional probability $P\{X_0 = i_0\}$, and the conditional probabilities

$$P\{X_1 = i_1 \mid X_0 = i_0\},$$

(2.2) $$P\{X_2 = i_2 \mid X_1 = i_1, X_0 = i_0\},$$

$$P\{X_3 = i_3 \mid X_2 = i_2, X_1 = i_1, X_0 = i_0\},$$

and so on, by making use of the Law of Multiplication [Chapter 3, Equation (3.5)]. Knowledge of the conditional probabilities (2.2) gives the *dynamic probabilistic structure* of the given stochastic process.

In general, if we want to construct a probability model for a given stochastic process, there are just as many probabilities of the form (2.2) to be specified as there are probabilities of the form (2.1). However, there is a large and very important class of discrete-time stochastic processes in which

(2.3)
$$P\{X_k = i_k | X_{k-1} = i_{k-1}, X_{k-2} = i_{k-2}, \ldots, X_1 = i_1, X_0 = i_0\}$$
$$= P\{X_k = i_k | X_{k-1} = i_{k-1}\},$$

for all times k, $k = 1, 2, 3, \ldots$, and for all state indices i_0, i_1, i_2, and so on. Put into words, the present state of the process depends probabilistically on past states of the process only through the state observed in the most immediate past. This type of probabilistic dependence is called a *Markov structure*, and a discrete-time stochastic process with conditional probabilities of the form (2.3) is called a *Markov chain*, in honor of the Russian probabilist A. A. Markov (1856–1922), who developed the probability theory for such processes.

To see what a Markov structure means, suppose that a process has only four states: states S_1, S_2, S_3, and S_4. If we wish to determine the probability of the event that the process is in state S_3 at time 0, state S_4 at time 1, state S_1 at time 2, and state S_2 at time 3, then from the Law of Multiplication for conditional probabilities,

$$P\{X_0 = 3, X_1 = 4, X_2 = 1, \text{ and } X_3 = 2\}$$
$$= P\{X_3 = 2 | X_2 = 1, X_1 = 4, X_0 = 3\}P\{X_2 = 1 | X_1 = 4, X_0 = 3\}$$
$$\times P\{X_1 = 4 | X_0 = 3\}P\{X_0 = 3\}.$$

However, the Markov structure of the process tells us that

$$P\{X_3 = 2 | X_2 = 1, X_1 = 4, X_0 = 3\} = P\{X_3 = 2 | X_2 = 1\}$$

and

$$P\{X_2 = 1 | X_1 = 4, X_0 = 3\} = P\{X_2 = 1 | X_1 = 4\},$$

so that

$$P\{X_3 = 2, X_2 = 1, X_1 = 4, \text{ and } X_0 = 3\}$$
$$= P\{X_3 = 2 | X_2 = 1\}P\{X_2 = 1 | X_1 = 4\}P\{X_1 = 4 | X_0 = 3\}P\{X_0 = 3\}.$$

More generally, the probability of any event concerning the outcome of a Markov chain can be determined through knowledge of the initial probabilities $P\{X_0 = i_0\}$, and the conditional probabilities

(2.4) $P\{X_1 = i_1 | X_0 = i_0\}, P\{X_2 = i_2 | X_1 = i_1\}, P\{X_3 = i_3 | X_2 = i_2\}, \ldots,$

for all integers $i_0, i_1, i_2, i_3, \ldots$.

The *transition probability* $P\{X_k = j \mid X_{k-1} = i\}$ gives the conditional probability that the process will be in (change to) state S_j at time k given that that the process is known to have been in state S_i at time $k - 1$. For a general Markov chain, the transition probabilities $P\{X_k = j \mid X_{k-1} = i\}$ have values depending on the state indices i and j and on the time index k. Thus, when $k > 1$, $P\{X_k = 1 \mid X_{k-1} = 2\}$ and $P\{X_1 = 1 \mid X_0 = 2\}$ are not the same. When the transition probabilities $P\{X_k = j \mid X_{k-1} = i\}$ of a Markov chain are independent of the time index k, so that

$$(2.5) \qquad P\{X_k = j \mid X_{k-1} = i\} = P\{X_1 = j \mid X_0 = i\},$$

for all $k = 1, 2, 3, \ldots$ and all states S_i and S_j, then we say that these transition probabilities are *stationary* (over time).

If a Markov chain has stationary transition probabilities, then in comparison to more general kinds of Markov chains, the task of determining a probability model for such a process is greatly simplified. The probability model for a Markov chain having N possible states and having stationary transition probabilities is completely determined in terms of the transition probabilities

$$p_{ij} = P\{X_1 = j \mid X_0 = i\} = P\{X_k = j \mid X_{k-1} = i\},$$

for $i, j = 1, 2, 3, \ldots, N$, and the initial probabilities

$$a_i = P\{X_0 = i\},$$

for $i = 1, 2, \ldots, N$. Once the initial probabilities $a_1, a_2, \ldots, a_N$ are known, the Markov chain can be represented by the array of transition probabilities given in Table 2.1.

Table 2.1: *Array of Transition Probabilities*

State at time 1

		S_1	S_2	S_3	$\cdots$	S_N
	S_1	p_{11}	p_{12}	p_{13}	$\cdots$	p_{1N}
	S_2	p_{21}	p_{22}	p_{23}	$\cdots$	p_{2N}
State at time 0	S_3	p_{31}	p_{32}	p_{33}	$\cdots$	p_{3N}
	$\vdots$	$\vdots$	$\vdots$	$\vdots$	$\vdots$	$\vdots$
	S_N	p_{N1}	p_{N2}	p_{N3}	$\cdots$	p_{NN}

The array in Table 2.1 is known as a *transition matrix*. Note that since the process must be in one of the N states at time 1,

(2.6) $$\sum_{j=1}^{N} p_{ij} = P\{X_1 = 1, 2, 3, \ldots, \text{or } N \,|\, X_0 = i\} = 1, \quad \text{all } i.$$

Thus, the row sums of any transition matrix are always equal to 1.

In our discussion we have gradually restricted the generality of the discrete-time stochastic processes considered to concentrate on Markov chains with a finite number N of states with stationary transition probabilities. Despite their special structure, such models have an astonishing breadth of application, as illustrated in the next section.

3. APPLICATIONS

The following examples serve to illustrate the use and importance of Markov chain models.

Example 3.1 (Mother–Infant Vocalization). In a study of joint infant–mother behavior conducted by Freedle and Lewis (1971), 6 distinct (and exhaustive) states of mother–infant vocalization were defined:

S_1: neither mother nor infant vocalize,

S_2: infant vocalizes alone,

S_3: mother vocalizes alone to infant,

(3.1) S_4: mother vocalizes alone to some other person,

S_5: mother and infant both vocalize (the mother vocalizes to infant),

S_6: mother vocalizes to another person and the infant vocalizes.

Observations were made at 10-second intervals and only one of the above 6 states was in effect during any given interval of time. Assume that the state of mother–infant vocalization at time $t + 1$ is influenced by past states only through the vocalization state at the immediately previous time, time t, of observation, and that identical psychological mechanisms cause the vocalization state at time t to influence the vocalization state at time $t + 1$ for all times $t = 0, 1, 2, \ldots$. In this case, the process of mother–infant vocalization can be modeled as a Markov chain having stationary transition probabilities.

To estimate the transition probabilities p_{ij} of this process from observation of one mother–infant pair, we can proceed as follows. For every pair of states S_i and S_j, count the number n_{ij} of times in which the process is in state

S_i at one time, say time t, and in state S_j at the very next time, time $t + 1$. Let

$$(3.2) \qquad n_{i\cdot} = \sum_{j=1}^{6} n_{ij}, \quad i = 1, 2, \ldots, 6.$$

The quantity $n_{i\cdot}$ is the number of times (not counting the very last time observed) at which the process is in state S_i, for $i = 1, 2, 3, 4, 5, 6$. Let

$$(3.3) \qquad \hat{p}_{ij} = \frac{n_{ij}}{n_{i\cdot}}.$$

Remark. If the process is not in state S_i at any time during the period of observation, then $n_{i\cdot} = 0$ and we define $\hat{p}_{ij} = n_{ij}/n_{i\cdot}$ to be $\frac{1}{6}$ for all $j = 1, 2, 3, 4, 5, 6$.

Because $\hat{p}_{ij}$ measures the proportion of transitions from state S_i to state S_j (in one time period) among all transitions in which state S_i is the initial state of the transition, it is intuitively reasonable to use $\hat{p}_{ij}$ to estimate p_{ij}.

As an example of the calculation of $\hat{p}_{ij}$, suppose that a given mother-infant pair is observed at 41 time periods, as in Table 3.1. Then, direct enumeration yields Table 3.2, and using (3.3), we obtain the estimated p_{ij}'s shown in Table 3.3.

In one of the actual experiments performed by Freedle and Lewis (1971), a mother-infant pair was observed for a total of 720 successive

Table 3.1: *Hypothetical Sequence of Observed States of Vocalization for a Given Mother-Infant Pair*

Time	State	Time	State	Time	State
0	S_2	14	S_4	28	S_6
1	S_1	15	S_3	29	S_5
2	S_1	16	S_1	30	S_3
3	S_3	17	S_4	31	S_2
4	S_4	18	S_1	32	S_2
5	S_5	19	S_6	33	S_6
6	S_3	20	S_2	34	S_1
7	S_6	21	S_3	35	S_5
8	S_4	22	S_3	36	S_4
9	S_2	23	S_5	37	S_1
10	S_5	24	S_2	38	S_6
11	S_1	25	S_4	39	S_6
12	S_2	26	S_4	40	S_3
13	S_2	27	S_6		

Table 3.2: Table of Values of n_{ij}, $i, j = 1, 2, 3,$
4, 5, 6 as Calculated from the Data in Table 3.1

(The entry in the ith row and jth column is n_{ij})

			j				
i	1	2	3	4	5	6	$n_i.$
1	1	1	1	1	1	2	7
2	1	2	1	2	1	1	8
3	1	1	1	1	1	1	6
4	2	1	1	1	1	1	7
5	1	1	2	1	0	0	5
6	1	1	1	1	1	2	7
Total							40

Table 3.3: Estimated Transition Probabilities for the
Mother–Infant Vocalization Process as Calculated
from the Hypothetical Data of Table 3.1

State at time $t + 1$

	S_1	S_2	S_3	S_4	S_5	S_6
S_1	$\frac{1}{7}$	$\frac{1}{7}$	$\frac{1}{7}$	$\frac{1}{7}$	$\frac{1}{7}$	$\frac{2}{7}$
S_2	$\frac{1}{8}$	$\frac{2}{8}$	$\frac{1}{8}$	$\frac{2}{8}$	$\frac{1}{8}$	$\frac{1}{8}$
S_3	$\frac{1}{6}$	$\frac{1}{6}$	$\frac{1}{6}$	$\frac{1}{6}$	$\frac{1}{6}$	$\frac{1}{6}$
S_4	$\frac{2}{7}$	$\frac{1}{7}$	$\frac{1}{7}$	$\frac{1}{7}$	$\frac{1}{7}$	$\frac{1}{7}$
S_5	$\frac{1}{5}$	$\frac{1}{5}$	$\frac{2}{5}$	$\frac{1}{5}$	0	0
S_6	$\frac{1}{7}$	$\frac{1}{7}$	$\frac{1}{7}$	$\frac{1}{7}$	$\frac{1}{7}$	$\frac{2}{7}$

State at time t (labels for rows S_3, S_4)

10-second time periods. The resulting estimated transition probabilities
appear in Table 3.4.

Because we assume that the properties of the Markov chain are stationary in time, we are able to estimate the transition probabilities of the vocalization process between any two successive time points t and $t + 1$ from the
observed transitions of the process between all observed successive time

Table 3.4: *Estimated Transition Probabilities for Vocalization States for Highly Vocal Participants*

State at time $t + 1$

		S_1	S_2	S_3	S_4	S_5	S_6
	S_1	0.42	0.09	0.13	0.22	0.02	0.12
	S_2	0.22	0.46	0.00	0.08	0.02	0.22
State at	S_3	0.18	0.04	0.51	0.12	0.05	0.10
time t	S_4	0.05	0.01	0.05	0.71	0.01	0.17
	S_5	0.27	0.13	0.20	0.07	0.07	0.26
	S_6	0.05	0.06	0.01	0.33	0.02	0.53

points, regarding each transition of the process from state to state at adjacent times as being a repetition of the experiment.

To estimate the initial probabilities a_i of the states S_i, $i = 1, 2, \ldots, 6$, several repetitions of the same mother–infant vocalization process could be utilized. If m similar mother–infant pairs are observed, and if m_i of these pairs are initially (at time 0) in state S_i, where $m = m_1 + m_2 + m_3 + m_4 + m_5 + m_6$, then we estimate the initial probability a_i by $\hat{a}_i = m_i/m$. Thus, if 10 similar mother–infant pairs were observed, and of these 10 pairs, 2 pairs began in state S_1, then we would estimate the initial probability of nonvocalization (state S_1) to be $\hat{a}_1 = \frac{2}{10} = 0.20$.

The estimates of the initial probabilities in the Freedle–Lewis experiment described above were actually obtained by a quite different method. Their estimates of the initial probabilities are

$$\hat{a}_1 = 0.13, \quad \hat{a}_2 = 0.07, \quad \hat{a}_3 = 0.09, \quad \hat{a}_4 = 0.44, \quad \hat{a}_5 = 0.02, \quad \hat{a}_6 = 0.25.$$

Example 3.2 (Conformity). The following experiment was first performed by Asch (1952). A subject is seated at the end of a row of pretrained confederates of the experimenter. The subject believes that he and the confederates of the experimenter are all part of an experiment in which certain perceptual judgments are to be made. A series of questions, each with only one clearly correct response, are put to the group. Responses to each question are given by one participant at a time, with the subject always responding last. Furthermore, the responses of the confederates of the experimenter are always identical and *incorrect*. The subject is thus faced with a choice between giving that answer to the question which he or she knows is correct, or conforming to the unanimous and incorrect answer of the group. Cohen (1958) carried out several experiments and compared results with those predicted by a Markov chain model in which at each trial t of the experiment, one of the following four states could be in effect:

S_1: The subject is motivated to answer correctly, independent of the answers of the group, at all trials.

S_2: The subject is motivated to answer correctly, but is still indecisive as to whether or not to yield to group pressure and conform in the future.

S_3: The subject is motivated to conform to the incorrect answer of the group, but is still indecisive as to whether or not to yield to group pressure and conform in the future.

S_4: The subject is motivated to conform to the unanimous answer of the group at all trials.

Using the definitions of the states S_1 and S_4, we can argue that $p_{11} = 1$, $p_{12} = p_{13} = p_{14} = 0$ and that $p_{44} = 1$, $p_{41} = p_{42} = p_{43} = 0$. In constructing a Markov chain model for the type of experiment described above, Cohen (1958) also assumes that it is not possible for a subject to make an abrupt change from an indecisively conforming state of mind (i.e., state S_3) at a trial t to a resolutely independent state of mind (state S_1) at the very next trial (trial $t + 1$). Thus, Cohen assumes that $p_{31} = 0$. Similarly, he assumes that $p_{24} = 0$. The other transition probabilities are not specified. Thus, it is hypothesized that the transition probabilities for a Markov chain which describes the process of individual reactions to group pressures have the form shown in Table 3.5.

Table 3.5: *Transition Probabilities for a Conformity Model*

State at trial $t + 1$

		S_1	S_2	S_3	S_4
	S_1	1	0	0	0
State at trial t	S_2	p_{21}	p_{22}	p_{23}	0
	S_3	0	p_{32}	p_{33}	p_{34}
	S_4	0	0	0	1

In one experiment reported by Cohen (1958), 33 subjects were used, and the deliberate errors made by the confederates did not deviate greatly from the correct answer. The estimated transition probabilities were obtained by a fairly complex method necessitated by the fact that states S_2 and S_3 are not directly observable. [Cohen (1958)]. The results were as follows:

State at trial $t + 1$

	S_1	S_2	S_3	S_4
S_1	1	0	0	0
S_2	0.06	0.76	0.18	0
S_3	0	0.27	0.69	0.04
S_4	0	0	0	1

State at trial t (rows: S_1, S_2, S_3, S_4)

In a second experiment with 27 subjects, the errors made by the confederates deviated from the correct answer as widely as possible. The transition probabilities obtained were

State at trial $t + 1$

	S_1	S_2	S_3	S_4
S_1	1	0	0	0
S_2	0.13	0.48	0.39	0
S_3	0	0.39	0.595	0.015
S_4	0	0	0	1

State at trial t (rows: S_1, S_2, S_3, S_4)

Example 3.3 (Breeding Experiments). In the simplest form of an inheritance model, each individual has a pair of genes, which may be

$$AA, \quad Aa \text{ (or } aA\text{)}, \quad \text{or} \quad aa,$$

that govern his or her inheritable characteristics. These three pairs of genes are called *genotypes*. In the mating of two parents, the offspring inherits one gene from each of its parents. The usual inheritance model assumes that the child has equal probability $\frac{1}{2}$ of getting either one of its parent's genes. For example, a given offspring of an $Aa \times Aa$ mating receives genotype AA from its parents with probability $\frac{1}{4}$, since the event that the offspring receives gene A from its father has probability $\frac{1}{2}$, and the event that the offspring receives gene A from its mother has probability $\frac{1}{2}$, and we assume that these events are

Table 3.6: *Probability Distributions for Genotypes of an Offspring Resulting from the Matings of Parents of Various Genotypes*

Mating	Probability assigned to offspring having genotype		
	AA	Aa	aa
$AA \times AA$	1	0	0
$AA \times Aa$	$\frac{1}{2}$	$\frac{1}{2}$	0
$AA \times aa$	0	1	0
$Aa \times AA$	$\frac{1}{2}$	$\frac{1}{2}$	0
$Aa \times Aa$	$\frac{1}{4}$	$\frac{1}{2}$	$\frac{1}{4}$
$Aa \times aa$	0	$\frac{1}{2}$	$\frac{1}{2}$
$aa \times AA$	0	1	0
$aa \times Aa$	0	$\frac{1}{2}$	$\frac{1}{2}$
$aa \times aa$	0	0	1

statistically independent. An offspring of the mating $Aa \times Aa$ can also receive genotype aa with probability $\frac{1}{4}$, or genotype Aa with probability $\frac{1}{2}$. The probability distributions of the genotypes of a given offspring resulting from the matings of various genotypes (for the parents) are shown in Table 3.6.

A very simple type of genetic mating experiment is the following. One parent is required to possess a specified genotype, say Aa. The other parent is selected at random from a population in which all genotypes AA, Aa, and aa are represented. From the offspring of the mating, one individual is selected at random and mated with an Aa parent. One offspring of that mating is selected at random, and again this offspring is mated with an Aa parent. Such a process is continued over many generations. At a given stage of the experiment the state of the process is the genotype of the offspring selected for the next mating. This process is a Markov chain with states $S_1 =$ genotype AA, $S_2 =$ genotype Aa, and $S_3 =$ genotype aa, and (see rows 4 to 6 of Table 3.6) with transition probabilities as given in Table 3.7. The "time t" mentioned in Table 3.7 refers to the matings of the experiment ($t = 0$ refers to the initial mating, $t = 1$ to the next mating, etc.); the transition probabilities are stationary over time. If at state $t = 0$, the randomly selected parent comes from a population of individuals in which each genotype is equally represented, then the initial probabilities a_1, a_2, and a_3 are all equal to $\frac{1}{3}$.

Various other breeding experiments are used in genetics. If a Markov chain model does describe the generation-to-generation variation in genotypes, such a model can be used to predict the genotypes of offspring several generations in the future of a given mating. For this reason, Markov chain

Table 3.7: *Transition Probabilities for Genotypes Arising from a Mating Where One Parent Always Has Genotype Aa, the Other Parent Is Randomly Selected from Offspring of a Previous Such Mating, and One Offspring Is Selected at Random from Their Children*

	Genotype of randomly selected offspring (at time $t + 1$)		
	State S_1	State S_2	State S_3
Genotype of S_1: AA	$\frac{1}{2}$	$\frac{1}{2}$	0
parent with unspecified S_2: Aa	$\frac{1}{4}$	$\frac{1}{2}$	$\frac{1}{4}$
genotype (at time t) S_3: aa	0	$\frac{1}{2}$	$\frac{1}{2}$

models can serve as a guide to agriculturists interested in breeding methods designed to improve crops and herds.

4. CALCULATION OF k-STEP TRANSITION PROBABILITIES

Recall that a Markov chain with stationary transition probabilities is completely determined once we know the values of the initial probabilities $a_i = P\{X_0 = i\}$, for all states S_i, and the values of the transition probabilities

$$p_{ij} = P\{X_{t+1} = j \mid X_t = i\},$$

for all states S_i and S_j. In the present section, we discuss the conditional probability that a Markov chain will undergo a transition to state S_j in k time units (k time steps) given that this transition begins in state S_i. Such a conditional probability is called a *k-step transition probability* and is denoted by $p_{ij}^{(k)}$. Thus,

(4.1) $$p_{ij}^{(k)} = P\{X_{t+k} = j \mid X_t = i\}.$$

As we will see, the conditional probability $P\{X_{t+k} = j \mid X_t = i\}$ does not depend for its value on the time t at which the transition from state S_i to state S_j begins. Thus,

$$p_{ij}^{(k)} = P\{X_k = j \mid X_0 = i\} = P\{X_{k+1} = j \mid X_1 = i\} = P\{X_{k+2} = j \mid X_2 = i\},$$

and so on.

Note that the transition probability p_{ij} is a special case of a k-step transition probability (namely, the case $k = 1$); for simplicity we write p_{ij} for $p_{ij}^{(1)}$.

2-Step Transition Probabilities

Suppose that we are interested in a Markov chain with N states $S_1, S_2, \ldots, S_N$, and that we know the (1-step) transition probabilities $p_{\alpha\beta}$, $1 \leq \alpha$, $\beta \leq N$, of this process. Then, the 2-step transition probability, $p_{ij}^{(2)} = P\{X_{t+2} = j | X_t = i\}$, that the process will be in state S_j at time $t+2$ given that it is observed to be in state S_i at time t can be calculated as follows:

$$(4.2) \qquad\qquad p_{ij}^{(2)} = \sum_{r=1}^{N} p_{ir}\, p_{rj}.$$

Note that it follows from (4.2) that the 2-step transition probability $p_{ij}^{(2)}$ is independent of the time index t.

The reasoning leading to (4.2) proceeds as follows. First, note that to pass from state S_i at time t to state S_j at time $t+2$, the process must follow exactly one of the N mutually exclusive transition paths $S_i \to S_r \to S_j$, where the intermediate state, state S_r, of such a transition is the state of the process at time $t+1$ (see Figure 4.1). Thus, the event $\{X_t = i \text{ and } X_{t+2} = j\}$ is the union of the N mutually exclusive events $\{X_t = i \text{ and } X_{t+1} = r \text{ and } X_{t+2} = j\}$, $r = 1, 2, \ldots, N$, and

$$(4.3) \quad P\{X_t = i \text{ and } X_{t+2} = j\} = \sum_{r=1}^{N} P\{X_t = i \text{ and } X_{t+1} = r \text{ and } X_{t+2} = j\}.$$

Next, note that since the process is a Markov chain with stationary transition probabilities,

$$
\begin{aligned}
(4.4) \quad & P\{X_t = i \text{ and } X_{t+1} = r \text{ and } X_{t+2} = j\} \\
&= P\{X_{t+2} = j | X_{t+1} = r \text{ and } X_t = i\} P\{X_{t+1} = r | X_t = i\} P\{X_t = i\} \\
&= p_{rj}\, p_{ir}\, P\{X_t = i\} = P\{X_t = i\} p_{ir}\, p_{rj}.
\end{aligned}
$$

Figure 4.1: *The N possible transitions $(S_i \to S_1 \to S_j, \ldots, S_i \to S_N \to S_j)$ that permit the process to go from state S_i to state S_j in two time steps.*

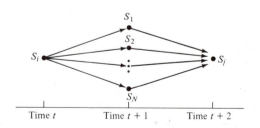

Now, substituting (4.4) into (4.3), we obtain

$$p_{ij}^{(2)} = P\{X_{t+2} = j \,|\, X_t = i\} = \frac{P\{X_t = i \text{ and } X_{t+2} = j\}}{P\{X_t = i\}}$$

$$= \sum_{r=1}^{N} p_{ir} p_{rj},$$

which verifies (4.2).

Example 4.1 (Learning Models). There are many variations of the original formulation of statistical learning theory by Estes (1950). Here we describe a simple learning process that is assumed to be a Markov chain with stationary transition probabilities. There are two states:

S_1: the subject is conditioned to make the correct response (for example, a rat is conditioned to press a bar),

S_2: the subject is not conditioned to make the correct response.

Once a subject has been conditioned (S_1), the subject remains conditioned from that time on; that is, at any time t,

$$p_{11} = P\{X_{t+1} = 1 \,|\, X_t = 1\} = 1, \qquad p_{12} = P\{X_{t+1} = 2 \,|\, X_t = 1\} = 0.$$

If a subject is unconditioned (S_2) at time t, it may become conditioned (move to state S_1) at time $t + 1$ with some probability α, $0 \le \alpha \le 1$. Thus,

$$p_{21} = P\{X_{t+1} = 1 \,|\, X_t = 2\} = \alpha,$$

$$p_{22} = P\{X_{t+1} = 2 \,|\, X_t = 2\} = 1 - p_{21} = 1 - \alpha.$$

The transition matrix for this simple Markov chain is shown in Table 4.1.

Suppose that at time t, the subject is unconditioned (S_2). The conditional probability $p_{21}^{(2)}$ that the subject will be conditioned (S_1) at time $t + 2$ is

$$p_{21}^{(2)} = p_{21} p_{11} + p_{22} p_{21} = (\alpha)(1) + (1 - \alpha)(\alpha) = 2\alpha - \alpha^2.$$

Table 4.1: *Transition Probabilities for a Two-State Markov Chain Learning Model*

		State at time $t + 1$	
		S_1	S_2
State at time t	S_1	1	0
	S_2	α	$1 - \alpha$

The conditional probability that the subject will be unconditioned (S_2) at time $t + 2$ is

$$p_{22}^{(2)} = p_{21}p_{12} + p_{22}p_{22} = (\alpha)(0) + (1 - \alpha)(1 - \alpha) = (1 - \alpha)^2.$$

Example 4.2 (Occupational Mobility). The process by which the male wage earners in a given family change in occupational status from generation to generation is of considerable interest.

Berger and Snell (1957) proposed a Markov chain model for occupational mobility in which the times 0, 1, 2, ... of observation are generations within families, and three occupation levels (states) are distinguished:

S_1: upper-level occupations (professional, high administrative, managerial, executive),

S_2: middle-level occupations (higher-grade supervisory and non-manual, skilled manual and nonmanual),

S_3: lower-level occupation (semiskilled manual, unskilled manual).

Glass and Hall (1954) obtained data on the occupational status of male residents of England and Wales and on the occupational status of their fathers. Using these data, a table (Table 4.2) of estimated transition probabilities for occupational mobility can be obtained. Here, we are implicitly assuming that the forces acting to cause transitions in occupational levels between generations are the same from generation to generation. This assumption implies that the Markov chain has stationary transition probabilities.

Suppose that the male wage earner of the tth generation of a certain family has a lower-level occupation (state S_3). From (4.2) and Table 4.2, the (conditional) probability that this individual's (eldest) grandson [the $(t + 2)$nd generation of the family] will be in the highest occupational level (S_1) is

$$p_{31}^{(2)} = p_{31}p_{11} + p_{32}p_{21} + p_{33}p_{31}$$

$$= 0.01(0.45) + 0.50(0.05) + 0.49(0.01) = 0.0344.$$

Table 4.2: Transition Probabilities for Level of Occupation

		Occupational level of sons (time $t + 1$)		
		S_1	S_2	S_3
Occupational level of fathers (time t)	S_1	0.45	0.48	0.07
	S_2	0.05	0.70	0.25
	S_3	0.01	0.50	0.49

Thus, the (conditional) probability of a two-generation rise "from rags to riches" is 0.0344.

Matrix Formulation of 2-Step Transition Probabilities

We have earlier stated that the rectangular array of transition probabilities (see Table 2.1) is called a transition matrix. Matrix theory deals with operations on matrices and, in particular, with the multiplication of such matrices. Let

$$A = \begin{pmatrix} a_{11} & a_{12} & \cdots & a_{1N} \\ a_{21} & a_{22} & \cdots & a_{2N} \\ \vdots & \vdots & & \vdots \\ a_{N1} & a_{N2} & \cdots & a_{NN} \end{pmatrix}, \qquad B = \begin{pmatrix} b_{11} & b_{12} & \cdots & b_{1N} \\ b_{21} & b_{22} & \cdots & b_{2N} \\ \vdots & \vdots & & \vdots \\ b_{N1} & b_{N2} & \cdots & b_{NN} \end{pmatrix}$$

be two $N \times N$ arrays of numbers (matrices). Then the product $C = AB$ is defined to be a matrix of the form

$$C = \begin{pmatrix} c_{11} & c_{12} & \cdots & c_{1N} \\ c_{21} & c_{22} & \cdots & c_{2N} \\ \vdots & \vdots & & \vdots \\ c_{N1} & c_{N2} & \cdots & c_{NN} \end{pmatrix},$$

where the element in the ith row, jth column of this array is given by the expression

$$c_{ij} = a_{i1}b_{1j} + a_{i2}b_{2j} + \cdots + a_{iN}b_{Nj}.$$

As a shorthand notation for a matrix, we write $A = (a_{ij})$, $B = (b_{ij})$, $C = (c_{ij})$. Thus, if $A = (a_{ij})$ and $B = (b_{ij})$, then $C = AB = (\sum_{k=1}^{N} a_{ik}b_{kj})$.

The matrix $P = (p_{ij})$ of one-step transition probabilities has entries p_{ij}. The square of this matrix P is the matrix P^2, which has the (i, j)th element

$$p_{i1}p_{1j} + p_{i2}p_{2j} + \cdots + p_{iN}p_{Nj}.$$

But from (4.2), we see that it is exactly this sum that provides the 2-step transition probabilities. Thus, if we let $P^{(2)}$ denote the matrix whose (i, j)th element is $p_{ij}^{(2)}$, then (4.2) shows that

(4.5)
$$P^{(2)} = P^2.$$

3-Step Transition Probabilities

Having shown how to compute 2-step probabilities $p_{ij}^{(2)}$, we now calculate the 3-step transition probabilities $p_{ij}^{(3)} = P\{X_{t+3} = j \mid X_t = i\}$. To go from state S_i at time t to state S_j at time $t + 3$, the process must go through two intermediate states, S_h and S_r, at times $t + 1$ and $t + 2$, respectively. For example, if there are only two possible states, S_1 and S_2, for the

$$
\begin{array}{ccccccc}
S_i & \rightarrow & S_1 & \rightarrow & S_1 & \rightarrow & S_j \\
S_i & \rightarrow & S_1 & \rightarrow & S_2 & \rightarrow & S_j \\
S_i & \rightarrow & S_2 & \rightarrow & S_1 & \rightarrow & S_j \\
S_i & \rightarrow & S_2 & \rightarrow & S_2 & \rightarrow & S_j
\end{array}
$$

Time: t $t+1$ $t+2$ $t+3$

Figure 4.2: *The four possible transitions that permit the process to go from state S_i at time t to state S_j three time units later.*

process, then in order for the process to go from S_i at time t to S_j at time $t+3$, one of the four transitions shown in Figure 4.2 must occur. Thus, using arguments similar to those used to derive (4.2), we can show that

$$(4.6) \qquad p_{ij}^{(3)} = p_{i1}p_{11}p_{1j} + p_{i1}p_{12}p_{2j} + p_{i2}p_{21}p_{1j} + p_{i2}p_{22}p_{2j}.$$

In general, when the Markov chain has N states, we can show that

$$
\begin{aligned}
p_{ij}^{(3)} = \; & p_{i1}p_{11}p_{1j} + p_{i1}p_{12}p_{2j} + \cdots + p_{i1}p_{1N}p_{Nj} \\
& + p_{12}p_{21}p_{1j} + p_{12}p_{22}p_{2j} + \cdots + p_{i2}p_{2N}p_{Nj} + \cdots \\
& + p_{iN}p_{N1}p_{1j} + p_{iN}p_{N2}p_{2j} + \cdots + p_{iN}p_{NN}p_{Nj}.
\end{aligned}
$$

(4.7)

The sum on the right-hand side of (4.7) has N^2 terms, each term consisting of a product of three numbers. If N is even moderately large, calculation of $p_{ij}^{(3)}$ by means of (4.7) can become tedious. In many cases where we are interested in calculating $p_{ij}^{(3)}$, however, we may have already calculated the 2-step transition probabilities $p_{ij}^{(2)}$. A formula that expresses $p_{ij}^{(3)}$ in terms of 2-step transition probabilities and 1-step transition probabilities would be of considerable help to us in such situations. Grouping terms in (4.7) and then applying (4.2), we find that

$$
\begin{aligned}
p_{ij}^{(3)} = \; & p_{i1}\left(p_{11}p_{1j} + p_{12}p_{2j} + \cdots + p_{1N}p_{Nj}\right) \\
& + p_{i2}\left(p_{21}p_{1j} + p_{22}p_{2j} + \cdots + p_{2N}p_{Nj}\right) \\
& + \cdots + p_{iN}\left(p_{N1}p_{1j} + p_{N2}p_{2j} + \cdots + p_{NN}p_{Nj}\right) \\
= \; & p_{i1}p_{1j}^{(2)} + p_{12}p_{2j}^{(2)} + \cdots + p_{iN}p_{Nj}^{(2)}.
\end{aligned}
$$

Thus,

$$(4.8) \qquad p_{ij}^{(3)} = \sum_{r=1}^{N} p_{ir}p_{rj}^{(2)}.$$

We can justify (4.8) heuristically by noting that to get from state S_i at time t to state S_j at time $t+3$, the process can move from state S_i at time t to some state, state S_h, at time $t+1$, and then from state S_h at time $t+1$ to state

S_j two time steps after that (at time $t + 3$). The rth term in the sum on the right-hand side of (4.8) gives the conditional probability that the process moves from state S_i to state S_r in one time step and then from state S_r to state S_j in two time steps, given that the process is initially in state S_i. The probability $p_{ij}^{(3)}$ is the sum of these N terms, since at time $t + 1$ the process must be in one and only one of the states $S_1, S_2, \ldots, S_N$.

We can restate (4.8) in matrix notation. If $P^{(3)} = (p_{ij}^{(3)})$ is the matrix of 3-step transition probabilities, then from (4.8),

$$(4.9) \qquad\qquad P^{(3)} = P^{(1)}P^{(2)},$$

where $P^{(2)}$ is the matrix of 2-step transition probabilities, and where for symmetry of notation we have written $P^{(1)} = P$ to denote the matrix of (1-step) transition probabilities. Recall that $P^{(2)} = P^2$. Hence,

$$(4.10) \qquad\qquad P^{(3)} = PP^2 = P^3.$$

However, it follows from (4.10) that

$$P^{(3)} = P^3 = P^2 P = P^{(2)}P^{(1)},$$

so that it is also true that

$$(4.11) \qquad\qquad p_{ij}^{(3)} = \sum_{h=1}^{N} p_{ih}^{(2)} p_{hj}.$$

General k-Step Transition Probabilities

The kinds of arguments needed to calculate k-step transition probabilities $p_{ij}^{(k)} = P\{X_{t+k} = j \mid X_t = i\}$ are the same as those used to calculate 2-step and 3-step transition probabilities. We may think of the process moving from state S_i at time t to state S_j at time $t + k$ in the following manner:

(1) First, the process moves from state S_i at time t to some state, state S_r, at time $t + 1$.
(2) Then the process moves from state S_r at time $t + 1$ to state S_j at time $t + k$.

The same kind of argument that led to (4.2) and (4.8) now yields the result:

$$(4.12) \qquad\qquad p_{ij}^{(k)} = p_{i1} p_{1j}^{(k-1)} + p_{i2} p_{2j}^{(k-1)} + \cdots + p_{iN} p_{Nj}^{(k-1)}.$$

We may also think of going from state S_i to state S_j in k time units by (i) going from state S_i to some state, state S_r, in $(k - 1)$ time units; and then (ii) going from state S_r to state S_j in one time unit. Hence, we can show that

$$(4.13) \qquad\qquad p_{ij}^{(k)} = p_{i1}^{(k-1)} p_{1j} + p_{i2}^{(k-1)} p_{2j} + \cdots + p_{iN}^{(k-1)} p_{Nj}.$$

Using either (4.12) or (4.13), all k-step transition probabilities can be recursively calculated.

Example 4.1 (continued). Recall that in the simple learning model described in this example, there are two states to the learning process:

S_1:   the subject is conditioned to make the correct response,

S_2:   the subject is not conditioned to make the correct response,

and the learning process is a Markov chain with the following transition matrix:

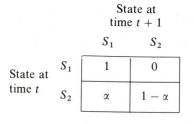

where α is a number between 0 and 1.

Using (4.2), we earlier showed that $p_{21}^{(2)} = 2\alpha - \alpha^2$ and $p_{22}^{(2)} = (1 - \alpha)^2$. From (4.2), we can also show that

$$p_{11}^{(2)} = p_{11}p_{11} + p_{12}p_{21} = 1,$$

$$p_{12}^{(2)} = p_{11}p_{12} + p_{12}p_{22} = 0.$$

The 2-step transition probabilities are given in Table 4.3.

Table 4.3: *Table of 2-Step Transition Probabilities*

State at time $t + 2$

		S_1	S_2
State at time t	S_1	1	0
	S_2	$2\alpha - \alpha^2$	$(1 - \alpha)^2$

Table 4.4: *Table of 3-Step Transition Probabilities*

State at time $t + 3$

		S_1	S_2
State at time t	S_1	1	0
	S_2	$3\alpha - 3\alpha^2 + \alpha^3$	$(1 - \alpha)^3$

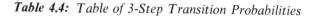

Table 4.5: *Table of 4-Step Transition Probabilities*

State at time $t + 4$

		S_1	S_2
State at time t	S_1	1	0
	S_2	$4\alpha - 6\alpha^2 + 4\alpha^3 - \alpha^4$	$(1 - \alpha)^4$

If we want to find the conditional probability $p_{21}^{(3)}$ that a subject will be conditioned (S_1) three time units after a time, time t, in which the subject was observed to be unconditioned (S_2), then from (4.8) with $N = 2$, $i = 2$, $j = 1$, we obtain

$$p_{21}^{(3)} = p_{21}p_{11}^{(2)} + p_{22}p_{21}^{(2)} = (\alpha)(1) + (1 - \alpha)(2\alpha - \alpha^2) = \alpha^3 - 3\alpha^2 + 3\alpha.$$

If $\alpha = \frac{1}{2}$, then $p_{21}^{(3)} = \frac{7}{8}$. We can also compute $p_{21}^{(3)}$ using (4.11):

$$p_{21}^{(3)} = p_{21}^{(2)}p_{11} + p_{22}^{(2)}p_{21} = (2\alpha - \alpha^2)(1) + (1 - \alpha)^2(\alpha)$$
$$= \alpha^3 - 3\alpha^2 + 3\alpha.$$

The other 3-step transition probabilities can be computed using either (4.8) or (4.11), whichever is more convenient. These 3-step transition probabilities are summarized in Table 4.4.

Turning next to the 4-step transition probabilities $p_{ij}^{(4)}$, these can be computed from the 1-step and 3-step transition probabilities by use of (4.12) or (4.13). Using (4.12) with $N = 2$, $k = 4$,

$$p_{11}^{(4)} = p_{11}p_{11}^{(3)} + p_{12}p_{21}^{(3)} = (1)(1) + (0)(3\alpha - 3\alpha^2 + \alpha^3) = 1,$$
$$p_{12}^{(4)} = p_{11}p_{12}^{(3)} + p_{12}p_{22}^{(3)} = (1)(0) + (0)(1 - \alpha)^3 = 0,$$
$$p_{21}^{(4)} = p_{21}p_{11}^{(3)} + p_{22}p_{21}^{(3)} = (\alpha)(1) + (1 - \alpha)(3\alpha - 3\alpha^2 + \alpha^3)$$
$$= 4\alpha - 6\alpha^2 + 4\alpha^3 - \alpha^4,$$
$$p_{22}^{(4)} = p_{21}p_{12}^{(3)} + p_{22}p_{22}^{(3)} = (\alpha)(0) + (1 - \alpha)(1 - \alpha)^3 = (1 - \alpha)^4,$$

and these 4-step transition probabilities are summarized in Table 4.5.

Because a process must be in some state, state S_j, at time $t + k$, it must always be the case that the row sums of tables of k-step transition probabilities (such as Tables 4.3, 4.4, and 4.5) are all equal to 1. That is, if a Markov chain has N states, $S_1, S_2, \ldots, S_N$, then for any number of time steps k and for any initial state i,

(4.14) $$p_{i1}^{(k)} + p_{i2}^{(k)} + \cdots + p_{iN}^{(k)} = 1.$$

Example 4.2 (continued). We have described a 3-state Markov chain model of occupational mobility in which the states are

$$S_1: \quad \text{upper occupation level,}$$

$$S_2: \quad \text{middle occupation level,}$$

$$S_3: \quad \text{lower occupation level,}$$

and the transition probability matrix is given by Table 4.6. We previously calculated the conditional probability $p_{31}^{(2)} = 0.0344$ that a son will be in the upper occupation level given that his grandfather was in the lower occupation level. Applying (4.2), we find that

$$p_{11}^{(2)} = p_{11}p_{11} + p_{12}p_{21} + p_{13}p_{31}$$

$$= (0.45)(0.45) + (0.48)(0.05) + (0.07)(0.01)$$

$$= 0.2272,$$

$$p_{21}^{(2)} = p_{21}p_{11} + p_{22}p_{21} + p_{23}p_{31}$$

$$= (0.05)(0.45) + (0.70)(0.05) + (0.25)(0.01)$$

$$= 0.0600,$$

and similarly,

$$p_{12}^{(2)} = 0.5870, \qquad p_{22}^{(2)} = 0.6390, \qquad p_{32}^{(2)} = 0.5998.$$

Instead of using (4.2) to compute $p_{31}^{(2)}$, $p_{32}^{(2)}$, and $p_{33}^{(2)}$, we can use (4.14), with $k = 2$, and obtain

$$p_{13}^{(2)} = 1 - p_{11}^{(2)} - p_{12}^{(2)} = 1 - 0.2272 - 0.5870 = 0.1858,$$

$$p_{23}^{(2)} = 1 - p_{21}^{(2)} - p_{22}^{(2)} = 1 - 0.0600 - 0.6390 = 0.3010,$$

$$p_{33}^{(2)} = 1 - p_{31}^{(2)} - p_{32}^{(2)} = 1 - 0.0344 - 0.5998 = 0.3658.$$

Table 4.6: Transition Probabilities for Level of Occupation

		Occupation level of sons (time $t + 1$)		
		S_1	S_2	S_3
Occupational level of fathers (time t)	S_1	0.45	0.48	0.07
	S_2	0.05	0.70	0.25
	S_3	0.01	0.50	0.49

Table 4.7: *Table of 2-Step Transition Probabilities for the Process of Occupational Mobility*

State at time $t + 2$

		S_1	S_2	S_3
State at time t	S_1	0.2272	0.5870	0.1858
	S_2	0.0600	0.6390	0.3010
	S_3	0.0344	0.5998	0.3658

Table 4.8: *Table of 3-Step Transition Probabilities for the Process of Occupational Mobility*

State at time $t + 3$

		S_1	S_2	S_3
State at time t	S_1	0.133	0.613	0.254
	S_2	0.062	0.627	0.311
	S_3	0.049	0.619	0.332

The 2-step transition probabilities of this Markov chain are summarized in Table 4.7. Using Tables 4.6 and 4.7 and either (4.12) or (4.13), we find the 3-step transition probabilities of this Markov chain (Table 4.8).

From these probabilities, Table 4.6, and either (4.12) or (4.13), recursive calculations yield the 4-step transition probabilities $p_{ij}^{(4)}$ and 5-step transition probabilities $p_{ij}^{(5)}$ summarized in Table 4.9. This process can be continued to

Table 4.9: *Table of 4-Step and 5-Step Transition Probabilities for the Process of Occupational Mobility*

		State at time $t + 4$			State at time $t + 5$		
		S_1	S_2	S_3	S_1	S_2	S_3
State at time t	S_1	0.093	0.620	0.287	0.076	0.622	0.302
	S_2	0.062	0.624	0.314	0.062	0.624	0.314
	S_3	0.056	0.623	0.321	0.060	0.623	0.317
		$p_{ij}^{(4)}$			$p_{ij}^{(5)}$		

the k-step transition probabilities $p_{ij}^{(k)}$, for $k = 6, 7, 8$, and so on. Since doing such calculations by hand is tedious, a computer program can be utilized to perform the task.

An Alternative Method for Calculating k-Step Transition Probabilities

We have seen that the k-step transition probabilities of a Markov chain can be determined from the $(k - 1)$-step and 1-step transition probabilities. Actually, for $k \geq 4$, the k-step transition probabilities can also be determined from the $(k - 2)$-step and 2-step transition probabilities. Indeed, if there are two positive integers k_1 and k_2 such that $k_1 + k_2 = k$, then we can obtain the k-step transition probabilities from the following equation:

(4.15)
$$p_{ij}^{(k)} = \sum_{r=1}^{N} p_{ir}^{(k_1)} p_{rj}^{(k_2)}.$$

For example, if $k = 4$, and $k_1 = k_2 = 2$, then

(4.16)
$$p_{ij}^{(4)} = p_{i1}^{(2)} p_{1j}^{(2)} + p_{i2}^{(2)} p_{2j}^{(2)} + \cdots + p_{iN}^{(2)} p_{Nj}^{(2)}.$$

Thus, in Example 4.2, rather than obtain $p_{ij}^{(4)}$ using Tables 4.6 and 4.8, we can obtain $p_{ij}^{(4)}$ from the use of Table 4.7 alone. For example, using (4.16), we obtain

$$p_{31}^{(4)} = p_{31}^{(2)} p_{11}^{(2)} + p_{32}^{(2)} p_{21}^{(2)} + p_{33}^{(2)} p_{31}^{(2)}$$

$$= (0.0344)(0.2272) + (0.5998)(0.0600) + (0.3658)(0.0344)$$

$$= 0.056,$$

in agreement with the result for $p_{31}^{(4)}$ given in Table 4.9.

Calculating k-Step Transition Probabilities by Matrix Multiplication

Let $P^{(k)} = (p_{ij}^{(k)})$ denote the matrix of k-step transition probabilities. From the definition of matrix multiplication and (4.12) and (4.13),

$$P^{(k)} = P^{(1)} P^{(k-1)} = P^{(k-1)} P^{(1)},$$

where $P^{(k-1)} = (p_{ij}^{(k-1)})$ is the matrix of $(k-1)$-step transition probabilities, and $P^{(1)} = P = (p_{ij})$ is the matrix of one-step transition probabilities. We have already seen that $P^{(2)} = P^2$ and $P^{(3)} = P^3$. In general,

(4.17)
$$P^{(k)} = P^k.$$

We note in passing that since

$$P^{k_1} P^{k_2} = P^{k_1 + k_2}$$

and $P^{(k_1)} = P^{k_1}$, $P^{(k_2)} = P^{k_2}$, (4.15) is a direct consequence of (4.17).

5. *ABSOLUTE OR UNCONDITIONAL PROBABILITIES*

In the previous section, we discussed how to calculate k-step transition probabilities $p_{ij}^{(k)}$ for Markov chains. These transition probabilities are conditional probabilities, in that $p_{ij}^{(k)}$ gives the probabilility that the process moves from state S_i to state S_j in k units of time given that the process is initially in state S_i at some time t. That is,

$$p_{ij}^{(k)} = P\{X_{t+k} = j \,|\, X_t = i\},$$

and in particular (when $t = 0$),

(5.1) $$p_{ij}^{(k)} = P\{X_k = j \,|\, X_0 = i\}.$$

Suppose instead that we are interested in the unconditional probability, $P\{X_k = j\}$, that the process is in state S_j at time k. Assume that the process has N possible states, $S_1, S_2, \ldots, S_N$. Recall from Chapter 3 that if B is any event, and $E_1, E_2, \ldots, E_N$ are N mutually exclusive events such that $E_1 \cup E_2 \cup \cdots \cup E_N$ is the whole sample space, then

(5.2) $$P(B) = P(B\,|\,E_1)P(E_1) + P(B\,|\,E_2)P(E_2) + \cdots + P(B\,|\,E_N)P(E_N).$$

We can apply this rule to the events

$$B = \{X_k = j\},\ E_1 = \{X_0 = 1\},\ E_2 = \{X_0 = 2\},\ \ldots,\ E_N = \{X_0 = N\},$$

obtaining the result

$$P\{X_k = j\} = \sum_{i=1}^{N} P\{X_k = j \,|\, X_0 = i\}P\{X_0 = i\} = \sum_{i=1}^{N} p_{ij}^{(k)} a_i$$

or

(5.3) $$P\{X_k = j\} = \sum_{i=1}^{N} a_i p_{ij}^{(k)}.$$

Example 5.1 (Brand Switching). Suppose that there are two brands, brand 1 and brand 2, of a given product. An individual purchases one of these brands each day. The sequence of the individual's brand choices from day to day is a discrete-time stochastic process. We assume that only the individual's choice of brand on the immediately preceding day, day t, affects his or her choices on day $t + 1$. Thus, the process of choice of brand is a Markov chain with possible states:

S_1: the individual chooses brand 1,

S_2: the individual chooses brand 2.

Table 5.1: *Transition Probabilities for*
Choice-of-Brand Process

State at day $t + 1$

		S_1	S_2
State at day t	S_1 (brand 1)	$\frac{1}{2}$	$\frac{1}{2}$
	S_2 (brand 2)	$\frac{3}{4}$	$\frac{1}{4}$

If the factors that influence the individual's choice of brand remain uniform in nature over time, we can assume that this choice-of-brand process has stationary transition probabilities.

Suppose that initially (day 0), brands 1 and 2 are valued equally by the individual. Consequently, the initial probabilities of the process are

$$a_1 = \tfrac{1}{2}, \qquad a_2 = \tfrac{1}{2}.$$

Based on survey data, the transition matrix (Table 5.1) of the process is estimated.

What is of interest are the unconditional probabilities $P\{X_k = 1\}$, $P\{X_k = 2\}$ that individuals will buy brand 1 or brand 2, respectively, k days after they begin to purchase the product of interest.

Using the methods of Section 4, 2-step, 3-step, 4-step, and 5-step transition probabilities are calculated (Table 5.2).

Table 5.2: *The 2-Step, 3-Step, 4-Step, and 5-Step Transition Probabilities for the Choice-of-Brand Process*

		State at day $t + 2$		State at day $t + 3$		State at day $t + 4$		State at day $t + 5$	
		S_1	S_2	S_1	S_2	S_1	S_2	S_1	S_2
State at day t	S_1	$\frac{10}{16}$	$\frac{6}{16}$	$\frac{38}{64}$	$\frac{26}{64}$	$\frac{154}{256}$	$\frac{102}{256}$	$\frac{614}{1024}$	$\frac{410}{1024}$
	S_2	$\frac{9}{16}$	$\frac{7}{16}$	$\frac{39}{64}$	$\frac{25}{64}$	$\frac{153}{256}$	$\frac{103}{256}$	$\frac{615}{1024}$	$\frac{409}{1024}$
		$p_{ij}^{(2)}$		$p_{ij}^{(3)}$		$p_{ij}^{(4)}$		$p_{ij}^{(5)}$	

Table 5.3: *Unconditional Probabilities (accurate to three decimal places) of Brand Choice for days* 0, 1, 2, 3, 4, *and* 5

Probability	Day k						
	0	1	2	3	4	5	$\cdots$
$P\{X_k = 1\}$	0.500	0.625	0.594	0.602	0.600	0.600	$\cdots$
$P\{X_k = 2\}$	0.500	0.375	0.406	0.398	0.400	0.400	$\cdots$

Now, from the initial probabilities, Table 5.1, and (5.3), the probability that the brand purchased by an individual on day 1 is brand 1 is

$$P\{X_1 = 1\} = a_1 p_{11}^{(1)} + a_2 p_{21}^{(1)} = a_1 p_{11} + a_2 p_{21}$$
$$= (\tfrac{1}{2})(\tfrac{1}{2}) + (\tfrac{1}{2})(\tfrac{3}{4}) = \tfrac{5}{8} = 0.625.$$

Using the initial probabilities, Table 5.2, and (5.3), the probability that the individual purchases brand 1 on day 2 is

$$P\{X_2 = 1\} = a_1 p_{11}^{(2)} + a_2 p_{21}^{(2)} = (\tfrac{1}{2})(\tfrac{10}{16}) + (\tfrac{1}{2})(\tfrac{9}{16}) = \tfrac{19}{32} = 0.594.$$

Comparable calculations of probabilities can be obtained for days 3, 4, 5, and so on, by making use of Table 5.2, the initial probabilities, and (5.3). These results are summarized in Table 5.3.

We see from Table 5.3 that the unconditional probability of choosing brand 1 appears by day 4 to have stabilized at 0.600, and that the unconditional probability of choosing brand 2 has stabilized at 0.400. Hence, if on any given day we observe the choice of brands only of individuals who have been purchasing the given product for 4 or more days, we can expect that approximately 60% of these individuals will have purchased brand 1.

6. FIRST-PASSAGE TIMES

Suppose that we are observing a discrete-time stochastic process which we have modeled as a Markov chain, and that one state is of special interest to us. For example, in Example 4.1 (the learning model), the state of interest may be S_1, the state in which the subject has been conditioned to make a correct response. In Example 3.3 (the genetical breeding experiments) the state S_j of interest may be some desirable (or undesirable) genotype for the offspring. In Example 4.2 (the occupational mobility process), the state of interest may be S_1, the highest and most desirable occupation level. Suppose that at time 0

Table 6.1: *Obtained Sequence of States over Time for
One Trial of the Random Experiment in Which a
Given Markov Chain Process Is Observed*

(The initial state is S_4)

Time	State	Time	State	Time	State
0	S_4	5	S_8	10	S_4
1	S_2	6	S_8	11	S_7
2	S_1	7	S_6	12	S_9
3	S_2	8	S_1	13	S_5
4	S_5	9	S_3	$\vdots$	$\vdots$

the process is observed to be in some state, state S_i, other than the state S_j of interest (i.e., $i \neq j$). We now consider the first time K_{ij}, after time 0, at which the process is in state S_j. Such variables K_{ij} are called *first-passage times*, since they tell us the time at which the process first passes into state S_j after having initially been in state S_i.

The observed value of the first-passage time K_{ij} depends upon the particular sequence of states obtained for the process during any particular trial. Thus, if we observe the sequence of states shown in Table 6.1, then $K_{49} = 12$, $K_{47} = 11$, $K_{42} = 1$, and so on. If we were to repeat the random experiment, and if once again S_4 is the state of the process at time 0, but otherwise the sequence of obtained states is as in Table 6.2, then $K_{49} = 1$, $K_{47} = 3$, $K_{42} = 10$, and so on.

Because the first-passage times K_{ij} depend on the sequence of states obtained for the random process, the K_{ij} are random variables. Since the properties of any random variable can be determined from knowledge of its distribution, we need to obtain the distributions of the first-passage times K_{ij}.

Table 6.2: *Obtained Sequence of States over Time for
Another Trial of the Random Experiment in Which
a Given Markov Chain Process Is Observed*

(The Initial State is S_4)

Time	State	Time	State	Time	State
0	S_4	5	S_1	10	S_2
1	S_9	6	S_3	11	S_2
2	S_3	7	S_9	12	S_1
3	S_7	8	S_6	13	S_5
4	S_5	9	S_8	$\vdots$	$\vdots$

Note that the first-passage time K_{ij} is defined for a given process only when the process is known to be initially (at time 0) in state S_i. It therefore seems appropriate to consider the conditional distribution of K_{ij} given that the process is in state S_i at time 0. Because K_{ij} is a discrete random variable, the conditional distribution of K_{ij} can be found by determining the values of its (conditional) probability mass function $P\{X_{ij} = k \mid X_0 = i\}$ for $k = 1, 2, 3$, and so on.

The definition of the first-passage time K_{ij} implies that K_{ij} equals the positive integer k if the process is in state S_j at time k, and has not been in state S_j at any time previous to time k. Let $f_{ij}^{(k)}$ represent the conditional probability that K_{ij} equals k given that the process is in state S_i at time 0. It follows from the definition of K_{ij} that

$$\begin{aligned}
(6.1) \quad f_{ij}^{(k)} &= P\{K_{ij} = k \mid X_0 = i\} \\
&= P\{X_0 = i \text{ and } X_1 \neq j \text{ and } \cdots \text{ and } X_{k-1} \neq j \text{ and } X_k = j \mid X_0 = i\}.
\end{aligned}$$

Although the notations we have used for k-step transition probabilities and first-passage probabilities are similar in form, it is important to note that in general $p_{ij}^{(k)}$ and $f_{ij}^{(k)}$ are not equal. Indeed, it is always the case that

$$(6.2) \qquad\qquad f_{ij}^{(k)} \leq p_{ij}^{(k)}.$$

To verify this assertion, note that if the process has entered state S_j for the first time at time k (i.e., if the event $\{K_{ij} = k\}$ has occurred), then it must be the case that the process is in state S_j at time k. Thus, the event $\{K_{ij} = k\}$ is included in the event $\{X_k = j\}$, and it follows from the Law of Inclusion for conditional probabilities (see Chapter 3) that $f_{ij}^{(k)} = P\{K_{ij} = k \mid X_0 = i\} \leq p_{ij}^{(k)} = P\{X_k = j \mid X_0 = i\}$. As we see later, for $k > 1$, $f_{ij}^{(k)}$ is in general strictly less than $p_{ij}^{(k)}$.

Calculation of the Probability Distribution of the First-Passage Time K_{ij}

The probability distribution of the first-passage time K_{ij} is determined by the quantities $f_{ij}^{(1)}, f_{ij}^{(2)}, f_{ij}^{(3)}$, and so on, where for any positive integer k,

$$f_{ij}^{(k)} = P\{K_{ij} = k \mid X_0 = i\}.$$

We now show that the quantities $f_{ij}^{(1)}, f_{ij}^{(2)}, f_{ij}^{(3)}$, and so on, can be computed solely from knowledge of the transition probabilities p_{ij} for the given Markov chain. We illustrate such computations for the genetical breeding experiment described in Example 3.3. Recall that in this experiment, one parent in a given mating always has a fixed genotype (here, the fixed genotype is assumed to be the genotype Aa), while the other parent is randomly selected from the offspring of a previous such mating. The state of the mating process at the kth such mating ("time $k - 1$") is the genotype of the offspring that is selected at random for the $(k + 1)$st mating. Thus, the possible states of this process are

S_1:   the offspring selected has genotype AA,

S_2:   the offspring selected has genotype Aa,

S_3:   the offspring selected has genotype aa.

As we observed in Section 3, the genetical mating process described above is a Markov chain with stationary transition probabilities p_{ij} summarized in Table 6.3.

Suppose that at the initial mating, both parents have genotype Aa; that is, the state of the mating process at time 0 is S_2. We determine the probability distribution of K_{21}, the number of matings after the initial mating that are required to obtain an offspring of genotype AA (state S_1) for the first time.

The conditional probability that the first-passage time K_{21} from state S_2 to state S_1 is equal to 1, given that the process is initially in state S_2, is the same as the conditional probability p_{21} that the process is in state S_1 at time 1 given that the process is in state S_2 at time 0. That is,

(6.3) $$f_{21}^{(1)} = P\{X_1 = 1 \mid X_0 = 2\} = p_{21} = \tfrac{1}{4}.$$

To find the value of $f_{21}^{(2)}$, first note that for the process to be in state S_1 for the *first* time at time 2, the process cannot have been in state S_1 at time 1. Thus, the only transitions that need be considered are the transitions $S_2 \to S_2 \to S_1$ and $S_2 \to S_3 \to S_1$. Given that the process is in state S_2 at time 0, we obtain

$$P\{S_2 \to S_2 \to S_1 \mid \text{the process is in state } S_2 \text{ at time 0}\}$$
$$= P\{X_0 = 2 \text{ and } X_1 = 2 \text{ and } X_2 = 1 \mid X_0 = 2\}$$
$$= p_{22} p_{21},$$
$$P\{S_2 \to S_3 \to S_1 \mid \text{the process is in state } S_2 \text{ at time 0}\}$$
$$= P\{X_0 = 2 \text{ and } X_1 = 3 \text{ and } X_2 = 1 \mid X_0 = 2\}$$
$$= p_{23} p_{31}.$$

Because $\{K_{21} = 2\} = \{X_0 = 2 \text{ and } X_1 = 2 \text{ and } X_2 = 1\} \cup \{X_0 = 2 \text{ and } X_1 = 3 \text{ and } X_2 = 1\}$, it follows that

$$f_{21}^{(2)} = P\{K_{21} = 2 \mid X_0 = 2\}$$
(6.4)
$$= p_{22} p_{21} + p_{23} p_{31} = (\tfrac{1}{2})(\tfrac{1}{4}) + (\tfrac{1}{2})(0) = \tfrac{1}{8}.$$

To determine $f_{21}^{(3)}$, we need to find the conditional probability of moving from state S_2 at time 0 to state S_1 for the first time at time 3, given that the process is initially (at time 0) in state S_2. The process moves to state S_1 for

Table 6.3: *Transition Probabilities for Genotypes Arising from a Mating Where One Parent Always Has Genotype Aa, the Other Parent Is Randomly Selected from the Offspring of a Previous Such Mating, and One Offspring Is Selected at Random from Their Children*

		State at time $t+1$ (mating $t+2$)		
		S_1	S_2	S_3
State at time t (mating $t+1$)	S_1: genotype AA	$\frac{1}{2}$	$\frac{1}{2}$	0
	S_2: genotype Aa	$\frac{1}{4}$	$\frac{1}{2}$	$\frac{1}{4}$
	S_3: genotype aa	0	$\frac{1}{2}$	$\frac{1}{2}$

the first time at time 3 if any one of the sequences of states over time shown in Figure 6.1 occurs.

Using conditional probability arguments,

$P\{S_2 \to S_2 \to S_2 \to S_1 \,|\, \text{the process is in state } S_2 \text{ at time } 0\}$

$\quad = P\{X_0 = 2 \ and \ X_1 = 2 \ and \ X_2 = 2 \ and \ X_3 = 1 | X_0 = 2\}$

$\quad = p_{22}p_{22}p_{21},$

$P\{S_2 \to S_2 \to S_3 \to S_1 \,|\, \text{the process is in state } S_2 \text{ at time } 0\}$

$\quad = p_{22}p_{23}p_{31},$

$P\{S_2 \to S_3 \to S_2 \to S_1 \,|\, \text{the process is in state } S_2 \text{ at time } 0\}$

$\quad = p_{23}p_{32}p_{21},$

$P\{S_2 \to S_3 \to S_3 \to S_1 \,|\, \text{the process is in state } S_2 \text{ at time } 0\}$

$\quad = p_{23}p_{33}p_{31}.$

Figure 6.1: *Any of these four sequences of states results in the process moving from state S_2 at time 0 to state S_1 for the first time at time 3.*

$$
\begin{array}{ccccccc}
S_2 & \to & S_2 & \to & S_2 & \to & S_1 \\
S_2 & \to & S_2 & \to & S_3 & \to & S_1 \\
S_2 & \to & S_3 & \to & S_2 & \to & S_1 \\
S_2 & \to & S_3 & \to & S_3 & \to & S_1 \\
\end{array}
$$

Time: 0   1   2   3

Since $\{K_{21} = 3\} = \{S_2 \to S_2 \to S_2 \to S_1\} \cup \{S_2 \to S_2 \to S_3 \to S_1\} \cup \{S_2 \to S_3 \to S_2 \to S_1\} \cup \{S_2 \to S_3 \to S_3 \to S_1\}$, it follows that

$$f_{21}^{(3)} = P\{K_{21} = 3 \mid X_0 = 2\}$$

(6.5)
$$= p_{22}p_{22}p_{21} + p_{22}p_{23}p_{31} + p_{23}p_{32}p_{21} + p_{23}p_{33}p_{31}$$

$$= (\tfrac{1}{2})(\tfrac{1}{2})(\tfrac{1}{4}) + (\tfrac{1}{2})(\tfrac{1}{4})(0) + (\tfrac{1}{4})(\tfrac{1}{2})(\tfrac{1}{4}) + (\tfrac{1}{4})(\tfrac{1}{2})(0) = \tfrac{3}{32}.$$

Computing $f_{21}^{(k)}$ for $k = 4, 5, \ldots$ by the method illustrated above can become cumbersome as k increases. Thus, we are led to look for an alternative way of calculating the first-passage probabilities $f_{21}^{(k)}$ for $k = 1, 2, 3, \ldots$. One way is a recursive method similar to the recursive method used in Section 4 to compute the k-step transition probabilities $p_{ij}^{(k)}$.

We earlier pointed out that in general, for $k > 1, f_{ij}^{(k)} < p_{ij}^{(k)}$. In particular, for the mating example that we have been considering in the present section, $f_{21}^{(2)} = p_{22}p_{21} + p_{23}p_{31} = \tfrac{1}{8}$, while

$$p_{21}^{(2)} = p_{21}p_{11} + p_{22}p_{21} + p_{23}p_{31} = \tfrac{2}{8}.$$

The difference between $p_{21}^{(2)}$ and $f_{21}^{(2)}$ is that the transition $S_2 \to S_1 \to S_1$ is counted when computing $p_{21}^{(2)}$, but is not counted when computing $f_{21}^{(2)}$. Indeed,

$$p_{21}^{(2)} = f_{21}^{(2)} + P\{X_0 = 2 \text{ and } X_1 = 1 \text{ and } X_2 = 1 \mid X_0 = 2\}$$

$$= f_{21}^{(2)} + p_{21}p_{11},$$

and since we have already seen that $f_{21}^{(1)} = p_{21}$,

(6.6) $$p_{21}^{(2)} = f_{21}^{(2)} + f_{21}^{(1)}p_{11}.$$

We can interpret (6.6) in the following way. To make the transition from state S_2 at time 0 to state S_1 at time 2, the process can either go from state S_2 at time 0 to state S_1 for the first time at time 2; or else go from state S_2 at time 0 to state S_1 for the first time at time 1, and then remain in state S_1 at time 2.

Let us try to obtain a relationship similar to (6.6) between the transition probabilities $p_{11}, p_{11}^{(2)}$, and $p_{21}^{(3)}$, and the first-passage probabilities $f_{21}^{(1)}, f_{21}^{(2)}$, and $f_{21}^{(3)}$. To go from state S_2 at time 0 to state S_1 at time 3, one of the following mutually exclusive events must occur:

(1) The process goes from state S_2 at time 0 to state S_1 for the first time at time 3. The conditional probability of this event, given that the process is at state S_2 at time 0, is $f_{21}^{(3)}$.
(2) The process goes from state S_2 at time 0 to state S_1 for the first time at time 2, and then from state S_1 at time 2 to state S_1 once again at

time 3. The conditional probability of this event, given that the process is in state S_2 at time 0, is equal to

$$P\{X_2 = 1, X_1 \neq 1 \,|\, X_0 = 2\} P\{X_3 = 1 \,|\, X_2 = 1 \text{ and } X_1 \neq 1 \text{ and } X_0 = 2\}$$
$$= f_{21}^{(2)} P\{X_3 = 1 \,|\, X_2 = 1\} = f_{21}^{(2)} p_{11}.$$

To obtain this result, we have used the fact that the process is a Markov chain; this fact implies that

$$P\{X_3 = 1 \,|\, X_2 = 1 \text{ and } X_1 \neq 1 \text{ and } X_0 = 2\} = P\{X_3 = 1 \,|\, X_2 = 1\}.$$

(3) The process goes from state S_2 at time 0 to state S_1 for the first time at time 1, and then from state S_1 back to state S_1 in two time steps (from time 1 to time 3). The conditional probability of this event, given that the process is in state S_2 at time 0, is equal to

$$P\{X_1 = 1 \,|\, X_0 = 2\} P\{X_3 = 1 \,|\, X_1 = 1 \text{ and } X_0 = 2\}$$
$$= p_{21}^{(1)} P\{X_3 = 1 \,|\, X_1 = 1\} = f_{21}^{(1)} p_{11}^{(2)}.$$

Thus, to obtain the conditional probability $p_{21}^{(3)}$ that the process is in state S_1 at time 3, given that at time 0 the process is in state S_2, we add the conditional probabilities of the events described in (1), (2), and (3) above. We obtain the result

(6.7) $$p_{21}^{(3)} = f_{21}^{(3)} + f_{21}^{(2)} p_{11} + f_{21}^{(1)} p_{11}^{(2)}.$$

Using arguments similar to those above, we can express the conditional probability $p_{21}^{(k)}$ that the process is in state S_1 at time k, given that the process is in state S_2 at time 0, as the sum of the conditional probability $f_{21}^{(k)}$ that the process is in state S_1 for the first time at time k; the conditional probability $f_{21}^{(k-1)} p_{11}$ that the process is in state S_1 for the first time at time $k-1$ and then moves from state S_1 back to state S_1 in one time step (from time $k-1$ to time k); the conditional probability $f_{21}^{(k-2)} p_{11}^{(2)}$ that the process is in state S_1 for the first time at time $k-2$ and then moves from state S_1 back to state S_1 in two time steps; and so on. The resulting equation is

(6.8) $$p_{21}^{(k)} = f_{21}^{(k)} + f_{21}^{(k-1)} p_{11}^{(1)} + f_{21}^{(k-2)} p_{11}^{(2)} + \cdots + f_{21}^{(1)} p_{11}^{(k-1)}.$$

From (6.8), we obtain

(6.9) $$f_{21}^{(k)} = p_{21}^{(k)} - f_{21}^{(k-1)} p_{11}^{(1)} - f_{21}^{(k-2)} p_{11}^{(2)} - \cdots - f_{21}^{(1)} p_{11}^{(k-1)}.$$

If we have already computed the necessary transition probabilities, $p_{ij}^{(k)}$, then (6.9) allows us to recursively calculate the first-passage probabilities $f_{21}^{(2)}, f_{21}^{(3)}, f_{21}^{(4)}$, and so on. That is, first calculate $f_{21}^{(1)} = p_{21}$; then, using $p_{21}^{(2)}$ and p_{11} already obtained, calculate

(6.10) $$f_{21}^{(2)} = p_{21}^{(2)} - f_{21}^{(1)} p_{11};$$

next, using the values of $p_{21}^{(3)}$, p_{11}, $p_{11}^{(2)}$, and the values of $f_{21}^{(1)}$ and $f_{21}^{(2)}$ already calculated, compute $f_{21}^{(3)}$ from

(6.11) $$f_{21}^{(3)} = p_{21}^{(3)} - f_{21}^{(2)}p_{11} - f_{21}^{(1)}p_{11}^{(2)};$$

and so forth.

In our genetical mating example, we have already obtained $f_{21}^{(2)}$ and $f_{21}^{(3)}$ by direct methods. Let us now calculate these same quantities by the recursive method based on (6.9). As a first step, we determine all of the 2-step and 3-step transition probabilities for the genetical mating process, using the methods of Section 4 (Table 6.4).

To carry out the computations of $f_{21}^{(2)}$ and $f_{21}^{(3)}$, recall from (6.3) that

$$f_{21}^{(1)} = p_{21}^{(1)} = p_{21} = \tfrac{1}{4}.$$

Next, from Table 6.3 note that $p_{11} = \tfrac{1}{2}$, and from Table 6.4 note that $p_{21}^{(2)} = \tfrac{1}{4}$. Thus, (6.10),

$$f_{21}^{(2)} = p_{21}^{(2)} - f_{21}^{(1)}p_{11} = \tfrac{1}{4} - (\tfrac{1}{4})(\tfrac{1}{2}) = \tfrac{1}{8},$$

which agrees with the answer for $f_{21}^{(2)}$ obtained by our earlier method. To compute $f_{21}^{(3)}$, we make use of (6.11). Since $p_{11}^{(2)} = \tfrac{3}{8}$, $p_{21}^{(3)} = \tfrac{1}{4}$, it follows that

$$f_{21}^{(3)} = p_{21}^{(3)} - f_{21}^{(2)}p_{11} - f_{21}^{(1)}p_{11}^{(2)} = \tfrac{1}{4} - (\tfrac{1}{8})(\tfrac{1}{2}) - (\tfrac{1}{4})(\tfrac{3}{8}) = \tfrac{3}{32}.$$

Using (6.9), we can also calculate $f_{21}^{(4)}$, $f_{21}^{(5)}$, $f_{21}^{(6)}$, and so on. For example, since from Table 6.4 we know that $p_{11}^{(3)} = \tfrac{5}{16}$, and since

$$p_{21}^{(4)} = p_{21}p_{11}^{(3)} + p_{22}p_{21}^{(3)} + p_{23}p_{31}^{(3)}$$
$$= (\tfrac{1}{4})(\tfrac{5}{16}) + (\tfrac{1}{2})(\tfrac{8}{32}) + (\tfrac{1}{4})(\tfrac{3}{16}) = \tfrac{16}{64},$$

Table 6.4: *The 2-Step and 3-Step Transition Probabilities for the Genetical Mating Process Whose Transition Probabilities Are Given in Table 6.3*

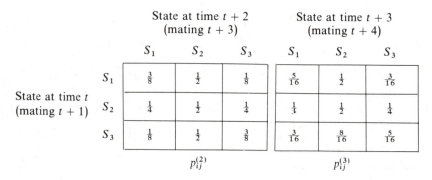

		State at time $t+2$ (mating $t+3$)			State at time $t+3$ (mating $t+4$)		
		S_1	S_2	S_3	S_1	S_2	S_3
	S_1	$\tfrac{3}{8}$	$\tfrac{1}{2}$	$\tfrac{1}{8}$	$\tfrac{5}{16}$	$\tfrac{1}{2}$	$\tfrac{3}{16}$
State at time t (mating $t+1$)	S_2	$\tfrac{1}{4}$	$\tfrac{1}{2}$	$\tfrac{1}{4}$	$\tfrac{1}{4}$	$\tfrac{1}{2}$	$\tfrac{1}{4}$
	S_3	$\tfrac{1}{8}$	$\tfrac{1}{2}$	$\tfrac{3}{8}$	$\tfrac{3}{16}$	$\tfrac{8}{16}$	$\tfrac{5}{16}$
			$p_{ij}^{(2)}$			$p_{ij}^{(3)}$	

it follows from (6.9) that

$$f_{21}^{(4)} = p_{21}^{(4)} - f_{21}^{(3)}p_{11} - f_{21}^{(2)}p_{11}^{(2)} - f_{21}^{(1)}p_{11}^{(3)}$$

$$= \tfrac{16}{64} - (\tfrac{3}{32})(\tfrac{1}{2}) - (\tfrac{1}{8})(\tfrac{3}{8}) - (\tfrac{1}{4})(\tfrac{5}{16}) = \tfrac{5}{64}.$$

Equations (6.9), (6.10), and (6.11) provide a method for recursively calculating the first-passage probabilities $f_{21}^{(k)}$, for $k = 1, 2, 3, 4, \ldots$. These involve states S_1 and S_2. In general, the result for any two states S_i and S_j are

(6.12)
$$f_{ij}^{(1)} = p_{ij},$$
$$f_{ij}^{(2)} = p_{ij}^{(2)} - f_{ij}^{(1)}p_{jj},$$
$$f_{ij}^{(3)} = p_{ij}^{(3)} - f_{ij}^{(2)}p_{jj} - f_{ij}^{(1)}p_{jj}^{(2)},$$
$$\vdots \quad \vdots \quad \vdots \qquad \vdots$$
$$f_{ij}^{(k)} = p_{ij}^{(k)} - f_{ij}^{(k-1)}p_{jj} - f_{ij}^{(k-2)}p_{jj}^{(2)} - \cdots - f_{ij}^{(2)}p_{jj}^{(k-2)} - f_{ij}^{(1)}p_{jj}^{(k-1)}.$$

For the genetical mating experiment whose 1-step transition probabilities appear in Table 6.3, we can use (6.12) to recursively calculate the values of the conditional probability mass function

$$f_{23}^{(k)} = P\{K_{23} = k \mid X_0 = 2\}, \quad k = 1, 2, 3, \ldots,$$

where K_{23} is the number of matings (after the initial mating) needed to obtain an offspring of genotype aa for the first time. Thus, from Tables 6.3 and 6.4, and from (6.12) with $i = 2$, $j = 3$, we find that

$$f_{23}^{(1)} = p_{23} = \tfrac{1}{4},$$
$$f_{23}^{(2)} = p_{23}^{(2)} - f_{23}^{(1)}p_{33} = \tfrac{1}{4} - (\tfrac{1}{4})(\tfrac{1}{2}) = \tfrac{1}{8},$$
$$f_{23}^{(3)} = p_{23}^{(3)} - f_{23}^{(2)}p_{33} - f_{23}^{(1)}p_{33}^{(2)} = \tfrac{1}{4} - (\tfrac{1}{8})(\tfrac{1}{2}) - (\tfrac{1}{4})(\tfrac{3}{8}) = \tfrac{3}{32};$$

and since

$$p_{23}^{(4)} = p_{21}p_{13}^{(3)} + p_{22}p_{23}^{(3)} + p_{23}p_{33}^{(3)} = (\tfrac{1}{4})(\tfrac{3}{16}) + (\tfrac{1}{2})(\tfrac{1}{4}) + (\tfrac{1}{4})(\tfrac{5}{16}) = \tfrac{1}{4},$$

we derive the result

$$f_{23}^{(4)} = p_{23}^{(4)} - f_{23}^{(3)}p_{33} - f_{23}^{(2)}p_{33}^{(2)} - f_{23}^{(1)}p_{33}^{(3)}$$

$$= \tfrac{1}{4} - (\tfrac{3}{32})(\tfrac{1}{2}) - (\tfrac{1}{8})(\tfrac{3}{8}) - (\tfrac{1}{4})(\tfrac{5}{16}) = \tfrac{5}{64}.$$

We have discussed calculation of the first-passage probabilities $f_{ij}^{(k)}$ only for situations in which the Markov chain under consideration has exactly three possible states. However, our derivation of (6.12) made no use of this fact. Indeed, (6.12) is valid for any Markov chain whatsoever. Assuming that the values of all the n-step transition probabilities (for $n = 1, 2, 3, \ldots$) are available, it is not difficult to use (6.12) to obtain the first few (say, four or five) first-passage probabilities by hand. However, if the entire conditional probability mass function

(6.13) $$f_{ij}^{(k)} = P\{K_{ij} = k \mid X_0 = i\}, \quad k = 1, 2, 3, \ldots$$

is desired, it is usually more convenient to make use of computer programs written for this purpose.

Example 6.1 (Political Preference). In a certain state, a voter has a chance to register every year as a member of one of three political parties, and to vote for representatives (delegates) to the state caucus of his party. Since voters change their party registration from year to year in unpredictable fashion, we might model this phenomenon as a discrete-time stochastic process. If, further, we assume that a voter's choice of political party in a given year depends on his past political preferences only in terms of his political choice in the immediately previous year, then we may model the process of political choice as a Markov chain. The states of this process are the three parties, and in addition, since a voter can also choose not to register, a fourth state of being "not registered" is required. Thus, the states of the process are the following:

$$S_1: \quad \text{the voter registers in party 1,}$$

$$S_2: \quad \text{the voter registers in party 2,}$$

$$S_3: \quad \text{the voter registers in party 3,}$$

$$S_4: \quad \text{the voter is not registered.}$$

The transition probabilities are given in Table 6.5.

If, in a given year, a voter is not registered (state S_4), the length of time K_{41} required before the voter registers in party 1 for the first time may be of

Table 6.5: Transition Probabilities for the Political Preference Process

		State of process (party preference) in year $t + 1$			
		S_1	S_2	S_3	S_4
State of process (party preference) in year t	S_1	0.43	0.22	0.10	0.25
	S_2	0.21	0.41	0.11	0.27
	S_3	0.30	0.02	0.33	0.35
	S_4	0.33	0.23	0.10	0.34

Table 6.6: *Values of the Conditional Probability Mass Function $f_{41}^{(k)}$ of the First-Passage Time K_{41} for the Political Preference Process of Example 6.1*

[The values of $p_{41}^{(k)}$ and $p_{11}^{(k)}$ are also given; all entries are to three significant figures]

k	$f_{41}^{(k)}$	$p_{41}^{(k)}$	$p_{11}^{(k)}$	k	$f_{41}^{(k)}$	$p_{41}^{(k)}$	$p_{11}^{(k)}$
1	0.330	0.330	0.430	11	0.010	0.330	0.330
2	0.190	0.332	0.344	12	0.007	0.330	0.330
3	0.134	0.330	0.331	13	0.005	0.330	0.330
4	0.097	0.330	0.330	14	0.004	0.330	0.330
5	0.070	0.330	0.330	15	0.003	0.330	0.330
6	0.050	0.330	0.330	16	0.002	0.330	0.330
7	0.036	0.330	0.330	17	0.001	0.330	0.330
8	0.026	0.330	0.330	18	0.001	0.330	0.330
9	0.019	0.330	0.330	19	0.001	0.330	0.330
10	0.013	0.330	0.330				

interest. Using a computer program, the conditional probability mass function

$$f_{41}^{(k)} = P\{K_{41} = k \mid X_0 = 4\}, \quad k = 1, 2, 3, \ldots$$

has been obtained in Table 6.6. Also in Table 6.6, we have summarized the values of $p_{41}^{(k)}$ and $p_{11}^{(k)}$, $k = 1, 2, 3, \ldots$, needed to compute $f_{41}^{(k)}$ by means of (6.12). The values of $p_{41}^{(k)}$ and $p_{11}^{(k)}$ were in turn computed recursively by means of the methods described in Section 4. After $k = 19$, the values of $f_{41}^{(k)}$ are (to three significant figures) equal to 0.

Infinite First-Passage Times

It is conceivable that if we start the Markov chain process at state S_i, the process never reaches state S_j, no matter how long the process is observed. In such a situation, the first-passage time K_{ij} between state S_i and state S_j is infinite (written $K_{ij} = \infty$). Thus, in general, the possible values of the first-passage time K_{ij} from state S_i to state S_j are the finite positive integers $1, 2, 3, \ldots$, and the additional value ∞. To compute the conditional probability

$$f_{ij}^{(\infty)} = P\{K_{ij} = \infty \mid X_0 = i\}$$

that K_{ij} is finite, note that if K_{ij} is not equal to one of the finite positive integers $1, 2, 3, \ldots$, then K_{ij} must be infinite. Let

$$(6.14) \qquad F_{ij} = f_{ij}^{(1)} + f_{ij}^{(2)} + f_{ij}^{(3)} + \cdots = \sum_{1 \leq k < \infty} f_{ij}^{(k)},$$

then

(6.15) $f_{ij}^{(\infty)} = P\{K_{ij} = \infty \,|\, X_0 = i\} = 1 - F_{ij}.$

Since $F_{ij} = P\{K_{ij} = 1 \text{ or } 2 \text{ or } 3 \text{ or } \cdots \,|\, X_0 = i\},$

(6.16) $F_{ij} \leq 1.$

If $F_{ij} = 1$, then $P\{K_{ij} = \infty \,|\, X_0 = i\} = 1 - F_{ij} = 0$. Thus, if $F_{ij} = 1$, the event that state S_j cannot be reached by the process, given that the process is initially in state S_i, has a (conditional) probability of 0, and the first-passage time K_{ij} from state S_i to state S_j is finite with probability equal to 1. For example, looking back at Example 6.1, we see (from Table 6.6) that the first-passage time K_{41} is finite with (conditional) probability equal to 1, since $F_{41} = f_{41}^{(1)} + f_{41}^{(2)} + f_{41}^{(3)} + \cdots = 1$.

It is possible, however, to exhibit Markov chains in which there exist two states, S_i and S_j, such that

(6.17) $F_{ij} < 1.$

If inequality (6.17) holds, then the conditional probability that state S_j is never reached, given that the process is initially in state S_i, is a nonzero number, and the first-passage time K_{ij} has a positive (conditional) probability of being infinite.

As an example of a Markov chain for which (6.17) holds, consider the Markov chain which has two states S_1 and S_2, and which has transition probabilities

		State at time $t+1$	
		S_1	S_2
State at time t	S_1	1	0
	S_2	$\frac{1}{2}$	$\frac{1}{2}$

The learning model of Example 4.1 (with $\alpha = \frac{1}{2}$) is an example of such a Markov chain. If this process begins initially in state S_1, then the process cannot leave this state to go to state S_2. To verify this assertion, recall that for any positive integer $k \neq 1$,

$$p_{12}^{(k)} = p_{11}^{(k-1)} p_{12} + p_{12}^{(k-1)} p_{22}.$$

However, for our particular Markov chain, $p_{12} = 0$ and $p_{22} = \frac{1}{2}$. Thus,

$$p_{12}^{(k)} = \frac{1}{2}\, p_{12}^{(k-1)} = \frac{1}{2^2}\, p_{12}^{(k-2)} = \frac{1}{2^3}\, p_{12}^{(k-3)} = \cdots = \frac{1}{2^{k-1}}\, p_{12}.$$

Since $p_{12} = 0$, $p_{12}^{(k)} = 0$ for all positive integers k. Since $f_{12}^{(k)}$ is a probability,

$f_{12}^{(k)} \geq 0$, and from (6.2), $f_{12}^{(k)} \leq p_{12}^{(k)} = 0$, $k = 1, 2, 3, \ldots$. Thus, $f_{12}^{(k)} = 0$, for all $k = 1, 2, \ldots$. Consequently,

$$F_{12} = f_{12}^{(1)} + f_{12}^{(2)} + f_{12}^{(3)} + \cdots = 0,$$

and therefore

$$P\{K_{12} = \infty \mid X_0 = 1\} = 1 - F_{12} = 1.$$

The conditional probabilities $f_{ij}^{(\infty)} = P\{K_{ij} = \infty \mid X_0 = i\}$ can be of practical interest. For example, in the learning process described in Example 4.1 (with $\alpha = \frac{1}{2}$), we have shown above that once a subject is conditioned, the conditional probability, $P\{K_{12} = \infty \mid X_0 = 1\}$, that the subject never becomes unconditioned is equal to 1. For this process, we can also verify that if a subject starts in an unconditioned state (state S_2), then the first-passage time K_{21} to the conditioned state (S_1) is finite with conditional probability equal to 1. First, recall from (4.17) that for any $k = 1, 2, 3, \ldots$,

(6.18) $$p_{11}^{(k)} = 1 - p_{12}^{(k)}.$$

and that for $k = 2, 3, 4, \ldots$,

(6.19) $$p_{21}^{(k-1)} = 1 - p_{22}^{(k-1)}.$$

We have already shown that $p_{12}^{(k)} = 0$ for $k = 1, 2, 3, \ldots$. Thus, (6.18) implies that $p_{11}^{(k)} = 1$ for all $k = 1, 2, 3, \ldots$. Next, using (6.19) and the fact that

$$p_{21}^{(k)} = p_{21}^{(k-1)}p_{11} + p_{22}^{(k-1)}p_{21} = (p_{21}^{(k-1)})(1) + (1 - p_{21}^{(k-1)})(\tfrac{1}{2}) = \tfrac{1}{2} + \tfrac{1}{2}p_{21}^{(k-1)},$$

it follows that $p_{21}^{(1)} = \tfrac{1}{2}$, $p_{21}^{(2)} = \tfrac{3}{4}$, and in general

$$p_{21}^{(k)} = \tfrac{1}{2} + (\tfrac{1}{2})^2 + (\tfrac{1}{2})^3 + \cdots + (\tfrac{1}{2})^k = 1 - (\tfrac{1}{2})^k,$$

for $k = 1, 2, 3, \ldots$. Finally, since $p_{11}^{(m)} = 1$ for $m = 1, 2, 3, \ldots$,

$$f_{21}^{(k)} = p_{21}^{(k)} - f_{21}^{(k-1)}p_{11} - f_{21}^{(k-2)}p_{11}^{(2)} - \cdots - f_{21}^{(2)}p_{11}^{(k-2)} - f_{21}^{(1)}p_{11}^{(k-1)}$$
$$= 1 - (\tfrac{1}{2})^k - f_{21}^{(k-1)} - f_{21}^{(k-2)} - \cdots - f_{21}^{(2)} - f_{21}^{(1)},$$

or

$$f_{21}^{(1)} + f_{21}^{(2)} + \cdots + f_{21}^{(k)} = 1 - (\tfrac{1}{2})^k.$$

Therefore,

$$F_{21} = f_{21}^{(1)} + f_{21}^{(2)} + f_{21}^{(3)} + \cdots = 1,$$

and it follows that once a subject enters the unconditioned state (S_2), the length of time K_{21} required for the subject to enter the conditioned state S_1 for the first time is finite with (conditional) probability equal to 1.

For general Markov chains, it is seldom an easy task to determine the conditional probability F_{ij} that the first-passage time K_{ij} between state S_i and state S_j is finite, or even to determine whether $F_{ij} < 1$ or $F_{ij} = 1$. The recursion relations for calculating $f_{ij}^{(k)}$ and (6.14) permit us, in theory, to calculate F_{ij} by simply computing $f_{ij}^{(1)}, f_{ij}^{(2)}, f_{ij}^{(3)}$, and so on, recursively, and then adding these quantities. However, unless only a finite number of the $f_{ij}^{(k)}$ are nonzero, computing F_{ij} in this manner requires summation of an infinite number of terms. For this reason, it becomes necessary in most cases to use mathematical arguments (such as those used above to verify that $F_{12} = 0$ and $F_{21} = 1$ in the learning-model example) either to determine the mathematical form of the $f_{ij}^{(k)}$'s or to directly determine the value of F_{ij}.

7. RECURRENT AND TRANSIENT STATES

In discussing first-passage times K_{ij} from a state S_i to a state S_j, we have up to now assumed that S_i and S_j are not the same state. However, no use of this assumption was made when we verified (6.12). Consequently, if we are interested in the random length K_{jj} of time required for a given state, state S_j, to recur or reappear for the first time, given that the process which we are observing was initially in state S_j, we can repeat the arguments that we used to obtain (6.12), and in this manner derive the following recursive equations for calculating the conditional probability mass function

$$f_{jj}^{(k)} = P\{K_{jj} = k \mid X_0 = j\}$$

of K_{jj}:

$$
\begin{aligned}
&f_{jj}^{(1)} = p_{jj}, \\
&f_{jj}^{(2)} = p_{jj}^{(2)} - f_{jj}^{(1)} p_{jj}, \\
&f_{jj}^{(3)} = p_{jj}^{(3)} - f_{jj}^{(2)} p_{jj} - f_{jj}^{(1)} p_{jj}^{(2)}, \\
&\quad \vdots \\
&f_{jj}^{(k)} = p_{jj}^{(k)} - f_{jj}^{(k-1)} p_{jj} - f_{jj}^{(k-2)} p_{jj}^{(2)} - \cdots - f_{jj}^{(2)} p_{jj}^{(k-2)} - f_{jj}^{(1)} p_{jj}^{(k-1)},
\end{aligned}
$$

(7.1)

and so on.

The random variable K_{jj} is a first-passage time between state S_j and itself. Alternatively, we can think of K_{jj} as being a (first) recurrence time for state S_j.

For example, in the genetical mating experiment described in Section 6, suppose that the initial parent chosen to mate with the parent of fixed genotype Aa has genotype aa. If genotype aa is an undesirable genotype, we might be interested in the number K_{33} of matings that can be performed before an offspring is chosen for mating which again has genotype aa. From Tables 6.3 and 6.4, and from (7.1) with $j = 3$, we can determine the first three values $f_{33}^{(1)}$, $f_{33}^{(2)}$, and $f_{33}^{(3)}$ of the conditional probability mass function $f_{33}^{(k)} = P\{K_{33} = k \mid X_0 = 3\}$ of K_{33}. These values are

$$f_{33}^{(1)} = p_{33} = \tfrac{1}{2} = 0.500,$$

$$f_{33}^{(2)} = p_{33}^{(2)} - f_{33}^{(1)}p_{33} = \left(\tfrac{3}{8}\right) - \left(\tfrac{1}{2}\right)\left(\tfrac{1}{2}\right) = \tfrac{1}{8} = 0.125,$$

and

$$f_{33}^{(3)} = p_{33}^{(3)} - f_{33}^{(2)}p_{33} - f_{33}^{(1)}p_{33}^{(2)}$$
$$= \left(\tfrac{5}{16}\right) - \left(\tfrac{1}{8}\right)\left(\tfrac{1}{2}\right) - \left(\tfrac{1}{2}\right)\left(\tfrac{3}{8}\right) = \tfrac{1}{16} = 0.0625.$$

The other values of this conditional probability mass function have been determined using a computer. These values are displayed in Table 7.1. From Table 7.1, it appears that

$$F_{33} = f_{33}^{(1)} + f_{33}^{(2)} + f_{33}^{(3)} + \cdots = 1,$$

so that

$$f_{33}^{(\infty)} = P\{K_{33} = \infty \mid X_0 = 3\} = 1 - F_{33} = 0.$$

However, although Table 7.1 is accurate enough for most practical purposes, the values of $f_{33}^{(k)}$ shown in that table are rounded off to three decimal places, and hence a roundoff error could have affected the sum F_{33}. Thus, it is possible that $f_{33}^{(\infty)}$ is actually not 0, but instead is a very small positive number. If it is important to know which is the case, then mathematical analysis is needed to find the exact values of $f_{33}^{(k)}$, $k = 1, 2, 3, \ldots$.

In some cases (such as in the example above) the recurrence time K_{jj} can never be infinite; that is,

$$f_{jj}^{(\infty)} = P\{K_{jj} = \infty \mid X_0 = j\} = 0,$$

or, equivalently,

(7.2) $$F_{jj} = f_{jj}^{(1)} + f_{jj}^{(2)} + f_{jj}^{(3)} + \cdots = 1.$$

Table 7.1: *Values of the Conditional Probability Mass Function $f_{33}^{(k)}$ of the (First) Recurrence Time K_{33} for the Genetical Mating Process*

k	$f_{33}^{(k)}$	k	$f_{33}^{(k)}$	k	$f_{33}^{(k)}$
1	0.500	6	0.033	11	0.015
2	0.125	7	0.028	12	0.013
3	0.063	8	0.024	13	0.011
4	0.047	9	0.021	14	0.009
5	0.039	10	0.018	:	:

In this case, the state S_j recurs in a finite length of time with conditional probability equal to 1, and we say that state S_j is *recurrent*. Thus, in the genetical mating experiment described in Section 1, state S_3 (genotype aa) is recurrent, since $F_{33} = 1$ (see Table 7.1). Because a recurrent state S_j has the property that the time K_{jj} between its recurrences is always finite, it follows that once the process has entered state S_j, it must eventually return to that state.

On the other hand, a state, S_j, is called a *transient state* if, once the process has entered state S_j, there is a positive probability, $P\{K_{jj} = \infty \mid X_0 = j\}$, that the process will never return to state S_j. Put another way, a state S_j is transient if it is not recurrent, that is, if

(7.3) $$F_{jj} = f_{jj}^{(1)} + f_{jj}^{(2)} + f_{jj}^{(3)} + \cdots < 1,$$

so that

$$f_{jj}^{(\infty)} = P\{K_{jj} = \infty \mid X_0 = j\} = 1 - F_{jj} > 0.$$

Within a given Markov chain, some states can be recurrent and some states can be transient. Classification of the states of a Markov chain in terms of whether they are recurrent or transient is a helpful qualitative way of analyzing such processes. If we know that certain states are recurrent, we know that once we have observed the process to be in one of those states, we will, at a future time, again observe the process to be in that state. On the other hand, once the process has been observed to enter a transient state, we might never again observe that state; transient states are, in other words, transitory.

Remark. The terms recurrent and transient are the terms most commonly used to distinguish between states S_j for which $F_{jj}^{(\infty)} = 1$ and states S_i for which $F_{ii}^{(\infty)} < 1$. However, various synonyms have been used in the literature. For example, recurrent states are sometimes called *persistent*, and transient states have been called *unessential*.

Example 7.1 (Clothing Styles). In a longitudinal study of changes in clothing styles for males, three alternative styles of trousers are identified:

> State S_1:   tapered trousers,
>
> State S_2:   straight-legged trousers,
>
> State S_3:   flared trousers.

The choice of dress style for each of several men is recorded daily. The process in which a given man daily chooses the style of dress that he will wear is assumed to be a Markov chain with stationary transition probabilities. Suppose that as a result of the longitudinal study, the transition probabilities shown in Table 7.2 are assigned to the process. We now determine which of the states S_1, S_2, or S_3 are recurrent and which are transient.

Table 7.2: *Table of Transition Probabilities for the Process of Choice of Dress Style*

		Style on day $t+1$		
		S_1	S_2	S_3
State on day t	S_1	$\frac{1}{2}$	$\frac{1}{2}$	0
	S_2	$\frac{1}{2}$	$\frac{1}{2}$	0
	S_3	$\frac{1}{4}$	$\frac{1}{4}$	$\frac{1}{2}$

Using the recursion equations for k-step transition probabilities discussed in Section 4, it can be shown that

$$(7.4) \qquad \begin{aligned} p_{11}^{(k)} &= \tfrac{1}{2}, & p_{12}^{(k)} &= \tfrac{1}{2}, & p_{13}^{(k)} &= 0, \\ p_{21}^{(k)} &= \tfrac{1}{2}, & p_{22}^{(k)} &= \tfrac{1}{2}, & p_{23}^{(k)} &= 0, \end{aligned}$$

and that

$$(7.5) \qquad p_{31}^{(k)} = \tfrac{1}{4} + \tfrac{1}{2}p_{31}^{(k-1)}, \qquad p_{33}^{(k)} = \tfrac{1}{4} + \tfrac{1}{2}p_{32}^{(k-1)}, \qquad p_{33}^{(k)} = \tfrac{1}{2}p_{33}^{(k-1)},$$

for all positive integers $k = 2, 3, 4, \ldots$. Thus, looking first at state S_1, we find that

$$f_{11}^{(1)} = p_{11} = \tfrac{1}{2}, \qquad f_{11}^{(2)} = p_{11}^{(2)} - f_{11}^{(1)}p_{11} = \tfrac{1}{4},$$

$$f_{11}^{(3)} = p_{11}^{(3)} - f_{11}^{(2)}p_{11} - f_{11}^{(1)}p_{11}^{(2)} = \tfrac{1}{8},$$

and making repeated use of (7.1), it can be shown in general that

$$(7.6) \qquad f_{11}^{(k)} = P\{K_{11} = k \mid X_0 = 1\} = (\tfrac{1}{2})^k, \quad k = 1, 2, 3, \ldots .$$

Now recall that if the discrete random variable X has a geometric distribution with parameter $p = \tfrac{1}{2}$, then the probability mass function of X has the form

$$p_X(k) = \tfrac{1}{2}(\tfrac{1}{2})^{k-1} = (\tfrac{1}{2})^k, \quad k = 1, 2, 3, \ldots .$$

Thus, we see that the recurrence time K_{11} for state S_1 has a (conditional) geometric distribution with parameter $p = \tfrac{1}{2}$. Because for a geometrically distributed random variable X, we have shown that

$$\sum_{k=1}^{\infty} p_X(k) = 1.$$

it follows from (7.6) that

$$F_{11} = f_{11}^{(1)} + f_{11}^{(2)} + f_{11}^{(3)} + \cdots = (\tfrac{1}{2}) + (\tfrac{1}{2})^2 + (\tfrac{1}{2})^3 + \cdots = 1.$$

Therefore, state S_1 is a recurrent state.

Because $p_{11}^{(k)} = p_{22}^{(k)} = \frac{1}{2}$ for all $k = 1, 2, 3, \ldots$, an analysis exactly similar to the above shows that $f_{22}^{(k)} = (\frac{1}{2})^k$, for $k = 1, 2, 3, \ldots$ and thus state S_2 is a recurrent state.

We now turn to state S_3. From (7.5), we find that

$$p_{33} = \tfrac{1}{2}, \qquad p_{33}^{(2)} = \tfrac{1}{2}p_{33}^{(1)} = (\tfrac{1}{2})^2, \qquad p_{33}^{(3)} = \tfrac{1}{2}p_{33}^{(2)} = (\tfrac{1}{2})^3,$$

and so on. Thus, $p_{33}^{(k)} = (\frac{1}{2})^k$ for $k = 1, 2, 3, \ldots$, and

$$f_{33} = p_{33} = \tfrac{1}{2}, \qquad f_{33}^{(2)} = p_{33}^{(2)} - f_{33}^{(1)}p_{33} = 0,$$
$$f_{33}^{(3)} = p_{33}^{(3)} - f_{33}^{(2)}p_{33} - f_{33}^{(1)}p_{33}^{(2)} = 0,$$

and in general,

$$f_{33}^{(k)} = 0, \quad \text{for } k = 2, 3, 4, 5, \ldots.$$

It follows that

$$F_{33} = f_{33}^{(1)} + f_{33}^{(2)} + f_{33}^{(3)} + f_{33}^{(4)} + f_{33}^{(5)} + \cdots = \tfrac{1}{2} < 1,$$

so that state S_3 is a transient state.

In Example 7.1, it is intuitively reasonable that state S_3 should be transient. Looking at Table 7.2, we see that starting in state S_3 at time 0 the process can either return to state S_3 for the first time at time 1, or leave state S_3. If the process leaves state S_3, it can never return because the probabilities p_{13} and p_{23} of going to state S_3 from state S_1 or state S_2, respectively, are 0. Thus, state S_3 either recurs at time 1, or state S_3 never recurs. Hence, the (conditional) probability that state S_3 ever recurs is less than 1, and state S_3 is transient.

8. MEAN FIRST-PASSAGE TIMES

As we have noted earlier in this section, a first-passage time K_{ij} from a state S_i to a state S_j is a random variable. Although all the probabilistic properties of K_{ij} can be determined once we know the distribution of K_{ij}, the distribution can be difficult to obtain. In Chapter 5, we demonstrated that certain descriptive indices of the distribution of a random variable can be used to provide a gross picture of the variation of that random variable. In particular, we showed how for a positive random variable X (such a first-passage time), the approximate magnitudes of the probabilities of certain events could be determined from knowledge of the mean of X.

First-passage times K_{ij} differ from the random variables previously discussed in that in some cases it is possible for a first-passage time K_{ij} to be infinite. In order to avoid more complicated mathematical concepts, assume that

$$f_{ij}^{(\infty)} = P\{K_{ij} = \infty \mid X_0 = i\} = 0,$$

or, equivalently, that

$$F_{ij} = f_{ij}^{(1)} + f_{ij}^{(2)} + f_{ij}^{(3)} + \cdots = 1,$$

so that

$$f_{ij}^{(k)} = P\{K_{ij} = k \mid X_0 = i\}, \quad k = 1, 2, 3, \ldots$$

is a probability mass function. In this case, we have the following (conditional) probability model for the first-passage time K_{ij}:

k	1	2	3	4	$\cdots$
$P\{K_{ij} = k \mid X_0 = i\}$	$f_{ij}^{(1)}$	$f_{ij}^{(2)}$	$f_{ij}^{(3)}$	$f_{ij}^{(4)}$	$\cdots$

The mean first-passage time (or expected first-passage time) m_{ij} from state S_i to state S_j is defined to be

$$(8.1) \qquad m_{ij} = \sum_{k=1}^{\infty} k f_{ij}^{(k)}.$$

Example 6.1 (continued). We earlier obtained (see Table 6.6) the conditional probability mass function $f_{41}^{(k)}$, $k = 1, 2, 3, \ldots$, for the number of years K_{41} required until a voter who is not now registered (state S_4) registers for the first time in party 1 (state S_1). From Table 6.6 and (8.1), the mean (expected) number of years m_{41} required before a voter who is not now registered chooses to register for the first time in party 1 is equal to

$$m_{41} = 1 f_{41}^{(1)} + 2 f_{41}^{(2)} + 3 f_{41}^{(3)} + \cdots + 18 f_{41}^{(18)} + 19 f_{41}^{(19)} + 0 = 3.357.$$

Thus, on the average, a voter who is not now registered will take 3.357 years before he first chooses to register in party 1.

The recurrence time K_{jj} for a state S_j is also a first-passage time. Thus, if state S_j is a recurrent state, so that $f_{jj}^{(k)} = P\{K_{jj} = k \mid X_0 = j\}$ is a probability mass function, the mean recurrence time m_{jj} is obtained from

$$(8.2) \qquad m_{jj} = \sum_{k=1}^{\infty} k f_{jj}^{(k)}.$$

Example 7.1 (continued). In the context of the process of choice of trouser style which was described earlier, we found that state S_1 (choice of a tapered trouser) is a recurrent state, and we determined that the conditional probability mass function $f_{11}^{(k)} = P\{K_{11} = k \mid X_0 = 1\}$ of the recurrence time K_{11} is given by

$$f_{11}^{(k)} = (\tfrac{1}{2})^k, \quad k = 1, 2, 3, \ldots.$$

Thus, if we see a man with tapered trousers on a given day, we know that the mean (expected) number of days m_{11} until for the first time that man once again chooses to wear tapered trousers is given by the following expression:

$$m_{11} = \sum_{k=1}^{\infty} k(\tfrac{1}{2})^k,$$

which is the mean, μ_X, of a random variable X having a geometric distribution with parameter $p = \tfrac{1}{2}$. It follows that

$$m_{11} = \frac{1}{\frac{1}{2}} = 2.$$

Thus, on the average, a man with tapered trousers will choose tapered trousers again for the first time 2 days after he initially wears tapered trousers.

If we want to find m_{22}, the mean number of days until a man with straight-legged trousers for the first time again wears straight-legged trousers, identical arguments show that $m_{22} = 2$ days. Because state S_3 is a transient state, we cannot compute m_{33} for this state. Note, however, that because S_3 is a transient state, there is a positive conditional probability [namely, $f_{33}^{(\infty)} = 1 - F_{33} = \tfrac{1}{2}$] that a man with flared trousers may never again wear flared trousers. (This shows how hypothetical our model actually is!)

Alternative Computation for Mean First-Passage Times

The foregoing computation of the mean first-passage time m_{ij} requires the first-passage probabilities $f_{ij}^{(1)}$, $f_{ij}^{(2)}$, $f_{ij}^{(3)}$, and so on, before we can compute the desired mean.

There is an alternative method of computation which determines all of the mean first-passage times m_{ij} simultaneously, and which requires knowledge only of the transition probabilities p_{ij} of the process. This method involves the simultaneous solution of a system of linear equations. For the sake of convenience and specificity, we first show how to apply this method to the calculation of the expected first-passage times for a three-state Markov chain, and then show how to extend our results to N-state Markov chains.

The Three-State Case. Suppose that we have a Markov chain with three states S_1, S_2, and S_3. Consider how the process can move from state S_1 to, say, state S_3 for the first time. The process could, of course, go from state S_1 to state S_3 in one time period; this event has (conditional) probability p_{13}. Alternatively, the process could move from state S_1 to state S_1 in one time step, and then move from state S_1 to state S_3 for the first time at some later time (taking, on the average, m_{13} time periods to do so); this event has (conditional) probability p_{11}. Finally, the process could move from state S_1 to state S_2 in one time period, and then move from state S_2 to state S_3 for the first time at some later time (taking, on the average, m_{23} time periods to do so); this event has conditional probability p_{12}. The three possibilities that we have indicated are illustrated in Figure 8.1.

Let W be a random variable defined as follows. If the process goes from state S_1 at time 0 to state S_j at time 1, then W equals the total additional length of time required on the average for the process to then go from state S_j to state S_3 for the first time. Thus,

$$W = \begin{cases} 0, & \text{if the process enters state } S_3 \text{ at time 1,} \\ m_{23}, & \text{if the process enters state } S_2 \text{ at time 1,} \\ m_{13}, & \text{if the process remains in state } S_1 \text{ at time 1.} \end{cases}$$

The total length of time required, on the average, for the process to go from state S_1 to state S_3 is then equal to 1 plus the conditional expected value of W, given that the process begins in state S_1. That is,

$$(8.3) \qquad m_{13} = 1 + E[W \mid X_0 = 1].$$

From Figure 8.1 and the definition of W,

$$E[W \mid X_0 = 1] = (0)P\{W = 0 \mid X_0 = 1\} + (m_{23})P\{W = m_{23} \mid X_0 = 1\}$$
$$+ (m_{13})P\{W = m_{13} \mid X_0 = 1\}$$
$$= (0)(p_{13}) + (m_{23})(p_{12}) + (m_{13})(p_{11})$$
$$= p_{11}m_{13} + p_{12}m_{23}.$$

Figure 8.1: *Figure illustrating ways that the process can move from state S_1 to state S_3 for the first time, the conditional probabilities of each of these ways, and the number of additional time steps after the first step required "on the average" for each way.*

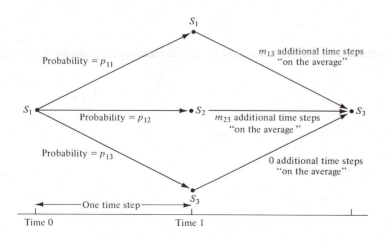

Hence, from (8.3),

$$m_{13} = 1 + p_{11}m_{13} + p_{12}m_{23}.$$

We have thus (somewhat heuristically) obtained an equation that relates the mean first-passage times m_{13} and m_{23} to the transition probabilities p_{11} and p_{12}. Proceeding in a similar fashion,

$$m_{23} = 1 + p_{21}m_{13} + p_{22}m_{23},$$
$$m_{33} = 1 + p_{31}m_{13} + p_{32}m_{23}.$$

Summarizing these three equations, we have

$$
\begin{align}
m_{13} &= 1 + p_{11}m_{13} + p_{12}m_{23}, \\
(8.4) \qquad m_{23} &= 1 + p_{21}m_{13} + p_{22}m_{23}, \\
m_{33} &= 1 + p_{31}m_{13} + p_{32}m_{23}.
\end{align}
$$

If we are given the values of the transition probabilities p_{11}, p_{12}, p_{21}, p_{22}, p_{31}, and p_{32}, we can solve (8.4) for the unknowns m_{13}, m_{23}, and m_{33}.

By a similar argument

$$
\begin{align}
m_{11} &= 1 + p_{12}m_{21} + p_{13}m_{31}, \\
(8.5) \qquad m_{21} &= 1 + p_{22}m_{21} + p_{23}m_{31}, \\
m_{31} &= 1 + p_{32}m_{21} + p_{33}m_{31},
\end{align}
$$

and

$$
\begin{align}
m_{12} &= 1 + p_{11}m_{12} + p_{13}m_{32}, \\
(8.6) \qquad m_{22} &= 1 + p_{21}m_{12} + p_{23}m_{32}, \\
m_{32} &= 1 + p_{31}m_{12} + p_{33}m_{32}.
\end{align}
$$

Remark. It should be noted that if $F_{ij} < 1$ for any pair of states S_i and S_j, then not all of (8.4), (8.5), and (8.6) will have finite solutions. Thus, to use these equations for finding the values of the expected first-passage times m_{ij}, it is helpful to know that $F_{ij} = 1$ for all states S_i and S_j. A condition under which $F_{ij} = 1$ for all i, j is given in Section 10.

As an example of the use of (8.4) through (8.6), consider the genetical mating process discussed in Section 6. The states of this process are

$$S_1: \quad \text{genotype } AA,$$
$$S_2: \quad \text{genotype } Aa,$$
$$S_3: \quad \text{genotype } aa,$$

and the transition probabilities are summarized in Table 6.3. From Table 6.3 and (8.6),

$$m_{12} = 1 + (\tfrac{1}{2})m_{12} + (0)m_{22},$$
$$m_{22} = 1 + (\tfrac{1}{4})m_{12} + (\tfrac{1}{4})m_{22},$$
$$m_{32} = 1 + (0)m_{12} + (\tfrac{1}{2})m_{22}.$$

The solution of this set of equations is

$$m_{12} = m_{22} = m_{32} = 2.$$

Similarly, from Table 6.3 and (8.5),

$$m_{11} = 1 + (\tfrac{1}{2})m_{21} + (0)m_{31},$$
$$m_{21} = 1 + (\tfrac{1}{2})m_{21} + (\tfrac{1}{4})m_{31},$$
$$m_{31} = 1 + (\tfrac{1}{2})m_{21} + (\tfrac{1}{2})m_{31},$$

from which

$$m_{11} = 4, \qquad m_{21} = 6, \qquad m_{31} = 8.$$

Finally, from Table 6.3 and (8.4),

$$m_{13} = 1 + (\tfrac{1}{2})m_{13} + (\tfrac{1}{2})m_{23},$$
$$m_{23} = 1 + (\tfrac{1}{4})m_{13} + (\tfrac{1}{2})m_{23},$$
$$m_{33} = 1 + (0)m_{13} + (\tfrac{1}{2})m_{23},$$

from which

$$m_{13} = 8, \qquad m_{23} = 6, \qquad m_{33} = 4.$$

The complete collection of mean first-passage times m_{ij} for the genetical mating process is summarized in Table 8.1.

Table 8.1: *Mean First-Passage Times m_{ij}, from State S_i to State S_j for the Genetical Mating Process Whose Transition Probabilities Are Given in Table 6.3*

Final state S_j

	S_1	S_2	S_3
S_1	4	2	8
S_2	6	2	6
S_3	8	2	4

Initial state S_i

The N-State Case. The method that we have just described for finding the expected first-passage times m_{ij} for a three-state Markov chain can be generalized to more general Markov chain contexts. Suppose that a Markov chain has a finite number N of states $S_1, S_2, \ldots, S_N$. Then, if we are given the values of the transition probabilities p_{ij} of this Markov chain (and if we know that $F_{ij} = 1$ for all states S_i and S_j, including where $i = j$), then, for example, the mean first-passage times $m_{11}, m_{21}, \ldots, m_{N1}$ for this process are given as the solution of N equations:

$$
\begin{aligned}
m_{11} &= 1 + p_{12}m_{21} + p_{13}m_{31} + \cdots + p_{1N}m_{N1}, \\
m_{21} &= 1 + p_{22}m_{21} + p_{23}m_{31} + \cdots + p_{2N}m_{N1}, \\
(8.7) \qquad m_{31} &= 1 + p_{32}m_{21} + p_{33}m_{31} + \cdots + p_{3N}m_{N1}, \\
&\;\;\vdots \qquad \vdots \qquad \vdots \qquad\qquad \vdots \qquad\qquad\quad \vdots \\
m_{N1} &= 1 + p_{N2}m_{21} + p_{N3}m_{31} + \cdots + p_{NN}m_{N1},
\end{aligned}
$$

for the unknowns $m_{11}, m_{21}, \ldots, m_{N1}$. Note that because the unknown m_{11} appears only in the first equation in (8.7), we can solve this set of equations by first solving the last $N-1$ equations for $m_{21}, m_{31}, \ldots, m_{N1}$, and then substituting our answers into the first equation in order to find m_{11}.

If we add and subtract $p_{i1}m_{11}$ from the right-hand side of the (i)th equation in (8.7), for $i = 1, 2, \ldots, N$, then (8.7) becomes

$$
\begin{aligned}
m_{11} &= (1 - p_{11}m_{11}) + p_{11}m_{11} + p_{12}m_{21} + \cdots + p_{1N}m_{N1}, \\
m_{21} &= (1 - p_{21}m_{11}) + p_{21}m_{11} + p_{22}m_{21} + \cdots + p_{2N}m_{N1}, \\
(8.8) \quad m_{31} &= (1 - p_{31}m_{11}) + p_{31}m_{11} + p_{32}m_{21} + \cdots + p_{3N}m_{N1}, \\
&\;\;\vdots \qquad \vdots \qquad \vdots \qquad\qquad \vdots \qquad\qquad\quad \vdots \\
m_{N1} &= (1 - p_{N1}m_{11}) + p_{N1}m_{11} + p_{N2}m_{21} + \cdots + p_{NN}m_{N1}.
\end{aligned}
$$

The collection of equations (8.8) exhibits a pattern that permits generalization to a collection of equations solvable for the expected first-passage times m_{1j}, $m_{2j}, m_{3j}, \ldots, m_{Nj}, j = 1, 2, \ldots, N$. This general set of equations is

$$
\begin{aligned}
m_{1j} &= (1 - p_{1j}m_{jj}) + p_{11}m_{1j} + p_{12}m_{2j} + \cdots + p_{1N}m_{Nj}, \\
m_{2j} &= (1 - p_{2j}m_{jj}) + p_{21}m_{1j} + p_{22}m_{2j} + \cdots + p_{2N}m_{Nj}, \\
(8.9) \quad m_{3j} &= (1 - p_{3j}m_{jj}) + p_{31}m_{1j} + p_{32}m_{2j} + \cdots + p_{3N}m_{Nj}, \\
&\;\;\vdots \qquad \vdots \qquad \vdots \qquad\qquad \vdots \qquad\qquad\quad \vdots \\
m_{Nj} &= (1 - p_{Nj}P_{jj}) + p_{N1}m_{1j} + p_{N2}m_{2j} + \cdots + p_{NN}m_{Nj}.
\end{aligned}
$$

For any state S_j, we can solve this set of N equations for the N unknowns $m_{1j}, m_{2j}, \ldots, m_{Nj}$.

9. PROBABILISTIC EQUILIBRIUM OF A MARKOV CHAIN

Some stochastic processes achieve a state of equilibrium in the sense that the conditional and unconditional probabilities of observing the various states of the process eventually become constant, ceasing to depend in value either on the condition (state) of the process at its inception, or on the length of time that the process has been in operation.

Consider, for example, the simple two-state Markov chain used to model the choice-of-brand process described in Section 5 with transition probabilities given in Table 9.1. Using the methods of computation described in Section 4, we can compute the 2-step, 3-step, 4-step, and so on, transition probabilities of this Markov chain. These transition probabilities are summarized in Table 9.2. (All entries in Table 9.2 are accurate to three decimal places.)

Looking across each row of Table 9.2, we see that for values of $k \geq 5$, the k-step transition probabilities $p_{ij}^{(k)}$ become independent of the time k. A comparison of rows 1 and 2 and of 3 and 4 shows that for $k \geq 5$, the k-step transition probabilities $p_{ij}^{(k)}$ also become independent of i.

Table 9.1: *Table of Transition Probabilities for a 2-State Markov Chain*

		State at time $t+1$	
		S_1	S_2
State at time t	S_1	0.500	0.500
	S_2	0.750	0.250

Table 9.2: *The k-Step Transition Probabilities $p_{ij}^{(k)}$ for the Markov Chain Whose Transition Probabilities Are Given in Table 9.1*

k-step transition probability to:	1	2	3	4	5	6	7	
State S_1 $\begin{cases} p_{11}^{(k)} \\ p_{21}^{(k)} \end{cases}$	0.500	0.625	0.594	0.602	0.600	0.600	0.600	$\cdots$
	0.750	0.562	0.609	0.598	0.600	0.600	0.600	$\cdots$
State S_2 $\begin{cases} p_{12}^{(k)} \\ p_{22}^{(k)} \end{cases}$	0.500	0.375	0.406	0.398	0.400	0.400	0.400	$\cdots$
	0.250	0.438	0.391	0.402	0.400	0.400	0.400	$\cdots$

From this analysis, we can conclude that the conditional probabilities $p_{ij}^{(k)}$ of observing state S_j at time k, given that the process starts in state S_i, eventually cease to depend either on the initial state S_i of the process, or on the length k of time that the process has been operating. Hence, after a sufficiently long time of operation, the probabilities of this Markov chain process exhibit equilibrium behavior, and we say that the Markov chain is in *probabilistic* (or *stochastic*) *equilibrium*.

Not all Markov chains achieve probabilistic equilibrium. However, if a Markov chain does achieve probabilistic equilibrium, then regardless of the state S_i in which the process begins, the conditional probability of observing a given state S_j at time k has (approximately) the same value π_j for all sufficiently distant times k. That is, for all initial states S_i and all sufficiently large values of k,

(9.1) $$p_{ij}^{(k)} \simeq \pi_j.$$

This common value π_j of the k-step transition probabilities $p_{ij}^{(k)}$ is called the *long-run* (or *steady-state*, or *equilibrium*) *probability* of state S_j. If we know that a Markov chain with states $S_1, S_2, \ldots, S_N$ can achieve probabilistic equilibrium, then we know that there is a probability distribution assigning probability π_j to state S_j, for $j = 1, 2, \ldots, N$, such that for all sufficiently large values of k, a table of the k-step transition probabilities of the Markov chain has the form shown in Table 9.3.

The probability distribution that assigns probability π_j to state S_j is called the *long-run* (or *steady-state*, or *equilibrium*) *distribution* of the Markov chain. Knowledge of this long-run distribution eliminates the need to calculate tables of k-step transition probabilities for large values of k, since all such tables have the form of Table 9.3.

Table 9.3: *The k-Step Transition Probabilities of an N-State Markov Chain, Where k Is Large Enough So That the Markov Chain Is in Probabilistic Equilibrium*

		State at time k			
		S_1	S_2	$\cdots$	S_N
	S_1	π_1	π_2	$\cdots$	π_N
State at time 0	S_2	π_1	π_2	$\cdots$	π_N
	$\vdots$	$\vdots$	$\vdots$		$\vdots$
	S_N	π_1	π_2	$\cdots$	π_N

Provided a long-run distribution exists for a given Markov chain, knowledge of this long-run distribution also enables us to calculate the values of the absolute probabilities $P\{X_k = j\}$ for all sufficiently large values of k. Recall from Section 5 that

$$(9.2) \qquad P\{X_k = j\} = a_1 p_{1j}^{(k)} + a_2 p_{2j}^{(k)} + \cdots + a_N p_{Nj}^{(k)},$$

where $a_1, a_2, \ldots, a_N$ are the initial probabilities (the probabilities at time 0) of the states $S_1, S_2, \ldots, S_N$, respectively. However, from (9.1) for k sufficiently large,

$$(9.3) \qquad p_{1j}^{(k)} \simeq p_{2j}^{(k)} \simeq \cdots \simeq p_{Nj}^{(k)} \simeq \pi_j.$$

Substituting (9.3) into (9.2), and remembering that $a_1 + a_2 + \cdots + a_N = 1$, we obtain

$$(9.4) \quad P\{X_k = j\} \simeq a_1 \pi_j + a_2 \pi_j + \cdots + a_N \pi_j = \pi_j(a_1 + a_2 + \cdots + a_N) = \pi_j.$$

Thus, for all sufficiently large values of k, the unconditional probability that the process is in state S_j at time k is (approximately) equal to the long-run probability π_j of state S_j. Note that this result is true regardless of the initial probabilities $a_1, a_2, \ldots, a_N$ of the states $S_1, S_2, \ldots, S_N$ of the Markov chain. Equation (9.4) is thus another illustration of the characteristic property of probabilistic equilibria mentioned earlier—that the probabilities of the states $S_1, S_2, \ldots, S_N$ in the long run cease to depend on the condition of the Markov chain at its start.

In the remainder of this chapter, we describe a large class of Markov chains that can achieve probabilistic equilibrium and show how to find the long-run distributions for such processes. In the course of our presentation, we also introduce certain new concepts concerning Markov chains that are of importance to our study of probabilistic equilibrium and also are of independent interest. Our study will not identify all Markov chains that achieve probabilistic equilibrium; however, the new concepts and techniques that we present can be used to accomplish that goal.

10. CLASSES OF STATES

In studying probabilistic equilibria and long-run probability distributions for Markov chains, it is useful to classify the states of a Markov chain according to various criteria. We have already discussed one such classification for the states of a Markov chain in Section 7: the classification of states according to whether such states are *recurrent* or *transient*. Before considering other, equally useful, classifications of states, let us make clear what we mean by a class of states and briefly indicate one important property that certain classes of states possess: the property of being closed.

In general, a class $\mathscr{C}$ of states of a Markov chain is any well-defined collection of such states. A class $\mathscr{C}$ of states can be defined by listing those states belonging to the class, or by indicating some property only met by all states in the class. Thus, in the Markov chain described in Example 7.1 (the process of choice of trouser style), one class of states is the class $\mathscr{C}$ containing the states S_1 (tapered trousers) and S_2 (straight-legged trousers). As we verified in Section 7, this class can be described as the class of recurrent states.

Suppose that we are interested in a particular class $\mathscr{C}$ of states for a given Markov chain. We say that such a class is *closed* or *absorbing* if once the process is in one of the states belonging to class $\mathscr{C}$ it can never leave the states in class $\mathscr{C}$ at any later time. For example, if a Markov chain has four states S_1, S_2, S_3, and S_4, and if the class $\mathscr{C}$ consists of states S_1 and S_2, then $\mathscr{C}$ is closed if once the process enters state S_1, it can then never go either to state S_3 or to state S_4; and if once the process enters state S_2, it then cannot go to states S_3 or S_4.

It can be shown that a class $\mathscr{C}$ of states is closed if for each state S_i belonging to class $\mathscr{C}$, and for every state S_j not belonging to class $\mathscr{C}$, the 1-step transition probability p_{ij} is equal to 0. Thus, if the Markov chain has four states S_1, S_2, S_3, and S_4, and if the class $\mathscr{C}$ consists of states S_1 and S_2, then $\mathscr{C}$ is closed if

$$p_{13} = p_{14} = p_{23} = p_{24} = 0,$$

and otherwise $\mathscr{C}$ is not closed. If the four-state Markov chain has transition probabilities of the form shown in Table 10.1(a), then the class $\mathscr{C}$ consisting of states S_1 and S_2 is closed. On the other hand, if the Markov chain has transition probabilities of the form shown in Table 10.1(b), then class $\mathscr{C}$ is not closed, since $p_{13} \neq 0$.

Table 10.1a: *Table of Transition Probabilities for a Markov Chain in Which Class $\mathscr{C}$ Is Closed*

(Here an * in any position in the table represents a transition probability that may or may not be equal to 0; class $\mathscr{C}$ consists of states S_1 and S_2)

State at time $t + 1$

	S_1	S_2	S_3	S_4
S_1	*	*	0	0
S_2	*	*	0	0
S_3	*	*	*	*
S_4	*	*	*	*

State at time t

Table 10.1b: *Table of Transition Probabilities for
a Markov Chain in Which Class $\mathscr{C}$ Is Not Closed*

(Class $\mathscr{C}$ consists of states S_1 and S_2)

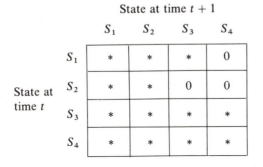

		State at time $t+1$			
		S_1	S_2	S_3	S_4
	S_1	*	*	*	0
State at	S_2	*	*	0	0
time t	S_3	*	*	*	*
	S_4	*	*	*	*

Classes of Communicating States

Two states, S_i and S_j, of a Markov chain process are said to *communicate* if the process can go from state S_i to state S_j and also from state S_j to state
S_i in a finite number of time periods. Formally, a state S_j *can be reached* from
another state S_i if, given that the process starts at state S_i, there is a nonzero
conditional probability that the process will be in state S_j after a finite
number of time periods. Two states, S_i and S_j, communicate if each can be
reached from the other. To demonstrate that state S_j and state S_i communicate, all we have to show is that there are finite integers n and m such that
$p_{ij}^{(n)} \neq 0$ and $p_{ji}^{(m)} \neq 0$. To show that two states, S_i and S_j, do not communicate,
we can try to show either that $p_{ij}^{(k)} = 0$ for $k = 1, 2, \ldots$, or that $p_{ji}^{(k)} = 0$, for
$k = 1, 2, \ldots$. However, in some cases it is easier to show that one of the two
states, say, S_i, belongs to a closed class to which the other state, S_j, does not
belong, since then, by the definition of a closed class, we know that S_j cannot
be reached from S_i.

For example, consider the four-state Markov chain that has the transition probabilities p_{ij} shown in Table 10.2. Let us first decide whether states S_1
and S_2 communicate. From Table 10.2, we see that $p_{12} = \frac{1}{2} \neq 0$, and thus
state S_2 can be reached from state S_1 in $n = 1$ time steps. Also from Table
10.2, we note that $p_{21} = 0$. Hence, the process cannot go from state S_2 to state
S_1 in one time step. However, we can compute the 2-step transition probabilities $p_{ij}^{(2)}$ (summarized in Table 10.3).

From Table 10.3, we see that $p_{21}^{(2)} = \frac{1}{4} \neq 0$, and thus state S_1 can be reached
from state S_2 in $m = 2$ time steps. From this analysis, we conclude that state
S_2 can be reached from state S_1, and that state S_1 can be reached from state
S_2. Hence, states S_1 and S_2 communicate. A similar analysis, using Tables
10.2 and 10.3, shows that states S_2 and S_3 communicate.

We could also use Tables 10.2 and 10.3 to prove that states S_1 and S_3 communicate. However, intuitively, this result should follow from the fact that state S_1 communicates with state S_2, and the fact that state S_2 communicates with state S_3. For, if states S_1 and S_2 communicate, then state S_2 can be reached from state S_1. In turn, the fact that states S_2 and S_3 communicate means that state S_3 can be reached from state S_2. Thus, starting from state S_1, the process can first reach state S_2; and from state S_2, the process can then reach state S_3. Hence, we seem to have shown that state S_3 can be reached from state S_1. A similar argument, using the fact that state S_1 can be reached from state S_2, and that state S_2 can be reached from state S_3, seems to indicate that state S_1 can be reached from state S_3. Hence, states S_1 and S_3 communicate.

Table 10.2: *Transition Probabilities for a Four-State Markov Chain*

State at time $t + 1$

		S_1	S_2	S_3	S_4
	S_1	$\frac{1}{2}$	$\frac{1}{2}$	0	0
State at time t	S_2	0	$\frac{1}{2}$	$\frac{1}{2}$	0
	S_3	$\frac{1}{2}$	0	$\frac{1}{2}$	0
	S_4	$\frac{1}{4}$	$\frac{1}{4}$	$\frac{1}{4}$	$\frac{1}{4}$

Table 10.3: *Two-Step Transition Probabilities for the Four-State Markov Chain Whose Transition Probabilities Appear in Table 10.2*

State at time $t + 2$

		S_1	S_2	S_3	S_4
	S_1	$\frac{1}{4}$	$\frac{1}{2}$	$\frac{1}{4}$	0
State at time t	S_2	$\frac{1}{4}$	$\frac{1}{4}$	$\frac{1}{2}$	0
	S_3	$\frac{1}{2}$	$\frac{1}{4}$	$\frac{1}{4}$	0
	S_4	$\frac{5}{16}$	$\frac{5}{16}$	$\frac{5}{16}$	$\frac{1}{16}$

In general (for any Markov chain) it is true that if state S_i communicates with state S_j, and if state S_j communicates with state S_h, then states S_i and S_h must communicate with each other. To verify this assertion, recall that for any two integers n_1 and n_2 such that $n = n_1 + n_2$,

$$(10.1) \qquad p_{ih}^{(n)} = p_{i1}^{(n_1)}p_{1h}^{(n_2)} + p_{i2}^{(n_1)}p_{2h}^{(n_2)} + \cdots + p_{ij}^{(n_1)}p_{jh}^{(n_2)} + \cdots + p_{iN}^{(n_1)}p_{Nh}^{(n_2)}.$$

Thus, if state S_j can be reached from state S_i in n_1 time steps, and state S_h can be reached from state S_j in n_2 time steps, then $p_{ij}^{(n_1)}p_{jh}^{(n_2)} \neq 0$, and thus $p_{ih}^{(n)} \neq 0$, so that state S_h can be reached from state S_i in $n = n_1 + n_2$ time steps. Using the equation

$$(10.2) \quad p_{hi}^{(m)} = p_{h1}^{(m_1)}p_{1i}^{(m_2)} + p_{h2}^{(m_1)}p_{2i}^{(m_2)} + \cdots + p_{hj}^{(m_1)}p_{ji}^{(m_2)} + \cdots + p_{hN}^{(m_1)}p_{Ni}^{(m_2)},$$

we can also show that if state S_j can be reached from state S_h in m_1 time steps, and if state S_i can be reached from state S_j in m_2 time steps, then $p_{hj}^{(m_1)}p_{ji}^{(m_2)} \neq 0$, so that $p_{hi}^{(m)} \neq 0$, and thus state S_i can be reached from state S_h in m time steps. Hence, if states S_i and S_j communicate, then states S_i and S_h each can be reached from the other, and thus must communicate.

Turning back to our four-state Markov chain example, we see that we have yet to determine whether or not any of states S_1, S_2, or S_3 communicate with state S_4. However, note from Table 10.2 that

$$p_{14} = p_{24} = p_{34} = 0.$$

It thus follows that the class $\mathscr{C}$ consisting of states S_1, S_2, and S_3 is a closed class to which state S_4 does not belong. Hence, state S_4 cannot be reached from any of the states S_1, S_2, or S_3, and therefore none of the pairs of states S_1 and S_4, S_2 and S_4, or S_3 and S_4 communicate with one another. It is worth noting, however, that although state S_4 cannot be reached from states S_1, S_2, or S_3, each of the states S_1, S_2, and S_3 can be reached in one time step from state S_4 (i.e., $p_{41} \neq 0$, $p_{42} \neq 0$, $p_{43} \neq 0$). This fact demonstrates that even though two states, S_i and S_j, do not communicate with each other, it may be possible for one of these states to be reached from the other state.

A class $\mathscr{C}$ of states which has the property that every pair of states in the class communicates with one another, and no state in the class communicates with any state outside of the class, is called a *communicating class*. A communicating class which is also closed is called a *closed, communicating class*. In a Markov chain that has a finite number N of states, communicating classes (and closed, communicating classes) have the following two important properties:

(1) The states in a communicating class are either all recurrent states, or all transient states. The states in a closed, communicating class are all recurrent.

(2) For every pair of states, S_i and S_j, in a closed, communicating class,

the conditional probability F_{ij} that the first-passage time K_{ij} between state S_i and state S_j is finite is equal to 1. That is, for every pair of states S_i and S_j in a closed, communicating class, $F_{ij} = 1$, and

$$f_{ij}^{(\infty)} = P\{K_{ij} = \infty \mid X_0 = i\} = 1 - F_{ij} = 0.$$

In most cases, it is quite straightforward to find the communicating classes of a Markov chain and to verify whether or not each such communicating class is closed. Thus, for example, in the four-state Markov chain whose transition probabilities are given in Table 10.2, we quite easily verified that the class $\mathscr{C}$ which consists of states S_1, S_2, and S_3 is a closed, communicating class. Hence, from the results above, we know that these three states $(S_1, S_2, \text{ and } S_3)$ are recurrent and that $F_{12} = F_{13} = F_{21} = F_{23} = F_{31} = F_{32} = 1$ (i.e., the first-passage times between any pair of the states S_1, S_2, and S_3 are finite with probability equal to 1). If we attempted to obtain these same results using the methods outlined in Section 6, we would have to first compute all of the k-step first-passage probabilities $f_{ij}^{(k)}$ for $k = 1, 2, 3, \ldots$. Then we would have to form the sums

$$F_{ij} = f_{ij}^{(1)} + f_{ij}^{(2)} + f_{ij}^{(3)} + \cdots, \quad i = 1, 2, 3, j = 1, 2, 3,$$

and check for each one of these nine sums whether or not the sum is equal to 1. However, using the properties of a closed, communicating class mentioned above, we immediately obtain the result that all nine of these sums are equal to 1. Thus, this example clearly illustrates the great utility of the concept of a closed, communicating class in the analysis of the first-passage times of a Markov chain.

There is one special case of a closed, communicating class that should be mentioned. Suppose that all the N states of an N-state Markov chain communicate with one another. In this case, all the states of the Markov chain form one communicating class. Since there are no states of the Markov chain outside this class, it follows that once the process enters this class (as it must, since the process must enter one of the states at time 0), it can never leave this class. Thus, in this case, all the states of the Markov chain form one big, closed, communicating class. It therefore follows, from the properties of a closed, communicating class quoted above, that every state of this Markov chain is recurrent, and that every pair of states, S_i and S_j, have a first-passage time K_{ij} that is finite with (conditional) probability $F_{ij} = 1$.

A Markov chain which has the property that every pair of its states communicate is called an *irreducible Markov chain*. Note that the two-state Markov chain whose transition probabilities are summarized in Table 9.1 is an irreducible Markov chain (states S_1 and S_2 communicate since $p_{12} = 0.500 \neq 0$, and since $p_{21} = 0.750 \neq 0$). It was this Markov chain that we used earlier to exhibit the possibility that a Markov chain can attain probabilistic equilibrium.

11. APERIODIC, IRREDUCIBLE MARKOV CHAINS

In the previous section, we noted that all states of an irreducible Markov chain are recurrent, and that every state S_j can be reached from every other state S_i in a finite number of time steps with (conditional) probability F_{ij} equal to 1. From such considerations, it follows that we can use the "alternative method of computation" described in Section 8 to calculate the expected first-passage times m_{ij} between all states, S_i and S_j, of the process.

We also noted that the two-state Markov chain whose transition probabilities are summarized in Table 9.1 is an irreducible Markov chain, and that this Markov chain achieves a probabilistic equilibrium. Although it is tempting to generalize from this example and state that all irreducible Markov chains achieve a probabilistic equilibrium, this assertion is, unfortunately, false. Consider, for example, the two-state Markov chain whose transition probabilities are summarized in Table 11.1. Since $p_{12} = p_{21} = 1 \neq 0$, it follows that states S_1 and S_2 of this Markov chain communicate. Since the Markov chain only has the two states S_1 and S_2, it is by definition irreducible. However, inspection of Table 11.1 reveals that if the process starts in state S_1, then it must go to state S_2 at time 1, return to state S_1 at time 2, go to state S_2 at time 3, and in this manner alternate between state S_1 and state S_2 at all future time points. Similarly, if the process begins in state S_2, then the process must visit state S_2 at all even times, and must visit state S_1 at all odd times. We conclude from this analysis that a table of k-step transition probabilities of this Markov chain has the form

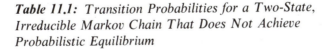
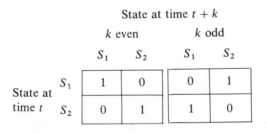

		k even		k odd	
		S_1	S_2	S_1	S_2
State at time t	S_1	1	0	0	1
	S_2	0	1	1	0

with heading "State at time $t + k$".

Table 11.1: *Transition Probabilities for a Two-State, Irreducible Markov Chain That Does Not Achieve Probabilistic Equilibrium*

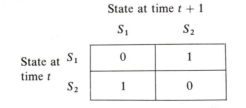

		State at time $t + 1$	
		S_1	S_2
State at time t	S_1	0	1
	S_2	1	0

Table 11.2: *The k-Step Transition Probabilities* $p_{ij}^{(k)}$
*for the Markov Chain Whose Transition Probabilities
Are Given in Table 11.1*

k-step transition probability to:		k									
		1	2	3	4	5	6	7	8	9	$\cdots$
State S_1	$p_{11}^{(k)}$	0	1	0	1	0	1	0	1	0	$\cdots$
	$p_{21}^{(k)}$	1	0	1	0	1	0	1	0	1	$\cdots$
State S_2	$p_{12}^{(k)}$	1	0	1	0	1	0	1	0	1	$\cdots$
	$p_{22}^{(k)}$	0	1	0	1	0	1	0	1	0	$\cdots$

We summarize these results in a table (Table 11.2) of k-step transition probabilities similar to Table 9.2.

From Table 11.2, we see that no matter how large k becomes, the k-step transition probability $p_{ij}^{(k)}$ always depends on the time k and the initial state S_i. Thus, the irreducible Markov chain whose 1-step transition probabilities appear in Table 11.1 does not achieve probabilistic equilibrium. This example shows that not all irreducible Markov chains achieve a probabilistic equilibrium.

In examining Table 11.2, we see that the k-step transition probabilities exhibit a "back-and-forth" regularity. The k-step transition probabilities $p_{ij}^{(k)}$ go from 0 (or 1) to 1 (or 0), and then back to 0 (or 1), periodically. If the process begins in state S_1 at time 0, it has nonzero (conditional) probability of returning to state S_1 at time k only when k is divisible by 2. Thus, the process can only periodically return to state S_1. Similarly, if the process starts at state S_2, the probability that it returns to state S_2 at time k is nonzero only if k is divisible by 2, and we see that the process can only periodically return to state S_2. These considerations motivate definition of the period (and periodicity) of a state, S_j of a Markov chain.

Periodic States and Periodic Chains

A state S_j of a Markov chain is periodic if there exists an integer $h > 1$ such that $p_{jj}^{(k)} = 0$ whenever k is not an integer multiple of h. [Note, however, that we are not asserting that $p_{jj}^{(k)}$ is always nonzero when k is an integer multiple of h.] The period T_j of a periodic state S_j is the smallest integer h $(h > 1)$ such that $p_{jj}^{(k)} = 0$ whenever k is not an integer multiple of h. A periodic state can have one and only one period.

For the two-state Markov chain whose transition probabilities are summarized in Table 11.1, we have already shown that $p_{11}^{(k)} = 0$ whenever k is not an integer multiple of the positive integer $h = 2$. Since $h = 2$ is also the smallest integer that is greater than 1, the period T_1 of state S_1 must be equal to 2. Similarly, we can show that state S_2 has period 2.

As another example of a periodic state, consider the four-state Markov chain whose transition probabilities p_{ij} are summarized in Table 11.3. Using the methods of computation described in Section 4, it can be shown for this Markov chain that

$$p_{11}^{(k)} = \begin{cases} \frac{1}{2}, & \text{if } k = 2, 4, 6, 8, 10, \ldots, \\ 0, & \text{if } k = 1, 3, 5, 7, 9, \ldots. \end{cases}$$

Since $p_{11}^{(k)} = 0$ whenever k is not an integer multiple of $h = 2$, it follows that state S_1 is periodic, and that the period T_1 of state S_1 is equal to 2.

It can be shown that if any single state in a communicating class $\mathscr{C}$ is periodic and has period T, then all states in that class are periodic and have period T. Thus, it is possible to talk about the periodicity (and the period) of a communicating class. A communicating class is periodic of period T if any one state in the class is periodic and has period T. From this fact it follows that if a Markov chain is irreducible, and if one state is periodic with period T, then all states of that irreducible Markov chain are periodic with period T. If any state of an irreducible Markov chain is periodic with period T, then we say that the Markov chain itself is periodic with period T.

For the four-state Markov chain whose transition probabilities are summarized in Table 11.3, we already know that state S_1 is periodic with period $T = 2$. Since for this Markov chain, $p_{12} = p_{21} = \frac{1}{2} \neq 0$, $p_{23} = p_{32} = \frac{1}{2} \neq 0$, and $p_{34} = p_{43} = \frac{1}{2} \neq 0$, it follows that states S_1 and S_2 communicate, that states S_2 and S_3 communicate, and that states S_3 and S_4 communicate. Thus, every pair of states S_i and S_j communicate, and the Markov chain is irreducible. Since state S_1 has period 2, it follows that all the states (S_1, S_2, S_2, and S_4) of the Markov chain have period $T = 2$, and thus the Markov chain itself is, by definition, periodic with period $T = 2$.

Table 11.3: *Table of Transition Probabilities for a Four-State Markov Chain*

		State at time $t + 1$			
		S_1	S_2	S_3	S_4
State at time t	S_1	0	$\frac{1}{2}$	0	$\frac{1}{2}$
	S_2	$\frac{1}{2}$	0	$\frac{1}{2}$	0
	S_3	0	$\frac{1}{2}$	0	$\frac{1}{2}$
	S_4	$\frac{1}{2}$	0	$\frac{1}{2}$	0

Irreducible Markov chains that are periodic do not achieve probabilistic equilibrium. We have already exhibited an example of this fact when we discussed the two-state, periodic, irreducible Markov chain whose transition probabilities are given in Table 11.1. However, even though periodic, irreducible Markov chains do not achieve a probabilistic equilibrium in the sense defined in Section 9, such Markov chains do exhibit certain long-run probabilistic regularities. Because the discussion of such regularities can become rather complicated, we leave the description and investigation of such properties to more advanced textbooks [see, e.g., Feller (1968)].

If a state S_j of a Markov chain is not periodic, we say that such a state is an *aperiodic state*. A state S_j is not periodic if for every integer h greater than 1, there is another integer k such that k is not an integer multiple of h and such that $p_{jj}^{(k)} \neq 0$. That is, state S_j is not periodic if for every integer h that could be the period for state S_j, it is possible for the process to return to state S_j after a length k of time has passed which is not an integer multiple of h. Although this definition of an aperiodic state may seem awkward and difficult to verify, in most cases it is straightforward to determine whether or not a state S_j is aperiodic. One quick test for the aperiodicity of a state S_j is from the 1-step transition probability p_{jj}. If p_{jj} is not zero, then state S_j is aperiodic. This assertion follows from the definition of an aperiodic state; if p_{jj} is positive, then for every integer h greater than 1, there is another integer k (namely, $k = 1$) which is not an integer multiple of h and for which $p_{jj}^{(k)}$ is not 0. For example, a look at Table 9.1 reveals that both state S_1 and state S_2 are aperiodic since $p_{11} = \frac{1}{2} > 0$, and $p_{22} = \frac{1}{4} > 0$.

Another test for the aperiodicity of a state S_j is the following. If we can find two positive integers k_1 and k_2 such that $p_{jj}^{(k_1)} > 0$ and $p_{jj}^{(k_2)} > 0$, and such that k_1 and k_2 have no common divisors other than 1, then state S_j is aperiodic. In particular, if state S_j has the property that $p_{jj}^{(2)} > 0$ and $p_{jj}^{(3)} > 0$, then state S_j is aperiodic, since $k_1 = 2$ and $k_2 = 3$ have no common divisors. For example, if we are interested in the three-state Markov chain with transition probabilities as shown in Table 11.4, then even though $p_{11} = p_{22} = p_{33} = 0$, we still are able to show that each of the states S_1, S_2, and S_3 is aperiodic. To do so, we compute the 2-step and 3-step transition probabilities of this Markov chain (these appear in Tables 11.5 and 11.6, respectively), and then note that $p_{11}^{(2)} = \frac{1}{3} > 0$, $p_{11}^{(3)} = \frac{21}{72} > 0$; that $p_{22}^{(2)} = \frac{13}{24} > 0$, $p_{22}^{(3)} = \frac{28}{96} > 0$; and that $p_{33}^{(2)} = \frac{13}{24} > 0$, $p_{33}^{(3)} = \frac{14}{48} > 0$. Applying our test for aperiodicity, states S_1, S_2, and S_3 are all aperiodic.

If one state, say state S_i, in a class $\mathscr{C}$ of communicating states is aperiodic, then all the states in the class $\mathscr{C}$ must be aperiodic. For, if it were the case that some state, state S_j, is in class $\mathscr{C}$, and that state S_j is periodic of period T, then all the states in $\mathscr{C}$ would have to be periodic with period T. Thus, since state S_i belongs to class $\mathscr{C}$, state S_i would have to be periodic. Since state S_i is known to be aperiodic, we have a contradiction caused by assuming that some state in $\mathscr{C}$ is periodic. Thus, every state in $\mathscr{C}$ is aperiodic, proving our

Table 11.4: *Table of Transition Probabilities for a Markov Chain for Which All Three States, S_1, S_2, S_3, Are Periodic*

		State at time $t + 1$		
		S_1	S_2	S_3
State at time t	S_1	0	$\frac{2}{3}$	$\frac{1}{3}$
	S_2	$\frac{1}{4}$	0	$\frac{3}{4}$
	S_3	$\frac{1}{2}$	$\frac{1}{2}$	0

Table 11.5: *Table of 2-Step Transition Probabilities for the Markov Chain Whose Transition Probabilities Appear in Table 11.4*

		State at time $t + 2$		
		S_1	S_2	S_3
State at time t	S_1	$\frac{1}{3}$	$\frac{1}{6}$	$\frac{1}{2}$
	S_2	$\frac{3}{8}$	$\frac{13}{24}$	$\frac{1}{12}$
	S_3	$\frac{1}{8}$	$\frac{1}{3}$	$\frac{13}{24}$

Table 11.6: *Table of 3-Step Transition Probabilities for the Three-State Markov Chain Whose Transition Probabilities Are Given in Table 11.4*

		State at time $t + 3$		
		S_1	S_2	S_3
State at time t	S_1	$\frac{21}{72}$	$\frac{34}{72}$	$\frac{17}{72}$
	S_2	$\frac{17}{96}$	$\frac{28}{96}$	$\frac{51}{96}$
	S_3	$\frac{17}{48}$	$\frac{17}{48}$	$\frac{14}{48}$

original assertion. If a communicating class $\mathscr{C}$ contains one aperiodic state, then we say that $\mathscr{C}$ is an aperiodic communicating class.

The foregoing fact offers us another way of determining whether or not a given state is aperiodic: if state S_i belongs to an aperiodic communicating class $\mathscr{C}$, then state S_i is aperiodic. In the Markov chain whose transition probabilities appear in Table 11.3, we could show that states S_2 and S_3 are aperiodic by first noting that state S_2 and state S_3 each communicate with state S_1 (since p_{12}, p_{21}, p_{13}, and p_{31} are all nonzero), and then verifying that state S_1 is aperiodic [by showing that $p_{11}^{(2)} \neq 0$ and $p_{11}^{(3)} \neq 0$, or by some other method]. Note that state S_1 communicates with state S_2 and with state S_3, so that this Markov chain is irreducible. An irreducible Markov chain that has one aperiodic state is called an *aperiodic, irreducible Markov chain*. All the states of an aperiodic irreducible Markov chain are aperiodic states.

Looking back at the Markov chain whose transition probabilities appear in Table 9.1, we see that this Markov chain is irreducible and aperiodic. Recall again from Section 9 that this Markov chain achieves a probabilistic equilibrium. Although, as we have demonstrated earlier, not all irreducible Markov chains achieve probabilistic equilibrium, perhaps it is the case that all aperiodic irreducible Markov chains achieve an equilibrium. This is indeed the case, and in fact the following assertions are true for any aperiodic irreducible Markov chain:

(1) The Markov chain achieves probabilistic equilibrium.
(2) The long-run probability π_j of state S_j of the Markov chain is equal to $1/m_{jj}$, where m_{jj} is the mean (first) recurrence time of state S_j, $j = 1, 2, \ldots, N$.

Recall that an N-state Markov chain achieves probabilistic equilibrium if we can find probabilities $\pi_1, \pi_2, \ldots, \pi_N$ for the states $S_1, S_2, \ldots, S_n$, respectively, such that for every pair of states, S_i and S_j, and for all sufficiently large values of k.

$$(11.1) \qquad\qquad p_{ij}^{(k)} \simeq \pi_j.$$

If we can show that a given Markov chain is aperiodic and irreducible, then assertion (i) tells us that the long-run probabilities $\pi_1, \pi_2, \ldots, \pi_N$ can be found. We could determine the values of $\pi_1, \pi_2, \ldots, \pi_N$ recursively by calculating the k-step transition probabilities $p_{ij}^{(k)}$ for $k = 1, 2, 3$, and so on, stopping when the rows of a table of k-step transition probabilities, for some value of k, are all the same.

However, since our calculation of k-step transition probabilities have roundoff errors, we cannot be certain that the values of the long-run probabilities $\pi_1, \pi_2, \ldots, \pi_N$ obtained by this method are sufficiently accurate. Further, this procedure could force us to compute k-step transition probabilities for k large. Assertion (ii) provides us with an alternative method of computing the long-run probabilities $\pi_1, \pi_2, \ldots, \pi_N$. Using this approach, first calculate the expected (first) recurrence times $m_{11}, m_{22}, \ldots, m_{NN}$ by one of the methods described in Section 8, from which

(11.2) $$\pi_j = \frac{1}{m_{jj}}, \quad j = 1, 2, \ldots, N.$$

For example, consider the two-state Markov chain whose transition probabilities appear in Table 9.1. As we have noted, this aperiodic, irreducible Markov chain achieves a probabilistic equilibrium; from Table 9.2, the long-run probabilities of the states S_1 and S_2 appear to be $\pi_1 = 0.600$ and $\pi_2 = 0.400$, respectively. From (8.9), with $N = 2, j = 1$,

$$m_{11} = (1 - p_{11}m_{11}) + p_{11}m_{11} + p_{12}m_{21} = 1 + (\tfrac{1}{2})m_{21},$$

$$m_{21} = (1 - p_{21}m_{11}) + p_{21}m_{11} + p_{22}m_{21} = 1 + (\tfrac{1}{4})m_{21},$$

so that $m_{21} = \tfrac{4}{3}$ and $m_{11} = \tfrac{5}{3}$. Again, from (8.9), with $N = 2, j = 2$,

$$m_{12} = (1 - p_{12}m_{22}) + p_{11}m_{12} + p_{12}m_{22} = 1 + (\tfrac{1}{2})m_{12},$$

$$m_{22} = (1 - p_{22}m_{22}) + p_{21}m_{12} + p_{22}m_{22} = 1 + (\tfrac{3}{4})m_{12},$$

so that $m_{12} = 2$ and $m_{22} = \tfrac{5}{2}$. Hence, from (11.2),

$$\pi_1 = \frac{1}{m_{11}} = \frac{1}{\tfrac{5}{3}} = 0.600,$$

$$\pi_2 = \frac{1}{m_{22}} = \frac{1}{\tfrac{5}{2}} = 0.400.$$

As another example of an aperiodic, irreducible Markov chain, consider the genetical mating process whose transition probabilities appear in Table 6.3. Recall that the states of this process are S_1: genotype AA, S_2: genotype Aa, and S_3: genotype aa. From Table 6.4, we see that $p_{12}^{(2)} = \tfrac{1}{2} > 0$, $p_{21}^{(2)} = \tfrac{1}{4} > 0$, and that $p_{23}^{(2)} = \tfrac{1}{4} > 0$, $p_{32}^{(2)} = \tfrac{1}{2} > 0$. Thus, from the discussion in Section 10, the three states S_1, S_2, and S_3 form a communicating class. Since the genetical mating process has only three states, it follows that this Markov chain is irreducible, and since $p_{11} = \tfrac{1}{2}$, it follows that the process is an aperiodic, irreducible Markov chain. From assertion (ii), we conclude that this Markov chain achieves a probabilistic equilibrium.

The expected first-passage times m_{ij} of this genetical mating process have been obtained in Section 8, and are summarized in Table 8.2. From Table 8.2, we see that $m_{11} = 4$, $m_{22} = 2$, and $m_{33} = 4$. Hence, from (11.2), the long-run probabilities of this process are

$$\pi_1 = \text{long-run probability of genotype } AA = \frac{1}{m_{11}} = \frac{1}{4},$$

$$\pi_2 = \text{long-run probability of genotype } Aa = \frac{1}{m_{22}} = \frac{1}{2},$$

$$\pi_3 = \text{long-run probability of genotype } aa = \frac{1}{m_{33}} = \frac{1}{4}.$$

Steady-State Equations

Useful as it is, (11.2) still is a somewhat indirect way of calculating the long-run probabilities π_1, π_2, ..., π_N of an aperiodic, irreducible Markov chain. Since we know that the nature of the probabilistic equilibrium of a Markov chain is determined solely by the 1-step transition probabilities that generate the Markov chain, we should be able to obtain the long-run probabilities π_1, π_2, ..., π_N of a Markov chain directly from knowledge of the 1-step transition probabilities p_{ij}. As we now demonstrate, this is indeed possible.

Suppose that we are observing an N-state Markov chain which achieves probabilistic equilibrium. However, instead of having observed this process from its beginning, we observe the process at a time t at which the process is already in probabilistic equilibrium. Because the process is in probabilistic equilibrium, the unconditional probability, $P\{X_t = j\}$, that the process is in state S_j at time t is equal to π_j. That is,

$$(11.3) \qquad P\{X_t = j\} = \pi_j, \quad j = 1, 2, ..., N.$$

However, since the Markov chain is in probabilistic equilibrium, we also know that at the next time, time $t + 1$, at which we observe the process, the unconditional probability, $P\{X_{t+1} = j\}$, that the process is in state S_j is the same as it is at time t; that is,

$$(11.4) \qquad P\{X_{t+1} = j\} = \pi_j, \quad j = 1, 2, ..., N.$$

If we think of the Markov chain process as beginning at time t, then we can think of the unconditional probabilities in (11.3) as if they were initial probabilities for the process, starting at time t. That is, we can think of time t as if it were time 0 on a new clock for the process, a clock that starts when we start observing the process. On this new clock, time $t + 1$ (measured on the old clock) becomes time 1, time $t + 2$ becomes time 2, and so on. The initial probabilities for the process now become $a_1 = \pi_1$, $a_2 = \pi_2$, ..., $a_N = \pi_N$. We know that the unconditional probability of state S_j at time 1 on the new clock can be found from the initial probabilities a_1, a_2, ..., a_N, and the 1-step transition probabilities p_{ij} through the equation

$$(11.5) \qquad P\{X_1 = j\} = a_1 p_{1j} + a_2 p_{2j} + \cdots + a_N p_{Nj}.$$

But time 1 on the new clock is time $t + 1$ on the old clock, and from (11.4) the unconditional probability that the process is in a given state, say state S_j, at time $t + 1$ is π_j. Thus, substituting π_j for $P\{X_1 = j\}$, and π_1, π_2, ..., π_N for a_1, a_2, ..., a_N, respectively, in (11.5), we obtain

$$(11.6) \qquad \pi_j = \pi_1 p_{1j} + \pi_2 p_{2j} + \cdots + \pi_N p_{Nj}.$$

Equation (11.6) is true for each state S_j; consequently,

$$\pi_1 = \pi_1 p_{11} + \pi_2 p_{21} + \cdots + \pi_N p_{N1},$$

$$\pi_2 = \pi_1 p_{12} + \pi_2 p_{22} + \cdots + \pi_N p_{N2},$$

$$\vdots \quad \vdots \qquad \vdots \qquad \qquad \vdots$$

(11.7)

$$\pi_j = \pi_1 p_{1j} + \pi_2 p_{2j} + \cdots + \pi_N p_{Nj},$$

$$\vdots \quad \vdots \qquad \vdots \qquad \qquad \vdots$$

$$\pi_N = \pi_1 p_{1N} + \pi_2 p_{2N} + \cdots + \pi_N p_{NN}.$$

Hence, if we know the transition probabilities of the given Markov chain, (11.7) provides us with a set of linear equations called the *steady-state equations*, which can be solved for the values of the unknowns $\pi_1, \pi_2, \ldots, \pi_N$.

Note that if we add the N equations in (11.7) we obtain the identity $\sum_1^N \pi_i = \sum_1^N \pi_i$, which implies that at least one of the equations is redundant. Thus, to solve (11.7) we can solve any subcollection of $N - 1$ of these equations and use the fact that $\sum_1^N \pi_i = 1$.

For the three-state Markov chain (the genetical mating process) whose transition probabilities appear in Table 6.3, we can find the long-run probabilities π_1, π_2, π_3 by taking any two of the steady-state equations

$$\pi_1 = (\tfrac{1}{2})\pi_1 + (\tfrac{1}{4})\pi_2 + (0)\pi_3,$$

$$\pi_2 = (\tfrac{1}{2})\pi_1 + (\tfrac{1}{2})\pi_2 + (\tfrac{1}{2})\pi_3,$$

$$\pi_3 = (0)\pi_1 + (\tfrac{1}{4})\pi_2 + (\tfrac{1}{2})\pi_3,$$

and solving these equations for the unknowns π_1, π_2, π_3, under the additional requirement that

$$\pi_1 + \pi_2 + \pi_3 = 1.$$

The resulting solutions

$$\pi_1 = \tfrac{1}{4}, \qquad \pi_2 = \tfrac{1}{2}, \qquad \pi_3 = \tfrac{1}{4},$$

are the same as those obtained earlier for this process by use of (11.2).

Remark. A *row vector*

$$\mathbf{a} = (a_1, a_2, \ldots, a_N)$$

is a $1 \times N$ matrix. The product $\mathbf{a}B$ of a matrix

$$B = \begin{pmatrix} b_{11} & b_{12} & \cdots & b_{1N} \\ b_{21} & b_{22} & \cdots & b_{2N} \\ \vdots & \vdots & & \vdots \\ b_{N1} & b_{N2} & \cdots & b_{NN} \end{pmatrix} = (b_{ij})$$

and a row vector $\mathbf{a}$ is defined to be a row vector $\mathbf{c} = \mathbf{a}B$, where

$$\mathbf{c} = (c_1, c_2, \dots, c_N)$$

and

$$c_j = \sum_{k=1}^{N} a_k b_{kj}.$$

In terms of vector-matrix products, (11.7) can be written

(11.8) $$\pi = \pi P,$$

where

$$\pi = (\pi_1, \pi_2, \dots, \pi_N)$$

is the row vector of long-run probabilities and $P = (p_{ij})$ is the matrix of one-step transition probabilities. The equation (11.8) is a famous equation in matrix theory, since it says that the row vector π of long-term probabilities is a (left-hand) characteristic vector of the matrix P corresponding to the largest characteristic root 1 of P. The existence of a solution π of this *characteristic equation*, under the additional constraints that the π_j's are nonnegative and add to 1, is a consequence of the well-known Perron–Frobenius Theorem [Perron (1907), Frobenius (1912)]. More important for our purposes, computer programs exist to solve the characteristic equation (11.8), thus freeing us from the necessity of solving such an equation by hand in cases when the number N of unknowns π_j is large.

12. ABSORBING MARKOV CHAINS

If we observe an irreducible Markov chain over time, the process typically jumps from one state to another and only rarely stays in one state for very long. In contrast, there exist Markov chains that have one or more states, called *absorbing states*, such that if the process enters one of these states, then the process can never leave that state (it is "absorbed" into that state). For example, consider the simple learning process first discussed in Example 4.1. This two-state Markov chain has transition probabilities of the form shown in Table 12.1.

In studying Table 12.1, recall that state S_1 is the state in which a subject under observation is conditioned to make a correct response, while state S_2 is the state in which the subject is still unconditioned. Once the process enters state S_1, it remains in state S_1 forever. We verified this assertion in Section 6 (for the case where α, in Table 12.1, equals $\frac{1}{2}$) by showing that $p_{12}^{(k)} = 0$ for all $k = 1, 2, 3, \dots$. That is, given that the process is in state S_1 at some time, time t, the conditional probability of leaving state S_1 and going to state S_2 at any

Table 12.1: *Table of Transition Probabilities*
for a Simple Learning Process

[The constant α is a number between 0 and 1 (i.e., $0 \le \alpha \le 1$)]

State at time $t + 1$

		S_1	S_2
State at time t	S_1	1	0
	S_2	α	$1 - \alpha$

future time, time $t + k$, $k = 1, 2, 3, \ldots$, is equal to 0. Thus, for the learning process, state S_1 is an absorbing state. On the other hand, state S_2 is not an absorbing state when $\alpha \neq 0$, since in this case $p_{21} \neq 0$, and the process can leave state S_2 and go to state S_1 one time unit after a time in which the process has been in state S_2.

In general, a state, state S_i, of a Markov chain is absorbing if $p_{ii} = 1$. If p_{ii} is not equal to 1, then since

$$p_{i1} + p_{i2} + \cdots + p_{ii} + \cdots + p_{iN} = 1,$$

it follows that at least one of the 1-step transition probabilities $p_{ij}, j \neq i$, must be nonzero. Hence, if $p_{ii} \neq 1$, it is possible for the process to leave state S_i, and hence state S_i cannot be absorbing. We conclude that state S_i is an absorbing state if and only if $p_{ii} = 1$.

It follows directly from the definition of an absorbing state that every absorbing state is also a recurrent state. If the process enters an absorbing state at time t, it can never leave that state afterward. Thus, the process must remain in that state at the very next time, time $t + 1$. Consequently, the (first) recurrence time for that absorbing state is finite (indeed, is equal to 1) with probability equal to 1, and hence, by definition, it follows that the absorbing state is a recurrent state.

If a Markov chain has one or more absorbing states, it is called an *absorbing Markov chain*. If an absorbing Markov chain has just one absorbing state, and if this absorbing state can be reached from every other state of the Markov chain, then the Markov chain has exactly one recurrent state (namely, the one absorbing state) and all the other states of the process are transient. Such a Markov chain achieves a probabilistic equilibrium in which the long-run probability of the single absorbing state is equal to 1, and the long-run probabilities of all the other states are equal to 0.

For example, the learning process whose transition probabilities are tabled in Table 12.1 is (when $\alpha \neq 0$) an absorbing Markov chain with a single absorbing state, state S_1. Thus, state S_1 is a recurrent state and state S_2 is a transient state (we proved this fact by a direct method in Section 2). Further, this process achieves a probabilistic equilibrium in which state S_1 has long-run probability π_1 equal to 1, and in which the long-run probability π_2 of state S_2 is equal to 0. Hence, after a long enough period of time, k, has passed, the unconditional probability $P\{X_k = 1\}$ that the process is in state S_1 is approximately equal to 1, and the unconditional probability that the process is in state S_2 is approximately equal to 0. (Indeed, when $k = 13$, the assertion that $P\{X_k = 1\} = 1.000$ and $P\{X_k = 0\} = 0.000$ is accurate to three decimal places.)

It is possible for an absorbing Markov chain to have just one absorbing state and yet have many recurrent states. For example, suppose that a Markov chain of interest to us has three states S_1, S_2, and S_3, and suppose that the transition probabilities of this process are those shown in Table 12.2. Since state S_1 is absorbing ($p_{11} = 1$), this Markov chain is an absorbing Markov chain. Although neither state S_2 nor state S_3 is absorbing ($p_{22} = p_{33} = \frac{1}{2} \neq 1$), these two states together form a closed, communicating class (closed since $p_{21} = p_{31} = 0$, communicating since $p_{23} \neq 0$, $p_{32} \neq 0$). Hence, from assertion (i) of Section 10, states S_2 and S_3 are recurrent. We have thus exhibited an example of an absorbing Markov chain in which there is only one absorbing state, but in which there are three recurrent states. Note that this example also illustrates the fact that although all absorbing states are recurrent, not all recurrent states are absorbing.

Suppose that we are interested in an absorbing Markov chain that has more than one absorbing state. The conformity process described earlier in Example 3.2 is an example of such an absorbing process. Recall that this process has four states:

S_1: The subject in the conformity experiment is motivated to answer correctly, independent of the answers of the group, at all trials of the experiment.

S_2: The subject is motivated to answer correctly this time, but is still indecisive about whether or not to conform in the future.

S_3: The subject is motivated to conform to the incorrect answer of the group, but is still indecisive about future conformity.

S_4: The subject is motivated to conform to the group's answer on all trials.

The transition probabilities of such a process are of the form shown in Table 12.3. From Table 12.3, we see that state S_1 and state S_4 are absorbing states (since $p_{11} = 1$, $p_{44} = 1$). Once the process enters one of these states, it can never leave that state.

Suppose, however, that a subject is initially undecided between answering correctly or conforming (i.e., the process is either in state S_2 or in state S_3). If p_{21} and p_{34} are not equal to 0, then it can be shown that states S_2 and

Table 12.2: *Table of Transition Probabilities for an Absorbing Markov Chain That Has One Absorbing State and Three Recurrent States*

		State at time $t+1$		
		S_1	S_2	S_3
	S_1	1	0	0
State at time t	S_2	0	$\frac{1}{2}$	$\frac{1}{2}$
	S_3	0	$\frac{1}{2}$	$\frac{1}{2}$

S_3 are transient states, and that the process eventually must leave both of these states and be absorbed either in state S_1 or in state S_4.

Suppose that the conformity process initially is in state S_3. We know that eventually the process will leave state S_3 to be absorbed in one of the two states S_1 or S_4. However, we do not know which state, S_1 or S_4, will be the state that "absorbs" the process. Thus, it is of interest to find the conditional probabilities A_{3j} that the process will be absorbed in state S_j, for $j = 1$ and $j = 4$, given that the process initially is in state S_3. In general, if state S_j is an absorbing state of an (absorbing) Markov chain, and if state S_j can be reached from a nonabsorbing state, state S_i, then

(12.1) $A_{ij} = P\{\text{the process is eventually absorbed in state } S_j | X_0 = i\}.$

The quantities A_{ij} are called *absorption probabilities*. Although the symbol A_{ij} is mnemonically useful for remembering that the quantity (12.1) is an absorption probability, we have discussed this quantity before under another name.

Table 12.3: *Transition Probabilities for a Conformity Model*

		State at time $t+1$			
		S_1	S_2	S_3	S_4
	S_1	1	0	0	0
State at time t	S_2	p_{21}	p_{22}	p_{23}	0
	S_3	0	p_{32}	p_{33}	p_{34}
	S_4	0	0	0	1

Note that once the process reaches the absorbing state, state S_j, for the first time, it is absorbed in state S_j. Consequently, the absorption probability A_{ij} and the probability F_{ij} of a finite first-passage time from state S_i to state S_j (see Section 6) are the same quantity. That is, if state S_j is an absorbing state, then

$$(12.2) \qquad\qquad\qquad A_{ij} = F_{ij}.$$

Thus, when feasible, we can calculate the absorption probability A_{ij} using the methods of computation described in Section 6.

There is, however, a more direct method for calculating the probabilities A_{ij}, which is usually easier to apply. Suppose that an absorbing Markov chain has N states, $S_1, S_2, \ldots, S_N$, and that state S_j is an absorbing state that can be reached from at least one of the nonabsorbing states of the Markov chain. Since state S_j can never be reached from any (other) absorbing state, let us adopt the convention that $A_{hj} = 0$ whenever state S_h is an absorbing state, $h \neq j$. Further, since if the process starts in state S_j, it is already absorbed in state S_j, it follows that $A_{jj} = 1$. Under these notational conventions, it can be shown that for every nonabsorbing state S_i,

$$(12.3) \qquad\qquad A_{ij} = p_{i1} A_{1j} + p_{i2} A_{2j} + \cdots + p_{iN} A_{Nj}.$$

If there are M nonabsorbing states, $S_{i_1}, S_{i_2}, \ldots, S_{i_M}$ $(M < N)$, in the Markov chain, we can write down M simultaneous linear equations [one equation of the form (12.3) for each nonabsorbing state], and solve these equations for the M unknowns $A_{i_1 j}, A_{i_2 j}, \ldots, A_{i_M j}$.

For example, in the conformity process described earlier in this section, state S_2 and state S_3 are nonabsorbing states. If we want the absorption probabilities A_{21} and A_{31} (the probabilities of being absorbed in state S_1), then we can solve the pair of linear equations

$$A_{21} = p_{21} A_{11} + p_{22} A_{21} + p_{23} A_{31} + p_{24} A_{41},$$
$$A_{31} = p_{31} A_{11} + p_{32} A_{21} + p_{33} A_{31} + p_{34} A_{41},$$

for the unknowns A_{21} and A_{31}. Recalling that under our notational conventions $A_{11} = 1$ and $A_{41} = 0$, this pair of linear equations becomes

$$A_{21} = p_{21} + p_{22} A_{21} + p_{23} A_{31},$$
$$(12.4) \qquad\qquad A_{31} = p_{31} + p_{32} A_{21} + p_{33} A_{31}.$$

In one of the conformity experiments (see Example 3.2) conducted by Cohen (1958), the unspecified transition probabilities in Table 12.3 were found to be

$$p_{21} = 0.06, \qquad p_{22} = 0.76, \qquad p_{23} = 0.18,$$
$$p_{31} = 0.00, \qquad p_{32} = 0.27, \qquad p_{33} = 0.69,$$

and $p_{34} = 0.04$. Substituting these values into (12.4), we see that we must solve the pair of linear equations

$$A_{21} = 0.06 + (0.76)A_{21} + (0.18)A_{31},$$
$$A_{31} = 0.00 + (0.27)A_{21} + (0.69)A_{31},$$

for the unknowns A_{21} and A_{31}. Solving these equations yields $A_{21} = 0.721$ and $A_{31} = 0.629$.

NOTES AND REFERENCES

The theory of Markov chains is quite extensive. In this chapter, we provided coverage of some of the main concepts of this theory. Our treatment of these concepts, however, has by no means been exhaustive. For example, although we have mostly considered N-state Markov chains, the concepts introduced in our discussion are meaningful and valid for Markov chains that have an infinite number of possible states. Further, in our analysis of the long-run behavior of N-state Markov chains, we confined our investigations to Markov chains that are either irreducible or absorbing. However, the Markov chain described in Example 7.1 is an example of a Markov chain that is neither irreducible nor absorbing. The long-run behavior of such a process can be studied by using some of the concepts that we have introduced in this chapter.

The theory of Markov chains continues to develop. In the last few years, many novel approaches have been suggested and many new results have been obtained. For example, considerable progress has occurred in that area dealing with "limit theorems" for the k-step transition probabilities of Markov chains. Another area of importance has been the study of the long-run behavior of Markov chains whose 1-step transition probabilities are not stationary over time. A third area of continuing interest is the use of Markov chains to approximate the behavior of stochastic processes that are not Markov chains.

EXERCISES

1. Ship lanes are sufficiently wide so that each ship may vary its direction from time to time to meet various contingencies and still avoid collisions with other ships. Suppose that after each unit of time, a ship chooses its direction by choosing one of three states S_1, S_2, S_3, rep-

resenting specific courses. Further, suppose that the process of choosing such directions is a Markov chain with the following transition probabilities:

State at time $t + 1$

		S_1	S_2	S_3
State at time t	S_1	p_1	p_2	p_3
	S_2	p_3	p_1	p_2
	S_3	p_2	p_3	p_1

where $p_1 + p_2 + p_3 = 1$. Assume $a_1 = a_2 = a_3 = \frac{1}{3}$. Find formulas for (a) the k-step transition probabilities, and (b) the unconditional probabilities at time k for this process.

2. In a study of the epidemiology of mental disease, Marshall and Goldhamer (1955) consider various Markov chain models. One of these has as states $S_1 =$ alive, sane; $S_2 =$ alive, insane (mild), unhospitalized; $S_3 =$ alive, insane (severe), unhospitalized; $S_4 =$ alive, insane, hospitalized; and $S_5 =$ dead, outside of mental institution (hospital). The transition probabilities for this five-state Markov chain are as follows:

State at time $t + 1$

		S_1	S_2	S_3	S_4	S_5
State at time t	S_1	p_{11}	p_{12}	p_{13}	0	p_{15}
	S_2	0	p_{22}	0	p_{24}	p_{25}
	S_3	0	0	p_{33}	p_{34}	p_{35}
	S_4	0	0	0	1	0
	S_5	0	0	0	0	1

Find 2-step, 3-step, and 4-step transition probabilities for this model.

3. The simple two-state learning model of Example 4.1 can be extended to deal with more complicated experiments. For example, we can have a three-state learning model in which state S_1 denotes long-term memory, state S_2 denotes short-term memory, and state S_3 denotes guessing. [The details of such models are discussed in Coombs, Dawes, and Tversky

(1970), ch. 9.] The transition matrix of such a three-state learning model is assumed to have the following form:

State at time $t + 1$

		S_1	S_2	S_3
	S_1	1	0	0
State at time t	S_2	α	$\bar{\alpha}\bar{\delta}$	$\bar{\alpha}\delta$
	S_3	β	$\bar{\beta}\bar{\delta}$	$\bar{\beta}\delta$

where $\bar{\alpha} = 1 - \alpha$, $\bar{\beta} = 1 - \beta$, and $\bar{\delta} = 1 - \delta$.

(a) Verify that the table is a valid table of transition probabilities for a Markov chain.

(b) What models do we obtain when (i) $\alpha = 0$, (ii) $\delta = 0$, or (iii) $\alpha = \beta = 0$?

(c) If $\alpha = 0.4$, $\beta = 0$, and $\delta = 0.5$, find the 2-step, 3-step, and 4-step transition probabilities.

(d) Determine

$$P\{X_4 = 2 | X_3 = 2\}, \quad P\{X_4 = 2 | X_2 = 2\},$$
$$P\{X_4 = 2 | X_1 = 2\}, \quad P\{X_4 = 2 | X_0 = 2\}.$$

(e) Determine

$$P\{X_3 = 2 | X_2 = 2\}, \quad P\{X_3 = 2 | X_1 = 2\}, \quad P\{X_3 = 2 | X_0 = 2\}.$$

(f) Determine the unconditional probabilities $P\{X_1 = 2\}$, $P\{X_2 = 2\}$, $P\{X_3 = 2\}$, and $P\{X_4 = 2\}$. Here assume that $a_1 = a_2 = a_3 = \frac{1}{3}$.

4. Suppose we are studying [see Berger and Snell (1957)] intercity population movements among three cities, A, B, and C. Each city sends, within a 1-year time period, certain fractions of its population to itself and to other cities. Here the states represent the cities, and the model for the intercity movement of a randomly chosen individual is assumed to be a Markov chain with transition probabilities

State (city) at time $t + 1$

		A	B	C
	A	0.85	0.07	0.08
State (city) at time t	B	0.25	0.70	0.05
	C	0.03	0.02	0.95

The initial populations of the three cities are 100,000 for city A, 500,000 for city B, and 200,000 for city C.

(a) What are the 2-step transition probabilities for this Markov chain?

(b) Determine the unconditional probabilities for this Markov chain after one unit of time.

(c) Determine the unconditional probabilities after two units of time.

5. Consider the following tables of transition probabilities for a three-state Markov chain:

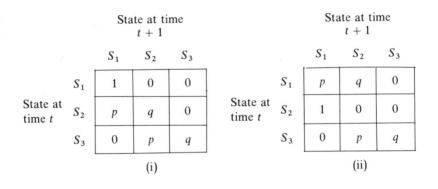

(i)   (ii)

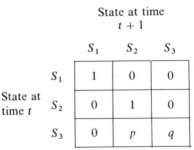

(iii)

where $p + q = 1$.

(a) In each case [(i), (ii), and (iii)] obtain formulas for all of the 2-step, 3-step, and 4-step transition probabilities in terms of the constants p and q.

(b) In each case, derive a formula for finding the higher-order transition probabilities in terms of p and q.

(c) If the initial probabilities are $a_1 = a_2 = a_3 = \frac{1}{3}$, find the unconditional probabilities $P\{X_3 = 1\}$, $P\{X_3 = 2\}$, and $P\{X_3 = 3\}$ in each of cases (i), (ii), and (iii).

6. The following transition probability structures arise in a variety of contexts:

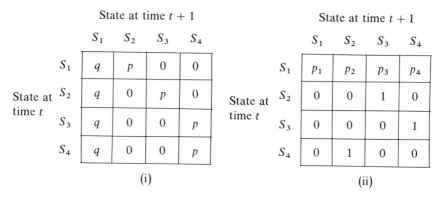

State at time $t + 1$

	S_1	S_2	S_3	S_4
S_1	q	p	0	0
S_2	q	0	p	0
S_3	q	0	0	p
S_4	q	0	0	p

State at time t

(i)

State at time $t + 1$

	S_1	S_2	S_3	S_4
S_1	p_1	p_2	p_3	p_4
S_2	0	0	1	0
S_3	0	0	0	1
S_4	0	1	0	0

State at time t

(ii)

State at time $t + 1$

	S_1	S_2	S_3	S_4	S_5
S_1	1	0	0	0	0
S_2	p	0	q	0	0
S_3	0	p	0	q	0
S_4	0	0	p	0	q
S_5	0	0	0	0	1

State at time t

(iii)

where $p + q = 1$ and $p_1 + p_2 + p_3 + p_4 = 1$.
(a) In each case, obtain formulas for the 2-step and 3-step transition probabilities in terms of the unspecified constants (i.e., p, q, p_1, p_2, p_3, and p_4).
(b) If the initial probabilities are $a_1 = a_2 = a_3 = a_4 = \frac{1}{4}$ for (i) and (ii), and $a_1 = a_2 = a_3 = a_4 = a_5 = \frac{1}{5}$ for (iii), find the unconditional probabilities of the various states at time $t = 2$.

7. Markov chain models have also been used in a spatial rather than temporal context [see American Geological Institute (1969)]. Suppose that we are interested in the study of rock formation and represent the different rock components as our states: $S_1 =$ sandstone, $S_2 =$ shale, $S_3 =$ siltstone, $S_4 =$ lignite. A vertical strip or section of rock surface, say a cliff or well, is observed upward from its base. Each layer of rock deposit can be thought of as corresponding to the passage of a unit of time. In a particular experiment, a table of transition probabilities is reported (Table E.1).

Table E.1: *Transition Probabilities for a Section in the Oficina Formation (Miocene) from a Well in Venezuela*

		State at layer $t + 1$			
		S_1	S_2	S_3	S_4
	S_1	0.79	0.07	0.07	0.07
State at layer t	S_2	0.05	0.79	0.06	0.10
	S_3	0.10	0.32	0.43	0.15
	S_4	0.18	0.39	0.13	0.30

(a) If we start our observations (layer 0) with lignite (state S_4), what is the conditional probability that lignite will again be present at layer 2 following immediately after siltstone at layer 1? What is the probability that lignite will again be present at layer 3 following after siltstone at layer 2? In other words, find $P\{X_2 = 4, X_1 = 3 \mid X_0 = 4\}$ and $P\{X_3 = 4, X_2 = 3 \mid X_0 = 4\}$.

(b) Determine the probability that lignite will be present at layer 2 or at layer 3 given that sandstone occurs at layer 1.

(c) It was found that the 40-step transition probabilities of this Markov chain model are

		State at layer 40			
		S_1	S_2	S_3	S_4
	S_1	0.27	0.49	0.12	0.12
State at layer 0	S_2	0.27	0.49	0.12	0.12
	S_3	0.27	0.49	0.12	0.12
	S_4	0.27	0.49	0.12	0.12

Interpret this finding in terms of the given experimental context.

8. Suppose that a three-state Markov chain has the following transition matrix:

State at time $t + 1$

	S_1	S_2	S_3
State at time t — S_1	$\frac{1}{3}$	$\frac{1}{3}$	$\frac{1}{3}$
S_2	0	$\frac{1}{2}$	$\frac{1}{2}$
S_3	0	0	1

Show that the k-step transition probabilities of this Markov chain are

State at time $t + k$

	S_1	S_2	S_3
State at time t — S_1	$(\frac{1}{3})^k$	$\frac{1}{3}p_{12}^{(k-1)} + \frac{1}{3}(\frac{1}{2})^{k-1}$	$1 - (\frac{1}{3})(\frac{1}{2})^{k-1} - \frac{1}{3}p_{12}^{(k-1)}$
S_2	0	$(\frac{1}{2})^k$	$1 - (\frac{1}{2})^k$
S_3	0	0	1

By recursive calculation find $p_{12}^{(2)}$, $p_{12}^{(3)}$, $p_{12}^{(4)}$, and $p_{12}^{(5)}$, and use these results to display the 2-step, 3-step, 4-step, and 5-step transition probability matrices of the Markov chain.

9. In Washington, D.C., taxicab fares are based on zones arranged in a pattern of concentric circles. A taxicab may start the day in one zone. The zone of destination of the first passenger then determines the zone in which the taxicab driver cruises for his next fare. The zone of destination of his next passenger then determines a new cruising zone in which the driver looks for his third fare, and so on. The process that we have just described might be modeled as a Markov chain. Suppose that there are four fare zones: $S_1 =$ the center zone; S_2, S_3, and $S_4 =$ the outer zone. Suppose that the transition probabilities among the zones are stationary over time, and that these transition probabilities are as follows:

Zone of destination of
$(t + 1)$st fare

	S_1	S_2	S_3	S_4
Zone of destination of tth fare — S_1	0.80	0.14	0.05	0.01
S_2	0.60	0.20	0.18	0.02
S_3	0.50	0.40	0.05	0.05
S_4	0.30	0.30	0.30	0.10

(a) If a taxicab driver lives in the center of town (state S_1) and starts looking for his first fare in state S_4 (the outermost zone), what is the (conditional) probability that he returns to his home zone (state S_1) for the first time after he has driven 4 fares to their destinations?

(b) If the driver starts looking for fares in the outermost zone (state S_4), what is the conditional probability that he must have 5 or more fares before he reaches his home zone for the first time?

(c) If the driver starts in the central zone (state S_1), what is the conditional probability that he will return to this zone for the first time by driving his second fare to that fare's zone of destination? That is, what is the value of $f_{11}^{(2)}$? What is the value of $f_{11}^{(3)}$?

(d) Given that the driver picks up his first fare in the central zone, what is the mean number of fares that the driver must take to their destinations before he returns to the central zone again for the first time?

(e) Does the above Markov chain process achieve a probabilistic equilibrium? Support your assertion.

(f) The taxicab driver generally stops for the day after he has driven 30 fares. What is the (approximate) unconditional probability that he will stop for the day in his home zone (state S_1) and thus not have to drive far to reach his house? Assume that the initial probabilities for the zones in which the driver finds his first fare of the day are $a_1 = 0.75$, $a_2 = 0.10$, $a_3 = 0.10$, and $a_4 = 0.05$.

10. In a study on the pattern of diseased and healthy trees, Pielou (1965) uses Markov chain models. As we walk along a randomly chosen path (transect) that leads through the trees in a given forested area, we successively encounter trees that are either healthy or diseased. Healthy trees may themselves be part of a subarea of noninfested, healthy trees, called a gap, or they may be part of a subarea of trees containing both diseased and healthy trees, called a patch. It is assumed that every subarea of trees can be uniquely labeled either as a patch or as a gap, and that every tree that we encounter can be assigned to one (and only one) of these subareas of trees. We model the succession of trees that we encounter as a Markov chain with stationary transition probabilities. The states of the Markov chain are S_1 = diseased tree, S_2 = healthy patch tree, and S_3 = healthy gap tree. From theoretical considerations, the transition probabilities of this Markov chain have the following form:

State of $(t + 1)$st tree encountered

	S_1	S_2	S_3
S_1	αv	$\alpha(1 - v)$	$1 - \alpha$
S_2	αw	$\alpha(1 - w)$	$1 - \alpha$
S_3	$\dfrac{\gamma w}{1 - v + w}$	$\dfrac{\gamma(1 - v)}{1 - v + w}$	$1 - \gamma$

State of tth tree encountered

(a) Show that for the entries in the table to be transition probabilities, we must have $0 \le \alpha \le 1$, $0 \le v \le 1$, $0 \le \gamma \le 1$, $0 \le w \le 1$, $0 \le 1 - v + w$.

(b) Suppose for a certain species of tree, $\alpha = 0.4$, $\gamma = 0.5$, $v = 0.8$, and $w = 0.2$. Find the first-passage probabilities $f_{11}^{(2)}, f_{12}^{(2)}, f_{13}^{(2)}, f_{21}^{(2)}, f_{22}^{(2)}, f_{23}^{(2)}, f_{31}^{(2)}, f_{32}^{(2)}$, and $f_{33}^{(2)}$.

(c) When $\alpha = 0.4$, $\gamma = 0.5$, $v = 0.8$, and $w = 0.2$, find the mean first-passage times $m_{11}, m_{12}, m_{13}, m_{21}, m_{22}, m_{23}, m_{31}, m_{32}$, and m_{33}.

(d) When $\alpha = 0.4$, $\gamma = 0.5$, $v = 0.8$, and $w = 0.2$, show that this Markov chain is irreducible. Does the Markov chain achieve probabilistic equilibrium? Is the Markov chain an absorbing chain?

(e) If $\gamma = 0.0$, $\alpha = 0.4$, $v = 0.8$, and $w = 0.2$, find $f_{33}^{(1)}, f_{33}^{(2)}, f_{33}^{(3)}$, and $f_{33}^{(4)}$. Also find $f_{13}^{(1)}, f_{13}^{(2)}, f_{13}^{(3)}$, and $f_{13}^{(4)}$.

11. In certain Markov chains not only do the row sums of a table of transition probabilities all equal 1, but each of the column sums of such a table equals 1. For example, consider the three-state Markov chain with the transition probabilities

State at time $t + 1$

	S_1	S_2	S_3
S_1	$\frac{6}{8}$	$\frac{2}{8}$	0
S_2	$\frac{1}{8}$	$\frac{4}{8}$	$\frac{3}{8}$
S_3	$\frac{1}{8}$	$\frac{2}{8}$	$\frac{5}{8}$

State at time t

Note that each column has a sum of 1.

(a) Show that a Markov chain with the transition matrix shown is irreducible and aperiodic. Thus, show that such a Markov chain achieves a probabilistic equilibrium and find the long-run probabilities π_1, π_2, and π_3.

(b) Show that any irreducible three-state Markov chain having a table of transition probabilities in which all the columns sum to 1 must achieve a probabilistic equilibrium. Further, show that the long-run probabilities of any such Markov chain are the same as the long-run probabilities found in part (a).

(c) Use the result in part (b) to analyze the long-run behavior of the Markov chain described in Exercise 1.

In each of the following Markov chains (Exercises 12–17) identify recurrent and transient states, identify absorbing states, and find the period of each state. Identify communicating classes and state whether each such class is closed or not. Verify whether the chain is aperiodic and irreducible. If it is, find the mean first-passage times and long-run probabilities. If it is not, see if it is an abosrbing chain, and calculate the absorption probabilities A_{ij} where these probabilities are appropriate.

12. The occupational mobility process described in Example 4.2.

13. The political preference process described in Example 6.1.

14. The learning process described in Exercise 3 when $\alpha = 0.4$, $\beta = 0.0$, and $\delta = 0.5$.

15. The intercity population mobility process described in Exercise 4.

16. Coleman (1964, p. 168) considers a Markov chain model in which there are two attributes, A and B, to which we may respond with a "+" or "−." These responses may mean "present–absent" in one context, "agree–disagree" in another, and so on. We then have four states of the system representing the responses on attributes A and B, respectively:

S_1: response is $(+, +)$, S_3: response is $(-, +)$,

S_2: response is $(+, -)$, S_4: response is $(-, -)$.

Thus, state S_2 is the state in which attribute A is present $(+)$ and attribute B is absent $(-)$. In one study of the social system of adolescents in 10 schools, 3260 (female) students were asked in a questionnaire about the "leading crowd." The attribute A was whether the individual was a member of the leading crowd, and B was whether she agreed or disagreed with the sentence: "If a girl wants to be part of the leading crowd around here, she sometimes has to go against her principles." The resulting (estimated) transition probabilities are given as follows:

State at time 2 (May 1958)

	S_1	S_2	S_3	S_4
S_1	0.676	0.130	0.149	0.045
S_2	0.376	0.369	0.101	0.154
S_3	0.103	0.032	0.610	0.255
S_4	0.075	0.076	0.307	0.542

State at time 1 (October 1957)

17. In a second study considered by Coleman (1964), the effects on sales of a favorable attitude toward a brand of a grocery item were studied. Here attribute A referred to the respondent's attitude toward the brand, and attribute B referred to whether the particular item was the respondent's "usual brand." A total of 1633 respondents were interviewed, yielding the following (estimated) transition probabilities:

State at time $t + 1$

	S_1	S_2	S_3	S_4
S_1	0.848	0.033	0.055	0.064
S_2	0.200	0.167	0.033	0.600
S_3	0.638	0.056	0.153	0.153
S_4	0.047	0.104	0.025	0.824

State at time t

Appendix A

The Gamma Function

The gamma function, $\Gamma(z)$, is one way of extending the factorial $n!$ computation to all positive numbers z in a continuous fashion. Perhaps the most straightforward way to define the gamma function is through *Euler's integral* (named after the Swiss mathematician Leonhard Euler (1707–1783); actually, this integral is Euler's second integral). Thus,

$$\text{(A.1)} \qquad \Gamma(z) = \int_0^\infty t^{z-1} e^{-t} \, dt,$$

where $z > 0$. To demonstrate the relationship of $\Gamma(z)$ to the factorial, let us integrate by parts in (A.1). Then, as long as $z \geq 1$,

$$\text{(A.2)} \quad \Gamma(z) = \int_0^\infty t^{z-1} e^{-t} \, dt = -e^{-t} t^{z-1} \Big|_0^\infty + (z-1) \int_0^\infty t^{z-2} e^{-t} \, dt$$

$$= (z-1)\Gamma(z-1),$$

since when $z \geq 1$, $\lim_{t \to \infty} e^{-t} t^{z-1} = \lim_{t \to 0} e^{-t} t^{z-1} = 0$, as can be seen using L'Hospital's Rule. Using (A.2) repeatedly, if $n \leq z < n+1$ for some positive integer n, then

$$\text{(A.3)} \quad \Gamma(z) = (z-1)\Gamma(z-1) = (z-1)(z-2)\Gamma(z-2) = \cdots$$

$$= (z-1)(z-2) \cdots (z-n+1)\Gamma(z-n+1).$$

Now, if $z = n$, then by (A.3),

$$\Gamma(n) = (n-1)(n-2) \cdots 2(1)\Gamma(1) = (n-1)! \, \Gamma(1).$$

However,

$$\Gamma(1) = \int_0^\infty e^{-t} \, dt = -e^{-t} \Big|_0^\infty = 1.$$

Thus, $\Gamma(n) = (n-1)!$, or equivalently,

(A.4) $$\Gamma(n+1) = n!.$$

In obtaining (A.2) and (A.3), we implicitly assumed that the definite integral in (A.1) is well defined and finite for positive values of z. This is, in fact, the case, and indeed it can be shown [see, e.g., Artin (1964)] that the integral in Equation (A.1) is defined and finite for all real numbers z other than $0, -1, -2, -3, \ldots$, in which case the integral in (A.1) is either $+\infty$ or $-\infty$. The graph of $\Gamma(z)$ is given in Figure A.1. From Figure A.1 it is apparent that $\Gamma(z)$ is continuous (and indeed differentiable) at all values z other than $z = 0, -1, -2, \ldots$.

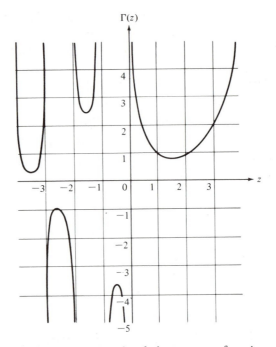

Figure A.1: *Graph of the gamma function* $\Gamma(z)$ *for* $-4 \le z \le 4$. *For* $z > 4$, *the graph continues to rise as shown. For* $z < -4$, *the graph continues to alternate between positive and negative curves of parabolic shape, where the minima of the positive curves and the maxima of the negative curves come closer and closer to* 0 *as* $z \to -\infty$.

Once we have values of $\Gamma(a)$ for $1 \le a < 2$, the value of $\Gamma(z)$ for any $z \ge 2$ can be obtained from (A.3). On the other hand, if $0 < z < 1$, then since

$$\Gamma(1 + z) = \int_0^\infty t^z e^{-t}\, dt = -e^{-t}t^z\Big|_0^\infty + z \int_0^\infty t^{z-1} e^{-t}\, dt,$$

and $\lim_{t\to\infty} e^{-t}t^z = \lim_{t\to 0} e^{-t}t^z = 0$, when $0 < z < 1$, it follows that

(A.5) $$\Gamma(z) = \frac{\Gamma(1 + z)}{z},$$

so that again we can obtain $\Gamma(z)$ from a table of values for $\Gamma(a)$, $1 \le a < 2$. The relation (A.5) can also be shown to hold when z is negative, but not an integer. Thus, we conclude that the entire gamma function $\Gamma(z)$ can be computed [using (A.3) and (A.5)] once we know the values of $\Gamma(a)$ for $1 \le a < 2$. These values are provided in Table A.1.

For example, to compute $\Gamma(3.45)$, using (A.3) with $n = 3$ and Table A.1, we obtain

$$\Gamma(3.45) = (2.45)(1.45)\Gamma(1.45) = (2.45)(1.45)(0.8857) = 3.1287.$$

Similarly, using (A.5) repeatedly,

$$\Gamma(-3.45) = \frac{\Gamma(-2.45)}{-3.45} = \frac{\Gamma(-1.45)}{(-3.45)(-2.45)} = \frac{\Gamma(0.45)}{(-3.45)(-2.45)(-1.45)}$$

$$= \frac{\Gamma(1.45)}{(-3.45)(-2.45)(-1.45)(0.45)} = -0.0016,$$

$$\Gamma(0.67) = \frac{\Gamma(1.67)}{0.67} = \frac{0.9033}{0.67} = 1.3482.$$

Table A.1: *Values of the Gamma Function $\Gamma(a)$, for $1 \le a < 2$*

a	0.00	0.01	0.02	0.03	0.04	0.05	0.06	0.07	0.08	0.09
1.00	1.0000	0.9943	0.9888	0.9835	0.9784	0.9735	0.9687	0.9642	0.9597	0.9555
1.10	0.9514	0.9474	0.9436	0.9399	0.9364	0.9330	0.9298	0.9267	0.9237	0.9209
1.20	0.9182	0.9156	0.9131	0.9108	0.9085	0.9064	0.9044	0.9025	0.9007	0.8990
1.30	0.8975	0.8960	0.8946	0.8934	0.8922	0.8912	0.8902	0.8893	0.8885	0.8879
1.40	0.8873	0.8868	0.8864	0.8860	0.8858	0.8857	0.8856	0.8856	0.8857	0.8859
1.50	0.8862	0.8866	0.8870	0.8876	0.8882	0.8889	0.8896	0.8905	0.8914	0.8924
1.60	0.8935	0.8947	0.8959	0.8972	0.8986	0.9001	0.9017	0.9033	0.9050	0.9068
1.70	0.9086	0.9106	0.9126	0.9147	0.9168	0.9191	0.9214	0.9238	0.9262	0.9288
1.80	0.9314	0.9341	0.9368	0.9397	0.9426	0.9456	0.9487	0.9518	0.9551	0.9384
1.90	0.9618	0.9652	0.9688	0.9724	0.9761	0.9799	0.9837	0.9877	0.9917	0.9948

By making various changes of variable in (A.1), we obtain other integral representations for $\Gamma(z)$ which frequently are useful. In (A.1) make the change of variable $u = \sqrt{t}$. Then, $dt = 2u\,du$, and

(A.6)
$$\Gamma(z) = 2\int_0^\infty u^{2z-1}e^{-u^2}\,du.$$

Now, changing variable from u to $v = \sqrt{2}\,u$, we obtain

(A.7)
$$\Gamma(z) = \frac{1}{2^{z-1}}\int_0^\infty v^{2z-1}e^{-(1/2)v^2}\,dv.$$

There is one value of z, namely $z = \frac{1}{2}$, which is a critical argument in the gamma function. We now show that

(A.8)
$$\Gamma(\tfrac{1}{2}) = \sqrt{\pi}.$$

There appears to be only one elementary proof of this fact. Using (A.6), we obtain

$$\Gamma(\tfrac{1}{2}) = 2\int_0^\infty e^{-u^2}\,du,$$

so that

(A.9)
$$[\Gamma(\tfrac{1}{2})]^2 = \left(2\int_0^\infty e^{-u^2}\,du\right)^2 = 4\int_0^\infty e^{-u^2}\,du \int_0^\infty e^{-v^2}\,dv$$

$$= 4\int_0^\infty \int_0^\infty e^{-(u^2+v^2)}\,du\,dv.$$

Now let $u = r\sin\theta$, $v = r\cos\theta$ be a transformation to polar coordinates. Then $du\,dv = r\,dr\,d\theta$, and (A.9) becomes

(A.10)
$$[\Gamma(\tfrac{1}{2})]^2 = 4\int_0^{\pi/2}\int_0^\infty re^{-r^2}\,dr\,d\theta$$

$$= 4\left(\int_0^{\pi/2} d\theta\right)\left(\int_0^\infty re^{-r^2}\,dr\right) = 4\left(\frac{\pi}{2}\right)\frac{\Gamma(1)}{2}$$

$$= \pi,$$

which completes the proof.

Stirling's Formula

The computation of $n!$ for large values of n [or of $\Gamma(z)$ for large values of z] is no easy task. We now try to find a more convenient mathematical expression to approximate $\Gamma(z)$.

For $x > 0$, a direct calculation shows that

$$\left(\frac{d}{dx}\right)^2 \left[x \log\left(1 + \frac{1}{x}\right)\right] = \frac{d}{dx}\left[\log\left(1 + \frac{1}{x}\right) - \frac{1}{1+x}\right]$$

$$= -\frac{1}{(x+1)^2 x} < 0,$$

$$\left(\frac{d}{dx}\right)^2 \left[(x+1) \log\left(1 + \frac{1}{x}\right)\right] = \frac{d}{dx}\left[\log\left(1 + \frac{1}{x}\right) - \frac{1}{x}\right]$$

$$= \frac{1}{x^2(1+x)} > 0,$$

so that

$$\frac{d}{dx}\left[x \log\left(1 + \frac{1}{x}\right)\right] = \log\left(1 + \frac{1}{x}\right) - \frac{1}{1+x}$$

is decreasing in x, and

$$\frac{d}{dx}\left[(x+1) \log\left(1 + \frac{1}{x}\right)\right] = \log\left(1 + \frac{1}{x}\right) - \frac{1}{x}$$

is increasing in x. It follows that

$$\frac{d}{dx}\left[x \log\left(1 + \frac{1}{x}\right)\right] \geq \lim_{x \to \infty}\left[\log\left(1 + \frac{1}{x}\right) - \frac{1}{1+x}\right] = 0,$$

$$\frac{d}{dx}\left[(x+1) \log\left(1 + \frac{1}{x}\right)\right] \leq \lim_{x \to \infty}\left[\log\left(1 + \frac{1}{x}\right) - \frac{1}{x}\right] = 0,$$

from which we conclude that $x \log(1 + 1/x)$ is increasing in x and $(x+1)\log(1 + 1/x)$ is decreasing in x, for $x \geq 0$. Consequently,

$$\left(1 + \frac{1}{x}\right)^x = e^{x \log(1 + 1/x)} \leq e^{\lim_{x \to \infty} x \log(1 + 1/x)} = e,$$

$$\left(1 + \frac{1}{x}\right)^{x+1} = e^{(x+1)\log(1 + 1/x)} \geq e^{\lim_{x \to \infty}(x+1)\log(1 + 1/x)} = e.$$

Putting these results together yields the well-known inequality

(A.11)
$$\left(1 + \frac{1}{x}\right)^x \leq e \leq \left(1 + \frac{1}{x}\right)^{x+1}.$$

Substituting $x = 1, 2, 3, \ldots, n - 1$, in (A.11), we obtain

$$\left(1 + \frac{1}{j}\right)^{j} \leq e \leq \left(1 + \frac{1}{j}\right)^{j+1}, \quad j = 1, \ldots, n - 1,$$

and multiplying these $(n - 1)$ inequalities together yields

$$\left(1 + \frac{1}{1}\right)^{1}\left(1 + \frac{1}{2}\right)^{2} \cdots \left(1 + \frac{1}{n-1}\right)^{n-1} \leq e^{n-1}$$

$$\leq \left(1 + \frac{1}{1}\right)^{2}\left(1 + \frac{1}{2}\right)^{3} \cdots \left(1 + \frac{1}{n-1}\right)^{n},$$

or

(A.12) $$\left(\frac{2}{1}\right)^{1}\left(\frac{3}{2}\right)^{2} \cdots \left(\frac{n}{n-1}\right)^{n-1} \leq e^{n-1} \leq \left(\frac{2}{1}\right)^{2}\left(\frac{3}{2}\right)^{3} \cdots \left(\frac{n}{n-1}\right)^{n}.$$

Canceling terms in (A.12), we are able to simplify to

(A.13) $$\frac{n^{n-1}}{(n-1)!} \leq e^{n-1} \leq \frac{n^{n}}{(n-1)!}.$$

Finally, multiplying all sides of (A.13) by $n!e^{-(n-1)}$, we find that

(A.14) $$n^{n}e^{-(n-1)} \leq n! \leq n^{n+1}e^{-(n-1)}.$$

Thus, we see that $n!$ is not free to increase with n at any rate, but must lie between $n^{n}e^{-(n-1)}$ and $n^{n+1}e^{-(n-1)}$.

Stirling's formula provides a compromise between these upper and lower bounds and approximates $n!$ by

(A.15) $$n! \simeq \sqrt{2\pi}\, n^{n+1/2}e^{-n} \equiv S(n).$$

Table A.2 shows the accuracy of the approximation $n! \cong S(n)$ for $n = 1(1)15$.

Note from Table A.2 that the difference $n! - S(n)$ is increasing as n increases. Indeed, $n! - S(n) \to \infty$ as $n \to \infty$. However, the *percent error* between $n!$ and $S(n)$ [i.e., $100(n! - S(n))/n!$] stays small, and in fact converges to 0 as $n \to \infty$. Thus, even for relatively small values of n, the Stirling approximation to $n!$ is proportionately quite accurate.

Remark. In Feller (1968), it is shown that $n!/S(n) = e^{r(n)/12n}$, where $(12n)/(12n + 1) < r(n) < 1$. Hence, as $n \to \infty$, $[n!/S(n)] \to 1$.

There is a similar Stirling approximation to $\Gamma(z + 1)$; namely, $S(z) = \sqrt{2\pi}\, z^{z+1/2}e^{-z}$. This approximation also has the property that the percentage error, $100(\Gamma(z + 1) - S(z))/\Gamma(z + 1)$, is small even for relatively small values of z, and converges to 0 as $z \to \infty$.

Table A.2: *Values of n! and S(n) for n = 1(1)15*

n	$n!$	$S(n)$	Percent error $= 100 \left[\dfrac{n! - S(n)}{n!} \right]$
1	1	0.9221370089	7.79
2	2	1.919004351	4.05
3	6	5.836209591	2.73
4	24	23.50617513	2.06
5	120	118.0191680	1.65
6	720	710.0781847	1.38
7	5,040	4,980.395833	1.18
8	40,320	39,902.39545	1.04
9	362,880	359,536.8729	0.92
10	3,628,800	3,598,695.619	0.83
11	39,916,800	39,615,625.05	0.75
12	479,001,600	475,687,486.5	0.69
13	6,227,020,800	6,187,239,475	0.64
14	87,178,291,200	86,661,001,780	0.59
15	1,307,674,368,000	1,300,430,722,000	0.55

The Beta Function

Earlier, we mentioned that the integral (A.1) is known as Euler's second integral. Euler's first integral is

(A.16)
$$B(x, y) = \int_0^1 t^{x-1}(1 - t)^{y-1}\, dt,$$

and, as a function of $x > 0$, $y > 0$, is known as the *beta function*. The beta function is related to the gamma function by the formula

(A.17)
$$B(x, y) = \frac{\Gamma(x)\Gamma(y)}{\Gamma(x + y)}.$$

To prove this relationship, we make use of (A.6):

$$\Gamma(x)\Gamma(y) = \left(2 \int_0^\infty u^{2x-1} e^{-u^2}\, du \right)\left(2 \int_0^\infty v^{2y-1} e^{-v^2}\, dv \right)$$

$$= 4 \int_0^\infty \int_0^\infty u^{2x-1} v^{2y-1} e^{-(u^2 + v^2)}\, du\, dv.$$

Let $u = r \sin \theta$, $v = r \cos \theta$. Then

$$\Gamma(x)\Gamma(y) = \left(2 \int_0^\infty r^{2(x+y)-1} e^{-r^2} \, dr\right)$$

(A.18)
$$\times \left(2 \int_0^{\pi/2} (\sin \theta)^{2x-1}(\cos \theta)^{2y-1} \, d\theta\right)$$

$$= \Gamma(x+y)\left(2 \int_0^{\pi/2} (\sin \theta)^{2x-1}(\cos \theta)^{2y-1} \, d\theta\right).$$

Now

$$B(x, y) = \int_0^1 t^{x-1}(1-t)^{y-1} \, dt$$

(A.19)
$$= 2 \int_0^{\pi/2} (\sin \theta)^{2x-1}(\cos \theta)^{2y-1} \, d\theta$$

as can be seen by making the change of variable from t to $\theta = \sin^{-1}[(t)^{1/2}]$, and noting that $t = \sin^2(\theta)$, so that $dt = 2 \sin \theta \cos \theta \, d\theta$. Putting (A.18) and (A.19) together yields (A.17).

Equation (A.19) gives one alternative integral expression for $B(x, y)$. Another expression can be obtained by making the change of variable from t to $s = t(1-t)^{-1}$ in (A.16). Since $t = s/(1+s)$, we have $dt = (1+s)^{-2} \, ds$, and

(A.20)
$$B(x, y) = \int_0^\infty \frac{s^{x-1}}{(1+s)^{x+y}} \, ds.$$

Still a third expression is obtained by letting $u = 1 - t$ in (A.16). We obtain

(A.21) $$B(x, y) = \int_0^1 t^{x-1}(1-t)^{y-1} \, dt = \int_0^1 (1-u)^{x-1}u^{y-1} \, du = B(y, x),$$

a relationship that could also have been inferred from the symmetry in the definition (A.16) of $B(x, y)$.

The beta function is closely related to the binomial coefficient $\binom{n}{k}$ through the expression

(A.22)
$$B(k+1, n-k+1) = \frac{1}{(n+1)\binom{n}{k}}.$$

Computation of $B(x, y)$ follows directly from (A.17) and computations with the gamma function. For example,

$$B(2.45, 2.67) = \frac{\Gamma(2.45)\Gamma(2.67)}{\Gamma(2.45 + 2.67)} = \frac{(1.45)(1.67)\Gamma(1.45)\Gamma(1.67)}{(4.12)(3.12)(2.12)\Gamma(1.12)}$$

$$= \frac{(1.45)(1.67)(0.8857)(0.9033)}{(4.12)(3.12)(2.12)(0.9436)} = 0.0753.$$

If one of the arguments is an integer n, then

$$B(n, y) = \frac{\Gamma(n)\Gamma(y)}{\Gamma(n + y)} = \frac{(n - 1)!\,\Gamma(y)}{(n - 1 + y)(n - 2 + y) \cdots (1 + y)y\Gamma(y)}$$

$$= \frac{(n - 1)!}{(n - 1 + y)(n - 2 + y) \cdots (1 + y)(y)}.$$

If both of the arguments are integers $(x = n, y = m)$, then

$$B(n, m) = \frac{\Gamma(n)\Gamma(m)}{\Gamma(n + m)} = \frac{(n - 1)!\,(m - 1)!}{(n + m - 1)!}.$$

Finally, the recursive formulas

$$B(x, y) = \frac{y - 1}{x} B(x + 1, y - 1),$$

$$B(x, y) = \frac{(x - 1)(y - 1)}{(x + y - 1)(x + y - 2)} B(x - 1, y - 1),$$

often are useful.

Appendix B

Tables

Table T.1: *Individual Terms,* $p_Z(k) = \binom{n}{k}p^k(1-p)^{n-k}$, *of the Probability Mass Function of a Binomially Distributed Random Variable Z*

					p			
n	k	0.01	0.02	0.03	0.04	0.05	0.10	0.15
1	0	0.99000	0.98000	0.97000	0.96000	0.95000	0.90000	0.85000
	1	0.01000	0.02000	0.03000	0.04000	0.05000	0.10000	0.15000
2	0	0.98010	0.96040	0.94090	0.92160	0.90250	0.81000	0.72250
	1	0.01980	0.03920	0.05820	0.07680	0.09500	0.18000	0.25500
	2	0.00010	0.00040	0.00090	0.00160	0.00250	0.01000	0.02250
3	0	0.97030	0.94119	0.91267	0.88474	0.85738	0.72900	0.61413
	1	0.02940	0.05762	0.08468	0.11059	0.13538	0.24300	0.32513
	2	0.00030	0.00118	0.00262	0.00461	0.00712	0.02700	0.05737
	3	0.00000	0.00001	0.00003	0.00006	0.00013	0.00100	0.00337
4	0	0.96060	0.92237	0.88529	0.84935	0.81451	0.65610	0.52201
	1	0.03881	0.07530	0.10952	0.14156	0.17148	0.29160	0.36848
	2	0.00059	0.00230	0.00508	0.00885	0.01354	0.04860	0.09754
	3	0.00000	0.00003	0.00010	0.00025	0.00047	0.00360	0.01148
	4	0.00000	0.00000	0.00000	0.00000	0.00001	0.00010	0.00051
5	0	0.95099	0.90392	0.85873	0.81537	0.77378	0.59049	0.44371
	1	0.04803	0.09224	0.13279	0.16987	0.20363	0.32805	0.39150
	2	0.00097	0.00376	0.00821	0.01416	0.02143	0.07290	0.13818
	3	0.00001	0.00008	0.00025	0.00059	0.00113	0.00810	0.02438
	4	0.00000	0.00000	0.00000	0.00001	0.00003	0.00045	0.00215
	5	0.00000	0.00000	0.00000	0.00000	0.00000	0.00001	0.00008
6	0	0.94148	0.88584	0.83297	0.78276	0.73509	0.53144	0.37715
	1	0.05706	0.10847	0.15457	0.19569	0.23213	0.35429	0.39933
	2	0.00144	0.00553	0.01195	0.02038	0.03054	0.09842	0.17618
	3	0.00002	0.00015	0.00049	0.00113	0.00214	0.01458	0.04145
	4	0.00000	0.00000	0.00001	0.00004	0.00008	0.00121	0.00549
	5	0.00000	0.00000	0.00000	0.00000	0.00000	0.00005	0.00039
	6	0.00000	0.00000	0.00000	0.00000	0.00000	0.00000	0.00001

Table T.1: (*continued*)

			p					
0.20	0.25	0.30	0.35	0.40	0.45	0.50	*n*	*k*
0.80000	0.75000	0.70000	0.65000	0.60000	0.55000	0.50000	1	0
0.20000	0.25000	0.30000	0.35000	0.40000	0.45000	0.50000		1
0.64000	0.56250	0.49000	0.42250	0.36000	0.30250	0.25000	2	0
0.32000	0.37500	0.42000	0.45500	0.48000	0.49500	0.50000		1
0.04000	0.06250	0.09000	0.12250	0.16000	0.20250	0.25000		2
0.51200	0.42188	0.34300	0.27463	0.21600	0.16638	0.12500	3	0
0.38400	0.42188	0.44100	0.44362	0.43200	0.40838	0.37500		1
0.09600	0.14063	0.18900	0.23888	0.28800	0.33413	0.37500		2
0.00800	0.01563	0.02700	0.04287	0.06400	0.09113	0.12500		3
0.40960	0.31641	0.24010	0.17851	0.12960	0.09151	0.06250	4	0
0.40960	0.42188	0.41160	0.38447	0.34560	0.29948	0.25000		1
0.15360	0.21094	0.26460	0.31054	0.34560	0.36754	0.37500		2
0.02560	0.04688	0.07560	0.11148	0.15360	0.20048	0.25000		3
0.00160	0.00391	0.00810	0.01501	0.02560	0.04101	0.06250		4
0.32768	0.23730	0.16807	0.11603	0.07776	0.05033	0.03125	5	0
0.40960	0.39551	0.36015	0.31239	0.25920	0.20589	0.15625		1
0.20480	0.26367	0.30870	0.33642	0.34560	0.33691	0.31250		2
0.05120	0.08789	0.13230	0.18115	0.23040	0.27565	0.31250		3
0.00640	0.01465	0.02835	0.04877	0.07680	0.11277	0.15625		4
0.00032	0.00098	0.00243	0.00525	0.01024	0.01845	0.03125		5
0.26214	0.17798	0.11765	0.07542	0.04666	0.02768	0.01563	6	0
0.39322	0.35596	0.30253	0.24366	0.18662	0.13589	0.09375		1
0.24576	0.29663	0.32413	0.32801	0.31104	0.27795	0.23438		2
0.08192	0.13184	0.18522	0.23549	0.27648	0.30322	0.31250		3
0.01536	0.03296	0.05953	0.09510	0.13824	0.18607	0.23438		4
0.00154	0.00439	0.01021	0.02048	0.03686	0.06089	0.09375		5
0.00006	0.00024	0.00073	0.00184	0.00410	0.00830	0.01563		6

Table T.1 (*continued*)

					p			
n	k	0.01	0.02	0.03	0.04	0.05	0.10	0.15
7	0	0.93207	0.86813	0.80798	0.75145	0.69834	0.47830	0.32058
	1	0.06590	0.12402	0.17492	0.21917	0.25728	0.37201	0.39601
	2	0.00200	0.00759	0.01623	0.02740	0.04062	0.12400	0.20965
	3	0.00003	0.00026	0.00084	0.00190	0.00356	0.02296	0.06166
	4	0.00000	0.00001	0.00003	0.00008	0.00019	0.00255	0.01088
	5	0.00000	0.00000	0.00000	0.00000	0.00001	0.00017	0.00115
	6	0.00000	0.00000	0.00000	0.00000	0.00000	0.00001	0.00007
	7	0.00000	0.00000	0.00000	0.00000	0.00000	0.00000	0.00000
8	0	0.92274	0.85076	0.78374	0.72139	0.66342	0.43047	0.27249
	1	0.07457	0.13890	0.19392	0.24046	0.27933	0.38264	0.38469
	2	0.00264	0.00992	0.02099	0.03507	0.05146	0.14880	0.23760
	3	0.00005	0.00040	0.00130	0.00292	0.00542	0.03307	0.08386
	4	0.00000	0.00001	0.00005	0.00015	0.00036	0.00459	0.01850
	5	0.00000	0.00000	0.00000	0.00001	0.00002	0.00041	0.00261
	6	0.00000	0.00000	0.00000	0.00000	0.00000	0.00002	0.00023
	7	0.00000	0.00000	0.00000	0.00000	0.00000	0.00000	0.00001
	8	0.00000	0.00000	0.00000	0.00000	0.00000	0.00000	0.00000
9	0	0.91352	0.83375	0.76023	0.69253	0.63025	0.38742	0.23162
	1	0.08305	0.15314	0.21161	0.25970	0.29854	0.38742	0.36786
	2	0.00336	0.01250	0.02618	0.04328	0.06285	0.17219	0.25967
	3	0.00008	0.00060	0.00189	0.00421	0.00772	0.04464	0.10692
	4	0.00000	0.00002	0.00009	0.00026	0.00061	0.00744	0.02830
	5	0.00000	0.00000	0.00000	0.00001	0.00003	0.00083	0.00499
	6	0.00000	0.00000	0.00000	0.00000	0.00000	0.00006	0.00059
	7	0.00000	0.00000	0.00000	0.00000	0.00000	0.00000	0.00004
	8	0.00000	0.00000	0.00000	0.00000	0.00000	0.00000	0.00000
	9	0.00000	0.00000	0.00000	0.00000	0.00000	0.00000	0.00000
10	0	0.90438	0.81707	0.73742	0.66483	0.59874	0.34868	0.19687
	1	0.09135	0.16675	0.22807	0.27701	0.31512	0.38742	0.34743
	2	0.00415	0.01531	0.03174	0.05194	0.07463	0.19371	0.27590
	3	0.00011	0.00083	0.00262	0.00577	0.01048	0.05740	0.12983
	4	0.00000	0.00003	0.00014	0.00042	0.00096	0.01116	0.04010
	5	0.00000	0.00000	0.00001	0.00002	0.00006	0.00149	0.00849
	6	0.00000	0.00000	0.00000	0.00000	0.00000	0.00014	0.00125
	7	0.00000	0.00000	0.00000	0.00000	0.00000	0.00001	0.00013
	8	0.00000	0.00000	0.00000	0.00000	0.00000	0.00000	0.00001
	9	0.00000	0.00000	0.00000	0.00000	0.00000	0.00000	0.00000
	10	0.00000	0.00000	0.00000	0.00000	0.00000	0.00000	0.00000

Table T.1: (*continued*)

0.20	0.25	0.30	0.35	0.40	0.45	0.50	n	k
0.20972	0.13348	0.08235	0.04902	0.02799	0.01522	0.00781	7	0
0.36700	0.31146	0.24706	0.18478	0.13064	0.08719	0.05469		1
0.27525	0.31146	0.31765	0.29848	0.26127	0.21402	0.16406		2
0.11469	0.17303	0.22689	0.26787	0.29030	0.29185	0.27344		3
0.02867	0.05768	0.09724	0.14424	0.19354	0.23878	0.27344		4
0.00430	0.01154	0.02500	0.04660	0.07741	0.11722	0.16406		5
0.00036	0.00128	0.00357	0.00836	0.01720	0.03197	0.05469		6
0.00001	0.00006	0.00022	0.00064	0.00164	0.00374	0.00781		7
0.16777	0.10011	0.05765	0.03186	0.01680	0.00837	0.00391	8	0
0.33554	0.26697	0.19765	0.13726	0.08958	0.05481	0.03125		1
0.29360	0.31146	0.29648	0.25869	0.20902	0.15695	0.10938		2
0.14680	0.20764	0.25412	0.27859	0.27869	0.25683	0.21875		3
0.04588	0.08652	0.13614	0.18751	0.23224	0.26266	0.27344		4
0.00918	0.02307	0.04668	0.08077	0.12386	0.17192	0.21875		5
0.00115	0.00385	0.01000	0.02175	0.04129	0.07033	0.10938		6
0.00008	0.00037	0.00122	0.00335	0.00786	0.01644	0.03125		7
0.00000	0.00002	0.00007	0.00023	0.00066	0.00168	0.00391		8
0.13422	0.07508	0.04035	0.02071	0.01008	0.00461	0.00195	9	0
0.30199	0.22525	0.15565	0.10037	0.06047	0.03391	0.01758		1
0.30199	0.30034	0.26683	0.21619	0.16124	0.11099	0.07031		2
0.17616	0.23360	0.26683	0.27162	0.25082	0.21188	0.16406		3
0.06606	0.11680	0.17153	0.21939	0.25082	0.26004	0.24609		4
0.01652	0.03893	0.07351	0.11813	0.16722	0.21276	0.24609		5
0.00275	0.00865	0.02100	0.04241	0.07432	0.11605	0.16406		6
0.00029	0.00124	0.00386	0.00979	0.02123	0.04069	0.07031		7
0.00002	0.00010	0.00041	0.00132	0.00354	0.00832	0.01758		8
0.00000	0.00000	0.00002	0.00008	0.00026	0.00076	0.00195		9
0.10737	0.05631	0.02825	0.01346	0.00605	0.00253	0.00098	10	0
0.26844	0.18771	0.12106	0.07249	0.04031	0.02072	0.00977		1
0.30199	0.28157	0.23347	0.17565	0.12093	0.07630	0.04395		2
0.20133	0.25028	0.26683	0.25222	0.21499	0.16648	0.11719		3
0.08808	0.14600	0.20012	0.23767	0.25082	0.23837	0.20508		4
0.02642	0.05840	0.10292	0.15357	0.20066	0.23403	0.24609		5
0.00551	0.01622	0.03676	0.06891	0.11148	0.15957	0.20508		6
0.00079	0.00309	0.00900	0.02120	0.04247	0.07460	0.11719		7
0.00007	0.00039	0.00145	0.00428	0.01062	0.02289	0.04395		8
0.00000	0.00003	0.00014	0.00051	0.00157	0.00416	0.00977		9
0.00000	0.00000	0.00001	0.00003	0.00010	0.00034	0.00098		10

Table T.2: Values for the Probability Mass Function
of the Hypergeometric Distribution for N = 2(1)9 and Selected Values of n and p

n	Np	k	$p_X(k)$	n	Np	k	$p_X(k)$	n	Np	k	$p_X(k)$
		$N = 2$						5	1	0	0.1667
1	1	0	0.5000	3	3	2	0.6000			1	0.8333
		1	0.5000			3	0.1000	5	2	1	0.3333
		$N = 3$		4	1	0	0.2000			2	0.6667
1	1	0	0.6667			1	0.8000	5	3	2	0.5000
		1	0.3333	4	2	1	0.4000			3	0.5000
						2	0.6000	5	4	3	0.6667
2	1	0	0.3333	4	3	2	0.6000			4	0.3333
		1	0.6667			3	0.4000	5	5	4	0.8333
2	2	1	0.6667	4	4	3	0.8000			5	0.1667
		2	0.3333			4	0.2000			$N = 7$	
		$N = 4$				$N = 6$		1	1	0	0.8571
1	1	0	0.7500	1	1	0	0.8333			1	0.1429
		1	0.2500			1	0.1667	2	1	0	0.7143
2	1	0	0.5000							1	0.2857
		1	0.5000	2	1	0	0.6667	2	2	0	0.4762
2	2	0	0.1667			1	0.3333			1	0.4762
		1	0.6667	2	2	0	0.4000			2	0.0476
		2	0.1667			1	0.5333				
						2	0.0667	3	1	0	0.5714
3	1	0	0.2500							1	0.4286
		1	0.7500	3	1	0	0.5000	3	2	0	0.2857
3	2	1	0.5000			1	0.5000			1	0.5714
		2	0.5000	3	2	0	0.2000			2	0.1429
3	3	2	0.7500			1	0.6000	3	3	0	0.1143
		3	0.2500			2	0.2000			1	0.5143
		$N = 5$		3	3	0	0.0500			2	0.3429
1	1	0	0.8000			1	0.4500			3	0.0286
		1	0.2000			2	0.4500				
						3	0.0500	4	1	0	0.4286
2	1	0	0.6000							1	0.5714
		1	0.4000	4	1	0	0.3333	4	2	0	0.1429
2	2	0	0.3000			1	0.6667			1	0.5714
		1	0.6000	4	2	0	0.0667			2	0.2857
		2	0.1000			1	0.5333	4	3	0	0.0286
						2	0.4000			1	0.3429
3	1	0	0.4000	4	3	1	0.2000			2	0.5143
		1	0.6000			2	0.6000			3	0.1143
3	2	0	0.1000			3	0.2000	4	4	1	0.1143
		1	0.6000	4	4	2	0.4000			2	0.5143
		2	0.3000			3	0.5333			3	0.3429
3	3	1	0.3000			4	0.0667			4	0.0286

Table T.2 (continued)

n	Np	k	$p_X(k)$	n	Np	k	$p_X(k)$	n	Np	k	$p_X(k)$
5	1	0	0.2857	3	3	1	0.5357	6	3	2	0.5357
		1	0.7143			2	0.2679			3	0.3571
5	2	0	0.0476			3	0.0179	6	4	2	0.2143
		1	0.4762							3	0.5714
		2	0.4762	4	1	0	0.5000			4	0.2143
5	3	1	0.1429			1	0.5000	6	5	3	0.3571
		2	0.5714	4	2	0	0.2143			4	0.5357
		3	0.2857			1	0.5714			5	0.1071
5	4	2	0.2857			2	0.2143	6	6	4	0.5357
		3	0.5714	4	3	0	0.0714			5	0.4286
		4	0.1429			1	0.4286			6	0.0357
5	5	3	0.4762			2	0.4286				
		4	0.4762			3	0.0714	7	1	0	0.1250
		5	0.0476	4	4	0	0.0143			1	0.8750
						1	0.2286	7	2	1	0.2500
6	1	0	0.1429			2	0.5143			2	0.7500
		1	0.8571			3	0.2286	7	3	2	0.3750
6	2	1	0.2857			4	0.0143			3	0.6250
		2	0.7143					7	4	3	0.5000
6	3	2	0.4286	5	1	0	0.3750			4	0.5000
		3	0.5714			1	0.6250	7	5	4	0.6250
6	4	3	0.5714	5	2	0	0.1071			5	0.3750
		4	0.4286			1	0.5357	7	6	5	0.7500
6	5	4	0.7143			2	0.3571			6	0.2500
		5	0.2857	5	3	0	0.0179	7	7	6	0.8750
6	6	5	0.8571			1	0.2679			7	0.1250
		6	0.1429			2	0.5357		N = 9		
	N = 8					3	0.1786	1	1	0	0.8889
1	1	0	0.8750	5	4	1	0.0714			1	0.1111
		1	0.1250			2	0.4286				
						3	0.4286	2	1	0	0.7778
2	1	0	0.7500			4	0.0714			1	0.2222
		1	0.2500	5	5	2	0.1786	2	2	0	0.5833
2	2	0	0.5357			3	0.5357			1	0.3889
		1	0.4286			4	0.2679			2	0.0278
		2	0.0357			5	0.0179				
								3	1	0	0.6667
3	1	0	0.6250	6	1	0	0.2500			1	0.3333
		1	0.3750			1	0.7500	3	2	0	0.4167
3	2	0	0.3571	6	2	0	0.0357			1	0.5000
		1	0.5357			1	0.4286			2	0.0833
		2	0.1071			2	0.5357	3	3	0	0.2381
3	3	0	0.1786	6	3	1	0.1071			1	0.5357

Table T.2 (continued)

n	Np	k	$p_X(k)$	n	Np	k	$p_X(k)$	n	Np	k	$p_X(k)$
	N = 9			5	5	1	0.0397	7	3	1	0.0833
3	3	2	0.2143			2	0.3175			2	0.5000
		3	0.0119			3	0.4762			3	0.4167
						4	0.1587	7	4	2	0.1667
4	1	0	0.5556			5	0.0079			3	0.5556
		1	0.4444							4	0.2778
4	2	0	0.2778	6	1	0	0.3333	7	5	3	0.2778
		1	0.5556			1	0.6667			4	0.5556
		2	0.1667	6	2	0	0.0833			5	0.1667
4	3	0	0.1190			1	0.5000	7	6	4	0.4167
		1	0.4762			2	0.4167			5	0.5000
		2	0.3571	6	3	0	0.0119			6	0.0833
		3	0.0476			1	0.2143	7	7	5	0.5833
4	4	0	0.0397			2	0.5357			6	0.3889
		1	0.3175			3	0.2381			7	0.0278
		2	0.4762	6	4	1	0.0476				
		3	0.1587			2	0.3571	8	1	0	0.1111
		4	0.0079			3	0.4762			1	0.8889
5	1	0	0.4444			4	0.1190	8	2	1	0.2222
		1	0.5556	6	5	2	0.1190			2	0.7778
5	2	0	0.1667			3	0.4762	8	3	2	0.3333
		1	0.5556			4	0.3571			3	0.6667
		2	0.2778			5	0.0476	8	4	3	0.4444
5	3	0	0.0476	6	6	3	0.2381			4	0.5556
		1	0.3571			4	0.5357	8	5	4	0.5556
		2	0.4762			5	0.2143			5	0.4444
		3	0.1190			6	0.0119	8	6	5	0.6667
5	4	0	0.0079							6	0.3333
		1	0.1587	7	1	0	0.2222	8	7	6	0.7778
		2	0.4762			1	0.7778			7	0.2222
		3	0.3175	7	2	0	0.0278	8	8	7	0.8889
		4	0.0397			1	0.3889			8	0.1111
						2	0.5833				

Table T.3: Individual Terms, $p_X(x)$, of the Probability Mass Function of the Poisson Distribution with Parameter λ

					λ					
x	0.1	0.2	0.3	0.4	0.5	0.6	0.7	0.8	0.9	1.0
0	.9048	.8187	.7408	.6703	.6065	.5488	.4966	.4493	.4066	.3679
1	.0905	.1637	.2222	.2681	.3033	.3293	.3476	.3595	.3659	.3679
2	.0045	.0164	.0333	.0536	.0758	.0988	.1217	.1438	.1647	.1839
3	.0002	.0011	.0033	.0072	.0126	.0198	.0284	.0383	.0494	.0613
4	.0000	.0001	.0003	.0007	.0016	.0030	.0050	.0077	.0111	.0153
5	.0000	.0000	.0000	.0001	.0002	.0004	.0007	.0012	.0020	.0031
6	.0000	.0000	.0000	.0000	.0000	.0000	.0001	.0002	.0003	.0005
7	.0000	.0000	.0000	.0000	.0000	.0000	.0000	.0000	.0000	.0001

					λ					
x	1.1	1.2	1.3	1.4	1.5	1.6	1.7	1.8	1.9	2.0
0	.3329	.3012	.2725	.2466	.2231	.2019	.1827	.1653	.1496	.1353
1	.3662	.3614	.3543	.3452	.3347	.3230	.3106	.2975	.2842	.2707
2	.2014	.2169	.2303	.2417	.2510	.2584	.2640	.2678	.2700	.2707
3	.0738	.0867	.0998	.1128	.1255	.1378	.1496	.1607	.1710	.1804
4	.0203	.0260	.0324	.0395	.0471	.0551	.0636	.0723	.0812	.0902
5	.0045	.0062	.0084	.0111	.0141	.0176	.0216	.0260	.0309	.0361
6	.0008	.0012	.0018	.0026	.0035	.0047	.0061	.0078	.0098	.0120
7	.0001	.0002	.0003	.0005	.0008	.0011	.0015	.0020	.0027	.0034
8	.0000	.0000	.0001	.0001	.0001	.0002	.0003	.0005	.0006	.0009
9	.0000	.0000	.0000	.0000	.0000	.0000	.0001	.0001	.0001	.0002

					λ					
x	2.1	2.2	2.3	2.4	2.5	2.6	2.7	2.8	2.9	3.0
0	.1225	.1108	.1003	.0907	.0821	.0743	.0672	.0608	.0550	0.498
1	.2572	.2438	.2306	.2177	.2052	.1931	.1815	.1703	.1596	.1494
2	.2700	.2681	.2652	.2613	.2565	.2510	.2450	.2384	.2314	.2240
3	.1890	.1966	.2033	.2090	.2138	.2176	.2205	.2225	.2237	.2240
4	.0992	.1082	.1169	.1254	.1336	.1414	.1488	.1557	.1622	.1680
5	.0417	.0476	.0538	.0602	.0668	.0735	.0804	.0872	.0940	.1008
6	.0146	.0174	.0206	.0241	.0278	.0319	.0362	.0407	.0455	.0504
7	.0044	.0055	.0068	.0083	.0099	.0118	.0139	.0163	.0188	.0216
8	.0011	.0015	.0019	.0025	.0031	.0038	.0047	.0057	.0068	.0081
9	.0003	.0004	.0005	.0007	.0009	.0011	.0014	.0018	.0022	.0027
10	.0001	.0001	.0001	.0002	.0002	.0003	.0004	.0005	.0006	.0008
11	.0000	.0000	.0000	.0000	.0000	.0001	.0001	.0001	.0002	.0002
12	.0000	.0000	.0000	.0000	.0000	.0000	.0000	.0000	.0000	.0001

Table T.3 (*continued*)

					λ					
x	3.1	3.2	3.3	3.4	3.5	3.6	3.7	3.8	3.9	4.0
0	.0450	.0408	.0369	.0334	.0302	.0273	.0247	.0224	.0202	.0183
1	.1397	.1304	.1217	.1135	.1057	.0984	.0915	.0850	.0789	.0733
2	.2165	.2087	.2008	.1929	.1850	.1771	.1692	.1615	.1539	.1465
3	.2237	.2226	.2209	.2186	.2158	.2125	.2087	.2046	.2001	.1954
4	.1734	.1781	.1823	.1858	.1888	.1912	.1931	.1944	.1951	.1954
5	.1075	.1140	.1203	.1264	.1322	.1377	.1429	.1477	.1522	.1563
6	.0555	.0608	.0662	.0716	.0771	.0826	.0881	.0936	.0989	.1042
7	.0246	.0278	.0312	.0348	.0385	.0425	.0466	.0508	.0551	.0595
8	.0095	.0111	.0129	.0148	.0169	.0191	.0215	.0241	.0269	.0298
9	.0033	.0040	.0047	.0056	.0066	.0076	.0089	.0102	.0116	.0132
10	.0010	.0013	.0016	.0019	.0023	.0028	.0033	.0039	.0045	.0053
11	.0003	.0004	.0005	.0006	.0007	.0009	.0011	.0013	.0016	.0019
12	.0001	.0001	.0001	.0002	.0002	.0003	.0003	.0004	.0005	.0006
13	.0000	.0000	.0000	.0000	.0001	.0001	.0001	.0001	.0002	.0002
14	.0000	.0000	.0000	.0000	.0000	.0000	.0000	.0000	.0000	.0001

					λ					
x	4.1	4.2	4.3	4.4	4.5	4.6	4.7	4.8	4.9	5.0
0	.0166	.0150	.0136	.0123	.0111	.0101	.0091	.0082	.0074	.0067
1	.0679	.0630	.0583	.0540	.0500	.0462	.0427	.0395	.0365	.0337
2	.1393	.1323	.1254	.1188	.1125	.1063	.1005	.0948	.0894	.0842
3	.1904	.1852	.1798	.1743	.1687	.1631	.1574	.1517	.1460	.1404
4	.1951	.1944	.1933	.1917	.1898	.1875	.1849	.1820	.1789	.1755
5	.1600	.1633	.1662	.1687	.1708	.1725	.1738	.1747	.1753	.1755
6	.1093	.1143	.1191	.1237	.1281	.1323	.1362	.1398	.1432	.1462
7	.0640	.0686	.0732	.0778	.0824	.0869	.0914	.0959	.1002	.1044
8	.0328	.0360	.0393	.0428	.0463	.0500	.0537	.0575	.0614	.0653
9	.0150	.0168	.0188	.0209	.0232	.0255	.0280	.0307	.0334	.0363
10	.0061	.0071	.0081	.0092	.0104	.0118	.0132	.0147	.0164	.0181
11	.0023	.0027	.0032	.0037	.0043	.0049	.0056	.0064	.0073	.0082
12	.0008	.0009	.0011	.0014	.0016	.0019	.0022	.0026	.0030	.0034
13	.0002	.0003	.0004	.0005	.0006	.0007	.0008	.0009	.0011	.0013
14	.0001	.0001	.0001	.0001	.0002	.0002	.0003	.0003	.0004	.0005
15	.0000	.0000	.0000	.0000	.0001	.0001	.0001	.0001	.0001	.0002

Table T.3 (*continued*)

	λ									
x	5.1	5.2	5.3	5.4	5.5	5.6	5.7	5.8	5.9	6.0
0	.0061	.0055	.0050	.0045	.0041	.0037	.0033	.0030	.0027	.0025
1	.0311	.0287	.0265	.0244	.0225	.0207	.0191	.0176	.0162	.0149
2	.0793	.0746	.0701	.0659	.0618	.0580	.0544	.0509	.0477	.0446
3	.1348	.1293	.1239	.1185	.1133	.1082	.1033	.0985	.0938	.0892
4	.1719	.1681	.1641	.1600	.1558	.1515	.1472	.1428	.1383	.1339
5	.1753	.1748	.1740	.1728	.1714	.1697	.1678	.1656	.1632	.1606
6	.1490	.1515	.1537	.1555	.1571	.1584	.1594	.1601	.1605	.1606
7	.1086	.1125	.1163	.1200	.1234	.1267	.1298	.1326	.1353	.1377
8	.0692	.0731	.0771	.0810	.0849	.0887	.0925	.0962	.0998	.1033
9	.0392	.0423	.0454	.0486	.0519	.0552	.0586	.0620	.0654	.0688
10	.0200	.0220	.0241	.0262	.0285	.0309	.0334	.0359	.0386	.0413
11	.0093	.0104	.0116	.0129	.0143	.0157	.0173	.0190	.0207	.0225
12	.0039	.0045	.0051	.0058	.0065	.0073	.0082	.0092	.0102	.0113
13	.0015	.0018	.0021	.0024	.0028	.0032	.0036	.0041	.0046	.0052
14	.0006	.0007	.0008	.0009	.0011	.0013	.0015	.0017	.0019	.0022
15	.0002	.0002	.0003	.0003	.0004	.0005	.0006	.0007	.0008	.0009
16	.0001	.0001	.0001	.0001	.0001	.0002	.0002	.0002	.0003	.0003
17	.0000	.0000	.0000	.0000	.0000	.0000	.0001	.0001	.0001	.0001

	λ									
x	6.1	6.2	6.3	6.4	6.5	6.6	6.7	6.8	6.9	7.0
0	.0022	.0020	.0018	.0017	.0015	.0014	.0012	.0011	.0010	.0009
1	.0137	.0126	.0116	.0106	.0098	.0090	.0082	.0076	.0070	.0064
2	.0417	.0390	.0364	.0340	.0318	.0296	.0276	.0258	.0240	.0223
3	.0848	.0806	.0765	.0726	.0688	.0652	.0617	.0584	.0552	.0521
4	.1294	.1249	.1205	.1162	.1118	.1076	.1034	.0992	.0952	.0912
5	.1579	.1549	.1519	.1487	.1454	.1420	.1385	.1349	.1314	.1277
6	.1605	.1601	.1595	.1586	.1575	.1562	.1546	.1529	.1511	.1490
7	.1399	.1418	.1435	.1450	.1462	.1472	.1480	.1486	.1489	.1490
8	.1066	.1099	.1130	.1160	.1188	.1215	.1240	.1263	.1284	.1304
9	.0723	.0757	.0791	.0825	.0858	.0891	.0923	.0954	.0985	.1014
10	.0441	.0469	.0498	.0528	.0558	.0588	.0618	.0649	.0679	.0710
11	.0245	.0265	.0285	.0307	.0330	.0353	.0377	.0401	.0426	.0452
12	.0124	.0137	.0150	.0164	.0179	.0194	.0210	.0227	.0245	.0264
13	.0058	.0065	.0073	.0081	.0089	.0098	.0108	.0119	.0130	.0142
14	.0025	.0029	.0033	.0037	.0041	.0046	.0052	.0058	.0064	.0071
15	.0010	.0012	.0014	.0016	.0018	.0020	.0023	.0026	.0029	.0033
16	.0004	.0005	.0005	.0006	.0007	.0008	.0010	.0011	.0013	.0014
17	.0001	.0002	.0002	.0002	.0003	.0003	.0004	.0004	.0005	.0006
18	.0000	.0001	.0001	.0001	.0001	.0001	.0001	.0002	.0002	.0002
19	.0000	.0000	.0000	.0000	.0000	.0000	.0000	.0001	.0001	.0001

Table T.3 (*continued*)

					λ					
x	7.1	7.2	7.3	7.4	7.5	7.6	7.7	7.8	7.9	8.0
0	.0008	.0007	.0007	.0006	.0006	.0005	.0005	.0004	.0004	.0003
1	.0059	.0054	.0049	.0045	.0041	.0038	.0035	.0032	.0029	.0027
2	.0208	.0194	.0180	.0167	.0156	.0145	.0134	.0125	.0116	.0107
3	.0492	.0464	.0438	.0413	.0389	.0366	.0345	.0324	.0305	.0286
4	.0874	.0836	.0799	.0764	.0729	.0696	.0663	.0632	.0602	.0573
5	.1241	.1204	.1167	.1130	.1094	.1057	.1021	.0986	.0951	.0916
6	.1468	.1445	.1420	.1394	.1367	.1339	.1311	.1282	.1252	.1221
7	.1489	.1486	.1481	.1474	.1465	.1454	.1442	.1428	.1413	.1396
8	.1321	.1337	.1351	.1363	.1373	.1382	.1388	.1392	.1395	.1396
9	.1042	.1070	.1096	.1121	.1144	.1167	.1187	.1207	.1224	.1241
10	.0740	.0770	.0800	.0829	.0858	.0887	.0914	.0941	.0967	.0993
11	.0478	.0504	.0531	.0558	.0585	.0613	.0640	.0667	.0695	.0722
12	.0283	.0303	.0323	.0344	.0366	.0388	.0411	.0434	.0457	.0481
13	.0154	.0168	.0181	.0196	.0211	.0227	.0243	.0260	.0278	.0296
14	.0078	.0086	.0095	.0104	.0113	.0123	.0134	.0145	.0157	.0169
15	.0037	.0041	.0046	.0051	.0057	.0062	.0069	.0075	.0083	.0090
16	.0016	.0019	.0021	.0024	.0026	.0030	.0033	.0037	.0041	.0045
17	.0007	.0008	.0009	.0010	.0012	.0013	.0015	.0017	.0019	.0021
18	.0003	.0003	.0004	.0004	.0005	.0006	.0006	.0007	.0008	.0009
19	.0001	.0001	.0001	.0002	.0002	.0002	.0003	.0003	.0003	.0004
20	.0000	.0000	.0001	.0001	.0001	.0001	.0001	.0001	.0001	.0002
21	.0000	.0000	.0000	.0000	.0000	.0000	.0000	.0000	.0001	.0001

					λ					
x	8.1	8.2	8.3	8.4	8.5	8.6	8.7	8.8	8.9	9.0
0	.0003	.0003	.0002	.0002	.0002	.0002	.0002	.0002	.0001	.0001
1	.0025	.0023	.0021	.0019	.0017	.0016	.0014	.0013	.0012	.0011
2	.0100	.0092	.0086	.0079	.0074	.0068	.0063	.0058	.0054	.0050
3	.0269	.0252	.0237	.0222	.0208	.0195	.0183	.0171	.0160	.0150
4	.0544	.0517	.0491	.0466	.0443	.0420	.0398	.0377	.0357	.0337
5	.0882	.0849	.0816	.0784	.0752	.0722	.0692	.0663	.0635	.0607
6	.1191	.1160	.1128	.1097	.1066	.1034	.1003	.0972	.0941	.0911
7	.1378	.1358	.1338	.1317	.1294	.1271	.1247	.1222	.1197	.1171
8	.1395	.1392	.1388	.1382	.1375	.1366	.1356	.1344	.1332	.1318
9	.1256	.1269	.1280	.1290	.1299	.1306	.1311	.1315	.1317	.1318
10	.1017	.1040	.1063	.1084	.1104	.1123	.1140	.1157	.1172	.1186
11	.0749	.0776	.0802	.0828	.0853	.0878	.0902	.0925	.0948	.0970
12	.0505	.0530	.0555	.0579	.0604	.0629	.0654	.0679	.0703	.0728
13	.0315	.0334	.0354	.0374	.0395	.0416	.0438	.0459	.0481	.0504
14	.0182	.0196	.0210	.0225	.0240	.0256	.0272	.0289	.0306	.0324

Table T.3 (*continued*)

					λ					
x	8.1	8.2	8.3	8.4	8.5	8.6	8.7	8.8	8.9	9.0
15	.0098	.0107	.0116	.0126	.0136	.0147	.0158	.0169	.0182	.0194
16	.0050	.0055	.0060	.0066	.0072	.0079	.0086	.0093	.0101	.0109
17	.0024	.0026	.0029	.0033	.0036	.0040	.0044	.0048	.0053	.0058
18	.0011	.0012	.0014	.0015	.0017	.0019	.0021	.0024	.0026	.0029
19	.0005	.0005	.0006	.0007	.0008	.0009	.0010	.0011	.0012	.0014
20	.0002	.0002	.0002	.0003	.0003	.0004	.0004	.0005	.0005	.0006
21	.0001	.0001	.0001	.0001	.0001	.0002	.0002	.0002	.0002	.0003
22	.0000	.0000	.0000	.0000	.0001	.0001	.0001	.0001	.0001	.0001

					λ					
x	9.1	9.2	9.3	9.4	9.5	9.6	9.7	9.8	9.9	10
0	.0001	.0001	.0001	.0001	.0001	.0001	.0001	.0001	.0001	.0000
1	.0010	.0009	.0009	.0008	.0007	.0007	.0006	.0005	.0005	.0005
2	.0046	.0043	.0040	.0037	.0034	.0031	.0029	.0027	.0025	.0023
3	.0140	.0131	.0123	.0115	.0107	.0100	.0093	.0087	.0081	.0076
4	.0319	.0302	.0285	.0269	.0254	.0240	.0226	.0213	.0201	.0189
5	.0581	.0555	.0530	.0506	.0483	.0460	.0439	.0418	.0398	.0378
6	.0881	.0851	.0822	.0793	.0764	.0736	.0709	.0682	.0656	.0631
7	.1145	.1118	.1091	.1064	.1037	.1010	.0982	.0955	.0928	.0901
8	.1302	.1286	.1269	.1251	.1232	.1212	.1191	.1170	.1148	.1126
9	.1317	.1315	.1311	.1306	.1300	.1293	.1284	.1274	.1263	.1251
10	.1198	.1210	.1219	.1228	.1235	.1241	.1245	.1249	.1250	.1251
11	.0991	.1012	.1031	.1049	.1067	.1083	.1098	.1112	.1125	.1137
12	.0752	.0776	.0799	.0822	.0844	.0866	.0888	.0908	.0928	.0948
13	.0526	.0549	.0572	.0594	.0617	.0640	.0662	.0685	.0707	.0729
14	.0342	.0361	.0380	.0399	.0419	.0439	.0459	.0479	.0500	.0521
15	.0208	.0221	.0235	.0250	.0265	.0281	.0297	.0313	.0330	.0347
16	.0118	.0127	.0137	.0147	.0157	.0168	.0180	.0192	.0204	.0217
17	.0063	.0069	.0075	.0081	.0088	.0095	.0103	.0111	.0119	.0128
18	.0032	.0035	.0039	.0042	.0046	.0051	.0055	.0060	.0065	.0071
19	.0015	.0017	.0019	.0021	.0023	.0026	.0028	.0031	.0034	.0037
20	.0007	.0008	.0009	.0010	.0011	.0012	.0014	.0015	.0017	.0019
21	.0003	.0003	.0004	.0004	.0005	.0006	.0006	.0007	.0008	.0009
22	.0001	.0001	.0002	.0002	.0002	.0002	.0003	.0003	.0004	.0004
23	.0000	.0001	.0001	.0001	.0001	.0001	.0001	.0001	.0002	.0002
24	.0000	.0000	.0000	.0000	.0000	.0000	.0000	.0001	.0001	.0001

Table T.3 (*continued*)

x	11	12	13	14	15	16	17	18	19	20
0	.0000	.0000	.0000	.0000	.0000	.0000	.0000	.0000	.0000	.0000
1	.0002	.0001	.0000	.0000	.0000	.0000	.0000	.0000	.0000	.0000
2	.0010	.0004	.0002	.0001	.0000	.0000	.0000	.0000	.0000	.0000
3	.0037	.0018	.0008	.0004	.0002	.0001	.0000	.0000	.0000	.0000
4	.0102	.0053	.0027	.0013	.0006	.0003	.0001	.0001	.0000	.0000
5	.0224	.0127	.0070	.0037	.0019	.0010	.0005	.0002	.0001	.0001
6	.0411	.0255	.0152	.0087	.0048	.0026	.0014	.0007	.0004	.0002
7	.0646	.0437	.0281	.0174	.0104	.0060	.0034	.0018	.0010	.0005
8	.0888	.0655	.0457	.0304	.0194	.0120	.0072	.0042	.0024	.0013
9	.1085	.0874	.0661	.0473	.0324	.0213	.0135	.0083	.0050	.0029
10	.1194	.1048	.0859	.0663	.0486	.0341	.0230	.0150	.0095	.0058
11	.1194	.1144	.1015	.0844	.0663	.0496	.0355	.0245	.0164	.0106
12	.1094	.1144	.1099	.0984	.0829	.0661	.0504	.0368	.0259	.0176
13	.0926	.1056	.1099	.1060	.0956	.0814	.0658	.0509	.0378	.0271
14	.0728	.0905	.1021	.1060	.1024	.0930	.0800	.0655	.0514	.0387
15	.0534	.0724	.0885	.0989	.1024	.0992	.0906	.0786	.0650	.0516
16	.0367	.0543	.0719	.0866	.0960	.0992	.0963	.0884	.0772	.0646
17	.0237	.0383	.0550	.0713	.0847	.0934	.0963	.0936	.0863	.0760
18	.0145	.0256	.0397	.0554	.0706	.0830	.0909	.0936	.0911	.0844
19	.0084	.0161	.0272	.0409	.0557	.0699	.0814	.0887	.0911	.0888
20	.0046	.0097	.0177	.0286	.0418	.0559	.0692	.0798	.0866	.0888
21	.0024	.0055	.0109	.0191	.0299	.0426	.0560	.0684	.0783	.0846
22	.0012	.0030	.0065	.0121	.0204	.0310	.0433	.0560	.0676	.0769
23	.0006	.0016	.0037	.0074	.0133	.0216	.0320	.0438	.0559	.0669
24	.0003	.0008	.0020	.0043	.0083	.0144	.0226	.0328	.0442	.0557
25	.0001	.0004	.0010	.0024	.0050	.0092	.0154	.0237	.0336	.0446
26	.0000	.0002	.0005	.0013	.0029	.0057	.0101	.0164	.0246	.0343
27	.0000	.0001	.0002	.0007	.0016	.0034	.0063	.0109	.0173	.0254
28	.0000	.0000	.0001	.0003	.0009	.0019	.0038	.0070	.0117	.0181
29	.0000	.0000	.0001	.0002	.0004	.0011	.0023	.0044	.0077	.0125
30	.0000	.0000	.0000	.0001	.0002	.0006	.0013	.0026	.0049	.0083
31	.0000	.0000	.0000	.0000	.0001	.0003	.0007	.0015	.0030	.0054
32	.0000	.0000	.0000	.0000	.0001	.0001	.0004	.0009	.0018	.0034
33	.0000	.0000	.0000	.0000	.0000	.0001	.0002	.0005	.0010	.0020
34	.0000	.0000	.0000	.0000	.0000	.0000	.0001	.0002	.0006	.0012
35	.0000	.0000	.0000	.0000	.0000	.0000	.0000	.0001	.0003	.0007
36	.0000	.0000	.0000	.0000	.0000	.0000	.0000	.0001	.0002	.0004
37	.0000	.0000	.0000	.0000	.0000	.0000	.0000	.0000	.0001	.0002
38	.0000	.0000	.0000	.0000	.0000	.0000	.0000	.0000	.0000	.0001
39	.0000	.0000	.0000	.0000	.0000	.0000	.0000	.0000	.0000	.0001

Source: Reprinted with permission from William H. Beyer (ed.), *Handbook of Tables for Probability and Statistics*, 2nd ed., pp. 207–11. Copyright 1964 The Chemical Rubber Co., CRC Press, Inc.

Table T.4: *Values of* $p_Y(k)$ *for the Negative Binomial Distribution with* $r = 2, 3, 4, 5$ *and* $p = 0.2, 0.4, 0.5, 0.6, 0.8$

			$r = 2$		
			p		
k	0.2	0.4	0.5	0.6	0.8
0	0.0400	0.1600	0.2500	0.3600	0.6400
1	0.0640	0.1920	0.2500	0.2880	0.2560
2	0.0768	0.1728	0.1875	0.1728	0.0768
3	0.0819	0.1382	0.1250	0.0922	0.0205
4	0.0819	0.1037	0.0781	0.0461	0.0051
5	0.0786	0.0746	0.0469	0.0221	0.0012
6	0.0734	0.0523	0.0273	0.0103	0.0003
7	0.0671	0.0358	0.0156	0.0047	0.0001
8	0.0604	0.0242	0.0088	0.0021	
9	0.0537	0.0161	0.0049	0.0009	
10	0.0472	0.0106	0.0027	0.0004	
11	0.0412	0.0070	0.0015	0.0002	
12	0.0357	0.0045	0.0008	0.0001	
13	0.0308	0.0029	0.0004		
14	0.0264	0.0019	0.0002		
15	0.0225	0.0012	0.0001		
16	0.0191	0.0008	0.0001		
17	0.0162	0.0005			
18	0.0137	0.0003			
19	0.0115	0.0002			
20	0.0097	0.0001			
21	0.0081	0.0001			
22	0.0068				
23	0.0057				
24	0.0047				
25	0.0039				
26	0.0033				
27	0.0027				
28	0.0022				
29	0.0019				
30	0.0015				

Table T.4 (*continued*)

k	r = 3				
	0.2	0.4	0.5	0.6	0.8
0	0.0080	0.0640	0.1250	0.2160	0.5120
1	0.0192	0.1152	0.1875	0.2592	0.3072
2	0.0307	0.1382	0.1875	0.2074	0.1229
3	0.0410	0.1382	0.1562	0.1382	0.0410
4	0.0492	0.1244	0.1172	0.0829	0.0123
5	0.0551	0.1045	0.0820	0.0464	0.0034
6	0.0587	0.0836	0.0547	0.0248	0.0009
7	0.0604	0.0645	0.0352	0.0127	0.0002
8	0.0604	0.0484	0.0220	0.0064	0.0001
9	0.0591	0.0355	0.0134	0.0031	
10	0.0567	0.0255	0.0081	0.0015	
11	0.0536	0.0181	0.0048	0.0007	
12	0.0500	0.0127	0.0028	0.0003	
13	0.0462	0.0088	0.0016	0.0002	
14	0.0422	0.0060	0.0009	0.0001	
15	0.0383	0.0041	0.0005		
16	0.0345	0.0028	0.0003		
17	0.0308	0.0019	0.0002		
18	0.0274	0.0012	0.0001		
19	0.0242	0.0008	0.0001		
20	0.0213	0.0005			
21	0.0187	0.0004			
22	0.0163	0.0002			
23	0.0142	0.0002			
24	0.0123	0.0001			
25	0.0106	0.0001			
26	0.0091				
27	0.0079				
28	0.0067				
29	0.0058				
30	0.0049				

The *p* column header spans columns 0.2–0.8.

Table T.5 (continued)

k	0.00	0.01	0.02	0.03	0.04	0.05	0.06	0.07	0.08	0.09
2.0	0.97725	0.97778	0.97831	0.97882	0.97932	0.97982	0.98030	0.98077	0.98124	0.98169
2.1	0.98214	0.98257	0.98300	0.98341	0.98382	0.98422	0.98461	0.98500	0.98537	0.98574
2.2	0.98610	0.98645	0.98679	0.98713	0.98745	0.98778	0.98809	0.98840	0.98870	0.98899
2.3	0.98928	0.98956	0.98983	0.99010	0.99036	0.99061	0.99086	0.99111	0.99134	0.99158
2.4	0.99180	0.99202	0.99224	0.99245	0.99266	0.99286	0.99305	0.99324	0.99343	0.99361
2.5	0.99379	0.99396	0.99413	0.99430	0.99446	0.99461	0.99477	0.99492	0.99506	0.99520
2.6	0.99534	0.99547	0.99560	0.99573	0.99585	0.99598	0.99609	0.99621	0.99632	0.99643
2.7	0.99653	0.99664	0.99674	0.99683	0.99693	0.99702	0.99711	0.99720	0.99728	0.99736
2.8	0.99744	0.99752	0.99760	0.99767	0.99774	0.99781	0.99788	0.99795	0.99801	0.99807
2.9	0.99813	0.99819	0.99825	0.99831	0.99836	0.99841	0.99846	0.99851	0.99856	0.99861
3.0	0.99865	0.99869	0.99874	0.99878	0.99882	0.99886	0.99889	0.99893	0.99896	0.99900
3.1	0.99903	0.99906	0.99910	0.99913	0.99916	0.99918	0.99921	0.99924	0.99926	0.99929
3.2	0.99931	0.99934	0.99936	0.99938	0.99940	0.99942	0.99944	0.99946	0.99948	0.99950
3.3	0.99952	0.99953	0.99955	0.99957	0.99958	0.99960	0.99961	0.99962	0.99964	0.99965
3.4	0.99966	0.99968	0.99969	0.99970	0.99971	0.99972	0.99973	0.99974	0.99975	0.99976
3.5	0.99977	0.99978	0.99978	0.99979	0.99980	0.99981	0.99981	0.99982	0.99983	0.99983
3.6	0.99984	0.99985	0.99985	0.99986	0.99986	0.99987	0.99987	0.99988	0.99988	0.99989
3.7	0.99989	0.99990	0.99990	0.99990	0.99991	0.99991	0.99992	0.99992	0.99992	0.99992
3.8	0.99993	0.99993	0.99993	0.99994	0.99994	0.99994	0.99994	0.99995	0.99995	0.99995
3.9	0.99995	0.99995	0.99996	0.99996	0.99996	0.99996	0.99996	0.99996	0.99997	0.99997
4.0	0.99997	0.99997	0.99997	0.99997	0.99997	0.99997	0.99998	0.99998	0.99998	0.99998

Table T.4 (continued)

	r = 4				
	p				
k	0.2	0.4	0.5	0.6	0.8
0	0.0016	0.0256	0.0625	0.1296	0.4096
1	0.0051	0.0614	0.1250	0.2074	0.3277
2	0.0102	0.0922	0.1562	0.2074	0.1638
3	0.0164	0.1106	0.1562	0.1659	0.0655
4	0.0229	0.1161	0.1367	0.1161	0.0229
5	0.0294	0.1115	0.1094	0.0743	0.0073
6	0.0352	0.1103	0.0820	0.0446	0.0022
7	0.0403	0.0860	0.0586	0.0255	0.0006
8	0.0443	0.0709	0.0403	0.0140	0.0002
9	0.0472	0.0568	0.0269	0.0075	
10	0.0491	0.0443	0.0175	0.0039	
11	0.0500	0.0338	0.0111	0.0020	
12	0.0500	0.0254	0.0069	0.0010	
13	0.0493	0.0187	0.0043	0.0005	
14	0.0479	0.0136	0.0026	0.0002	
15	0.0459	0.0098	0.0016	0.0001	
16	0.0436	0.0070	0.0009	0.0001	
17	0.0411	0.0049	0.0005		
18	0.0383	0.0035	0.0003		
19	0.0355	0.0024	0.0002		
20	0.0327	0.0017	0.0001		
21	0.0299	0.0011			
22	0.0272	0.0008			
23	0.0246	0.0005			
24	0.0221	0.0004			
25	0.0198	0.0002			
26	0.0177	0.0002			
27	0.0157	0.0001			
28	0.0139	0.0001			
29	0.0123				
30	0.0108				

Table T.4 (continued)

			r = 5		
			p		
k	0.2	0.4	0.5	0.6	0.8
0	0.0003	0.0102	0.0312	0.0778	0.3277
1	0.0013	0.0307	0.0781	0.1555	0.3277
2	0.0031	0.0553	0.1172	0.1866	0.1966
3	0.0057	0.0774	0.1367	0.1742	0.0918
4	0.0092	0.0929	0.1367	0.1393	0.0367
5	0.0132	0.1003	0.1230	0.1003	0.0132
6	0.0176	0.1003	0.1025	0.0669	0.0044
7	0.0221	0.0946	0.0806	0.0420	0.0014
8	0.0266	0.0851	0.0604	0.0252	0.0004
9	0.0307	0.0738	0.0436	0.0146	0.0001
10	0.0344	0.0620	0.0305	0.0082	
11	0.0375	0.0507	0.0208	0.0045	
12	0.0400	0.0406	0.0139	0.0024	
13	0.0419	0.0318	0.0091	0.0012	
14	0.0431	0.0246	0.0058	0.0006	
15	0.0436	0.0187	0.0037	0.0003	
16	0.0436	0.0140	0.0023	0.0002	
17	0.0431	0.0104	0.0014	0.0001	
18	0.0422	0.0076	0.0009		
19	0.0408	0.0055	0.0005		
20	0.0392	0.0040	0.0003		
21	0.0373	0.0028			
22	0.0353	0.0020			
23	0.0332	0.0014			
24	0.0309	0.0010			
25	0.0287	0.0007			
26	0.0265	0.0005			
27	0.0243	0.0003			
28	0.0223	0.0002			
29	0.0203	0.0002			
30	0.0184	0.0001			

Table T.5: Table of the Cumulative Distribution Function of a Standard Normal Random Variable

k	0.00	0.01	0.02	0.03	0.04	0.05	0.06	0.07	0.08	0.09
0.0	0.50000	0.50399	0.50798	0.51197	0.51595	0.51994	0.52392	0.52790	0.53188	0.53586
0.1	0.53983	0.54380	0.54776	0.55172	0.55567	0.55962	0.56356	0.56749	0.57142	0.57535
0.2	0.57926	0.58317	0.58706	0.59095	0.59483	0.59871	0.60257	0.60642	0.61026	0.61409
0.3	0.61791	0.62172	0.62552	0.62930	0.63307	0.63683	0.64058	0.64431	0.64803	0.65173
0.4	0.65542	0.65910	0.66276	0.66640	0.67003	0.67364	0.67724	0.68082	0.68439	0.68793
0.5	0.69146	0.69497	0.69847	0.70194	0.70540	0.70884	0.71226	0.71566	0.71904	0.72240
0.6	0.72575	0.72907	0.73237	0.73565	0.73891	0.74215	0.74537	0.74857	0.75175	0.75490
0.7	0.75804	0.76115	0.76424	0.76730	0.77035	0.77337	0.77637	0.77935	0.78230	0.78524
0.8	0.78814	0.79103	0.79389	0.79673	0.79955	0.80234	0.80511	0.80785	0.81057	0.81327
0.9	0.81594	0.81859	0.82121	0.82381	0.82639	0.82894	0.83147	0.83398	0.83646	0.83891
1.0	0.84134	0.84375	0.84614	0.84849	0.85083	0.85314	0.85543	0.85769	0.85993	0.86214
1.1	0.86433	0.86650	0.86864	0.87076	0.87286	0.87493	0.87698	0.87900	0.88100	0.88298
1.2	0.88493	0.88686	0.88877	0.89065	0.89251	0.89435	0.89617	0.89796	0.89973	0.90147
1.3	0.90320	0.90490	0.90658	0.90824	0.90988	0.91149	0.91309	0.91466	0.91621	0.91774
1.4	0.91924	0.92073	0.92220	0.92364	0.92507	0.92647	0.92785	0.92922	0.93056	0.93189
						0.93943	0.94062	0.94179	0.94295	0.94408
										0.95449

Table T.6: *Table of the Exponential Function, e^{-x}, for $x > 0$*

	0.00	0.01	0.02	0.03	0.04	0.05	0.06	0.07	0.08	0.09
0.0	1.00000	0.99005	0.98020	0.97045	0.96079	0.95123	0.94176	0.93239	0.92312	0.91393
0.1	0.90484	0.89583	0.88692	0.87810	0.86936	0.86071	0.85214	0.84366	0.83527	0.82696
0.2	0.81873	0.81058	0.80252	0.79453	0.78663	0.77880	0.77105	0.76338	0.75578	0.74826
0.3	0.74082	0.73345	0.72615	0.71892	0.71177	0.70469	0.69768	0.69073	0.68386	0.67706
0.4	0.67032	0.66365	0.65705	0.65051	0.64404	0.63763	0.63128	0.62500	0.61878	0.61263
0.5	0.60653	0.60050	0.59452	0.58860	0.58275	0.57695	0.57121	0.56553	0.55990	0.55433
0.6	0.54881	0.54335	0.53794	0.53259	0.52729	0.52204	0.51685	0.51171	0.50662	0.50158
0.7	0.49659	0.49164	0.48675	0.48191	0.47711	0.47237	0.46767	0.46301	0.45841	0.45384
0.8	0.44933	0.44486	0.44043	0.43605	0.43171	0.42741	0.42316	0.41895	0.41478	0.41066
0.9	0.40657	0.40252	0.39852	0.39455	0.39063	0.38674	0.38289	0.37908	0.37531	0.37158

	0.0	0.1	0.2	0.3	0.4	0.5	0.6	0.7	0.8	0.9
1.0	0.36788	0.33287	0.30119	0.27253	0.24660	0.22313	0.20190	0.18268	0.16530	0.14957
2.0	0.13534	0.12246	0.11080	0.10026	0.09072	0.08208	0.07427	0.06721	0.06081	0.05502
3.0	0.04979	0.04505	0.04076	0.03688	0.03337	0.03020	0.02732	0.02472	0.02237	0.02024
4.0	0.01832	0.01657	0.01500	0.01357	0.01228	0.01111	0.01005	0.00910	0.00823	0.00745
5.0	0.00674	0.00610	0.00552	0.00499	0.00452	0.00409	0.00370	0.00335	0.00303	0.00274
6.0	0.00248	0.00224	0.00203	0.00184	0.00166	0.00150	0.00136	0.00123	0.00111	0.00101

Table T.7: *Values of the Incomplete Gamma Function* $I_r(\tau)$ *for Use in the Computation of the Cumulative Gamma Distribution Function*

τ	r				
	1	2	3	4	5
0.2	0.18127	0.01752	0.00115	0.00006	0.00000
0.4	0.32968	0.06155	0.00793	0.00078	0.00006
0.6	0.45119	0.12190	0.02312	0.00336	0.00039
0.8	0.55067	0.19121	0.04742	0.00908	0.00141
1.0	0.63212	0.26424	0.08030	0.01899	0.00366
1.2	0.69881	0.33737	0.12051	0.03377	0.00775
1.4	0.75340	0.40817	0.16650	0.05372	0.01425
1.6	0.79810	0.47507	0.21664	0.07881	0.02368
1.8	0.83470	0.53716	0.26938	0.10871	0.03641
2.0	0.86466	0.59399	0.32332	0.14288	0.05265
2.2	0.88920	0.64543	0.37729	0.18065	0.07250
2.4	0.90928	0.69156	0.43029	0.22128	0.09587
2.6	0.92573	0.73262	0.48157	0.26400	0.12258
2.8	0.93919	0.76892	0.53055	0.30806	0.15232
3.0	0.95021	0.80085	0.57681	0.35277	0.18474
3.2	0.95924	0.82880	0.62010	0.39748	0.21939
3.4	0.96663	0.85316	0.66026	0.44164	0.25582
3.6	0.97268	0.87431	0.69725	0.48478	0.29356
3.8	0.97763	0.89262	0.73110	0.52652	0.33216
4.0	0.98168	0.90842	0.76190	0.56653	0.37116
4.2	0.98500	0.92202	0.78976	0.60460	0.41017
4.4	0.98772	0.93370	0.81486	0.64055	0.44882
4.6	0.98995	0.94371	0.83736	0.67429	0.48677
4.8	0.99177	0.95227	0.85746	0.70577	0.52374
5.0	0.99326	0.95957	0.87535	0.73497	0.55951
5.2	0.99448	0.96580	0.89121	0.76193	0.59387
5.4	0.99548	0.97109	0.90524	0.78671	0.62669
5.6	0.99630	0.97559	0.91761	0.80938	0.65785
5.8	0.99697	0.97941	0.92849	0.83004	0.68728
6.0	0.99752	0.98265	0.93803	0.84880	0.71494

Table T.7 (*continued*)

τ	r				
	1	2	3	4	5
6.2	0.99797	0.98539	0.94638	0.86577	0.74082
6.4	0.99834	0.98770	0.95368	0.88108	0.76493
6.6	0.99864	0.98966	0.96003	0.89485	0.78730
6.8	0.99889	0.99131	0.96556	0.90719	0.80797
7.0	0.99909	0.99270	0.97036	0.91823	0.82701
7.2	0.99925	0.99388	0.97453	0.92808	0.84448
7.4	0.99939	0.99487	0.97813	0.93685	0.86047
7.6	0.99950	0.99570	0.98124	0.94463	0.87506
7.8	0.99959	0.99639	0.98393	0.95152	0.88833
8.0	0.99966	0.99698	0.98625	0.95762	0.90037
8.5	0.99980	0.99807	0.99072	0.96989	0.92564
9.0	0.99988	0.99877	0.99377	0.97877	0.94504
9.5	0.99993	0.99921	0.99584	0.98514	0.95974
10.0	0.99995	0.99950	0.99723	0.98966	0.97075
10.5	0.99997	0.99968	0.99817	0.99285	0.97891
11.0	0.99998	0.99980	0.99879	0.99508	0.98490
11.5	0.99999	0.99987	0.99920	0.99664	0.98925
12.0	0.99999	0.99992	0.99948	0.99771	0.99240
12.5	1.00000	0.99995	0.99966	0.99845	0.99465
13.0	1.00000	0.99997	0.99978	0.99895	0.99626
13.5	1.00000	0.99998	0.99986	0.99929	0.99740
14.0	1.00000	0.99999	0.99991	0.99953	0.99819
14.5	1.00000	0.99999	0.99994	0.99968	0.99875
15.0	1.00000	1.00000	0.99996	0.99979	0.99914

Table T.7 (*continued*)

τ	\(r \) 6	7	8	9	10
1.0	0.00059	0.00008	0.00001		
1.2	0.00150	0.00025	0.00004		
1.4	0.00320	0.00062	0.00011	0.00002	
1.6	0.00604	0.00134	0.00026	0.00005	0.00001
1.8	0.01038	0.00257	0.00056	0.00011	0.00002
2.0	0.01656	0.00453	0.00110	0.00024	0.00005
2.2	0.02491	0.00746	0.00198	0.00047	0.00010
2.4	0.03567	0.01159	0.00334	0.00086	0.00020
2.6	0.04904	0.01717	0.00533	0.00149	0.00038
2.8	0.06511	0.02441	0.00813	0.00243	0.00066
3.0	0.08392	0.03351	0.01190	0.00380	0.00110
3.2	0.10541	0.04462	0.01683	0.00571	0.00176
3.4	0.12946	0.05785	0.02307	0.00829	0.00271
3.6	0.15588	0.07327	0.03079	0.01167	0.00402
3.8	0.18444	0.09089	0.04011	0.01598	0.00580
4.0	0.21487	0.11067	0.05113	0.02136	0.00813
4.2	0.24686	0.13254	0.06394	0.02793	0.01113
4.4	0.28009	0.15635	0.07858	0.03580	0.01489
4.6	0.31424	0.18197	0.09505	0.04507	0.01953
4.8	0.34899	0.20920	0.11333	0.05582	0.02514
5.0	0.38404	0.23782	0.13337	0.06809	0.03183
5.2	0.41909	0.26761	0.15508	0.08193	0.03967
5.4	0.45387	0.29833	0.17834	0.09735	0.04875
5.6	0.48814	0.32974	0.20302	0.11432	0.05913
5.8	0.52169	0.36161	0.22897	0.13281	0.07084
6.0	0.55432	0.39370	0.25602	0.15276	0.08392
6.2	0.58589	0.42579	0.28398	0.17409	0.09838
6.4	0.61626	0.45767	0.31268	0.19669	0.11420
6.6	0.64533	0.48916	0.34192	0.22044	0.13136
6.8	0.67302	0.52008	0.37151	0.24523	0.14982

Table T.7 (*continued*)

τ	\multicolumn{5}{c}{r}				
	6	7	8	9	10
7.0	0.69929	0.55029	0.40129	0.27091	0.16950
7.2	0.72410	0.57964	0.43106	0.29733	0.19035
7.4	0.74744	0.60804	0.46067	0.32435	0.21226
7.6	0.76932	0.63538	0.48996	0.35181	0.23515
7.8	0.78975	0.66159	0.51879	0.37956	0.25889
8.0	0.80876	0.68663	0.54704	0.40745	0.28338
8.5	0.85040	0.74382	0.61440	0.47689	0.34703
9.0	0.88431	0.79322	0.67610	0.54435	0.41259
9.5	0.91147	0.83505	0.73134	0.60818	0.47817
10.0	0.93291	0.86986	0.77978	0.66718	0.54207
10.5	0.94962	0.89837	0.82149	0.72059	0.60287
11.0	0.96248	0.92139	0.85681	0.76801	0.65949
11.5	0.97227	0.93973	0.88627	0.80941	0.71121
12.0	0.97966	0.95418	0.91050	0.84497	0.75761
12.5	0.98518	0.96543	0.93017	0.87508	0.79857
13.0	0.98927	0.97411	0.94597	0.90024	0.83419
13.5	0.99227	0.98075	0.95852	0.92100	0.86474
14.0	0.99447	0.98577	0.96838	0.93794	0.89060
14.5	0.99606	0.98955	0.97606	0.95162	0.91224
15.0	0.99721	0.99237	0.98200	0.96255	0.93015
15.5	0.99803	0.99446	0.98654	0.97121	0.94481
16.0	0.99862	0.99599	0.99000	0.97801	0.95670
16.5	0.99903	0.99712	0.99261	0.98331	0.96626
17.0	0.99933	0.99794	0.99457	0.98741	0.97388

Table T.7 (*continued*)

τ	11	12	13	14	15
			r		
4.0	0.00284	0.00091	0.00027	0.00008	0.00002
4.5	0.00667	0.00240	0.00081	0.00025	0.00007
5.0	0.01370	0.00545	0.00202	0.00070	0.00023
5.5	0.02525	0.01099	0.00445	0.00169	0.00060
6.0	0.04262	0.02009	0.00883	0.00363	0.00140
6.5	0.06684	0.03388	0.01603	0.00710	0.00296
7.0	0.09852	0.05335	0.02700	0.01281	0.00572
7.5	0.13776	0.07924	0.04267	0.02156	0.01026
8.0	0.18411	0.11192	0.06380	0.03418	0.01726
8.5	0.23664	0.15134	0.09092	0.05141	0.02743
9.0	0.29401	0.19699	0.12423	0.07385	0.04147
9.2	0.31797	0.21682	0.13926	0.08438	0.05999
9.4	0.34236	0.23743	0.15524	0.09581	0.05590
9.6	0.36705	0.25876	0.17212	0.10815	0.06428
9.8	0.39195	0.28072	0.18988	0.12139	0.07346
10.0	0.41696	0.30322	0.20844	0.13554	0.08346
10.2	0.44197	0.32618	0.22777	0.15055	0.09429
10.4	0.46687	0.34951	0.24779	0.16641	0.10596
10.6	0.49159	0.37310	0.26843	0.18309	0.11847
10.8	0.51603	0.39687	0.28963	0.20054	0.11318
11.0	0.54011	0.42073	0.31130	0.21871	0.14596
11.2	0.56376	0.44459	0.33337	0.23756	0.16090
11.4	0.58690	0.46837	0.35576	0.25702	0.17661
11.6	0.60949	0.49198	0.37839	0.27703	0.19305
11.8	0.63146	0.51535	0.40117	0.29754	0.21019
12.0	0.65277	0.53840	0.42403	0.31846	0.22798
12.2	0.67338	0.56108	0.44690	0.33974	0.24637
12.4	0.69327	0.58331	0.46968	0.36130	0.26531
12.6	0.71239	0.60504	0.49232	0.38307	0.28474
12.8	0.73075	0.62623	0.51475	0.40498	0.30462

Table T.7 (*continued*)

τ	11	12	13	14	15
			r		
13.0	0.74832	0.64684	0.53690	0.42696	0.32487
13.2	0.76510	0.66681	0.55870	0.44893	0.34543
13.4	0.78108	0.68614	0.58012	0.47084	0.36625
13.6	0.79627	0.70478	0.60110	0.49262	0.38725
13.8	0.81068	0.72273	0.62158	0.51421	0.40838
14.0	0.82432	0.73996	0.64154	0.53555	0.42956
14.2	0.83720	0.75647	0.66094	0.55659	0.45075
14.4	0.84934	0.77225	0.67975	0.57728	0.47188
14.6	0.86076	0.78731	0.69793	0.59756	0.49289
14.8	0.87149	0.80164	0.71549	0.61741	0.51373
15.0	0.88154	0.81525	0.73239	0.63678	0.53435
15.5	0.90388	0.84622	0.77173	0.68292	0.58459
16.0	0.92260	0.87301	0.80688	0.72549	0.63247
16.5	0.93813	0.89593	0.83790	0.76426	0.67746
17.0	0.95088	0.91533	0.86498	0.79913	0.71917
17.5	0.96126	0.93160	0.88835	0.83013	0.75736
18.0	0.96963	0.94511	0.90833	0.85740	0.79192
18.5	0.97635	0.95624	0.92525	0.88114	0.82286
19.0	0.98168	0.96533	0.93944	0.90160	0.85025
19.5	0.98589	0.97269	0.95125	0.91908	0.87427
20.0	0.98919	0.97861	0.96099	0.93387	0.89514
20.5	0.99176	0.98335	0.96897	0.94630	0.91310
21.0	0.99375	0.98710	0.97545	0.95664	0.92843
21.5	0.99528	0.99005	0.98069	0.96520	0.94141
22.0	0.99645	0.99237	0.98488	0.97222	0.95231
22.5	0.99735	0.99418	0.98823	0.97794	0.96140
23.0	0.99802	0.99557	0.99088	0.98257	0.96893
23.5	0.99853	0.99665	0.99297	0.98630	0.97512
24.0	0.99892	0.99748	0.99460	0.98928	0.98018
24.5	0.99920	0.99811	0.99587	0.99166	0.98428

Bibliography

Abramowitz, M., and **I. A. Stegun** (eds.) 1965. *Handbook of Mathematical Functions*. New York: Dover Publications, Inc.

Aczél, J. 1966. *Lectures on Functional Equations and Their Applications*. New York, London: Academic Press.

Adams, J. D. 1962. Failure time distribution estimation. *Semiconductor Reliability*, vol. 2, pp. 41–52.

Ahrens, L. H. 1954a. The lognormal distribution of the elements. *Geochimica et Cosmochimica Acta*, vol. 5, pp. 49–73.

Ahrens, L. H. 1954b. The lognormal distribution of the elements. *Geochimica et Cosmochimica Acta*, vol. 6, pp. 121–31.

Ahrens, L. H. 1957. The lognormal distribution of the elements. *Geochimica et Cosmochimica Acta*, vol. 11, pp. 205–12.

American Geological Institute. 1969. *Models of Geologic Processes—An Introduction to Mathematical Geology*. Washington, D.C.: American Geological Institute.

Artin, E. 1964. *The Gamma Function*. New York: Holt, Rinehart and Winston, Inc.

Asch, S. E. 1952. *Social Psychology*. Englewood Cliffs, N.J.: Prentice-Hall, Inc.

Ayer, A. J. 1965. Chance. *Scientific American*, vol. 213, October, p. 44.

Berger, J., and **J. L. Snell.** 1957. On the concept of equal exchange. *Behavioral Science*, vol. 2, pp. 111–18.

Birnbaum, Z. W., and **S. C. Saunders.** 1958. A statistical model for the lifelength of materials. *Journal of the American Statistical Association*, vol. 53, pp. 151–60.

Bliss, C. I. 1934. The method of probits. *Science*, vol. 79, pp. 38–39, 409–10.

Burgess, E. W., and **L. S. Cottrell, Jr.** 1955. The prediction of adjustment in marriage, in P. F. Lazarsfeld and M. Rosenberg (eds.). *The Language of Social Research*. Glencoe, Ill.: The Free Press, pp. 267–76.

Chatfield, C. 1970. Discrete distributions in market research, in G. P. Patil (ed.). *Random Counts in Physical Science, Geo Science, and Business*. University Park, Pa.: Pennsylvania State University Press.

Chayes, F. 1954. The lognormal distribution of the elements: A discussion. *Geochimica et Cosmochimica Acta*, vol. 6, pp. 119–20.

Cohen, B. 1958. A probability model for conformity. *Sociometry*, vol. 21, pp. 69–81.

Coleman, J. S. 1964. *Introduction to Mathematical Sociology*. London: Free Press of Glencoe, Collier-Macmillan Ltd.

Coombs, C. H., R. M. Dawes, and **A. Tversky.** 1970. *Mathematical Psychology, An Elementary Introduction.* Englewood Cliffs, N. J.: Prentice-Hall, Inc.

David, F. N. 1962. *Games, Gods, and Gambling.* New York: Hafner Publishing Company.

Davis, D. J. 1952. An analysis of some failure data. *Journal of the American Statistical Association,* vol. 47, pp. 113–50.

Davis, H. T. 1935. *Tables of the Higher Mathematical Functions,* 2 vols. Bloomington, Ind.: Principia Press.

Derman, C., L. L. J. Gleser, and **I. Olkin.** 1973. *A Guide to Probability Theory and Application.* New York: Holt, Rinehart and Winston, Inc.

Detlefsen, J. A. 1918. Fluctuations of sampling in a Mendelian population. *Genetics,* vol. 3, pp. 597–607.

Dhrymes, P. J. 1962. On devising unbiased estimators for the parameters of the Cobb–Douglas production function. *Econometrica,* vol. 30, pp. 297–304.

Dwass, M. 1967. *First Steps in Probability.* New York: McGraw-Hill Book Company, Inc.

Epstein, B. 1947. The mathematical description of certain breakage mechanisms leading to the logarithmico-normal distribution. *Journal of the Franklin Institute,* vol. 244, pp. 471–77.

Epstein, B. 1948. Statistical aspects of fracture problems. *Journal of Applied Physics,* vol. 19, pp. 140–47.

Estes, W. K. 1950. Toward a statistical theory of learning. *Psychological Review,* vol. 57, pp. 94–107.

Fechner, G. T. 1897. *Kollektivmasslehre.* Leipzig: W. Engelmann.

Feinlieb, M. 1960. A method of analyzing log-normally distributed survival data with incomplete follow-up. *Journal of the American Statistical Association,* vol. 55, pp. 534–45.

Feller, W. 1968. *An Introduction to Probability Theory and Its Applications,* vol. 1, 3rd ed. New York: John Wiley & Sons, Inc.

Fieller, E. C. 1932. The distribution of the index in a normal bivariate population. *Biometrika,* vol. 24, pp. 428–40.

Fienberg, S. E. 1971. Randomization and social affairs: The 1970 draft lottery. *Science,* vol. 171, pp. 255–61.

Finkelstein, M. O. 1966. The application of statistical decision theory to the jury discrimination cases. *Harvard Law Review,* vol. 80, pp. 338–76.

Fisher, R. A. 1950. *Statistical Methods for Research Workers,* 11th ed. Edinburgh: Oliver & Boyd Ltd.

Fisher, R. A., and **K. Mather.** 1936. A linkage test with mice. *Annals of Eugenics,* vol. 7, pp. 265–80.

Freedle, R. O., and **M. Lewis.** 1971. Application of Markov processes to the concept of state. *Research Bulletin,* 71–34. Princeton, N. J.: Educational Testing Service.

Frobenius, G. 1912. Über Matrizen aus nicht-negativen Elementen. *Sitzungsberichte der Koeniglichen Preussischen Akademie der Wissenschaften,* pp. 456–77.

Froggatt, P. 1970. Application of discrete distribution theory to the study of noncommunicable events in medical epidemiology, in G. P. Patil (ed.). *Random Counts in Biomedical and Social Sciences,* vol. 2. University Park, Pa.: Pennsylvania State University Press.

Gaddum, J. H. 1945. Lognormal distributions. *Nature,* vol. 156, pp. 463–66.

Galton, F. 1879. The geometric mean in vital and social statistics. *Proceedings of the Royal Society, London*, vol. 29, pp. 365–67.

Galton, F. 1892. *Finger Prints*. London: Macmillan & Co. Ltd.

Gantmacher, F. R. 1959. *Theory of Matrices*, vol. II. New York: Chelsea Publishing Company.

Geissler, A. 1889. Beiträge zur Frage des Geschlechts ver hältnisses der Gebarenen. *Zeitschrift des Königlich Sachsischen Statistischen Bureaus.*

Glass, D. V., and **J. R. Hall.** 1954. A study of inter-generation changes in status, in D. V. Glass (ed.). *Social Mobility in Britain*. Glencoe, Ill.: The Free Press, pp. 177–241.

Goldberg, S. 1960. *Probability, An Introduction*. Englewood Cliffs, N. J.: Prentice-Hall, Inc.

Goldthwaite, L. R. 1961. Failure rate study for the lognormal lifetime model. *Proceedings of the Seventh National Symposium on Reliability and Quality Control in Electronics*, pp. 208–13.

Greenwood, M., Jr. 1904. A first study of the weight, variability, and correlation of the human viscera, with special reference to the healthy and diseased heart. *Biometrika*, vol. 3, pp. 63–83.

Greenwood, M., and **G. U. Yule.** 1920. An inquiry into the nature of frequency distributions of multiple happenings with particular reference to the occurrence of multiple attacks of disease or repeated accidents. *Journal of the Royal Statistical Society*. Series A, vol. 83, pp. 255–79.

Gregory, S. 1963. *Statistical Methods and the Geographer*. London: Longmans, Green & Co. Ltd.

Griffiths, J. C. 1960. Frequency distributions in accessory mineral analysis. *The Journal of Geology*, vol. 68, pp. 353–65.

Grundy, P. M. 1951. The expected frequencies in a sample of an animal population in which the abundances of species are log-normally distributed. I. *Biometrika*, vol. 38, pp. 427–34.

Guenther, W. C. 1968. *Concepts of Probability*. New York: McGraw-Hill Book Company, Inc.

Gutenberg, B., and **C. F. Richter.** 1944. Frequency of earthquakes in California. *Bulletin of the Seismological Society of America*, vol. 34, pp. 185–88.

Hagstroem, K.-G. 1960. Remarks on Pareto distributions. *Skandinavisk Aktuarietidskrift*, vol. 43, pp. 59–71.

Haight, F. A. 1970. Group size distributions, with applications to vehicle occupancy, in G. P. Patil (ed.). *Random Counts in Physical Science, Geo Science, and Business*, vol. 3. University Park, Pa.: Pennsylvania State University Press, pp. 95–105.

Hasofer, A. M. 1970. Random mechanisms in Talmudic literature, in E. S. Pearson and M. G. Kendall (eds.). *Studies in the History of Statistics and Probability*. New York: Hafner Publishing Co., pp. 39–43.

Hatch, T., and **S. P. Choute.** 1929. Statistical description of the size properties of non-uniform particles. *Journal of the Franklin Institute*, vol. 207, pp. 369–80.

Herdan, G. 1958. The relation between the dictionary distribution and the occurrence distribution of word length and its importance for the study of quantitative linguistics. *Biometrika*, vol. 45, pp. 222–28.

Herdan, G. 1960. *Small Particle Statistics*, 2nd ed. London: Butterworth & Co. Ltd.

Herdan, G. 1964. *Quantitative Linguistics*. London: Butterworth & Co. Ltd.

Herdan, G. 1966. *The Advanced Theory of Language as Choice and Chance.* New York: Springer-Verlag.

Hodges, J. L., Jr., and **E. L. Lehmann.** 1970. *Elements of Finite Probability,* 2nd ed. San Francisco: Holden-Day, Inc.

Hollingshead, A. de B. 1949. *Elmtown's Youth: The Impact of Social Classes on Adolescents.* New York: John Wiley & Sons, Inc.

Indow, T. 1971. Models for responses of customers with a varying rate. *Journal of Marketing Research,* vol. 8, pp. 78–84.

Jaffe, J., and **S. Feldstein.** 1970. *Rhythms of Dialogue.* New York: Academic Press, Inc.

James, J. 1953. The distribution of free-forming small group size. *American Sociological Review,* vol. 18, p. 569.

Johnson, N. L., and **S. Kotz.** 1969. *Distributions in Statistics: Discrete Distributions.* Boston: Houghton Mifflin Company.

Johnson, N. L., and **S. Kotz.** 1970. *Distributions in Statistics: Continuous Univariate Distributions,* vols. 1, 2. Boston: Houghton Mifflin Company.

Kac, M. 1964. Probability. *Scientific American,* vol. 211, September, p. 92.

Kendall, M. G. 1963. Isaac Todhunter's history of the mathematical theory of probability. *Biometrika,* vol. 50, pp. 204–205.

Kitagawa, T. 1952. *Tables of Poisson Distribution.* Tokyo: Baifukan.

Kline, M. 1962. *Mathematics: A Cultural Approach.* Reading, Mass.: Addison-Wesley Publishing Company, Inc.

Koch, G. S., Jr. and **R. F. Link.** (1971). *Statistical Analysis of Geological Data,* vol. II. New York: John Wiley & Sons, Inc.

Kolmogorov, A. N. 1933/1950/1956. *Foundations of the Theory of Probability* (translation of the original). New York: Chelsea Publishing Company.

Krumbein, W. C. 1936. Application of logarithmic moments to size frequency distributions of sediments. *Journal of Sedimentary Petrology,* vol. 6, pp. 35–47.

Krumbein, W. C. 1954. Applications of statistical methods to sedimentary rocks. *Journal of the American Statistical Association,* vol. 49, pp. 51–66.

Kyburg, H. E., and **H. E. Smokler.** 1964. *Studies in Subjective Probability.* New York: John Wiley & Sons, Inc.

Laplace, P. S. de. 1951. *A Philosophical Essay on Probabilities* (translation with an introduction by E. T. Bell). New York: Dover Publications, Inc.

Latter, O. H. 1902. The egg of *Cuculus canorus. Biometrika,* vol. 1. pp. 164–76.

Lazarsfeld, P. F., and **W. Thielens, Jr.** 1958. *The Academic Mind.* Glencoe, Ill.: The Free Press.

Lieberman, G. J., and **D. B. Owen.** 1961. *Tables of the Hypergeometric Distribution.* Stanford, Cal.: Stanford University Press.

Life Insurance Fact Book. 1967. New York: Institute of Life Insurance.

Macdonell, W. R. 1902. On criminal anthropometry and the identification of criminals. *Biometrika,* vol. 1, pp. 177–227.

Marshall, A. W., and **H. Goldhamer.** 1955. An application of Markov processes to the study of the epidemiology of mental disease. *Journal of the American Statistical Association,* vol. 50, pp. 99–129.

Masuyama, M., and **Y. Kuroiwa.** 1952. Table for the likelihood solutions of gamma distribution and its medical applications. *Reports of Statistical Application Research (JUSE),* vol. 1, pp. 18–23.

McAlister, D. 1879. The law of the geometric mean. *Proceedings of the Royal Society, London*, vol. 29, pp. 367–75.

McGill, W. J. 1963. Stochastic latency mechanisms, in R. D. Luce, R. R. Bush, and E. Galanter (eds.). *Handbook of Mathematical Psychology*. New York: John Wiley & Sons, Inc.

Miller, R. L., and **J. S. Kahn.** 1962. *Statistical Analysis in the Geological Sciences*. New York: John Wiley & Sons, Inc.

Mises, R. von. 1957. *Probability, Statistics, and Truth*, 2nd rev. English ed. (prepared by H. Geiringer). New York: Macmillan Publishing Co., Inc.

Molina, E. C. 1942. *Poisson's Exponential Binomial Limit*. New York: D. Van Nostrand Company, Inc.

Mosteller, F., R. E. K. Rourke, and **G. B. Thomas.** 1961. *Probability with Statistical Applications*. Reading, Mass.: Addison-Wesley Publishing Company, Inc.

Mueller, C. G. 1950. Theoretical relationships among some measures of conditioning. *Proceedings of the National Academy of Sciences*, vol. 56, pp. 123–34.

National Bureau of Standards. 1950. *Tables of the Binomial Probability Distribution, Applied Mathematics Series 6*. Washington, D.C.: U.S. Government Printing Office.

National Bureau of Standards. 1951. *Tables of n! and $\Gamma(n + \frac{1}{2})$ for the First Thousand Values of n, Applied Mathematics Series 16*. Washington, D.C.: U.S. Government Printing Office.

National Bureau of Standards. 1959. *Tables of the Bivariate Normal Distribution Function and Related Functions, Applied Mathematics Series 50*. Washington, D.C.: U.S. Government Printing Office.

National Research Council. 1975. *Doctoral Scientists and Engineers in the United States, 1975 Profile*. Washington, D.C.

National Research Council. 1975. *Employment Status of Ph.D. Scientists and Engineers: 1973 and 1975*. Washington, D.C.

Newman, J. R. 1956. *The World of Mathematics*. New York: Simon and Schuster, Inc.

Ore, O. 1953. *Cardano, The Gambling Scholar*. Princeton, N.J.: Princeton University Press.

Ore, O. 1960. Pascal and the invention of probability theory. *American Mathematical Monthly*, vol. 67, pp. 409–19.

Patil, G. P. (ed.). 1965. *Classical and Contagious Distributions*. Calcutta, India: Statistical Publishing Society.

Pearl, R. 1905. Biometrical studies on man. I. Variation and correlation in brain-weight. *Biometrika*, vol. 4, pp. 13–104.

Pearson, E. S., and **M. G. Kendall.** 1970. *Studies in the History of Statistics and Probability*. New York: Hafner Publishing Company.

Pearson, K. 1924. On a certain double hypergeometrical series and its representation by continuous frequency surfaces. *Biometrika*, vol. 16, pp. 172–88.

Pearson, K. (ed.). 1934. *Tables of the Incomplete Beta-Function*. London: The "Biometrika" Office, University College.

Pearson, K., and **A. Lee.** 1903. On the laws of inheritance in man. I. Inheritance of physical characters. *Biometrika*, vol. 2, pp. 357–462.

Perron, O. 1907. Zur Theorie der Matrizen. *Mathematische Annalen*, vol. 64, pp. 248–63.

Pielou, E. C. 1965. The concept of segregation pattern in ecology: Some discrete distributions applicable to the run lengths of plants in narrow transects, in G. P. Patil (ed.). *Classical and Contagious Discrete Distributions.* Calcutta, India: Statistical Publishing Society, pp. 410–18.

Proschan, F. 1963. Theoretical explanation of observed decrease failure rate. *Technometrics,* vol. 5, pp. 375–84.

Rabinovitch, N. L. 1973. *Probability and Statistical Inference in Ancient and Medieval Jewish Literature.* Toronto: University of Toronto Press.

Rasch, G. 1960. *Probabilistic Models for Some Intelligence and Attainment Tests.* Studies in Mathematical Psychology I. Copenhagen, Denmark: Nielson and Lydiche.

Richardson, L. F. 1944. Distribution of wars in time. *Journal of the Royal Statistical Society,* vol. 107, pp. 242–50.

Romig, H. G. 1953. *50–100 Binomial Tables.* New York: John Wiley & Sons, Inc.

Rutherford, E., J. Chadwick, and **C. D. Ellis.** 1930. *Radiations from Radioactive Substances.* New York: The Macmillan Company.

Rutherford, E., and **H. Geiger.** 1910. The probability variations in the distribution of a particle. *Philosophical Magazine,* vol. 20, pp. 698–707.

Sartwell, P. E. 1950. The distribution of incubation periods of infectious diseases. *American Journal of Hygiene,* vol. 51, pp. 310–18.

Savage, L. J. 1954. *The Foundations of Statistics.* New York: John Wiley & Sons, Inc.

Schwartz, A. 1967. *Calculus and Analytic Geometry,* 2nd ed. New York: Holt, Rinehart and Winston.

Simon, H. A., and **C. P. Bonini.** 1958. *The Size Distribution of Business Firms.* Pittsburgh, Pa.: Reprint 20 of the Graduate School of Business Administration, Carnegie Institute of Technology.

Simpson, G. G., A. Roe, and **R. C. Lewontin.** 1960. *Quantitative Zoology.* New York: Harcourt, Brace and World, Inc.

Slack, H. A., and **W. C. Krumbein.** 1955. Measurement and statistical evaluation of low-level radioactivity in rocks. *Transactions, American Geophysical Union,* vol. 36, pp. 460–64.

Smith, E., and **E. A. Suchman.** 1955. Do people know why they buy? In P. F. Lazarsfeld and M. Rosenberg, (eds.). *The Language of Social Research.* Glencoe, Ill.: The Free Press, pp. 404–10.

Spencer, H. 1877. *The Principles of Sociology.* New York: Appleton and Co.

Statistical Abstract of the United States. 1965 (86th ed.). Washington, D.C.: U.S. Bureau of the Census.

Svedberg, T. 1912. *Existenz der Moleküle.* Leipzig: Akademische Verlagsgesellschaft m.b.H.

Terman, L. M. 1919. *The Intelligence of School Children.* Boston: Houghton Mifflin Company.

Thorndike, F. 1926. Applications of Poisson's probability summation. *Bell System Technical Journal,* vol. 5, pp. 604–24.

Todhunter, I. 1865. *A History of the Mathematical Theory of Probability from the Time of Pascal to That of Laplace* (reprint). New York: Chelsea Publishing Company, 1949.

Trumpler, R. J., and **H. F. Weaver.** 1953. *Statistical Astronomy.* Berkeley: University of California Press.

Urban, F. M. 1909. Die psychophysischen Massmethoden als Grundlage empirischer Messungen. *Archiv für die gesamte Psychologie,* vol. 15, pp. 261–415.

Weaver, W. 1950. Probability. *Scientific American*, vol. 183, October, p. 44.

Weaver, W. 1952. Statistics. *Scientific American*, vol. 186, January, p. 60.

Weibull, W. 1951. A statistical distribution function of wide applicability. *Journal of Applied Mechanics*, vol. 18, pp. 293–97.

Whitaker, L. 1914. On the Poisson law of small numbers. *Biometrika*, vol. 10, pp. 36–71.

Wilk, M. B., R. Gnanadesikan, and **M. J. Huyett.** 1962. Estimation of parameters of the gamma distribution using order statistics. *Biometrika*, vol. 49, pp. 525–46.

Williams, C. B. 1940. A note on the statistical analysis of sentence length as a criterion of literary style. *Biometrika*, vol. 31, pp. 356–61.

Williams, C. B. 1956. Studies in the history of probability and statistics. IV. A note on an early statistical study of literary style. *Biometrika*, vol. 43, pp. 248–56.

Williamson, E., and **M. K. Bretherton.** 1963. *Tables of the Negative Binomial Probability Distribution.* New York: John Wiley & Sons, Inc.

Winkler, R. L. 1972. *Introduction to Bayesian Inference and Decision.* New York: Holt, Rinehart and Winston, Inc.

Wright, S. 1968. *Genetic and Biometric Foundations*, vol. 1. Chicago, Ill.: University of Chicago Press.

Zipf, G. K. 1949. *Human Behavior and the Principle of Least Effort.* Reading, Mass.: Addison-Wesley Publishing Company, Inc.

Name Index

565

Subject Index

569